Engineering Electromagnetics

Engineering Electromagnetics

Edited by **Edgar Wilson**

NY RESEARCH PRESS

New York

Published by NY Research Press,
23 West, 55th Street, Suite 816,
New York, NY 10019, USA
www.nyresearchpress.com

Engineering Electromagnetics
Edited by Edgar Wilson

International Standard Book Number: 978-1-63238-500-0 (Hardback)

Printed in China.

Contents

Preface

In my initial years as a student, I used to run to the library at every possible instance to grab a book and learn something new. Books were my primary source of knowledge and I would not have come such a long way without all that I learnt from them. Thus, when I was approached to edit this book; I became understandably nostalgic. It was an absolute honor to be considered worthy of guiding the current generation as well as those to come. I put all my knowledge and hard work into making this book most beneficial for its readers.

Electromagnetics refers to the study of magnetism and electricity and their inter-relation. It is a branch of physics which deals with studying electromagnetic fields produced by electrically charged objects. The applications of electromagnetic fields are used in various motors, CAT scanners, TVs, radio transmissions, magnetic resonance imaging, common speakers, etc. This book contains some path-breaking studies related to this field. It includes a detailed explanation of the various theories and applications of electromagnetism. The topics included in this text are of utmost significance and are bound to provide incredible insights to the readers. It includes contributions of experts and scientists from across the globe and thus, will serve as a reference guide to students, scientists, physicists, engineers, researchers, and all those who are interested in this subject.

I wish to thank my publisher for supporting me at every step. I would also like to thank all the authors who have contributed their researches in this book. I hope this book will be a valuable contribution to the progress of the field.

Editor

1

New Wavelet Thresholding Algorithm in Dropping Ambient Noise from Underwater Acoustic Signals

Mohammad Monirujjaman Khan, Ratil Hasnat Ashique, Badrun Naher Liya, Md. Mohsin Sajjad, Md. Anisur Rahman, M. T. Hasan Amin

Department of EEE, Primeasia University, Dhaka, Bangladesh
Email: monirkhan.qmul@gmail.com

Abstract

Underwater Wireless Communication, largely dependent on the acoustic communication between the machines, is largely affected by various types of noise in the shallow and deep water. However ambient noise which is due to multiple sources (e.g. shipping, wind) and no one source dominates. Ambient noise masks the acoustic signal to a large extent. Hence today it has drawn the attention of the experts to reduce its effect on the received signal. This paper discusses ambient noise problem and devises a new wavelet thresholding method to reduce its effect. Afterwards a comparative study on statistical parameters is shown to prove the efficiency of the devised method.

Keywords

Ambient Noise, Wavelet Transform, Thresholding, Signal to Noise Ratio (SNR), Root Mean Square Error (RMSE), Power Spectral Density (PSD), Percentage Root Mean Square Difference (PRD)

1. Introduction

Compared to radio waves, acoustic waves have become the most effective way in underwater wireless communication [1]. It is because radio waves are highly attenuated and spreading occurs due to high frequency. Hence they can propagate only over very short distances. On the other hand if acoustic waves are used, long distance communication can be established. However underwater wireless communication is still challenging due to frequency band limitation and underwater channel disturbances in the form of ambient noise. The disturbance is generated by both natural (seismic, wind marine animals, rain, breaking waves etc.) and manmade sources (shipping, other machineries etc.). We will discuss ambient noise properties in details and reducing algorithm afterwards [2]-[4].

2. Ambient Noise

Oceans are filled with various types of sound which may be generated by natural or man-made sources. Breaking waves, rain, marine, shrimp, volcanoes, earthquakes can be god examples of natural sources. Ships, military sonars exemplify man made sources.

This background noise present in the ocean is called ambient noise. All types of noise is not present everywhere all the time.

Category by Frequency

Ambient noise sources can be categorized by frequency. The typical frequency range of various types of ambient noise is shown by Wenz in the Wenz curve [5]. In the range of 20 Hz to 500 Hz, the noise is primarily due to distant shipping. In this range frequency content can be detected even after eliminating other types of shipping noise at the receiver. Moreover it is easy to understand that the noise level expressed in terms of signal power level (SPL) will be greater where shipping traffic is heavier. For example, low frequency ambient noise level is lower in the southern hemisphere as shipping traffic is very low. In the frequency range 0.5 KHz to 100 KHz the noise is mainly due to spray and bubbles associated with breaking waves. In this range noise is proportional to wind speed that is noise level increases with increasing wind speed. Above 100 KHz thermal noise dominates which arises due to random motion of water molecules. Thermal noise determines the ultimate minimum level of sound that can be used for communication purpose in a particular time and location [6].

3. Wavelet Transform

Wavelets are mathematical functions which break up the data into different frequency components and hence each frequency component can be studied at different resolution. Wavelet analysis is superior to Fourier analysis in the sense that Fourier analysis is well suited for frequency analysis but if we require time frequency analysis then Fourier analysis becomes useless. That is wavelet analysis provides us with time frequency data of the signal. That is we can determine through wavelet analysis where and at what frequency noise spikes or discontinuities occur. In fine wavelets are specially designed for non-stationary signals which are transient in behavior [7] [8].

4. Wavelet Thresholding

For many signals, the low-frequency content is the most important part as because it contains the information. The high-frequency content, on the other hand, imparts flavor or nuance. Considering the human voice, If we remove the high-frequency components, the voice sounds different, but we can still tell what's being said. However, if we remove enough of the low-frequency components, we hear gibberish. In wavelet analysis, we often speak of approximations and *details.* The approximations are the high-scale, low-frequency components of the signal. The details are the low-scale, high-frequency components. The thresholding process of wavelet coefficients can be divided into two steps.

4.1. Step 1

The first step is the policy choice, *i.e.* the choice of thresholding function T. Two standard choices are hard and soft thresholding mentioned as T_{hard} and T_{soft} respectively with corresponding transformations are given by,

$$T_{\text{hard}}\left(d_j,\lambda\right)=0 \quad \text{when} \quad \left|d_j\right|>\lambda \tag{1}$$

$$T_{\text{hard}}\left(d_j,\lambda\right)=d_j \quad \text{when} \quad \left|d_j\right|<\lambda \tag{2}$$

$$T_{\text{soft}}\left(d_j,\lambda\right)=0 \quad \text{when} \quad \left|d_j\right|\le\lambda \tag{3}$$

$$T_{\text{soft}}\left(d_j,\lambda\right)=d_j-\lambda \quad \text{when} \quad d_j>\lambda \tag{4}$$

$$T_{\text{soft}}\left(d_j,\lambda\right)=d_j+\lambda \quad \text{when} \quad d_j<-\lambda \tag{5}$$

The hyperbola function is as stated below,

$$T_{\text{hyper}}\left(d_j,\lambda\right)=\text{sgn}\left(d_j\right)*\sqrt{\left(d^2-\lambda^2\right)} \quad \text{when} \quad \left|d_j\right|>\lambda \tag{6}$$

$$T_{\text{hyper}}\left(d_j,\lambda\right)=0 \quad \text{when} \quad \left|d_j\right|>\lambda \tag{7}$$

The hyperbola function in (6) and (7) is a compromise between soft and hard thresholding functions. The function T_{hyper} is an almost hard thresholder with the continuity property [9]-[11].

4.2. Step 2

The second step is the choice of threshold value [12] [13]. Donho and Johnston (1993) proposed a threshold λ defined as follows:

$\lambda=\sigma*\sqrt{(2*\log N)}$ which is called universal threshold by the authors. This threshold is one of the oldest and provides fast and easy thresholding. There are several possibilities for the estimator σ. Some standard estimators are mentioned in (8) and (9):

$$\sigma^2=\frac{1}{\frac{N}{2}-1}*\sum_{i=1}^{N/2}\left(d_{n-1,i}-\bar{d}\right)^2 \tag{8}$$

$$\sigma^2=\frac{1}{0.6745}*\text{MAD}\left(d_{n-1,i}, i=1,2,3,\cdots,m\right) \tag{9}$$

where

n is the highest level of decomposition;

$\bar{d}=$ mean value of coefficients at level n;

$i=$ number of coefficients.

5. Proposed Thresholding Method

The new thresholding method consists of thresholding value estimation and thresholding function design both for detail and approximation coefficients at each level of wavelet decomposition. Detail coefficients carry the original information signal between 8 to 12 KHz range and approximation coefficients carry the noise component. The ambient noise component is dominating in between 1 to 4 KHz range [5] [11].

As the first step in our proposed thresholding method we offer the new threshold value for detail and approximation coefficients as follows

$$\lambda_a=\left\{\left(-\frac{2}{L}\right)+\mathrm{e}^{-j}\right\}*\sigma_1*\sqrt{\left(2*\log_{10}N_a\right)} \tag{10}$$

$$\lambda_d=\left(-\frac{2}{L}\right)+\mathrm{e}^{-j}+\mathrm{e}^{4*\sigma_1}*\sigma_1*\sqrt{\left(2*\log_{10}N_d\right)} \tag{11}$$

where,

$\lambda_a=$ Threshold value for detail coefficients;

$\lambda_d=$ Threshold value for approximation coefficients;

$L=$ Highest level of decomposition;

$j=$ Current level of decomposition;

$\sigma_1=\sigma/0.6745$;

$N_a=$ Number of approximation coefficients at current level;

$N_d=$ Number of detail coefficients at current level.

When data size is large the general thresholding estimation gives under fitted results [7] [8].

Proposed detail coefficient thresholding function,

$$T_{\text{proposed}}\left(a_j,\lambda_d\right)=0 \quad \text{when} \quad \left|a_j\right|>\lambda_d \tag{12}$$

$$T_{\text{proposed}}\left(a_j,\lambda_d\right)=a_j+0.001*\lambda_d \quad \text{when} \quad \left|a_j\right|>\lambda_d \tag{13}$$

Proposed approximation coefficient thresholding function,

$$T_{\text{proposed}}\left(a_j,\lambda\right)=0 \quad \text{when} \quad \left|a_j\right|>\lambda_a \tag{14}$$

$$T_{\text{proposed}}\left(a_j,\lambda\right)=a_j+0.001*\lambda_a \quad \text{when} \quad \left|a_j\right|<\lambda_a \tag{15}$$

where $a_j =$ Approximation coefficients at level j; $\lambda_a =$ Threshold value for approximation coefficients.

6. Proposed Noise Reduction Model

In our proposed method we assume a reference signal of 10 KHz frequency. The noise data is provided by MREG, Swansea University, UK. The data was recorded in high tidal current over the board 11/05/2013 at 07.06 am on the West coast of Wales.

The noise reduction algorithm can be described on a step by step basis as follows:

Step1: 10 KHz reference signal is used to produce the noisy signal (noisy signal = Ref. Signal + Noise).

Step 2: A level 8 wavelet Decomposition is performed using discrete wavelet transform (dwt) function.

Step 3: Detail coefficients at level 1 to 8 are modified using proposed thresholding method.

Step 4: Approximation coefficients are at each level are modified using proposed method.

Step 5: Inverse Wavelet Transform is performed to reconstruct the denoised signal.

The algorithm is applied on two sets of noise data to ensure algorithm efficiency.

7. MATLAB® Simulation Results

To proof the efficiency of the proposed thresholding value estimation method and thresholding function, we calculate the SNR, RMSE and PSD of the denoised signal. **Table 1** and **Table 2** show the result of simulation.

Table 1. Statistical parameter values using various wavelets (noise data set 1).

Wavelet Type					
haar					
Parameters	Universal/Soft	Universal/Hard	Proposed Soft	Proposed/ Hard	Proposed/Proposed
SNR	**11.8835**	**0.6798**	**13.8293**	**12.6516**	**16.2375**
RMSE	**103.8128**	**183.8403**	**57.0462**	**57.4009**	**56.5608**
PRD	**91.8910**	**115.4690**	**102.0446**	**102.4946**	**101.1777**
db2					
Parameters	Universal/Soft	Universal/Hard	Proposed Soft	Proposed Hard	Proposed/Proposed
SNR	**3.2517**	**0.4998**	**35.1749**	**17.0781**	**29.0811**
RMSE	**46.3308**	**50.6391**	**55.9077**	**56.4519**	**55.9362**
PRD	**57.5384**	**30.7650**	**100.0106**	**100.7891**	**100.0614**
coifl					
Parameters	Universal/Soft	Universal/Hard	Proposed Soft	Proposed Hard	Proposed/Proposed
SNR	**4.9903**	**0.5058**	**35.1130**	**23.0077**	**28.0949**
RMSE	**122.9994**	**131.0551**	**55.9132**	**56.0445**	**55.9541**
PRD	**154.4232**	**82.1063**	**100.0205**	**100.0665**	**100.0928**
sym4					
Parameters	Universal/Soft	Universal/Hard	Proposed Soft	Proposed Hard	Proposed/Proposed
SNR	**1.8481**	**0.4972**	**43.0030**	**29.5009**	**42.8615**
RMSE	**137.6016**	**126.3309**	**55.8981**	**55.9283**	**55.8982**
PRD	**172.8118**	**79.1153**	**99.9933**	**99.9017**	**99.9935**

Table 2. Statistical parameter values using various wavelets (noise data set 2).

Wavelet Type					
haar					
Parameters	Universal/Soft	Universal/Hard	Proposed Soft	Proposed/ Hard	Proposed/Proposed
SNR	**10.5025**	**8.2338**	**13.8333**	**8.6551**	**16.2377**
RMSE	**89.6563**	**91.8674**	**57.0452**	**60.3108**	**56.5608**
PRD	**108.3071**	**106.8285**	**101.8067**	**96.6354**	**101.0411**
db2					
Parameters	Universal/Soft	Universal/Hard	Proposed Soft	Proposed Hard	Proposed/Proposed
SNR	**4.3565**	**8.3479**	**35.1423**	**10.0502**	**29.0804**
RMSE	**51.6204**	**65.4517**	**55.9078**	**59.3309**	**55.9372**
PRD	**57.9868**	**73.5408**	**100.0070**	**94.9298**	**100.0572**
coifl					
Parameters	Universal/Soft	Universal/Hard	Proposed Soft	Proposed Hard	Proposed/Proposed
SNR	**5.9929**	**0.3022**	**35.2410**	**10.7386**	**28.0944**
RMSE	**115.0065**	**135.1313**	**55.9129**	**58.9470**	**55.9540**
PRD	**136.4476**	**80.0767**	**100.0153**	**94.2438**	**100.0841**
sym4					
Parameters	Universal/Soft	Universal/Hard	Proposed Soft	Proposed Hard	Proposed/Proposed
SNR	**16.5239**	**1.0773**	**39.8266**	**8.3438**	**42.8783**
RMSE	**84.4747**	**141.6139**	**55.8987**	**84.9418**	**55.8983**
PRD	**102.9858**	**167.9913**	**99.9936**	**100.3516**	**99.9954**

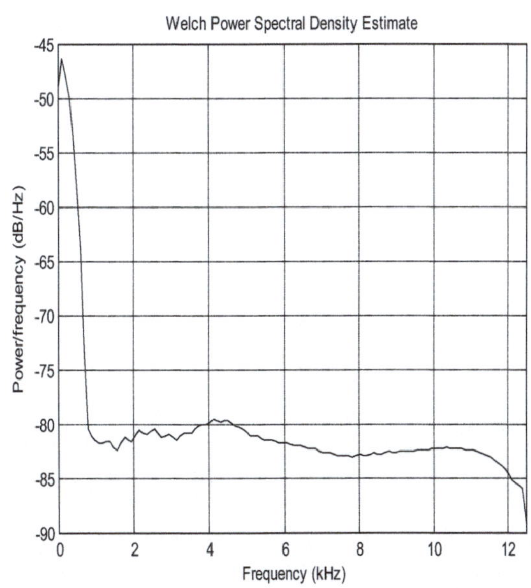

Figure 1. Noise PSD for data set 1.

In **Table 1** and **Table 2** we calculate SNR, RMSE and PSD respectively for wavelet filter types "db2", "Haar", "coifl" and "sym4" using hard, soft and proposed thresholding function. For each type of function we have calculated the statistical parameter values using universal and proposed threshold estimation method.

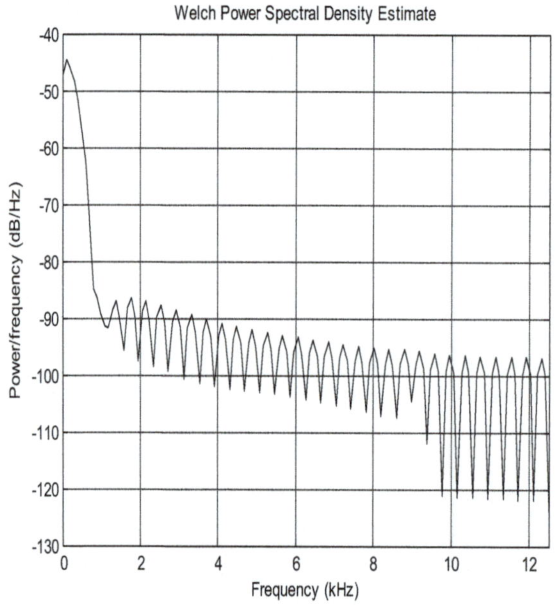

Figure 2. Noise PSD for data set 2.

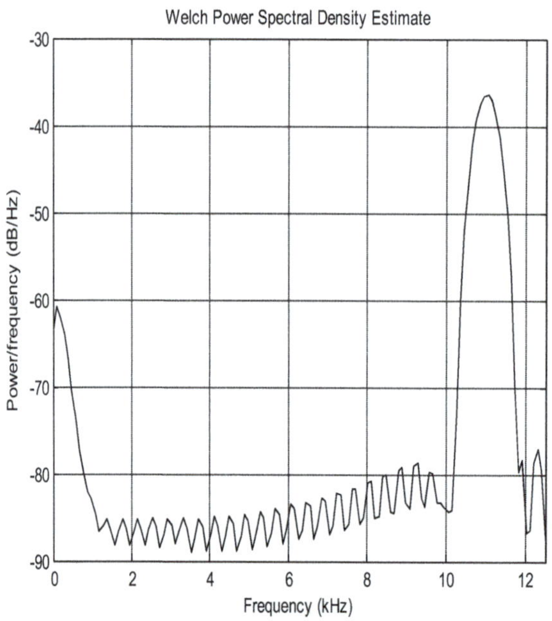

Figure 3. Noise signal (reference signal + noise) PSD using data set 1#.

8. Conclusion

From the above data it is clear that the devised method has produced improved SNR, PSD and lower RMSE for all noise data sets compared with global soft or hard thresholding methods. The PSD of noise signal being almost similar in frequency contents is showing that the noise spectrum have a peak in the lower frequency region as expected with the information signal at higher frequency region showed in **Figures 1-5**. The soft thresholding function with proposed estimation method has produced improved SNR with lower RMSE especially when noise signal strength is strong. The devised method seems to be efficient for all wavelet filter types applied to this experiment.

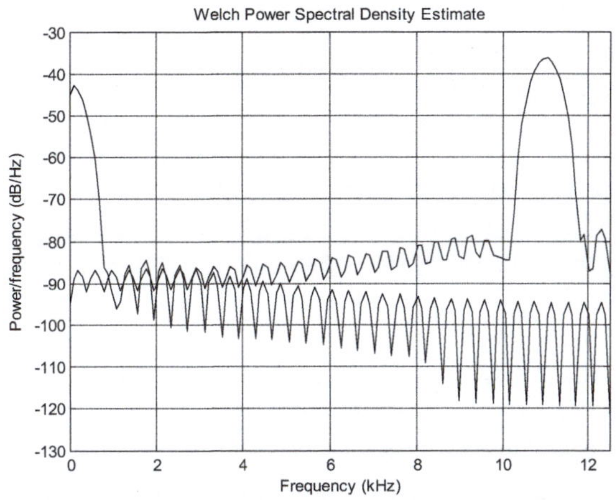

Figure 4. Noisy signal (reference signal + noise) PSD using data set 2.

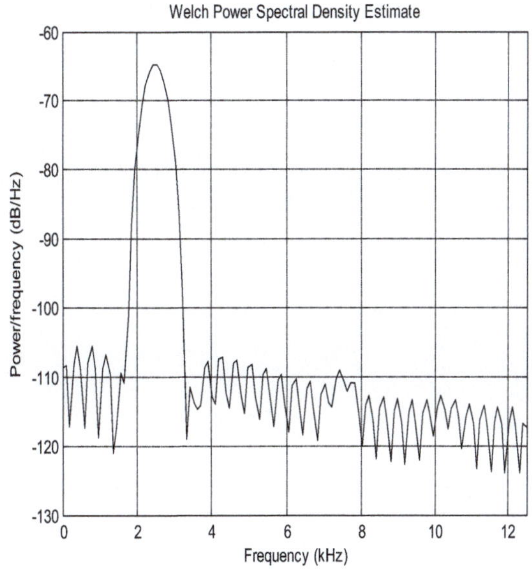

Figure 5. PSD of denoised signal using proposed method.

Acknowledgements

The authors would like to thank Mr Merin Broudic, Research Assistant Marine Renewables, Low Carbon Research Institute (LCRI Marine), School of Engineering, Swansea University, UK, for his perpetual support.

References

[1] Teng, D., Chen, H. and Zhu, N. (2010) Computer Simulation of Sound Field Formed around Transducer Source Used in Underwater Acoustic Communication. *3rd International Conference on Advanced Computer Theory and Engineering (ICACTE)*, Chengdu, 20-22 August 2010, V1-144-V1-148.

[2] Kinsler, L.E., Frey, A.R., Coppens, A.B. and Sanders, J.V. (1982) Fundamentals of Acoustics. 3rd Edition, John Wiley & Sons, New York.

[3] Dahl, P.H., Miller, J.H., Kato, D.H. and Andrew, R.K. (2007) Underwater Ambient Noise. Acoustics Today.

[4] Veni, K.S., Murugan, S.S. and Natarajan, V. (2011) Modified LMS Adaptive Algorithm for Detection of Underwater Acoustic Signals Against Ambient Noise in Shallow Water of Indian Sea. *IEEE International Conference on Recent Trends in Information Technology (ICRTIT)*, Chennai, 3-5 June 2011, 901-905.

[5] Raj, K.M., Murugan, S.S., Natarajan, V. and Radha, S. (2011) Denoising Algorithm using Wavelet for Underwater

Signal Affected by Wind Driven Ambient Noise. *IEEE-International Conference on Recent Trends in Information Technology* (*ICRTIT*), Chennai, 3-5 June 2011, 943-946.

[6] Stojanovic, M. (2006) Underwater Wireless Communications: Current Achievements and Research Challenges. IEEE Newsletter.

[7] Bassett, C. (2010) Underwater Ambient Noise at a Proposed Tidal Energy Site in Puget Sound. M.S Thesis, Department of Mechanical Engineering, University of Washington, WA.

[8] Chang, S.G., Yu, B. and Vetterli, M. (2000) Adaptive Wavelet Thresholding for Image Denoisingand Compression. *IEEE Transactions on Image Processing*, **9**, 1532-1546.

[9] Cohen, R. (2012) Signal Denoising Using Wavelets. Project Report, Department of Electrical Engineering Technion, Israel Institute of Technology, Haifa.

[10] Antoniadis, A., Leporini, D. and Pesquet, J.-C. (2002) Wavelet Thresholding for Some Classes of Non-Gaussian Noise. *Statistica Neerlandica*, **56**, 434-453.

[11] Discovery of Sound of the Sea. University of Rhode Island. www.dosits.org

[12] Daubechies, I. (1992) Ten Lectures on Wavelets. SLAM, Philadelphia. http://dx.doi.org/10.1137/1.9781611970104

[13] Stojanovic, M. (2014) Underwater Wireless Communications. Wiley-Blackwell, an Imprint of John Wiley & Sons Ltd.

2

Integrated and Explicit Boundary Conditions of Electromagnetic Fields at Arbitrary Interfaces between Two Anisotropic Media

Bing Zhou[1], Graham Heinson[2], Aixa Rivera-Rios[2]

[1]Petroleum Geosciences, The Petroleum Institute, Abu Dhabi, UAE
[2]Geology and Geophysics, The University of Adelaide, Adelaide, Australia
Email: bizhou@pi.ac.ae, graham.heinson@adelaide.edu.au, aixa-rivera-rios@adelaide.edu.au

Abstract

This paper derives two new integrated and explicit boundary conditions, named the "explicit normal version" and "explicit tangential versions" respectively for electromagnetic fields at an arbitrary interface between two anisotropic media. The new versions combine two implicit boundary equations into a single explicit matrix formula and reveal the boundary values linked by a 3 × 3 matrix, which depends on the interface topography and model property tensors. We analytically demonstrate the new versions equivalent to the common implicit boundary conditions and their application to transformation of the boundary values in the boundary integral equations. We also give two synthetic examples that show recovery of the boundary values on a hill and a ridge, and highlight the advantage of the new versions of being a simpler and more straightforward method to compute the electromagnetic boundary values.

Keywords

Electromagnetic Theory, Electromagnetism, Boubdary Condition, Electric Anisotropy

1. Introduction

The boundary conditions are often expressed in two equations—continuity of the tangential components and discontinuity of the normal components of electromagnetic field intensities $(\boldsymbol{E}, \boldsymbol{H})$ [1]. The former is yielded by applying Stokes' law to a differential line integral on the interface between two media, and the latter is obtained by applying Gauss' law to a differential sized cylinder surface containing a section of the interface. This gives two separate and implicit formulae that define "boundary equations" linking the boundary values of the fields in two anisotropic media. Two boundary equations are implicit functions of the interface normal $(\boldsymbol{n})$, electric conductivity and permittivity tensors $(\boldsymbol{\sigma},\boldsymbol{\varepsilon})$, or magnetic permeability tensor $(\boldsymbol{\mu})$. In isotropic cases, it is not difficult to obtain the explicit formulae of the boundary values because all these tensors reduce to scalars that make the explicit solution straightforward. The difficulty is increased in applying the separate and implicit formulae to anisotropic media and arbitrary interface topography as they do not explicitly give the solutions of the boundary values, so that they must be individually or successively employed in electromagnetic field modeling. In addition, most of numerical modeling techniques, such as finite-difference, finite-element and boundary element methods approximate the boundary values with some numerical schemes, e.g. the finite-difference method often replaces the interfaces with great gradients to produce the "strong solution" of electromagnetic fields [2]. The finite-element method employs combinations of the edge-vectors to approach the field intensities so that the boundary conditions are satisfied at the sampled points [3]. However, the accuracy of the edge-vector approximation depends on the number of the samples of the edge-vectors [4]. Also, these numerical approaches cannot simultaneously produce the complete set of boundary values due to only involving one-side boundary values in the assembled linear equations, and need an explicit formula to recover another side boundary values. In order to simplify the implementation of the boundary conditions or recover whole boundary values at an interface, it is desirable to combine the two separate and implicit equations into a single integrated and explicit formula so that it can be more directly and easily applied to theoretical and numerical electromagnetic anisotropy problems.

This paper derives two new integrated and explicit versions of the boundary conditions, called the explicit "*normal*" and "*tangential*" versions respectively. They successfully combine two common implicit boundary equations into a single explicit linear matrix formula without altering their applicability to interfaces that have arbitrary topography and two anisotropic media. These new versions consistently present the boundary values of electromagnetic field intensities $(\boldsymbol{E},\boldsymbol{H})$ linked by a 3 × 3 matrix, which can be calculated with the known interface topography $(\boldsymbol{n})$ and tensors of model electric permittivity $(\boldsymbol{\varepsilon})$, conductivity $(\boldsymbol{\sigma})$ and magnetic permeability $(\boldsymbol{\mu})$. We analytically demonstrate equivalence of the single matrix formula to two common implicit boundary equations, and show theoretical applications of the new versions to transformation of the boundary values from one-side to another in the boundary integral equation and boundary element approach. In addition, two synthetic experiments of utilizing the new versions are conducted, and show the advantage of the new versions of being a simpler and more straightforward method to recover the whole boundary values at arbitrary interfaces.

2. Boundary Conditions

In the frequency-domain, electric and magnetic field intensities $(\boldsymbol{E},\boldsymbol{H})$ in anisotropic media satisfy Maxwell's equations [5]

$$\begin{aligned}\nabla\times\boldsymbol{E}+i\omega\boldsymbol{\mu}\cdot\boldsymbol{H}&=-\boldsymbol{m}_e,\\ \nabla\times\boldsymbol{H}-\tilde{\boldsymbol{\sigma}}\cdot\boldsymbol{E}&=\boldsymbol{j}_e,\end{aligned}\tag{1}$$

where $\boldsymbol{m}_e$ and $\boldsymbol{j}_e$ represent the external magnetic and electric current sources supplied by human or natural existence, and $\tilde{\boldsymbol{\sigma}}$ is the complex-valued tensor defined by:

$$\tilde{\boldsymbol{\sigma}}=\boldsymbol{\sigma}+i\omega\boldsymbol{\varepsilon}\tag{2}$$

Here, ω represents an angular frequency and $\{\boldsymbol{\mu},\boldsymbol{\sigma},\boldsymbol{\varepsilon}\}$ are three tensors of magnetic permeability, electric conductivity and permittivity. The complex-valued conductivity tensor $(\tilde{\boldsymbol{\sigma}})$ implies that the electric current density $(\boldsymbol{J}=\tilde{\boldsymbol{\sigma}}\cdot\boldsymbol{E})$ consists of the conduction $(\boldsymbol{J}_c=\boldsymbol{\sigma}\cdot\boldsymbol{E})$ and displacement $(\boldsymbol{J}_d=\boldsymbol{\varepsilon}\cdot(i\omega\boldsymbol{E}))$ current densities. In this paper, $(\tilde{\mu},\tilde{\sigma})$ or (μ,σ,ε) are simply called the *model property tensors* because they define the electromagnetic properties of media. In isotropic cases, the model property tensors $(\tilde{\boldsymbol{\mu}},\tilde{\boldsymbol{\sigma}})$ or $(\boldsymbol{\mu},\boldsymbol{\sigma},\boldsymbol{\varepsilon})$ are scalars, *i.e.* $(\tilde{\mu},\tilde{\sigma})$ or (μ,σ,ε). In general, the field intensities $(\boldsymbol{E},\boldsymbol{H})$, model property tensors $(\boldsymbol{\mu},\boldsymbol{\sigma},\boldsymbol{\varepsilon})$ or scalars (μ,σ,ε), and external current sources $(\boldsymbol{m}_e,\boldsymbol{j}_e)$ are functions of the spatial coordinates $\mathbf{x}=(x_1,x_2,x_3)$.

Applying Equation (1) and its zero-divergences $\nabla\cdot(i\omega\boldsymbol{\mu}\cdot\boldsymbol{H}+\boldsymbol{m}_e)=0$ and $\nabla\cdot(\tilde{\boldsymbol{\sigma}}\cdot\boldsymbol{E}+\boldsymbol{j}_e)=0$ to a closed differential line integral and surface integral of a differential sized cylinder surface that contains a section of the interface between two media, respectively [1], the following boundary conditions of the electric and magnetic field intensities are obtained:

$$\left(\boldsymbol{E}^{+}-\boldsymbol{E}^{-}\right)\times\hat{\boldsymbol{n}}=\boldsymbol{0},$$
$$\left(\tilde{\boldsymbol{\sigma}}^{+}\cdot\boldsymbol{E}^{+}-\tilde{\boldsymbol{\sigma}}^{-}\cdot\boldsymbol{E}^{-}\right)\cdot\hat{\boldsymbol{n}}=\Delta J_{n}^{\mp}. \tag{3}$$

$$\left(\boldsymbol{H}^{+}-\boldsymbol{H}^{-}\right)\times\hat{\boldsymbol{n}}=\boldsymbol{0},$$
$$\left(\boldsymbol{\mu}^{+}\cdot\boldsymbol{H}^{+}-\boldsymbol{\mu}^{-}\cdot\boldsymbol{H}^{-}\right)\cdot\hat{\boldsymbol{n}}=\Delta M_{n}^{\mp}. \tag{4}$$

Here, the scalar quantities $\Delta J_n^{\mp}$ and $\Delta M_n^{\mp}$ are the normal components of the net external current densities at the interface:

$$\Delta J_n^{\mp}=\left(\boldsymbol{j}_e^{-}-\boldsymbol{j}_e^{+}\right)\cdot\hat{\boldsymbol{n}},$$
$$\Delta M_n^{\mp}=\left(\boldsymbol{m}_e^{-}-\boldsymbol{m}_e^{+}\right)\cdot\hat{\boldsymbol{n}}/i\omega. \tag{5}$$

The superscripts "–" and "+" stand for the boundary values on the two sides of the interface, and $\hat{\boldsymbol{n}}$ is a unit normal of the interface (see **Figure 1**).

In order to remove computational singularities (infinite value) of the external point sources $\boldsymbol{m}_e$ and $\boldsymbol{j}_e$, the field intensities are often expressed in two portions [2] [3] [6], *i.e.* $\{\boldsymbol{E},\boldsymbol{H}\}=\{\boldsymbol{E}^p+\boldsymbol{E}^s,\boldsymbol{H}^p+\boldsymbol{H}^s\}$, where $\{\boldsymbol{E}^p,\boldsymbol{H}^p\}$ are the primary fields generated by $\boldsymbol{m}_e$ and $\boldsymbol{j}_e$ in a reference model given by $\{\tilde{\boldsymbol{\sigma}}_p,\tilde{\boldsymbol{\mu}}_p\}$, and $\{\boldsymbol{E}^s,\boldsymbol{H}^s\}$ are the secondary fields governed by the following equations obtained by substitution of the field decomposition into Equation (1):

$$\nabla\times\boldsymbol{E}^s+i\omega\boldsymbol{\mu}\cdot\boldsymbol{H}^s=-i\omega\delta\boldsymbol{\mu}\cdot\boldsymbol{H}^p,$$
$$\nabla\times\boldsymbol{H}^s-\tilde{\boldsymbol{\sigma}}\cdot\boldsymbol{E}^s=\delta\tilde{\boldsymbol{\sigma}}\cdot\boldsymbol{E}^p. \tag{6}$$

These equations demonstrate that the source terms of the secondary fields are $i\omega\delta\boldsymbol{\mu}\cdot\boldsymbol{H}^p$ and $\delta\tilde{\boldsymbol{\sigma}}\cdot\boldsymbol{E}^p$ instead of $\boldsymbol{m}_e$ and $\boldsymbol{j}_e$, where $\{\delta\boldsymbol{\mu},\delta\tilde{\boldsymbol{\sigma}}\}=\{\boldsymbol{\mu}-\boldsymbol{\mu}_p,\tilde{\boldsymbol{\sigma}}-\tilde{\boldsymbol{\sigma}}_p\}$, Similarly, Applying Equation (6) and its zero divergences to an interface of two media, and appointing $\{\tilde{\boldsymbol{\sigma}}_p,\boldsymbol{\mu}_p\}=\{\tilde{\boldsymbol{\sigma}}^{+}\cup\tilde{\boldsymbol{\sigma}}_p^{-},\boldsymbol{\mu}^{+}\cup\boldsymbol{\mu}_p^{-}\}$, $\forall\{\tilde{\boldsymbol{\sigma}}_p^{-},\boldsymbol{\mu}_p^{-}\}$ or $\{\tilde{\boldsymbol{\sigma}}_p,\boldsymbol{\mu}_p\}=\{\tilde{\boldsymbol{\sigma}}_p^{+}\cup\tilde{\boldsymbol{\sigma}}^{-},\boldsymbol{\mu}_p^{+}\cup\boldsymbol{\mu}^{-}\}$, $\forall\{\tilde{\boldsymbol{\sigma}}_p^{+},\boldsymbol{\mu}_p^{+}\}$ in the cases of $\{\boldsymbol{E}^{s(+)},\boldsymbol{H}^{s(+)}\}$ or $\{\boldsymbol{E}^{s(-)},\boldsymbol{H}^{s(-)}\}$ respectively, the following boundary conditions of the secondary fields are obtained:

$$\left(\boldsymbol{E}^{s(+)}-\boldsymbol{E}^{s(-)}\right)\times\hat{\boldsymbol{n}}=\boldsymbol{0},$$
$$\left(\tilde{\boldsymbol{\sigma}}^{+}\cdot\boldsymbol{E}^{s(+)}-\tilde{\boldsymbol{\sigma}}^{-}\cdot\boldsymbol{E}^{s(-)}\right)\cdot\hat{\boldsymbol{n}}=\delta\tilde{\boldsymbol{\sigma}}^{\mp}\cdot\boldsymbol{E}^{p(-)}\cdot\hat{\boldsymbol{n}}. \tag{7}$$

$$\left(\boldsymbol{H}^{s(+)}-\boldsymbol{H}^{s(-)}\right)\times\hat{\boldsymbol{n}}=\boldsymbol{0},$$
$$\left(\boldsymbol{\mu}^{+}\cdot\boldsymbol{H}^{s(+)}-\boldsymbol{\mu}^{-}\cdot\boldsymbol{H}^{s(-)}\right)\cdot\hat{\boldsymbol{n}}=\delta\boldsymbol{\mu}^{\mp}\cdot\boldsymbol{H}^{p(-)}\cdot\hat{\boldsymbol{n}}. \tag{8}$$

Here, $\delta\tilde{\boldsymbol{\sigma}}^{\mp}=\tilde{\boldsymbol{\sigma}}^{-}-\tilde{\boldsymbol{\sigma}}^{+}$ and $\delta\boldsymbol{\mu}^{\mp}=\boldsymbol{\mu}^{-}-\boldsymbol{\mu}^{+}$. Equations (7) and (8) are also yielded by substituting $\boldsymbol{E}=\boldsymbol{E}^p+\boldsymbol{E}^s$ and $\boldsymbol{H}=\boldsymbol{H}^p+\boldsymbol{H}^s$ into Equations (3) and (4) respectively, and then applying the same boundary conditions to the primary fields. Equations (3) and (4) or Equations (7) and (8) are general and applicable to any interface between two media. Here, we named these boundary conditions as the "*implicit boundary equations*" because they consist of two separate and implicit equations that involve the boundary values of the field intensities $\{\boldsymbol{E}^{\pm},\boldsymbol{H}^{\pm}\}$, unit normal $\tilde{\boldsymbol{n}}$ of an interface and model property tensors $\{\tilde{\boldsymbol{\sigma}}^{\pm},\boldsymbol{\mu}^{\pm}\}$. By comparing Equation (3) with (4), or Equation (7) with (8), the similarities of the boundary conditions of magnetic fields to electric fields are observed. It is shown that the boundary conditions of magnetic fields can be obtained by simply replacing the electric field symbols $\{\tilde{\boldsymbol{\sigma}}^{\pm},\boldsymbol{E}^{\pm},\Delta J_n^{\mp}\}$ with the magnetic field symbols $\{\boldsymbol{\mu}^{\pm},\boldsymbol{H}^{\pm},\Delta M_n^{\mp}\}$. Therefore, derivations below will only deal with electric field whose result can be easily extended to magnetic field by the symbol replacements.

3. Explicit Normal Version

Equation (3) can be rewritten in the following matrix form

$$\boldsymbol{A}_{\tilde{\sigma}}^{+}\boldsymbol{E}^{+}=\boldsymbol{A}_{\tilde{\sigma}}^{-}\boldsymbol{E}^{-}+\Delta\boldsymbol{J}_n^{\mp}, \tag{9}$$

where the vectors $\boldsymbol{E}^{\pm}$ and $\Delta \boldsymbol{J}_n^{\mp}$ are defined by $\boldsymbol{E}^{\pm} = \left(E_x^{\pm}, E_y^{\pm}, E_z^{\pm}\right)^{\mathrm{T}}$ and $\Delta \boldsymbol{J}_n^{\mp} = \left(\Delta J_n^{\mp}, 0, 0\right)^{\mathrm{T}}$ respectively, and the matrices $\boldsymbol{A}_{\tilde{\sigma}}^{\pm}$ are given by:

$$\boldsymbol{A}_{\tilde{\sigma}}^{\pm} = \begin{cases} \begin{pmatrix} \tilde{\sigma}_{i1}^{\pm}\hat{n}_i & \tilde{\sigma}_{i2}^{\pm}\hat{n}_i & \tilde{\sigma}_{i3}^{\pm}\hat{n}_i \\ 0 & \hat{n}_3 & -\hat{n}_2 \\ -\hat{n}_3 & 0 & \hat{n}_1 \end{pmatrix}, & (\hat{n}_3 \neq 0); \\ \begin{pmatrix} \tilde{\sigma}_{i1}^{\pm}\hat{n}_i & \tilde{\sigma}_{i2}^{\pm}\hat{n}_i & \tilde{\sigma}_{i3}^{\pm}\hat{n}_i \\ 0 & \hat{n}_3 & -\hat{n}_2 \\ \hat{n}_2 & -\hat{n}_1 & 0 \end{pmatrix}, & (\hat{n}_2 \neq 0); \\ \begin{pmatrix} \tilde{\sigma}_{i1}^{\pm}\hat{n}_i & \tilde{\sigma}_{i2}^{\pm}\hat{n}_i & \tilde{\sigma}_{i3}^{\pm}\hat{n}_i \\ -\hat{n}_3 & 0 & -\hat{n}_1 \\ \hat{n}_2 & -\hat{n}_1 & 0 \end{pmatrix}, & (\hat{n}_1 \neq 0). \end{cases} \tag{10}$$

Here, the summation convention over the double subscripts i has been applied, and the redundant row arising from curl calculation has been removed in three cases. Accordingly, the determinant of the matrix cannot be zero$\left(\left|\boldsymbol{A}_{\tilde{\sigma}}^{\pm}\right| \neq 0\right)$, therefore, the matrix $\boldsymbol{A}_{\tilde{\sigma}}^{\pm}$ is invertible and its inverse matrix can be calculated by linear algebra:

$$\left(\boldsymbol{A}_{\tilde{\sigma}}^{\pm}\right)^{-1} = \begin{cases} \dfrac{1}{\Delta_{\tilde{\sigma}}^{\pm}\hat{n}_3} \begin{pmatrix} \hat{n}_1\hat{n}_3 & -\tilde{\sigma}_{i2}^{\pm}\hat{n}_i\hat{n}_1 & -\left(\tilde{\sigma}_{i2}^{\pm}\hat{n}_i\hat{n}_2 + \tilde{\sigma}_{i3}^{\pm}\hat{n}_i\hat{n}_3\right) \\ \hat{n}_2\hat{n}_3 & \tilde{\sigma}_{i1}^{\pm}\hat{n}_i\hat{n}_1 + \tilde{\sigma}_{i3}^{\pm}\hat{n}_i\hat{n}_3 & \tilde{\sigma}_{i1}^{\pm}\hat{n}_i\hat{n}_2 \\ \hat{n}_3\hat{n}_3 & -\tilde{\sigma}_{i2}^{\pm}\hat{n}_i\hat{n}_3 & \tilde{\sigma}_{i1}^{\pm}\hat{n}_i\hat{n}_3 \end{pmatrix}, & (\hat{n}_3 \neq 0); \\ \dfrac{1}{\Delta_{\tilde{\sigma}}^{\pm}\hat{n}_2} \begin{pmatrix} \hat{n}_1\hat{n}_2 & \tilde{\sigma}_{i3}^{\pm}\hat{n}_i\hat{n}_1 & \tilde{\sigma}_{i2}^{\pm}\hat{n}_i\hat{n}_2 + \tilde{\sigma}_{i3}^{\pm}\hat{n}_i\hat{n}_3 \\ \hat{n}_2\hat{n}_2 & \tilde{\sigma}_{i3}^{\pm}\hat{n}_i\hat{n}_2 & -\tilde{\sigma}_{i1}^{\pm}\hat{n}_i\hat{n}_2 \\ \hat{n}_2\hat{n}_3 & -\left(\tilde{\sigma}_{i1}^{\pm}\hat{n}_i\hat{n}_1 + \tilde{\sigma}_{i2}^{\pm}\hat{n}_i\hat{n}_2\right) & -\tilde{\sigma}_{i1}^{\pm}\hat{n}_i\hat{n}_3 \end{pmatrix}, & (\hat{n}_2 \neq 0); \\ \dfrac{1}{\Delta_{\tilde{\sigma}}^{\pm}\hat{n}_1} \begin{pmatrix} \hat{n}_1\hat{n}_1 & -\tilde{\sigma}_{i3}^{\pm}\hat{n}_i\hat{n}_1 & \tilde{\sigma}_{i2}^{\pm}\hat{n}_i\hat{n}_1 \\ \hat{n}_1\hat{n}_2 & -\tilde{\sigma}_{i3}^{\pm}\hat{n}_i\hat{n}_2 & -\left(\tilde{\sigma}_{i1}^{\pm}\hat{n}_i\hat{n}_1 + \tilde{\sigma}_{i3}^{\pm}\hat{n}_i\hat{n}_3\right) \\ \hat{n}_1\hat{n}_3 & \tilde{\sigma}_{i1}^{\pm}\hat{n}_i\hat{n}_1 + \tilde{\sigma}_{i2}^{\pm}\hat{n}_i\hat{n}_2 & \tilde{\sigma}_{i2}^{\pm}\hat{n}_i\hat{n}_3 \end{pmatrix}, & (\hat{n}_1 \neq 0), \end{cases} \tag{11}$$

where

$$\Delta_{\tilde{\sigma}}^{\pm} = \hat{n}_i \tilde{\sigma}_{ij}^{\pm} \hat{n}_j\,. \tag{12}$$

Multiplying $\left(\boldsymbol{A}_{\tilde{\sigma}}^{\pm}\right)^{-1}$ to Equation (9) gives

$$\boldsymbol{E}^{\pm} = \boldsymbol{C}_{\tilde{\sigma}}^{\mp}\boldsymbol{E}^{\mp} + \left(\Delta J_n^{\mp} / \Delta_{\tilde{\sigma}}^{\pm}\right)\hat{\boldsymbol{n}} \tag{13}$$

where

$$\boldsymbol{C}_{\tilde{\sigma}}^{\mp} = \frac{1}{\Delta_{\tilde{\sigma}}^{\pm}} \begin{pmatrix} \Delta_{\tilde{\sigma}}^{\pm} + \left(\tilde{\sigma}_{i1}^{\mp} - \tilde{\sigma}_{i1}^{\pm}\right)\hat{n}_i\hat{n}_1 & \left(\tilde{\sigma}_{i2}^{\mp} - \tilde{\sigma}_{i2}^{\pm}\right)\hat{n}_i\hat{n}_1 & \left(\tilde{\sigma}_{i3}^{\mp} - \tilde{\sigma}_{i3}^{\pm}\right)\hat{n}_i\hat{n}_1 \\ \left(\tilde{\sigma}_{i1}^{\mp} - \tilde{\sigma}_{i1}^{\pm}\right)\hat{n}_i\hat{n}_2 & \Delta_{\tilde{\sigma}}^{\pm} + \left(\tilde{\sigma}_{i2}^{\mp} - \tilde{\sigma}_{i1}^{\pm}\right)\hat{n}_i\hat{n}_2 & \left(\tilde{\sigma}_{i3}^{\mp} - \tilde{\sigma}_{i3}^{\pm}\right)\hat{n}_i\hat{n}_2 \\ \left(\tilde{\sigma}_{i1}^{\mp} - \tilde{\sigma}_{i1}^{\pm}\right)\hat{n}_i\hat{n}_3 & \left(\tilde{\sigma}_{i2}^{\mp} - \tilde{\sigma}_{i2}^{\pm}\right)\hat{n}_i\hat{n}_3 & \Delta_{\tilde{\sigma}}^{\pm} + \left(\tilde{\sigma}_{i3}^{\mp} - \tilde{\sigma}_{i3}^{\pm}\right)\hat{n}_i\hat{n}_3 \end{pmatrix}, \tag{14a}$$

or

$$c_{ij}\left(\tilde{\boldsymbol{\sigma}}^{\mp}\right) = \delta_{ij} + \hat{n}_i\hat{n}_k\left(\tilde{\sigma}_{kj}^{\mp} - \tilde{\sigma}_{kj}^{\pm}\right) / \Delta_{\tilde{\sigma}}^{\pm} \tag{14b}$$

Here, δ_{ij} is the Kronecker delta symbol. The above equation shows that the three cases given in Equations

(10) and (11) are unnecessary in the matrix $\boldsymbol{C}_{\tilde{\sigma}}^{\mp}$. In this paper, the matrix $\boldsymbol{C}_{\tilde{\sigma}}^{\mp}$ is called the *boundary matrix* because it is a function of the boundary conductivity tensors $\tilde{\boldsymbol{\sigma}}^{\pm}$ and the unit normal $\hat{\boldsymbol{n}}$ of the interface, and links the two boundary values of the field intensities. With the known interface normal $\hat{\boldsymbol{n}}$ and conductivity tensors $\tilde{\boldsymbol{\sigma}}^{\pm}$, Equation (13) directly give the solution of the boundary values and successfully combines two implicit boundary equations into a single explicit linear matrix formula. This integrated and explicit form of the boundary conditions is advantageous to application without altering its applicability to any interface between two media. Therefore, Equation (13) is termed the "*explicit normal versions*" of the boundary conditions.

Substituting $\boldsymbol{E}^{\pm} = \boldsymbol{E}^{p(\pm)} + \boldsymbol{E}^{s(\pm)}$ into Equation (13) and then applying the same boundary conditions to the primary fields $\{\boldsymbol{E}^{p}, \boldsymbol{H}^{p}\}$ in the reference conductivity model: $\tilde{\boldsymbol{\sigma}}_p = \{\tilde{\boldsymbol{\sigma}}^{+} \cup \tilde{\boldsymbol{\sigma}}_p^{-}\}$, $\forall \tilde{\boldsymbol{\sigma}}_p^{-}$ or $\tilde{\boldsymbol{\sigma}}_p = \{\tilde{\boldsymbol{\sigma}}_p^{+} \cup \tilde{\boldsymbol{\sigma}}^{-}\}$, $\forall \tilde{\boldsymbol{\sigma}}_p^{+}$, the integrated and explicit boundary conditions of the secondary electric fields are obtained:

$$\boldsymbol{E}^{s(\pm)} = \boldsymbol{C}_{\tilde{\sigma}}^{\mp}\boldsymbol{E}^{s(\mp)} + \left(\boldsymbol{C}_{\tilde{\sigma}}^{\mp} - \boldsymbol{C}_{\tilde{\sigma}_p}^{\mp}\right)\boldsymbol{E}^{p(\mp)} \tag{15}$$

This equation corresponds to Equation (7) but explicitly gives the boundary values of the secondary fields. It achieves transformation of the boundary values at an interface.

The explicit boundary conditions for magnetic fields can be obtained by replacing the electric symbols $\{\tilde{\boldsymbol{\sigma}}^{\pm}, \boldsymbol{E}^{\pm}, \Delta J_n^{\mp}\}$ with the magnetic symbols $\{\boldsymbol{\mu}^{\pm}, \boldsymbol{H}^{\pm}, \Delta M_n^{\mp}\}$ in Equations (13) and (15), *i.e.*

$$\boldsymbol{H}^{\pm} = \boldsymbol{C}_{\mu}^{\mp}\boldsymbol{H}^{\mp} + \left(\Delta M_n^{\mp} / \Delta_{\mu}^{\pm}\right)\hat{\boldsymbol{n}} \tag{16}$$

$$\boldsymbol{H}^{s(\pm)} = \boldsymbol{C}_{\mu}^{\mp}\boldsymbol{H}^{s(\mp)} + \left(\boldsymbol{C}_{\mu}^{\mp} - \boldsymbol{C}_{\mu_p}^{\mp}\right)\boldsymbol{H}^{p(\mp)} \tag{17}$$

From these explicit normal versions, it is apparent that the boundary matrices $\{\boldsymbol{C}_{\alpha}^{\pm}, \alpha \in (\boldsymbol{\mu}, \tilde{\boldsymbol{\sigma}})\}$ are crucial in solving the boundary values of the field intensities. With given model property tensors $\{\boldsymbol{\mu}^{\pm}, \tilde{\boldsymbol{\sigma}}^{\pm}\}$ and interface normal $(\hat{\boldsymbol{n}})$, the boundary values can be directly calculated through the boundary matrix. This mathematical merit is not possessed by the implicit boundary equations given in the previous section when dealing with the arbitrary interface between two anisotropic rocks.

In isotropic media, $\Delta_{\tilde{\sigma}}^{\pm} = \tilde{\sigma}^{\pm}$ and $\tilde{\sigma}_{ij}^{\pm}\hat{n}_i = \tilde{\sigma}^{\pm}\hat{n}_j$, and Equation (14b) is changed into

$$c_{ij}\left(\tilde{\sigma}^{\mp}\right) = \delta_{ij} + \hat{n}_i\hat{n}_j\left(\frac{\tilde{\sigma}^{\mp}}{\tilde{\sigma}^{\pm}} - 1\right). \tag{18}$$

This indicates that if there is no difference in model properties, the boundary matrix becomes a unit matrix $\boldsymbol{C}_{\tilde{\sigma}}^{\mp} = \boldsymbol{I}$ due to $\tilde{\sigma}^{\mp}/\tilde{\sigma}^{\pm} = 1$. It indicates that the field intensity maintains its continuity when the net external current source is zero at the interface $\left(\Delta J_n^{\mp} = \Delta M_n^{\mp} = 0\right)$.

At the air-earth interface, we have $\Delta_{\sigma}^{+} = \tilde{\sigma}_0 = i\omega\varepsilon_0$ (pure imaginary value) and $\tilde{\boldsymbol{\sigma}}^{+} = \tilde{\sigma}_0\boldsymbol{I}$, the boundary matrix Equation (14b) becomes

$$c_{ij}\left(\tilde{\sigma}^{-}\right) = \delta_{ij} + \hat{n}_i\hat{n}_k\left(\frac{\tilde{\sigma}_{kj}^{-}}{\tilde{\sigma}_0}\right) - \hat{n}_i\hat{n}_j. \tag{19}$$

Specifically, if the electric permittivity of the earth is the same as air, *i.e.* $\tilde{\sigma}_{kj}^{-} = \sigma_{kj}^{-} + \tilde{\sigma}_0\delta_{kj}$, Equation (19) is reduced to

$$c_{ij}\left(\sigma^{-}\right) = \delta_{ij} - \frac{i}{\omega\varepsilon_0}\hat{n}_i\hat{n}_k\sigma_{kj}^{-}. \tag{20}$$

It indicates that if the electric field $\boldsymbol{E}^{-}$ is real $\left(\mathrm{Im}\{\boldsymbol{E}^{-}\} = \boldsymbol{0}\right)$ and the net external current source continues at the interface, then the real and imaginary boundary values on the "+" side are given by $\mathrm{Re}\{E_i^{+}\} = \mathrm{Re}\{E_i^{-}\}$ and $\mathrm{Im}\{E_i^{+}\} = -\hat{n}_i\hat{n}_k\sigma_{kj}^{-}E_j^{-}/\omega\varepsilon_0$ respectively. This shows that the imaginary values of the field intensity on the "+" side are not zero cross the interface.

4. Explicit Tangential Version

In contrast to the implicit formulae given by Equations (3) and (6), the explicit normal versions of the boundary conditions, e.g. Equations (13) and (18), do not directly indicate continuity of the tangential components of electromagnetic field intensities $(\boldsymbol{E},\boldsymbol{H})$ at an interface due to absence of the tangential vectors of an interface. In order to overcome this weakness, three perpendicular interface vectors $\{\hat{\boldsymbol{\tau}}_1,\hat{\boldsymbol{\tau}}_2,\hat{\boldsymbol{n}}\}$ are introduced at a point of the interface (see **Figure 1**):

$$\begin{aligned}\hat{\boldsymbol{\tau}}_1 &= (\cos\alpha, 0, \sin\alpha),\\ \hat{\boldsymbol{\tau}}_2 &= (-\cos\alpha\sin\alpha\sin\beta, \cos\beta, \cos^2\alpha\sin\beta)\Big/\sqrt{\sin^2\alpha\cos^2\beta+\cos^2\alpha},\\ \hat{\boldsymbol{n}} &= (-\sin\alpha\cos\beta, -\cos\alpha\sin\beta, \cos\alpha\cos\beta)\Big/\sqrt{\sin^2\alpha\cos^2\beta+\cos^2\alpha}.\end{aligned} \tag{21}$$

Here, the angles $\{\alpha,\beta\}$ are calculated by

$$\alpha = \tan^{-1}\left[\partial_x z(x,y)\right],\quad \beta = \tan^{-1}\left[\partial_y z(x,y)\right], \tag{22}$$

where $z(x,y)$ defines topography of an arbitrary interface. According to spline theory [7], $z(x,y)$ may be approached by a 2-D spline interpolation:

$$z(x,y) = \sum_{k,l=1}^{4} a_{ijkl}\left(x-x_{i-1}\right)^{k-1}\left(y-y_{j-1}\right)^{l-1},\quad (x,y)\in\Omega_{ij}. \tag{23}$$

The coefficients a_{ijkl} are defined in the subdomain $\Omega_{ij} = [x_{i-1},x_i]\times[y_{j-1},y_j]$ and determined by the known regularly-gridded or scattered samples of $z(x_i,y_j)$ $(1\le i\le N_x, 1\le j\le N_y)$. According to the spline theory, Equation (23) guarantees the continuity of the interface vectors $\{\hat{\boldsymbol{\tau}}_1,\hat{\boldsymbol{\tau}}_2,\hat{\boldsymbol{n}}\}$ at every point of the interface. Equations (21) and (22) indicate that the interface vectors $\{\hat{\boldsymbol{\tau}}_1,\hat{\boldsymbol{\tau}}_2,\hat{\boldsymbol{n}}\}$ change with the interface topography $z(x_i,y_j)$. If it is flat $(\alpha=\beta=0)$, then the interface vectors $\{\hat{\boldsymbol{\tau}}_1,\hat{\boldsymbol{\tau}}_2,\hat{\boldsymbol{n}}\}$ become the Cartesian vectors $\{\boldsymbol{e}_x,\boldsymbol{e}_y,\boldsymbol{e}_z\}$ or $\{\boldsymbol{e}_1,\boldsymbol{e}_2,\boldsymbol{e}_3\}$, which are the constant directions of the *x*-, *y*- and *z*-axis. Consequently, the electromagnetic field intensities may be expressed by either the Cartesian or interface-vector forms, *i.e.*

$$\boldsymbol{E}^{\pm} = E_i^{\pm}\boldsymbol{e}_i = E_n^{\pm}\hat{\boldsymbol{n}} + E_{\tau_1}^{\pm}\hat{\boldsymbol{\tau}}_1 + E_{\tau_2}^{\pm}\hat{\boldsymbol{\tau}}_2. \tag{24}$$

Therefore, Equation (3) can be rewritten in the following forms:

$$\begin{aligned}&E_{\hat{\tau}_1}^{+} = E_{\hat{\tau}_1}^{-},\quad E_{\hat{\tau}_2}^{+} = E_{\hat{\tau}_2}^{-},\\ &\hat{n}_i\tilde{\sigma}_{ij}^{+}\left(\hat{n}_jE_n^{+} + \hat{\tau}_{1j}E_{\hat{\tau}_1}^{+} + \hat{\tau}_{2j}E_{\hat{\tau}_2}^{+}\right) = \hat{n}_i\tilde{\sigma}_{ij}^{-}\left(\hat{n}_jE_n^{-} + \hat{\tau}_{1j}E_{\hat{\tau}_1}^{-} + \hat{\tau}_{2j}E_{\hat{\tau}_2}^{-}\right) + \Delta J_n^{\mp}.\end{aligned} \tag{25}$$

Combining these two equations yields

$$E_n^{\pm} = Q_1(\tilde{\boldsymbol{\sigma}}^{\mp})E_n^{\mp} + Q_2(\tilde{\boldsymbol{\sigma}}^{\mp})E_{\hat{\tau}_1}^{\mp} + Q_3(\tilde{\boldsymbol{\sigma}}^{\mp})E_{\hat{\tau}_2}^{\mp} + \Delta J_n^{\mp}/\Delta_{\tilde{\sigma}}^{\pm}, \tag{26}$$

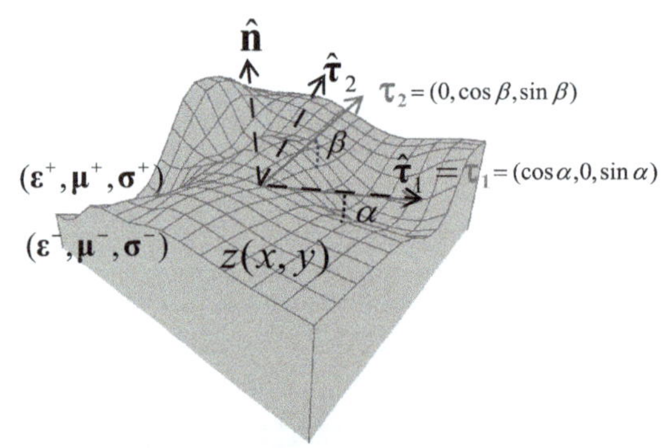

Figure 1. Illustration of three perpendicular vectors $(\hat{\boldsymbol{\tau}}_1,\hat{\boldsymbol{\tau}}_2,\hat{\boldsymbol{n}})$ at a point of an interface $z(x,y)$. $\boldsymbol{\tau}_1$ and $\boldsymbol{\tau}_2$ are the slope vectors of the interface and are employed to compute the normal $\hat{\boldsymbol{n}} = \boldsymbol{\tau}_1\times\boldsymbol{\tau}_2$ of the interface. The perpendicular tangential vectors $(\hat{\boldsymbol{\tau}}_1,\hat{\boldsymbol{\tau}}_2)$ are obtained by assigning $\hat{\boldsymbol{\tau}}_1 = \boldsymbol{\tau}_1$ and the cross product $\hat{\boldsymbol{\tau}}_2 = \hat{\boldsymbol{n}}\times\boldsymbol{\tau}_1$.

where

$$\begin{aligned} Q_1\left(\tilde{\boldsymbol{\sigma}}^{\mp}\right) &= \Delta_{\tilde{\sigma}}^{\mp}/\Delta_{\tilde{\sigma}}^{\pm}, \\ Q_2\left(\tilde{\boldsymbol{\sigma}}^{\mp}\right) &= \hat{n}_i\left(\tilde{\sigma}_{ij}^{\mp}-\tilde{\sigma}_{ij}^{\pm}\right)\hat{\tau}_{1j}/\Delta_{\tilde{\sigma}}^{\pm}, \\ Q_3\left(\tilde{\boldsymbol{\sigma}}^{\mp}\right) &= \hat{n}_i\left(\tilde{\sigma}_{ij}^{\mp}-\tilde{\sigma}_{ij}^{\pm}\right)\hat{\tau}_{2j}/\Delta_{\tilde{\sigma}}^{\pm}. \end{aligned} \tag{27}$$

Substituting Equation (26) into Equation (24) results in

$$E_i^{\pm}\boldsymbol{e}_i=\left\{\left[Q_1\left(\tilde{\boldsymbol{\sigma}}^{\mp}\right)\hat{n}_j+Q_2\left(\tilde{\boldsymbol{\sigma}}^{\mp}\right)\hat{\tau}_{1j}+Q_3\left(\tilde{\boldsymbol{\sigma}}^{\mp}\right)\hat{\tau}_{2j}\right]\hat{n}_i+\hat{\tau}_{1j}\hat{\tau}_{1i}+\hat{\tau}_{2j}\hat{\tau}_{2i}\right\}E_j^{\mp}\boldsymbol{e}_i+\left(\Delta J_n^{\mp}/\Delta_{\tilde{\sigma}}^{\pm}\right)\hat{\boldsymbol{n}}, \tag{28}$$

and

$$E_i^{\pm}=c_{ij}\left(\tilde{\boldsymbol{\sigma}}^{\mp}\right)E_j^{\mp}+\left(\Delta J_n^{\mp}/\Delta_{\tilde{\sigma}}^{\pm}\right)\hat{n}_i, \tag{29}$$

where

$$c_{ij}\left(\tilde{\boldsymbol{\sigma}}^{\mp}\right)=\hat{\tau}_{1i}\hat{\tau}_{1j}+\hat{\tau}_{2i}\hat{\tau}_{2j}+\hat{n}_i\left[Q_1\left(\tilde{\boldsymbol{\sigma}}^{\mp}\right)\hat{n}_j+Q_2\left(\tilde{\boldsymbol{\sigma}}^{\mp}\right)\hat{\tau}_{1j}+Q_3\left(\tilde{\boldsymbol{\sigma}}^{\mp}\right)\hat{\tau}_{2j}\right]. \tag{30}$$

Upon comparing Equations (29) and (30) with Equations (13) and (14), it is apparent that Equation (29) displays the same explicit linear matrix form as Equation (13) but with different boundary matrices $\boldsymbol{C}_{\tilde{\sigma}}^{\mp}$; the boundary matrix $\boldsymbol{C}_{\tilde{\sigma}}^{\mp}$ given by Equation (30) involves two tangential vectors $\{\hat{\boldsymbol{\tau}}_1,\hat{\boldsymbol{\tau}}_2\}$, whereas the previous matrix $\boldsymbol{C}_{\tilde{\sigma}}^{\mp}$ given *by* Equation (14) does not. Therefore, it can be deduced that Equation (30) is another form of Equation (14), and given the term "*explicit tangential versions*" of the boundary conditions to distinguish from the explicit normal versions.

Similarly, substituting $E_i^{\pm}=E_i^{s(\pm)}+E_i^{p(\pm)}$ into Equation (29) and then applying the boundary conditions $E_i^{p(\pm)}=c_{ij}^{\mp}E_j^{p(\pm)}+\left(\Delta J_n^{\mp}/\Delta_{\tilde{\sigma}_p}^{\pm}\right)\hat{n}_i$ in the reference model tensor $\tilde{\boldsymbol{\sigma}}_p=\left\{\tilde{\boldsymbol{\sigma}}^{+}\cup\tilde{\boldsymbol{\sigma}}_p^{-}\right\}$, $\forall\tilde{\boldsymbol{\sigma}}_p^{-}$ or $\tilde{\boldsymbol{\sigma}}_p=\left\{\tilde{\boldsymbol{\sigma}}_p^{+}\cup\tilde{\boldsymbol{\sigma}}^{+}\right\}$, $\forall\tilde{\boldsymbol{\sigma}}_p^{+}$, the following explicit tangential versions of the boundary conditions are obtained:

$$E_i^{s(\pm)}=c_{ij}\left(\tilde{\boldsymbol{\sigma}}^{\mp}\right)E_j^{s(\mp)}+\left[c_{ij}\left(\tilde{\boldsymbol{\sigma}}^{\mp}\right)-c_{ij}\left(\tilde{\boldsymbol{\sigma}}_p^{\mp}\right)\right]E_j^{p(\mp)} \tag{31}$$

Equations (29) and (31) can be changed for magnetic field intensity by symbol replacements:

$$H_i^{\pm}=c_{ij}\left(\boldsymbol{\mu}^{\mp}\right)H_j^{\mp}+\left(\Delta M_n^{\mp}/\Delta_{\mu}^{\pm}\right)\hat{n}_i \tag{32}$$

$$H_i^{s(\pm)}=c_{ij}\left(\boldsymbol{\mu}^{\mp}\right)H_j^{s(\mp)}+\left[c_{ij}\left(\boldsymbol{\mu}^{\mp}\right)-c_{ij}\left(\boldsymbol{\mu}_p^{\mp}\right)\right]H_j^{p(\mp)} \tag{33}$$

These equations correspond to Equations (4) and (8), or Equations (16) and (17).

At an isotropic interface, $Q_1\left(\tilde{\sigma}^{\mp}\right)=\tilde{\sigma}^{\mp}/\tilde{\sigma}^{\pm}$ and $Q_2\left(\tilde{\sigma}^{\mp}\right)=Q_3\left(\tilde{\sigma}^{\mp}\right)=0$. Thus, Equation (30) can be simplified to

$$c_{ij}\left(\tilde{\sigma}^{\mp}\right)=\hat{\tau}_{1i}\hat{\tau}_{1j}+\hat{\tau}_{2i}\hat{\tau}_{2j}+\left(\tilde{\sigma}^{\mp}/\tilde{\sigma}^{\pm}\right)\hat{n}_i\hat{n}_j, \tag{34}$$

At the air-earth interface, Equation (30) becomes

$$c_{ij}\left(\tilde{\boldsymbol{\sigma}}^{-}\right)=\left[\delta_{il}+\hat{n}_i\left(\tilde{\sigma}_{kl}^{-}\hat{n}_k\right)/\tilde{\sigma}_0\right]\left(\hat{n}_l\hat{n}_j+\hat{\tau}_{1l}\hat{\tau}_{1j}+\hat{\tau}_{2l}\hat{\tau}_{2j}\right)-\hat{n}_i\hat{n}_j, \tag{35}$$

and if the media possesses the same electric permittivity as air, *i.e.* $\tilde{\sigma}_{kj}^{-}=\sigma_{kl}^{-}+\tilde{\sigma}_0\delta_{ij}$, Equation (35) is changed into

$$c_{ij}\left(\tilde{\boldsymbol{\sigma}}^{-}\right)=\left[\delta_{il}+\hat{n}_i\left(\sigma_{kl}^{-}\hat{n}_k\right)/\tilde{\sigma}_0\right]\left(\hat{n}_l\hat{n}_j+\hat{\tau}_{1l}\hat{\tau}_{1j}+\hat{\tau}_{2l}\hat{\tau}_{2j}\right). \tag{36}$$

5. Equivalence of the Different Version

The two integrated and explicit boundary conditions formulated above demonstrate a matrix $\left\{\boldsymbol{C}_{\alpha}^{\pm},\boldsymbol{\alpha}\in\left(\tilde{\boldsymbol{\sigma}},\boldsymbol{\mu}\right)\right\}$ that can be calculated by either Equation (14) or Equation (30). Although the two versions are derived from the same implicit formulae, e.g. Equations (3) and (7), the boundary matrices $\{C_\alpha^\pm, \alpha \in (\tilde{\sigma}, \mu)\}$ appear to differ. From a mathematical perspective, the different versions, *i.e.* explicit normal and tangential versions, as well the original implicit equations should be equivalent to each other because of uniqueness of the boundary values.

Multiplying the matrix $A_{\tilde{\sigma}}^\pm$ to Equation (13), and then applying the factorization of the boundary matrix $C_{\tilde{\sigma}}^\pm = \left(A_{\tilde{\sigma}}^\pm\right)^{-1} A_{\tilde{\sigma}}^\mp$, the matrix form of Equation (3) is obtained from Equation (13):

$$\begin{aligned} A_{\tilde{\sigma}}^\pm E^\pm &= A_{\tilde{\sigma}}^\pm \left[C_{\tilde{\sigma}}^\mp E^\mp + \left(\Delta J_n^\mp / \Delta_{\tilde{\sigma}}^\pm\right) \hat{n} \right] \\ &= A_{\tilde{\sigma}}^\pm \left(A_{\tilde{\sigma}}^\mp\right)^{-1} A_{\tilde{\sigma}}^\mp E^\mp + \left(\Delta J_n^\mp / \Delta_{\tilde{\sigma}}^\pm\right) A_{\tilde{\sigma}}^\mp \hat{n} \\ &= A_{\tilde{\sigma}}^\mp E^\mp + \Delta J_n^\mp, \end{aligned} \tag{37}$$

Similarly, Equations (15), (16) and (17) can be changed into Equations (7), (4) and (8) respectively. These formulations show that the explicit normal versions are equivalent to two common implicit boundary equations.

Applying the perpendicular properties of the interface vectors to Equation (30), e.g. $\hat{n} \times \hat{n} = 0$, $\hat{\tau}_1 \cdot \hat{n} = 0$ and $\hat{\tau}_2 \cdot \hat{n} = 0$, the following equations are obtained:

$$\begin{aligned} &\hat{\tau}_{1i} c_{ij}\left(\tilde{\sigma}^\mp\right) = \hat{\tau}_{1j}, \quad \hat{\tau}_{2i} c_{ij}\left(\tilde{\sigma}^\mp\right) = \hat{\tau}_{2j}, \\ &\hat{n}_i c_{ij}\left(\tilde{\sigma}^\mp\right) = Q_1\left(\tilde{\sigma}^\mp\right)\hat{n}_j + Q_2\left(\tilde{\sigma}^\mp\right)\hat{\tau}_{1j} + Q_3\left(\tilde{\sigma}^\mp\right)\hat{\tau}_{1j}. \end{aligned} \tag{38}$$

Substituting these identities into Equation (29) yields

$$\begin{aligned} E_{\hat{\tau}_1}^\pm &= \hat{\tau}_{1i} E_i^\pm = \hat{\tau}_{1i} c_{ij}^\mp(\tilde{\sigma}) E_j^\mp = \hat{\tau}_{1j} E_j^\mp = E_{\hat{\tau}_1}^\mp, \\ E_{\hat{\tau}_2}^\pm &= \hat{\tau}_{2i} E_i^\pm = \hat{\tau}_{2i} c_{ij}^\mp(\tilde{\sigma}) E_j^\mp = \hat{\tau}_{2j} E_j^\mp = E_{\hat{\tau}_2}^\mp, \\ E_n^\pm &= \hat{n}_i E_i^\pm = \hat{n}_i c_{ij}^\mp(\tilde{\sigma}) E_j^\mp = Q_1\left(\tilde{\sigma}^\mp\right) E_n^\mp + Q_2\left(\tilde{\sigma}^\mp\right) E_{\hat{\tau}_1}^\mp + Q_3\left(\tilde{\sigma}^\mp\right) E_{\hat{\tau}_2}^\mp. \end{aligned} \tag{39}$$

These equations indicate continuity of the tangential components and discontinuity of the normal components of the electric field intensities. It proves that the explicit tangential versions are also equivalent to two common implicit boundary conditions.

Note that the three interface vectors given by Equation (21) satisfy the following equation

$$\delta_{ij} = \hat{\tau}_{1i}\hat{\tau}_{1j} + \hat{\tau}_{2i}\hat{\tau}_{2j} + \hat{n}_i\hat{n}_j. \tag{40}$$

Accordingly, equation (30) may be rewritten as follow

$$\begin{aligned} c_{ij}\left(\tilde{\sigma}^\mp\right) &= \left(\hat{n}_i\hat{n}_j + \hat{\tau}_{1i}\hat{\tau}_{1j} + \hat{\tau}_{2i}\hat{\tau}_{2j}\right) + \hat{n}_i\hat{n}_k\left(\tilde{\sigma}_{kl}^- - \tilde{\sigma}_{kl}^+\right)\left[\hat{n}_l\hat{n}_j + \hat{\tau}_{1l}\hat{\tau}_{1j} + \hat{\tau}_{2l}\hat{\tau}_{2j}\right] / \Delta_{\tilde{\sigma}}^\pm \\ &= \delta_{ij} + \hat{n}_i\left(\tilde{\sigma}_{kj}^- - \tilde{\sigma}_{kj}^+\right)\hat{n}_k / \Delta_{\tilde{\sigma}}^\pm, \end{aligned} \tag{41}$$

which is the same as Equation (14b). Similarly, substituting Equation (40) for Equations (34), (35) and (36) respectively, they become Equations (18), (19) and (20). Therefore, the explicit tangential versions are equivalent to the explicit normal versions and vice versa as Equation (41) are reversible. Specifically, when the two media have the same electric permittivity ε_0, *i.e.* $\tilde{\sigma}_{kj}^\pm = \sigma_{kj}^\pm + i\omega\varepsilon_0\delta_{kj}$, Equation (41) is changed into

$$c_{ij}\left(\tilde{\sigma}^\mp\right) = \delta_{ij} + \hat{n}_i\left(\sigma_{kj}^- - \sigma_{kj}^+\right)\hat{n}_k \left[\frac{\hat{n}_i\sigma_{kj}^+\hat{n}_k - i\omega\varepsilon_0}{\left(\hat{n}_i\sigma_{kj}^+\hat{n}_k\right)^2 + \left(\omega\varepsilon_0\right)^2}\right]. \tag{42}$$

This shows the small imaginary value $(\omega\varepsilon_0 \ll 1)$ when a low frequency is considered.

6. Transformation of Boundary Values

The boundary element theory has shown that if there is not any external current source j_e and m_e in a homogeneous medium, the electromagnetic field intensities $\{E^+, H^+\}$ in the medium domain Ω^+ may be expressed by the following boundary integral [8]:

$$\Theta\begin{pmatrix} \boldsymbol{E}^+(\boldsymbol{r}) \\ \boldsymbol{H}^+(\boldsymbol{r}) \end{pmatrix} = \int_\Gamma \left[G_0(\boldsymbol{r},\boldsymbol{r}') \begin{pmatrix} \dfrac{\partial \boldsymbol{E}^+}{\partial \hat{n}} \\ \dfrac{\partial \boldsymbol{H}^+}{\partial \hat{n}} \end{pmatrix} - \frac{\partial G_0(\boldsymbol{r},\boldsymbol{r}')}{\partial \hat{n}} \begin{pmatrix} \boldsymbol{E}^+ \\ \boldsymbol{H}^+ \end{pmatrix} \right] \mathrm{d}\Gamma, \quad (\boldsymbol{r} \in \Omega^+, \boldsymbol{r}' \in \Gamma). \tag{43}$$

Here, $G_0(\boldsymbol{r},\boldsymbol{r}')$ is the Greens function of the homogeneous medium, Θ takes the values of 1.0, 0.5 and $\theta/2\pi$ responses to $\boldsymbol{r} \in \Omega^+$, $\boldsymbol{r} \in \Gamma$ (smooth) and $\boldsymbol{r} \in \Gamma$ (not smooth) respectively, and θ is the corner angle at $\boldsymbol{r} \in \Gamma$. This equation indicates that calculation of the electromagnetic field intensities $\{\boldsymbol{E}^+, \boldsymbol{H}^+, \boldsymbol{r} \in \Omega^+\}$ in the homogeneous medium require not only the boundary values of the field intensities $\{\boldsymbol{E}^+, \boldsymbol{H}^+, \boldsymbol{r} \in \Gamma\}$ but also the normal derivatives $\{\partial \boldsymbol{E}^+/\partial \hat{n}, \partial \boldsymbol{H}^+/\partial \hat{n}, \boldsymbol{r} \in \Gamma\}$. The boundary element method based on Equation (43) [8] [9] offers a tool to find the boundary values $\{\partial \boldsymbol{E}^+/\partial \hat{n}, \partial \boldsymbol{H}^-/\partial \hat{n}, \boldsymbol{r} \in \Gamma\}$ or $\{\boldsymbol{E}^+, \boldsymbol{H}^+, \boldsymbol{r} \in \Omega^+\}$ with the known field intensities $\{\boldsymbol{E}^+, \boldsymbol{H}^+, \boldsymbol{r} \in \Omega^+\}$ or normal derivatives $\{\partial \boldsymbol{E}^+/\partial \hat{n}, \partial \boldsymbol{H}^+/\partial \hat{n}, \boldsymbol{r} \in \Gamma\}$. Unfortunately, in most of electromagnetic modeling cases, neither the boundary values of the field intensities $\{\boldsymbol{E}^+, \boldsymbol{H}^+, \boldsymbol{r} \in \Omega^+\}$ nor the normal derivatives $\{\partial \boldsymbol{E}^+/\partial \hat{n}, \partial \boldsymbol{H}^-/\partial \hat{n}, \boldsymbol{r} \in \Gamma\}$ are known. However, if the field intensities $\{\boldsymbol{E}^-, \boldsymbol{H}^-\}$ in the connected domain Ω^- are given or going to be solved, the normal derivatives $\{\partial \boldsymbol{E}^-/\partial \hat{n}, \partial \boldsymbol{H}^-/\partial \hat{n}\}$ at the interface can be calculated by numerical differentiations with the known or solved field intensities $\{\boldsymbol{E}^-, \boldsymbol{H}^-\}$. In this case, application of Equation (43) needs transformations of the boundary values from Ω^+ to Ω^-. Apparently, the integrated and explicit boundary conditions presented in the previous sections are directly applicable to these transformations, e.g. substituting Equations (13) and (16) for the second term of the right-hand-side surface integral of Equation (43) achieves the transformation of the boundary values $\{\boldsymbol{E}^+, \boldsymbol{H}^+\}$ into $\{\boldsymbol{E}^-, \boldsymbol{H}^-\}$. For transforming the normal derivatives $\{\partial \boldsymbol{E}^+/\partial \hat{n}, \partial \boldsymbol{H}^+/\partial \hat{n}\}$ into $\{\partial \boldsymbol{E}^-/\partial \hat{n}, \partial \boldsymbol{H}^-/\partial \hat{n}\}$, one may follow the same methodology as described in the previous sections and obtain the integrated and explicit boundary conditions of the normal derivatives.

We calculate $\partial/\partial \hat{n}$ on both sides of Equation (1) and obtain

$$\begin{aligned} &\nabla \times \frac{\partial \boldsymbol{E}}{\partial \hat{n}} + i\omega \frac{\partial}{\partial \hat{n}}(\boldsymbol{\mu} \cdot \boldsymbol{H}) = -\frac{\partial \boldsymbol{m}_e}{\partial \hat{n}}, \\ &\nabla \times \frac{\partial \boldsymbol{H}}{\partial \hat{n}} - \frac{\partial}{\partial \hat{n}}(\tilde{\boldsymbol{\sigma}} \cdot \boldsymbol{E}) = \frac{\partial \boldsymbol{j}_e}{\partial \hat{n}}, \end{aligned} \tag{44}$$

which give zero divergences

$$\begin{aligned} &\nabla \cdot \left[i\omega \frac{\partial}{\partial \hat{n}}(\boldsymbol{\mu} \cdot \boldsymbol{H}) + \frac{\partial \boldsymbol{m}_e}{\partial \hat{n}} \right] = 0, \\ &\nabla \cdot \left[\frac{\partial}{\partial \hat{n}} \cdot (\tilde{\boldsymbol{\sigma}} \cdot \boldsymbol{E}) + \frac{\partial \boldsymbol{j}_e}{\partial \hat{n}} \right] = 0. \end{aligned} \tag{45}$$

Applying Equations (44) and (45) to an interface of two anisotropic media, we obtain the boundary conditions of the partial derivatives:

$$\begin{aligned} &\left(\frac{\partial \boldsymbol{E}^+}{\partial \hat{n}} - \frac{\partial \boldsymbol{E}^-}{\partial \hat{n}} \right) \times \hat{\boldsymbol{n}} = \boldsymbol{0}, \\ &\left(\tilde{\boldsymbol{\sigma}}^+ \cdot \frac{\partial \boldsymbol{E}^+}{\partial \hat{n}} - \tilde{\boldsymbol{\sigma}}^- \cdot \frac{\partial \boldsymbol{E}^-}{\partial \hat{n}} \right) \cdot \hat{\boldsymbol{n}} = \left(\frac{\partial \tilde{\boldsymbol{\sigma}}^-}{\partial \hat{n}} \cdot \boldsymbol{E}^- - \frac{\partial \tilde{\boldsymbol{\sigma}}^+}{\partial \hat{n}} \cdot \boldsymbol{E}^+ \right) \cdot \hat{\boldsymbol{n}} + \frac{\partial \Delta J_n^{\mp}}{\partial \hat{n}}, \end{aligned} \tag{46}$$

and

$$\begin{aligned} &\left(\frac{\partial \boldsymbol{H}^+}{\partial \hat{n}} - \frac{\partial \boldsymbol{H}^-}{\partial \hat{n}} \right) \times \hat{\boldsymbol{n}} = \boldsymbol{0}, \\ &\left(\boldsymbol{\mu}^+ \cdot \frac{\partial \boldsymbol{H}^+}{\partial \hat{n}} - \boldsymbol{\mu}^- \cdot \frac{\partial \boldsymbol{H}^-}{\partial \hat{n}} \right) \cdot \hat{\boldsymbol{n}} = \left(\frac{\partial \boldsymbol{\mu}^-}{\partial \hat{n}} \cdot \boldsymbol{H}^- - \frac{\partial \boldsymbol{\mu}^+}{\partial \hat{n}} \cdot \boldsymbol{H}^+ \right) \cdot \hat{\boldsymbol{n}} + \frac{\partial \Delta M_n^{\mp}}{\partial \hat{n}}. \end{aligned} \tag{47}$$

Therefore, we have the following integrated and explicit versions of Equations (46) and (47):

$$\frac{\partial \boldsymbol{E}^{\pm}}{\partial \hat{n}} = \boldsymbol{C}_{\tilde{\sigma}}^{\mp} \frac{\partial \boldsymbol{E}^{\mp}}{\partial \hat{n}} + \boldsymbol{C}_{\tilde{\sigma}}^{\prime\mp} \boldsymbol{E}^{\mp} + \frac{1}{\Delta_{\tilde{\sigma}}^{\pm}} \left[\frac{\partial \Delta J_n^{\mp}}{\partial \hat{n}} - \frac{\Delta J_n^{\mp}}{\Delta_{\tilde{\sigma}}^{\pm}} \left(\hat{n}_k \frac{\partial \tilde{\sigma}_{kl}^{\pm}}{\partial \hat{n}} \hat{n}_l \right) \right] \hat{\boldsymbol{n}}, \tag{48}$$

$$\frac{\partial \boldsymbol{H}^{\pm}}{\partial \hat{n}} = \boldsymbol{C}_{\mu}^{\mp} \frac{\partial \boldsymbol{H}^{\mp}}{\partial \hat{n}} + \boldsymbol{C}_{\mu}^{\prime\mp} \boldsymbol{H}^{\mp} + \frac{1}{\Delta_{\mu}^{\pm}} \left[\frac{\partial \Delta M_n^{\mp}}{\partial \hat{n}} - \frac{\Delta M_n^{\mp}}{\Delta_{\mu}^{\pm}} \left(\hat{n}_k \frac{\partial \mu_{kl}^{\pm}}{\partial \hat{n}} \hat{n}_l \right) \right] \hat{\boldsymbol{n}}, \tag{49}$$

where the components of matrices $\boldsymbol{C}_{\tilde{\sigma}}^{\prime\mp}$ and $\boldsymbol{C}_{\mu}^{\prime\mp}$ are given by

$$c_{\alpha}^{\prime\pm} = \hat{n}_i \left[\hat{n}_k \frac{\partial \alpha_{kl}^{-}}{\partial \hat{n}} \delta_{lj} - \hat{n}_k \frac{\partial \alpha_{kl}^{+}}{\partial \hat{n}} c_{lj}(\boldsymbol{\alpha}) \right], \quad \boldsymbol{\alpha} \in \{\tilde{\boldsymbol{\sigma}}, \boldsymbol{\mu}\}. \tag{50}$$

Applying Equations (13), (16), (48) and (49) for Equation (43), one can fulfill the transformations of the boundary values from the domain Ω^{+} into Ω^{-}, in which the field intensities $(\boldsymbol{E}^{-}, \boldsymbol{H}^{-})$ are going to be solved and the normal derivatives $(\partial \boldsymbol{E}^{-}/\partial \hat{n}, \partial \boldsymbol{H}^{-}/\partial \hat{n})$ can be calculated with the given interface topography $z(x, y)$ and its nearby field intensities $(\boldsymbol{E}^{-}, \boldsymbol{H}^{-})$. Particularly, after achieving the transformations of the boundary values, the boundary integral $(\boldsymbol{r} \in \Gamma)$ in Equation (43) can be approached by the boundary element method [9] that results in N_{Γ} (total points of the interface) linear equations of the field intensity $\boldsymbol{E}^{-}$ or $\boldsymbol{H}^{-}$. These equations are considered as "*the boundary equations*" of the field intensities $(\boldsymbol{E}^{-}, \boldsymbol{H}^{-})$ and independently complementary to the linear equations yielded by other numerical approach applied to the domain Ω^{-}, e.g. finite-difference or finite-element method. Therefore, the numerical computations of the field intensities $(\boldsymbol{E}^{-}, \boldsymbol{H}^{-})$ are implemented only in the domain Ω^{-} and have nothing relating to Ω^{+}, so that the computational dimensions are significantly reduced. These developments of hybrid methods are beyond the topic of this paper and will be given in our future articles.

7. Synthetic Examples

In order to demonstrate possible applications of the integrated and explicit versions of the boundary conditions, synthetic experiments of a hill and a ridge model have been conducted (see **Figure 2** and **Figure 4**). These models may represent the Earth's surface, or seafloors, or subsurface interfaces of rocks. The synthetic experiments were only carried out using electric fields $\boldsymbol{E}^{\pm}$ with the explicit normal versions due to the similarity between magnetic fields $\boldsymbol{H}^{\pm}$ and electric fields $\boldsymbol{E}^{\pm}$, and the equivalence of the two explicit versions. In these experiments, the frequency of 0.1 Hz and an external plane-wave source at infinity were considered $(\Delta J_n^{\mp} = 0)$, and the hill and ridge interfaces were approximated by Equation (23) using regularly-gridded samples of the interface topographies. Above the interface, the conductivity tensor $\tilde{\boldsymbol{\sigma}}^{+}$ was assigned to the air $(\tilde{\boldsymbol{\sigma}}^{+} = i\omega\varepsilon_0 \boldsymbol{I})$ and an anisotropic medium $(\tilde{\boldsymbol{\sigma}}^{+} = (0.5, 0.1, 0.2, 0.7, 0.4.1.0) + i\omega\varepsilon_0\delta_{ij})$ respectively. Below the interface, a different anisotropic medium was applied $(\tilde{\boldsymbol{\sigma}}^{-} = (1.0, 0.5, 0.4, 2.0, 0.3, 3.0) + i\omega\varepsilon_0\delta_{ij})$. These two media have the same electric permittivity as air. In addition, we assumed the boundary values of the electric field intensity $\boldsymbol{E}^{+}$ in the air-domain are known, e.g. $\boldsymbol{E}_{ij}^{+} = \cos(ax_i)\sin(by_j)\hat{\boldsymbol{\tau}}_1 + \sin(ax_i)\cos(by_j)\hat{\boldsymbol{\tau}}_2 + \sin(ax_i)\sin(by_j)\hat{\boldsymbol{n}}$, $\forall(x_i, y_j) \in z(x, y)$, which represents the observed data on the Earth's surface or seafloor from a practical measurement [10], or the numerical solution from the boundary element method [8] [9]. The new integrated and explicit versions enable us to directly recover the boundary values $\boldsymbol{E}^{-}$ under the ground or seafloor. It is possible to combine the transformed boundary values with other numerical method in Ω^{-} and perform the forward modeling or tomographic inversion without the air or seawater domain.

Figure 2 displays the synthetic results at the air-earth interface of a hill. Three components of the boundary values $\{E_x^{\pm}, E_y^{\pm}, E_z^{\pm}\}$ are plotted and show discontinuities throughout the vertical components $E_z^{\pm}$, and continuity in the horizontal components $\{E_x^{\pm}, E_y^{\pm}\}$ at the flat portions of the interface due to $\hat{\boldsymbol{\tau}}_1 = \boldsymbol{e}_x$ and $\hat{\boldsymbol{\tau}}_2 = \boldsymbol{e}_y$. Discontinuity of $\{E_x^{\pm}, E_y^{\pm}\}$ in the hill area arises when $\hat{\boldsymbol{\tau}}_1 \neq \boldsymbol{e}_x$ and $\hat{\boldsymbol{\tau}}_2 \neq \boldsymbol{e}_y$. It also shows that the imaginary parts $\mathrm{Im}\{E_x^{\pm}, E_y^{\pm}, E_z^{\pm}\}$ are very small $(< 10^{-11}\ \mathrm{V/m})$ due to the low frequency (0.1 Hz) and same electric permittivity ε_0 of the two media. Therefore, these imaginary parts are often ignored in most magnetotelluric measurements [1].

Figure 3 demonstrates three components of the electric current density $\boldsymbol{J}^{\pm} = (J_{\hat{\tau}_1}^{\pm}, J_{\hat{\tau}_2}^{\pm}, J_{\hat{n}}^{\pm})$, whose real $(\mathrm{Re}\{\boldsymbol{J}^{\pm}\})$ and imaginary values $(\mathrm{Im}\{\boldsymbol{J}^{\pm}\})$ display the conduction and displacement current densities respectively. These diagrams indicate that the conduction current density disappears in air $(\mathrm{Re}\{\boldsymbol{J}^{+}\} = \boldsymbol{0})$ due to zero

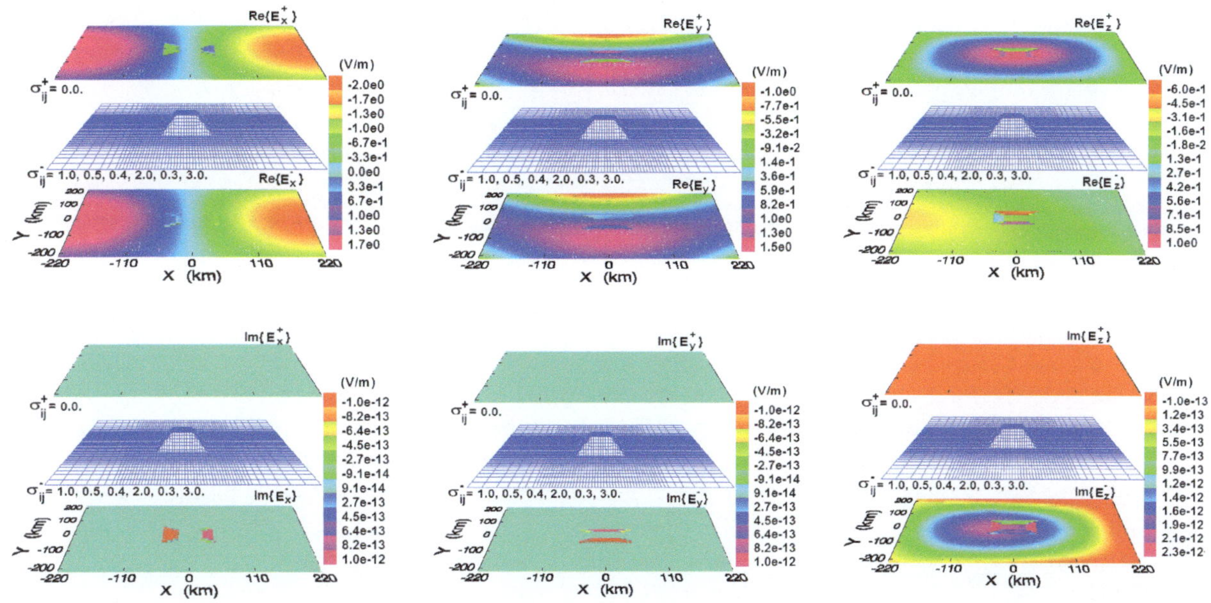

Figure 2. Synthetic results of electric fields $\boldsymbol{E}^{\pm}=\left(E_x^{\pm},E_y^{\pm},E_z^{\pm}\right)$ at the air-earth interface that has a hill topography and anisotropic ground. The images over and under the surface give the boundary values $\boldsymbol{E}^{+}=\left(E_x^{+},E_y^{+},E_z^{+}\right)$ and $\boldsymbol{E}^{-}=\left(E_x^{-},E_y^{-},E_z^{-}\right)$ computed by the explicit normal versions of the boundary conditions.

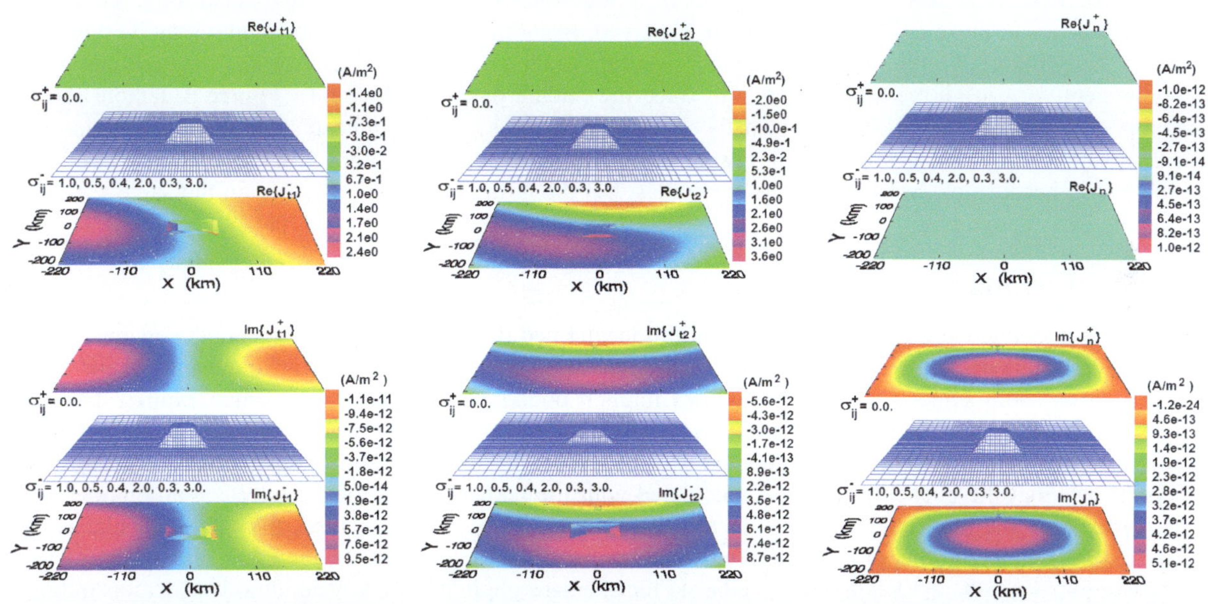

Figure 3. Synthetic results of electric current density $\boldsymbol{J}^{\pm}=\left(J_{\hat{t}_1}^{\pm},J_{\hat{t}_2}^{\pm},J_{\hat{n}}^{\pm}\right)$ at the air-earth interface that has a hill topography and anisotropic ground. The images over and under the surface are the boundary values $\boldsymbol{J}^{+}=\left(J_{\hat{t}_1}^{+},J_{\hat{t}_2}^{+},J_{\hat{n}}^{+}\right)$ and $\boldsymbol{J}^{-}=\left(J_{\hat{t}_1}^{-},J_{\hat{t}_2}^{-},J_{\hat{n}}^{-}\right)$ computed by the explicit normal version of the boundary conditions.

conductivity $\left(\boldsymbol{\sigma}^{+}=0\right)$ and displacement current density occurs $\left(\operatorname{Im}\left\{\boldsymbol{J}^{\pm}\right\}\neq\mathbf{0}\right)$ because of non-zero electric permittivity $\left(\boldsymbol{\varepsilon}^{\pm}=\varepsilon_0\boldsymbol{I}\right)$, whilst the normal total current densities remain unchanged $\left(\boldsymbol{J}_{\hat{n}}^{+}=\boldsymbol{J}_{n}^{-}\right)$ and the tangential total current densities vary $\left(\left\{\boldsymbol{J}_{\hat{t}_1}^{+},\boldsymbol{J}_{\hat{t}_2}^{+}\right\}\neq\left\{\boldsymbol{J}_{\hat{t}_1}^{-},\boldsymbol{J}_{\hat{t}_2}^{-}\right\}\right)$. These characteristics are predictable from the implicit boundary equations.

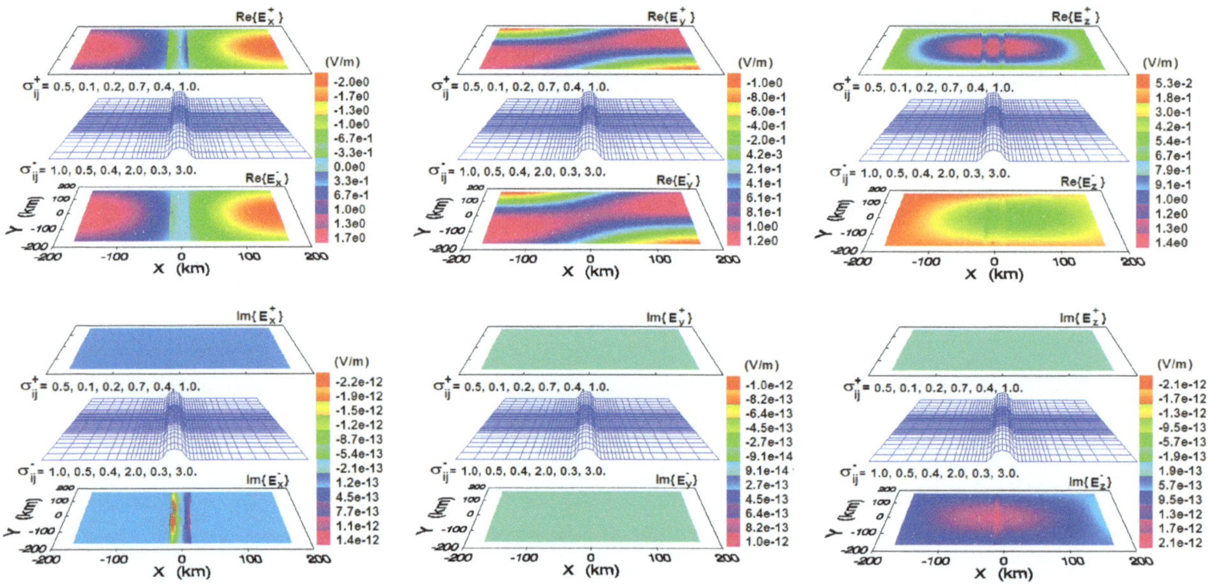

Figure 4. Synthetic results of electric fields $\boldsymbol{E}^{\pm}=\left(E_x^{\pm},E_y^{\pm},E_z^{\pm}\right)$ at a ridge interface between two anisotropic rocks. The images over and under the surface give the boundary values $\boldsymbol{E}^{+}=\left(E_x^{+},E_y^{+},E_z^{+}\right)$ and $\boldsymbol{E}^{-}=\left(E_x^{-},E_y^{-},E_z^{-}\right)$ computed by the explicit normal version of the boundary conditions.

Figure 4 demonstrates the synthetic results of a ridge interface that connects two anisotropic media. Similar characteristics to those in **Figure 2** are again observed, including discontinuities throughout the vertical components $E_z^{\pm}$, continuity in the *x*-components $E_x^{\pm}$ except in the ridge area where $\hat{\boldsymbol{\tau}}_1 \neq \boldsymbol{e}_x$, and continuity in the *y*-component $E_y^{\pm}$ in all areas due to $\hat{\boldsymbol{\tau}}_2=\boldsymbol{e}_y$ (see the middle panel in **Figure 4**). **Figure 5** demonstrates three components of the electric current density $\boldsymbol{J}^{\pm}=\left\{J_{\hat{\tau}_1}^{\pm},J_{\hat{\tau}_2}^{\pm},J_{\hat{n}}^{\pm}\right\}$, which indicate that the conduction and displacement current densities exist in the two media, and the tangential current densities $\left\{J_{\hat{\tau}_1}^{-},J_{\hat{\tau}_2}^{-}\right\}$ differ from $\left\{J_{\hat{\tau}_1}^{+},J_{\hat{\tau}_2}^{+}\right\}$ because of two different conductivities, but the normal currents $J_{\hat{n}}^{\pm}$ remain the same regardless of the interface topography.

8. Conclusions

Two new integrated and explicit boundary conditions, termed the "normal" and "tangential" versions, have been presented in this paper for electromagnetic fields at an arbitrary interface between two anisotropic media. These two versions both achieve combination of two implicit boundary equations into a single explicit linear matrix form, and consistently reveal that the boundary values are linked by a 3 × 3 boundary matrix dependent on the interface topography and electric conductivity or magnetic permeability tensors of the media. The normal version shows that the boundary matrix is calculated with the known normal of the interface and model property tensors; while the tangential version indicates that the boundary matrix requires two perpendicular tangential vectors besides the normal of the interface. However, despite these differences, the mathematical equivalence of the two new versions to each other, as well as to the standard implicit boundary conditions is demonstrated. With known normal $\hat{\boldsymbol{n}}$ of an interface, the explicit normal version is more compact and efficient compared to the explicit tangential version because the two perpendicular tangential vectors $\{\hat{\boldsymbol{\tau}}_1,\hat{\boldsymbol{\tau}}_2\}$ are not required. With a given interface $z(x,y)$, there is no difference between the two versions in computational efficiency as the tangential vectors $\{\hat{\boldsymbol{\tau}}_1,\hat{\boldsymbol{\tau}}_2\}$ and normal $\hat{\boldsymbol{n}}$ must be calculated from the interface topography function $z(x,y)$.

The synthetic examples of a hill and a ridge interface demonstrate possible applications in conversions of the boundary values, and capability of the new versions to arbitrary interfaces that may involve complex topography and anisotropic rocks. These results numerically show continuity of the tangential components and discontinuities of the normal components of electromagnetic field intensities, and continuity of the normal components and discontinuities of the tangential components of electric current densities across the air-earth interface and the boundary of two anisotropic rocks. These synthetic examples also demonstrate that the boundary values of the

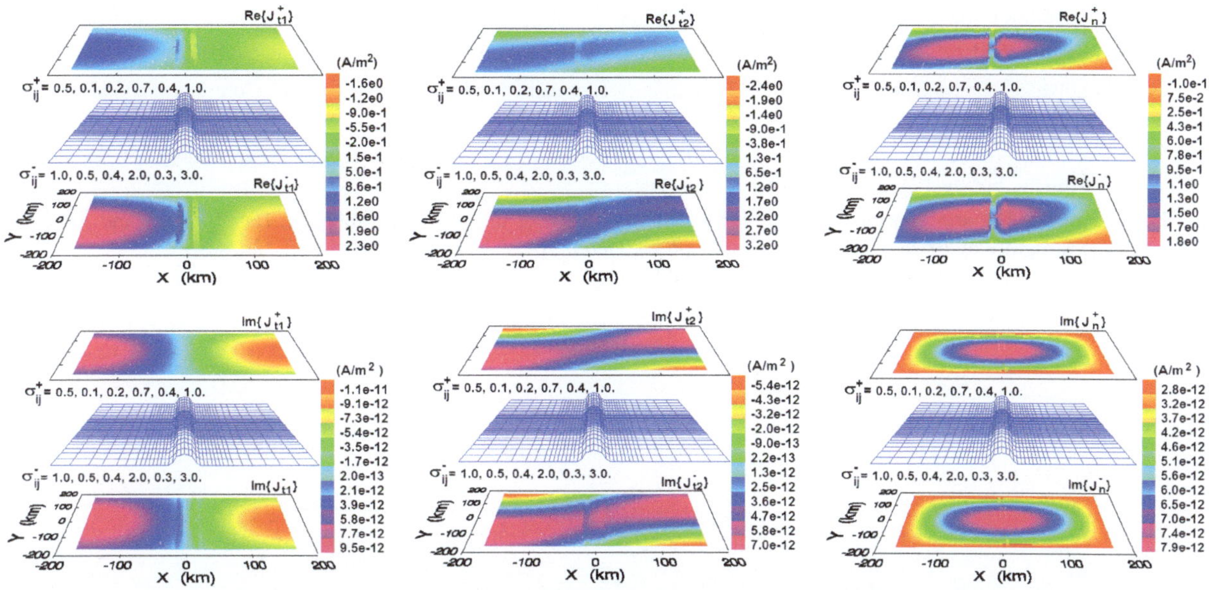

Figure 5. Synthetic results of electric current density $\boldsymbol{J}^{\pm}=\left(J_{\hat{t}_1}^{\pm},J_{\hat{t}_2}^{\pm},J_{\hat{n}}^{\pm}\right)$ at a ridge interface between two anisotropic rocks. The images over and under the surface show the boundary values $\boldsymbol{J}^{+}=\left(J_{\hat{t}_1}^{+},J_{\hat{t}_2}^{+},J_{\hat{n}}^{+}\right)$ and $\boldsymbol{J}^{-}=\left(J_{\hat{t}_1}^{-},J_{\hat{t}_2}^{-},J_{\hat{n}}^{-}\right)$ respectively, computed by the explicit version of the boundary conditions.

field intensities may change with alterations in topography of the interface, electric conductivity and permittivity tensors, or magnetic permeability tensors. It is shown that with help of the new integrated and explicit versions, the unknown boundary values can be obtained by simply multiplying a boundary matrix with the known boundary values. Therefore, it provides a more straightforward and easier method to transform the boundary values from one domain to another. It is greatly helpful to not only extrapolation of electromagnetic fields with the boundary element approach, but also combination of the boundary element approach with other numerical methods, such as finite-difference, finite-element and integral equation method, because the boundary element approach with the transformed boundary values can offer complementary linear equations to these numerical methods, so that the numerical computations remain in the interesting model domain and the computational dimensions are significantly reduced.

Acknowledgements

This work was supported by a Discovery Project (DP1093110) of the Australia Research Council. The authors thank Mr. Craig Patten for his assistance in using high-performance computing facility at e Research SA in Australia.

References

[1] Weaver, J.T. (1995) Mathematic Methods for Geo-Electromagnetic Induction. Research Studies Press Ltd., Taunton, Somerset.

[2] Key, K. and Weiss, C. (2006) Adaptive Finite-Element Modeling Using Unstructured Grids: The 2D Magnetotelluirc Example. *Geophysics*, **71**, G291-G299. http://dx.doi.org/10.1190/1.2348091

[3] Mukheriee, S. and Everett, M. (2011) 3D Controlled-Source Electromagnetic Edge-Based Finite Element Modeling of Conductive and Permeable Heterogeneities. *Geophysics*, **76**, F215-F226. http://dx.doi.org/10.1190/1.3571045

[4] Axia, R. (2014) Multi-Order Hexahedral Vector Finite Element Method for 3-D MT Modeling, Including Anisotropy and Complex Geometry. PhD Thesis, Adelaide University, Adelaide.

[5] Thide, B. (2004) Electromagnetic Field Theory. Upsilon Books, Communa AB, Uppsala.

[6] Zhdanov, M.S., Varentsov, I.M., Waever, J.T., Golubev, N.G. and Krylov, V.A. (1997) Methods for Modeling Electromagnetic Fields Results from COMMEMI—The International Project on the Comparison of Modeling Methods for Electromagnetic Induction. *Journal of Applied Geophysics*, **37**, 133-271.

http://dx.doi.org/10.1016/S0926-9851(97)00013-X

[7] Helmuth, S. (1995) Two Dimensional Spline Interpolation Algorithms. A. K. Peter Ltd, Wellesley.

[8] Brebbia, C.A. and Dominguez, J. (1992) Boundary Elements: An Introductory Course. Computational Mechanics Publications, Boston.

[9] Beer, G., Smith, I.M. and Duenser, C. (2008) The Boundary Element Method with Programming for Engineering and Scientists. Springer Wien, New York.

[10] Everett, M.E. and Constable, S. (1999) Electric Dipole Fields over an Anisotropic Seafloor: Theory and Application to the Structure of 40Ma Pacific Ocean Lithosphere. *Geophysical Journal International*, **136**, 41-56. http://dx.doi.org/10.1046/j.1365-246X.1999.00725.x

3

Performance Enhancement for Adaptive Beam-Forming Application Based Hybrid PSOGSA Algorithm

Ahmed Magdy[1], Osama M. EL-Ghandour[2], Hesham F. A. Hamed[1]

[1]Department of Electrical Engineering, Minia University, Minia, Egypt
[2]Department of Electronics, Communications and Computer Engineering, Helwan University, Cairo, Egypt
Email: eng_ahmedmagdy86@yahoo.com

Abstract

Recently researchers were interested in hybrid algorithms for optimization problems for several communication systems. In this paper, a novel algorithm based on hybrid PSOGSA technique (combination of Gravitational Search Algorithm and Particle Swarm Optimization) is presented to enhance the performance analysis of beam-forming for smart antennas systems using N elements for Uniform Circular Array (UCA) geometry. Complex excitations (phases) of the array radiation pattern are optimized using hybrid PSOGSA technique for a set of simultaneously incident signals. Our results have shown tremendous improvement over the previous work was done using Uniform Linear Array (ULA) geometry and standard GSA in terms of normalized array factor and computational speed for normalized fitness values.

Keywords

Smart Antennas, UCA, Optimization Techniques, Beam-Forming

1. Introduction

Adaptive beam-forming capabilities for smart antenna arrays are nowadays used in different applications such as suppression and reduction of interference in wireless mobile communication, besides its effects on the overall quality of service [1] [2]. There are different optimization techniques dealing with adaptive beam-forming for smart antennas. Particle Swarm Optimization (PSO), Central Force Optimization (CFO), and Bacterial Swarm Optimization (BSO) are well known global optimization techniques that are based on a nature-inspired heuristic [3]-[5]. It is proven that CFO requires higher computational complexity but on the other hand it has better performance than PSO in [6].

Recently, Gravitational Search Algorithm (GSA) is considered as a new optimization technique based on the law of gravity and mass interaction [7]. Where a set of various standard benchmark functions, synthesis of thinned scanned concentric ring array antenna and a fully digital controlled reconfigurable concentric ring array antenna problems were examined in [8] [9]. In most cases the GSA provided superior or at least comparable results with PSO and CFO. The GSA was proposed in [10] [11] for calculating the dimensions of a rectangular patch antenna, and for Direction of Arrival (DOA) estimation using a Uniform Circular Array (UCA) of 12 elements based on maximum likelihood (ML) criteria and showed better performance results over PSO and multiple signal classification (MUSIC) in terms of computational time for fitness function and RMSE. In [12], planar ultra-wide band (UWB) antennas with irregular radiator shapes were designed using GSA and compared with those obtained using CFO algorithm, it was found that the GSA gave better performance than CFO. In [13], GSA was implemented using Open-MP and its results outperformed the PSO by 65.09% in terms of normalized array factor. In [14], an algorithm based on collective animal behaviour (CAB) was used for finding the best optimal non-uniform excitation weights and optimal uniform inter-element spacing for hyper beam-forming of linear antenna arrays. In [15], a new algorithm based on Hybrid Particle Swarm Optimization with Gravitational Search Algorithm (Hybrid PSOGSA) technique was proposed and showed better performance than standard PSO and GSA in terms of computational speed.

In this paper, a novel algorithm that is based on the hybrid PSOGSA technique is developed for optimal beam-forming using ULA and UCA. The goal is to maximize the beam of the radiation pattern towards the intended user or Signal of Interest (SOI) and minimize the beam of the radiation pattern towards Signal Not of Interest (SNOI) based on controlling the complex weights (phase) of ULA or UCA. The paper is organized as follows. In Section 2, the system model and problem formulation for adaptive beam-forming are explained. In Section 3, two models for smart antenna array models are described. However, hybrid PSOGSA algorithm is proposed in Section 4. Simulation results and discussions for beam-forming are discussed in Section 5. Finally, the conclusion is presented in Section 6.

2. Problem Formulation

Smart antenna based on UCA topology by using N elements is showed in **Figure 1**. In this section, 12 elements in array structure are distributed uniformly along the circle of radius r, where θ is the azimuth angle, λ is the wavelength, the angle between adjacent elements is θ_0, and $d = 0.5\lambda$ is the space between two adjacent elements.

In the synthesis of beam-forming, the complex excitation for each element must be optimized to minimize radiation power intensity at certain directions and maximize the main-lobes to other directions. The following fitness function must be minimized to maximize the total output power toward the desired signal at θ_i and minimize the total output power in the direction of the interfering signals at θ_j.

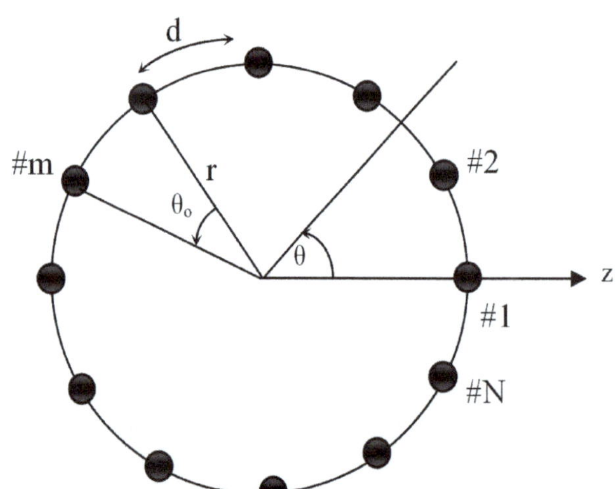

Figure 1. Geometry of the UCA with N elements.

$$\text{fitness function} = -\sum_{i=1}^{k} a_i \text{AF}(\theta_i) + \sum_{j=1}^{L} b_j \text{AF}(\theta_j) \quad (1)$$

where the number of SOI users is represented in constant k, and L represents the number of SNOI hackers. $\text{AF}(\theta_i)$ is the array factor that will be maximized or minimized in specific directions for ULA and UCA using evolutionary algorithms.

$$\text{AF}(\theta_i) = \sum_{n=1}^{N} \mathrm{e}^{j[\beta * r * \cos(\theta - \text{pos}_n) - \alpha_n]}, \quad \text{for UCA} \quad (2)$$

$$\text{AF}(\theta_i) = \sum_{n=1}^{N} \mathrm{e}^{j[\beta * (n-1) * d * \cos(\theta) - \alpha_n]}, \quad \text{for ULA} \quad (3)$$

where α_n represents the complex excitation phase of the n-th element, pos_n is the angular position of the n-th element, and β is the phase shift constant.

3. Smart Antennas Array Beam-Forming Models

As two models are proposed as shown in **Figure 2**, the first model assumes user #2 as transmitter, desired user #1 as receiver at desired angle $\text{SOI} = \theta_d$ from user #2, and hacker at angle $\text{SNOI} = \theta_h$ from user #2. In our work, smart antennas array using linear and circular topology are obtained. The second model assumes number of desired users M as SOI angles at θ_i.

4. Hybrid PSOGSA Algorithm Optimization Technique

Several different hybridization methods for heuristic algorithms was presented in [16], when two different algorithms can be hybridized in high-level or low-level with relay or co-evolutionary method as homogeneous or heterogeneous. In 1995, Kennedy and Eberhart surprised the world by PSO as shown in [3]. PSO was improved by several researchers when many interest variants were developed as shown in [4]. PSO depended on two main equations for updating the velocity and position as follow:

$$v_{iD}^{k+1} = w v_{iD}^{k} + \text{C1Rand 1}\left(\text{pbest}_{iD} - x_{iD}^{k}\right) + \text{C2Rand 2}\left(\text{gbest}_{D} - x_{iD}^{k}\right) \quad (4)$$

$$x_{iD}^{k+1} = x_{iD}^{k} + v_{iD}^{k+1} \quad (5)$$

where x_{iD} $(x_{i1}, x_{i2}, \cdots, x_{iD})$ and v_{iD} $(v_{i1}, v_{i2}, \cdots, v_{iD})$ are position and variation in position (velocity) of the i-th particle for an D-dimensional problem, respectively. pbest_{iD} $(\text{pbest}_{i1}, \text{pbest}_{i2}, \cdots, \text{pbest}_{iD})$ and gbest_D $(\text{gbest}_1, \text{gbest}_2, \cdots, \text{gbest}_D)$ represent for personal best position of the i-th particle and global best position vector, respectively. C1 and C2 are the acceleration constants, which represent the weighting of stochastic acceleration terms that pull each particle towards pbest and gbest positions. Rand 1 and Rand 2 are two random numbers in the range [0, 1], w is the inertia weight introduced to balance between the global and local search abilities

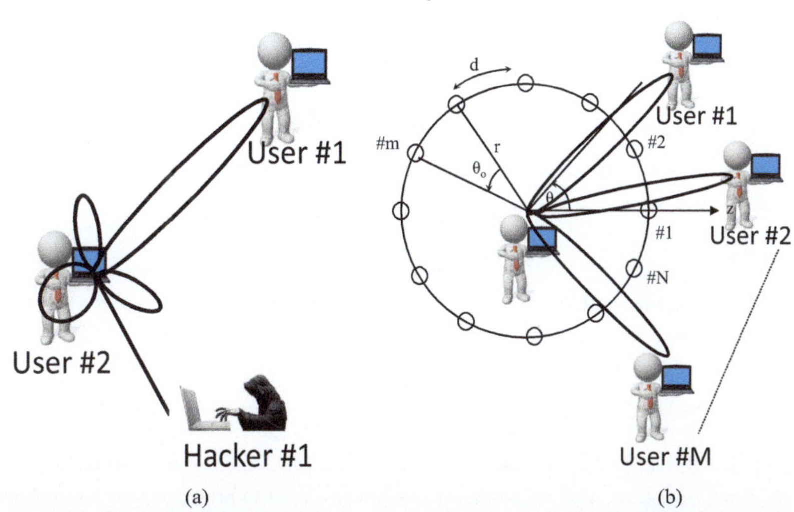

Figure 2. Two models for smart antennas array.

in terms of k iterations.

Recently, Gravitational Search Algorithm (GSA) was provided as an optimization problem based on the law of gravity and mass interaction [7]. The algorithm considered agents as objects consisting of different masses. Each agent in GSA was specified by four parameters: position of the mass in d-th dimension, inertia mass, active gravitational mass and passive gravitational mass. At each iteration, perform of the fitness evolution for all agents and also compute the best and worst fitness defined as below (for minimization problems):

$$\text{best}(t) = \min_{j\in\{1,\cdots,K\}} \text{fit}_j(t) \quad (6)$$

$$\text{worst}(t) = \max_{j\in\{1,\cdots,K\}} \text{fit}_j(t) \quad (7)$$

where $\text{fit}_j(t)$ represents the fitness of the j-th agent at iteration t, $\text{best}(t)$ and $\text{worst}(t)$ represents the best and worst fitness at generation t. Then the mass of the agents (gravitational and inertia masses) can be calculated as: $M_{ai} = M_{pi} = M_{ii} = M_i;\ i = 1,2,\cdots,K$

$$m_i(t) = \left[\text{fit}_i(t) - \text{worst}_i(t)\right] / \left[\text{best}(t) - \text{worst}(t)\right] \quad (8)$$

$$M_i(t) = m_i(t) / \sum_{j=1}^{K} m_i(t) \quad (9)$$

where M_{ai} is the active gravitational mass of the i-th agent, and M_{pi} is the passive gravitational mass of the i-th agent.

In this paper, PSO with GSA was hybridized using low-level co-evolutionary heterogeneous hybrid. The hybrid is low-level because the functionality of both algorithms is combined. It is co-evolutionary because both algorithms aren't used one after another but run in parallel. It is heterogeneous because there are two different algorithms that are involved to produce final results. The basic idea of hybrid PSOGSA is to combine the ability of social thinking (gbest) in PSO with the local search capability of GSA. In order to combine these algorithms, updating velocity is proposed as follow:

$$v_i(t+1) = w * v_i(t) + \text{C1} * \text{Rand} * a_i(t) + \text{C2} * \text{Rand} * (\text{gbest} - x_i(t)) \quad (10)$$

where $v_i(t)$ is the velocity of agent i at iteration t, w, Rand, gbest, C1 and C2 are taken from PSO algorithm. On the other hand, from GSA in [7], $a_i(t)$ is the acceleration of agent i at iteration t and can be calculated by

$$a_i^d(t) = F_i^d(t) / M_{ii}(t) \quad (11)$$

where $F_i^d(t)$ is the total force acting on i-th agent calculated as:

$$F_i^d(t) = \sum_{i=1,\, j\neq 1} \text{rand}_j F_{ij}^d(t) \quad (12)$$

where $F_{ij}^d(t)$ is the force acting on agent i from agent j at d-th dimension and t-th iteration and is computed as below:

$$F_{ij}^d(t) = G(t) M_{pi}(t) M_{aj}(t) \left(x_j^d(t) - x_i^d(t)\right) / \left[R_{ij}(t) + \epsilon\right] \quad (13)$$

where $R_{ij}(t)$ is the Euclidian distance between two agents i and j at iteration t, $G(t)$ is the computed gravitational constant at the same iteration, and ϵ is a small constant.

$$G(t) = G_0 e^{(\alpha t/T)} \quad (14)$$

In this problem G_0 is set to 100, α is set to 20 and T is the total number of iterations. In each iteration, the positions of particles are updated as follow:

$$x_i(t+1) = x_i(t) + v_i(t+1) \quad (15)$$

In hybrid PSOGSA [15], firstly, all agents are randomly initialized. Each agent is considered as a candidate solution. As can be seen in **Figure 3**, after initialization, evaluate the fitness function based on Equation (1). Gravitational force, gravitational constant, and resultant forces among agents are calculated using Equations (12), (14) and (13) respectively. After that, the accelerations of particles are defined as Equation (11). In each iteration,

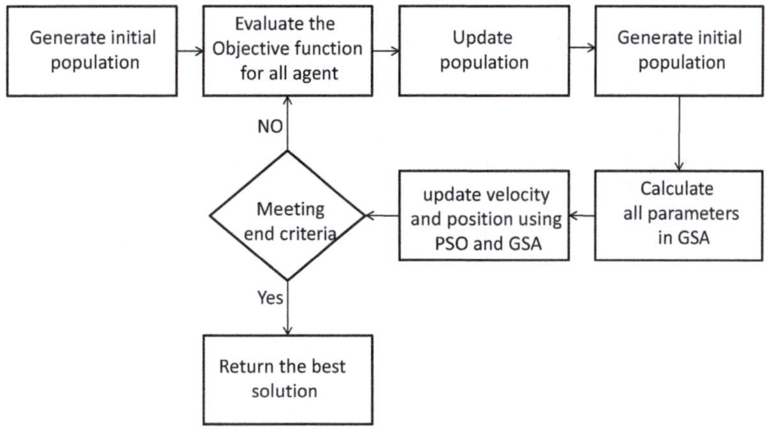

Figure 3. Flow chart for steps in hybrid PSOGSA algorithm.

the best solution so far (gbest) must be updated. After calculating the accelerations and with updating the best solution so far, the velocities and the positions of all agents can be updated using Equations (10) and (15) respectively. Finally, after agents are updated, the process of updating velocities and positions will be stopped by meeting an end criterion.

5. Simulation and Discussion Results

To validate the above analysis, we have developed a custom event driven simulator using Matlab package. In this section, the capability of hybrid PSOGSA technique for adaptive beam-forming with a UCA is studied.

In this section two models are discussed, the first model is shown in Section 3. **Figures 4-6** show beam patterns in comparison with smart antenna for different geometries (UCA and ULA) from user #2 to user #1 (SOI = 30°) and from user #2 to hacker #1 (SNOI = −30° (330°) or −20° (340°) or −40° (320°)). **Figure 4(a)** and **Figure 4(b)** is obtained for normalized array factor comparison using ULA and UCA by hybrid PSOGSA at SOI 30° and SNOI −30° (330°) in polar and rectangular representation respectively. It is found that only UCA topology has capability to direct the main beam toward user #1 (SOI) and null at hacker #1 (SNOI). It is clear that the directed power toward the intended direction (30°) using UCA is better than that obtained by ULA by approximately 55% as shown in **Figure 4(a)** (more than 6 dB as shown in **Figure 4(b)**). On the other hand, the directed null (zero power) toward the intended direction (−30°) using UCA is better than that obtained by ULA by approximately 35 dB as shown in **Figure 4(b)**.

Figure 5(a) and **Figure 5(b)** shows SOI at 30° and a changed value −40° (320°) for SNOI. Accordingly the results showed an improvement of 17% (more than 1.5 dB) and approximately 14 dB for SOI and SNOI respectively.

Figure 6(a) and **Figure 6(b)** shows SOI at 30° and a changed value −20° (340°) for SNOI. Accordingly the results showed an improvement of 35% (more than 3.5 dB) and approximately 5 dB for SOI and SNOI respectively.

Also, it can be noticed that an extra undesired main beam in the broadside direction is obtained in the ULA geometry. Therefore, the first case is the worst case in ULA because $\text{SNOI} = -30^{\circ}\,(330^{\circ}) = 360 - \text{SOI}$ (broadside direction). In general, the results obtained by UCA are better than those obtained from ULA which used in [14] for all directions in all scenarios.

The second model is shown in Section 3. **Figure 7** shows two scenarios normalized array factor comparison and normalized fitness functions versus iteration number using hybrid PSOGSA and GSA. The hybrid PSOGSA is used to adjust the weights of phase shift of the excitation of each element of the UCA array for beam synthesis to maximize the output power toward the desired signals at θ_i (scenario #1 two users at: $\theta_i = \left[30^{\circ}, 0^{\circ}\right]$, scenario #2 five users at: $\theta_i = \left[30^{\circ}, 0^{\circ}, -30^{\circ}, 90^{\circ}, 120^{\circ}\right]$). To validate the algorithm, the results are compared to those obtained using GSA which used in [13]. Both algorithms are employed with a population size of 30 and 150 iterations. First part in **Figure 7(a)** and **Figure 7(b)** shows the radiation pattern for the case of feeding each element of the UCA with the original precise values of phases optimized by hybrid PSOGSA and GSA for scenarios #1 and #2. Part two in **Figure 7(a)** and **Figure 7(b)** shows the normalized global best fitness of the hybrid PSOGSA and GSA algorithms for adaptive beam-forming of the array feeding. Comparing the global best

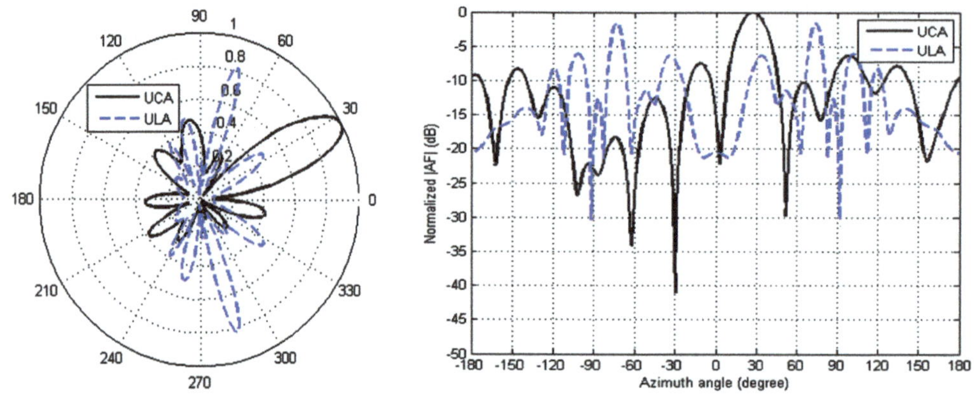

(a) (b)

Figure 4. Normalized array factor comparison using ULA and UCA by hybrid PSOGSA at SOI 30° and SNOI −30° (330°). (a) Polar representation; (b) Rectangular representation.

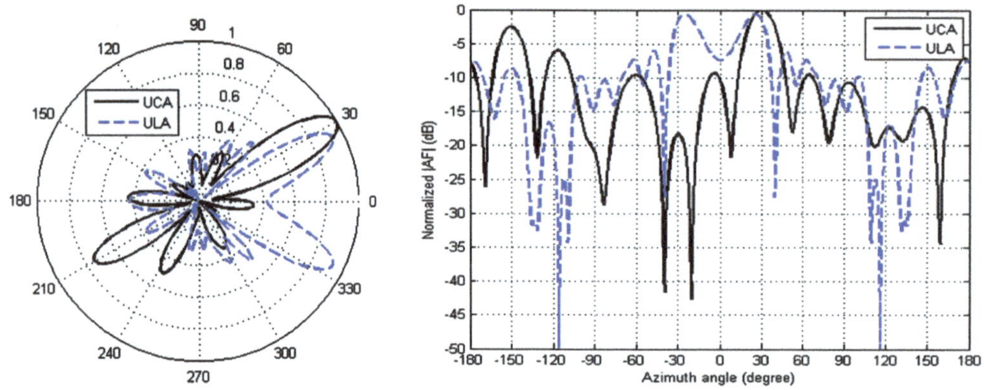

(a) (b)

Figure 5. Normalized array factor comparison using ULA and UCA by hybrid PSOGSA at SOI 30° and SNOI −40° (320°). (a) Polar representation; (b) Rectangular representation.

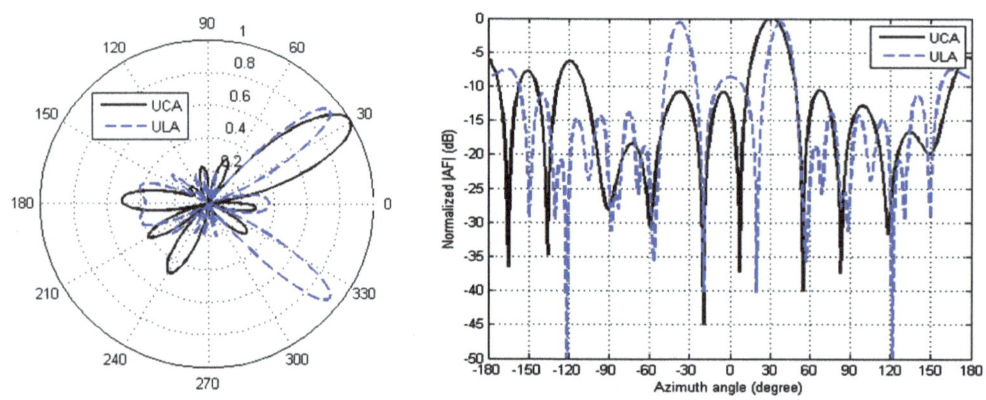

(a) (b)

Figure 6. Normalized array factor comparison using ULA and UCA by hybrid PSOGSA at SOI 30° and SNOI −20° (340°). (a) Polar representation; (b) Rectangular representation.

fitness of the two algorithms, it is found that the hybrid PSOGSA converges faster than GSA with a better performance in terms of computed final fitness values.

The pervious **Figure 7** indicates that the hybrid PSOGSA in general outperforms the GSA in precise feedings. **Table 1** shows the corresponding normalized |AF| values at the SOI directions for different scenarios. As shown,

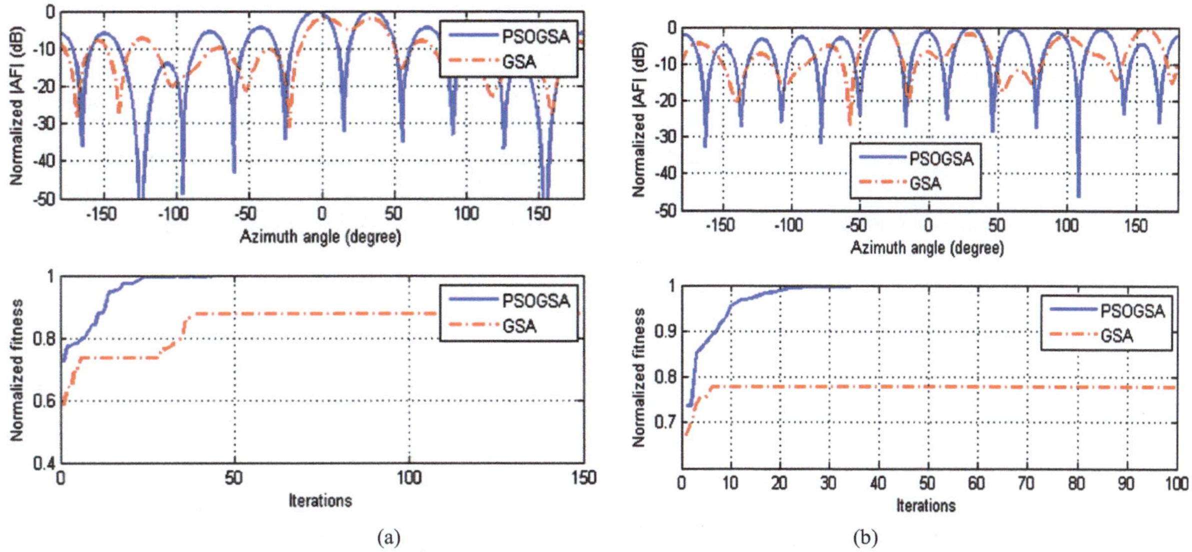

Figure 7. Two scenarios normalized array factor comparison and normalized fitness functions versus iteration number using Hybrid PSOGSA and GSA. (a) Scenario #1; (b) Scenario #2.

Table 1. The corresponding normalized |AF| values at the SOI directions for different scenarios.

		Scenario #1			Scenario #2					
	Desired Angles θ_i	0°	30°	Average Normalized |AF|	120°	90°	30°	0°	−30°	Average Normalized |AF|
Normalized |AF| in (dB)	Hybrid PSOGSA	0	−1	−0.5	−1.2	−1.9	0	−1.53	−0.6	−1.046
	Standard GSA	−3.2	−3.9	−3.55	−7.6	−3.96	−2	−6.98	−0.6	−4.228

the hybrid PSOGSA is better than GSA by −3.05 dB and −3.182 dB on the average for scenario #1 and scenario #2 respectively.

6. Conclusion

In this paper, a new novel (hybrid PSOGSA) technique is proposed with ULA and UCA antenna system for enhancing the performance of adaptive beam-forming in wireless communications applications. The technique is simple and appropriate for real time applications. It is clear that the directed power toward the intended direction (SOI) using UCA is better than that obtained by ULA by approximately 55% (more than 6 dB), 35% (more than 3.5 dB) and 17% (more than1.5 dB); on the other hand, directed null to SNOI better than ULA by approximately 35 dB, 5 dB and 14 dB in all cases in the first model in terms of normalized array factor. Simulations of beam-forming show accurate results even for a big set of simultaneously incident signals. Strategically pairing GSA with PSO has the desired advantages over GSA. It is found that hybrid PSOGSA is more attractive for beam-forming applications and better than GSA with approximately −3 dB in average. Via extensive simulation studies, it is demonstrated that hybrid PSOGSA achieves fast and robust global convergence over GSA.

References

[1] Lehne, P.H. and Pettersen, M. (1999) An Overview of Smart Antenna Technology for Mobile Communications Systems. *IEEE Communications Surveys and Tutorials*, **2**, 2-13.

[2] Chryssomallis, M. (2000) Smart Antennas. *IEEE Antennas and Propagation Magazine*, **42**, 129-136. http://dx.doi.org/10.1109/74.848965

[3] Kennedy, J. and Eberhart, R. (1995) Particle Swarm Optimization. *Proceedings of IEEE International Conference on Neural Networks*, Perth, 27 November-01 December 1995, **4**, 1942-1948. http://dx.doi.org/10.1109/ICNN.1995.488968

[4] Mahmoud, K.R., Eladawy, M., Bansal, R., Zainud-Deen, S.H. and Ibrahem, S.M.M. (2008) Analysis of Uniform Circular Arrays for Adaptive Beamforming Applications Using Particle Swarm Optimization Algorithm. *International Journal of RF and Microwave Computer-Aided Engineering*, **18**, 42-52. http://dx.doi.org/10.1002/mmce.20265

[5] Formato, R.A. (2007) Central Force Optimization: A New Metaheuristic with Applications in Applied Electromagnetics. *Progress in Electromagnetics Research*, **77**, 425-491. http://dx.doi.org/10.2528/PIER07082403

[6] Mahmoud, K.R. (2011) Central Force Optimization: Nelder-Mead Hybrid Algorithm for Rectangular Microstrip Antenna Design. *Electromagnetics*, **31**, 578-592. http://dx.doi.org/10.1080/02726343.2011.621110

[7] Rashedi, E., Nezamabadi-Pour, H. and Saryazdi, S. (2009) GSA: A Gravitational Search Algorithm. *Information Sciences*, **179**, 2232-2248. http://dx.doi.org/10.1016/j.ins.2009.03.004

[8] Chatterjee, A., Mahanti, G.K. and Pathak, N.N. (2010) Comparative Performance of Gravitational Search Algorithm and Modified Particle Swarm Optimization Algorithm for Synthesis of Thinned Scanned Concentric Ring Array Antenna. *Progress in Electromagnetics Research B*, **25**, 331-348. http://dx.doi.org/10.2528/PIERB10080405

[9] Chatterjee, A., Mahanti, G.K. and Mahapatra, P.R.S. (2011) Design of Fully Digital Controlled Reconfigurable Dual-Beam Concentric Ring Array Antenna Using Gravitational Search Algorithm. *Progress in Electromagnetics Research C*, **18**, 59-72. http://dx.doi.org/10.2528/PIERC10101806

[10] Altinoz, O.T. and Yilmaz, A.E. (2011) Calculation of Optimized Parameters Of Rectangular Patch Antenna Using Gravitational Search Algorithm. 2011 *International Symposium on Innovations in Intelligent Systems and Applications* (*INISTA*), Istanbul, 15-18 June 2011, 349-353. http://dx.doi.org/10.1109/INISTA.2011.5946090

[11] Magdy, A., Mahmoud, K.R., Abdel-Gawad, S.G. and Ibrahim, I.I. (2013) Direction of Arrival Estimation Based on Maximum Likelihood Criteria Using Gravitational Search Algorithm. *Progress in Electromagnetics Research Symposium Proceedings*, Taipei, 25-28 March 2013, 1162-1167.

[12] Mahmoud, K.R. (2013) UWB Antenna Using Gravitational Search Algorithm. *Journal of Engineering Sciences*, 41, 1890-1903.

[13] Mahmoud, K.R. and Hamad, S. (2014) Parallel Implementation of Hybrid GSA-NM Algorithm for Adaptive Beam-Forming Applications. *Progress in Electromagnetics Research B*, **58**, 47-57. http://dx.doi.org/10.2528/PIERB13102106

[14] Ram, G., Mandal, D., Kar, R. and Ghoshal, S.P. (2013) Optimized Hyper Beamforming of Linear Antenna Arrays Using Collective Animal Behaviour. *The Scientific World Journal*, **2013**, Article ID: 982017.

[15] Mirjalili, S. and Hashim, S.Z.M. (2010) A New Hybrid PSOGSA Algorithm for Function Optimization. *International Conference on Computer and Information Application* (*ICCIA*), Tianjin, 3-5 December 2010, 374-377.

[16] Talbi, E.G (2002) A Taxonomy of Hybrid Metaheuristic. *Journal of Heuristics*, **8**, 541-546. http://dx.doi.org/10.1023/A:1016540724870

4

Radiated Ultrashort High-Power Electromagnetic Pulses Induce ATP Release in B16F10 Murine Melanoma Cells

Sabrina Macaire[1], Alexandre Catrain[1], Stéphane Tortel[1], Jean-Christophe Joly[1], Sébastien Girard[2,3], Pierre Bonnet[2,3], Alain Vian[4]

[1]CEA, DAM, Gramat, France
[2]Institut Pascal, Université Blaise Pascal, Clermont-Ferrand, France
[3]CNRS, UMR 6602, Aubière, France
[4]ARCH-E. Université d'Angers, UMR 1345 IRHS, Angers, France
Email: pierre.bonnet@lasmea.univ-bpclermont.fr

Abstract

Ultrawideband electromagnetic pulses with high amplitude and short duration are reported to affect several aspects of cell physiology. They are usually delivered to the living material through electrodes in small dedicated chambers. Here we showed, using a totally different experimental setup, that radiated EM pulses illuminating the living material through a specialized antenna (without any direct contact) are able to trigger a rapid release of ATP in cultured murine cells that was concomitant with a drop of intracellular AEC. Despite this rapid and strong response, we found that cell viability and clonogenicity were only slightly affected by the EMF exposure.

Keywords

Radiated Electromagnetic Pulses, Koshelev Antenna, eATP, Cell Viability

1. Introduction

Electromagnetic-pulsed radiation is nowadays increasingly used for diverse military [1] and civilian (mainly medical) purposes [2], including cancer therapy and chronic pain treatment [3]-[5]. This technology is also giving rise to new kind of radar imagery both in the medical and military area [6], including generation of electromagnetic weapons based on high power ultrashort (nanosecond) electromagnetic pulses, commonly designated as Ultra-WideBand electromagnetic radiation (UWB). Their specific properties, namely a very high amplitude (several dozens of $kV \cdot m^{-1}$), a complex temporal signature and an ultrashort (nanosecond) duration, make them difficult to produce and little is known on their overall possible effects on living organism (including human). It is worth noting that the very low energy transfers from the electromagnetic waves to the living material, mainly due to the ultrashort duration, only allows for non-thermal effects to arise [7].

The most common method used to apply such electromagnetic radiation to living organisms consists of a direct injection of pulsed electric field using a small chamber surrounded by metallic electrodes [8]. This allows an effective technical set-up and an excellent control and transfer of the electromagnetic waves to the living cells or organisms that are maintained in an appropriate liquid medium. Indeed, numerous physiological modifications, including changes in signaling pathways, were reported after exposing cell or living tissues to various kinds of pulsed electromagnetic fields. The exposure could lead to DNA damage and cell viability issues [9] or cell apoptosis in human cells [10]. A cytosolic calcium increase followed by caspase activation arose in human lymphocytes after nanosecond pulse [11]. Exposure of pre-neoplastic CL-S1 mammary epithelial cells to 18 $kV \cdot m^{-1}$, 1 kHz repetition rate, 10 ns pulse during 4 - 6 h promote their proliferation through the mitogen-activated protein kinase (MAPK) signaling pathway [12]. The mitogenic activity in AML-12 Mouse Hepatocytes was increased after 2 h exposure to UWB radiations (pulse width of 10 ns, repetition rate of 1 kHz, amplitude of 5 - 20 $kV \cdot m^{-1}$) through the induction of cyclin A [13]. These physiological alterations are mainly understood as being the consequence of cell membrane permeabilization. Indeed, it is well known that electromagnetic pulses provided to biological samples using electrodes in a small chamber evoke the formation of pores both in the plasmalemma [14] [15] and in the nuclear membrane [16]. Experiments demonstrated that the pores were created very rapidly (within a few nano seconds) and that their number, conductance and size were proportional to the pulsed electric strength [17]-[19]. The increase in membrane conductance allows external ionic and polar molecules, normally efficiently blocked by the highly hydrophobic living membranes, to enter or exit the cell [20] [18], especially as the resealing of the cell membrane is a slow process that lasts for minutes [21]-[23]. Adenosine tri-phosphate (ATP) is a highly polar intracellular energy-related and signal molecule that is commonly used to report pore formation in the plasma membrane [24] because of the very low membrane permeability toward ATP. Thus, the so-called extracellular ATP (eATP) might appear as a major physiological event [25] especially in the context of electromagnetic or electric field stimulation [24] [26]. Aside from the formation of pores, an increasing number of reports highlight the existence of stretch or mechano-sensitive ATP-channels that could be involved in ATP release and indirect evidences suggest their involvement in cells responses to UWB exposure [27].

In contrast to the electrode-based electromagnetic pulses injection, the use of radiating antennas to perform, at a distance, stimulations of the biological samples (not by direct contact of electrodes with the culture medium) received only little attention. This kind of exposure is however quite commonly encountered in various kinds of environments, both in civilian and military areas. Very little information is indeed available at present time to decipher if living organisms 1) actually respond to radiated UWB radiation and 2) if these responses behave similarly or differently from those observed after direct UWB surge application by immersed electrodes. Our aim in the present work was to determine if high power UWB electromagnetic waves emitted by a Koshelev antenna induce physiological changes in cultured animal cells. Here we report evidences showing that radiated UWB electromagnetic pulses cause a rapid and important ATP release, significantly diminished cell clonogenicity, but only slightly affects overall cell viability.

2. Material and Methods

2.1. Cell Culture

Murine melanoma cells (B16F10) were maintained (37˚C, 5% CO_2) in Dulbecco's modified Eagle's medium (DMEM) high glucose containing 10% heat inactivated FBS and supplemented with penicillin (100 $U \cdot mL^{-1}$), streptomycin (100 $\mu g \cdot mL^{-1}$) and L-glutamine (2 mM). The cells were transferred on fresh medium every 2 days.

2.2. Electromagnetic Stimulation Design and Exposure Experiments

All radiation tests were performed in an anechoic chamber. A high pulsed power generator combined with an ad hoc antenna was used for the generation of high-power electromagnetic waves. **Figure 1(a)** summarizes the experimental set-up. A signal was first shaped using a high voltage pulse generator (Kentech model PBG5, with a

250 ps rise-time measured from 10% to 90% of the amplitude value, a 2.4 ns duration measured at 50% of the peak value, and a 2 kV peak value for a 50 Ω load). Then, a Koshelev-like ultra wide band antenna [28] radiated this transient electromagnetic signal in the form of repetitive pulses (100 Hz pulse repetition frequency). This compact antenna (38.5 × 30 × 30 cm^3) has been designed and optimized by CISTEME [29] to widen the bandwidth from 200 MHz to 2 GHz. The biological samples were located in the antenna heart between the two radiating elements (**Figure 1(b)**). At the same position but in the absence of the biological samples, the ultra wide band EM field was measured with an electric field surface sensor made of a SMA connector whose metallic shield was connected to the antenna and the inner conductor was introduced in the antenna. The length of the inner conductor is very short (less than 1mm) so that it is considered as a non-intrusive derivative sensor with a very high bandwidth (more than 10's of GHz): $E_{\text{sensor (in V/m)}}(t) = K\int V_{\text{sensor (in V)}}(t)\mathrm{d}t$. K is obtained from calibration using a network analyzer. The signal was monitored with a numerical oscilloscope (Lecroy-Waverunner, model 640 Zi, **Figure 2(a)**). The electric field radiated by the antenna was also simulated and verified using the temporal solver of CST Microwave Studio® (**Figure 2(b)**).

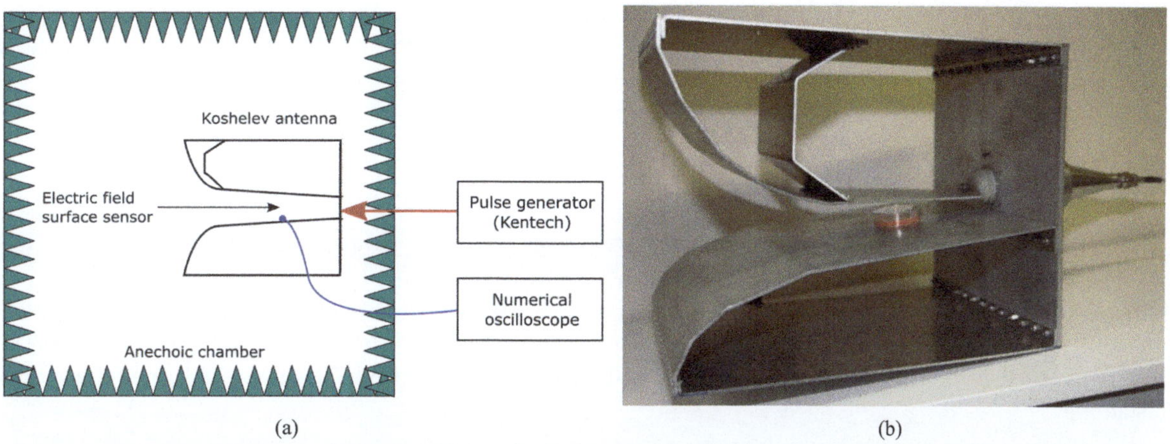

Figure 1. Experimental set-up. (a) General set-up. This scheme illustrates the Koshelev-like antenna powered by a Kentech generator and placed in an anechoic chamber in order to suppress interactions with undesired electromagnetic waves. A surface electric field sensor coupled to a numeric oscilloscope (Lecroy-Waverunner, model 640 Zi) measured the radiated electric field. (b) Koshelev antenna detail showing the position of the Petri dish containing the B16F10 murine melanoma cells.

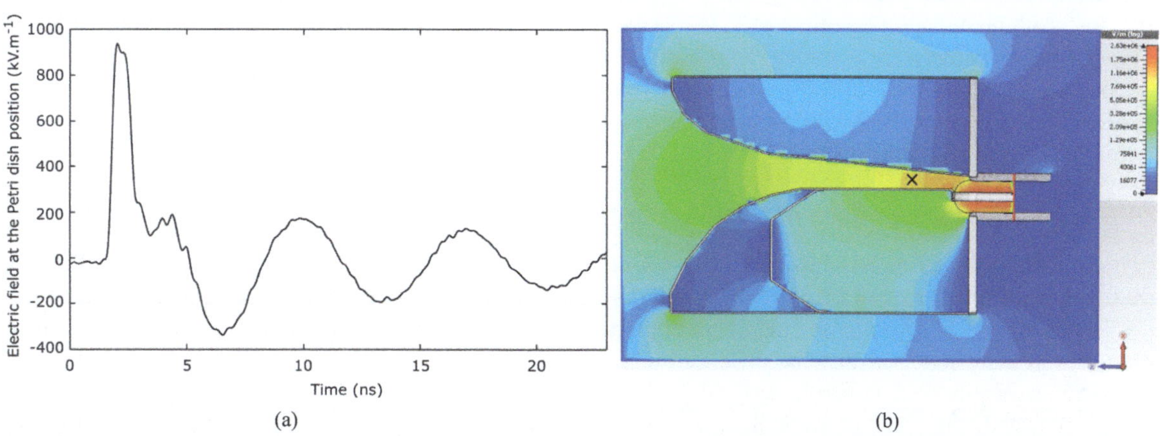

Figure 2. Radiated electric field characteristics. (a) Wave form of the transient radiated electric field. The signal was measured at the Petri dish location, in absence of the Petri dish. The main peak value was measured at 930 kV·m^{-1}. The rise-time is 0.25 ns, corresponding to an UWB frequency spectrum [15]. (b) Spatial electric field distribution in the Koshelev-like antenna in absence of the Petri dish. This numerical simulation performed with CST Microwave Studio® shows that the electric field amplitude is maximal at the antenna center, where the biological samples are located (cross marker). Numerical peak value is 845 kV·m^{-1}, for a vertical distance between the two radiating element of 2.8 cm.

We exposed the B16F10 cells plated in a 3 cm Petri dish in triplicate at room temperature to 60,000 electromagnetic pulses (2 distinct 5 min 30,000 pulses each, peak-intensity 930 $kV \cdot m^{-1}$, separated by 20 min interval, with 100 Hz pulse repetition frequency). The temperature of the culture medium was monitored with an optic fiber thermometer (Luxtron 812) and found to be not affected by the electromagnetic pulses. Shame (*i.e.* control) experiments were done exactly the same way, except that the Kentech amplifier was turned-off.

2.3. Measurement of Extracellular ATP (eATP) Concentration

ATP release in the culture medium was measured using the ATP bioluminescence assay kit HS II (Roche Diagnostic) as described by the manufacturer. Briefly, 5×10^5 cells were subjected to the electromagnetic stimulation in the Petri dish and 110 µL of the suspension were collected and centrifuged. The supernatant (50 µl) was used for ATP determination immediately after the addition of an equal volume of Luciferin/luciferase mixture. The bioluminescence was measured in opaque 96-wells plate to avoid optical leaks using a Berthold LB-96V microplate luminometer. The ATP concentration in sample was determined using a standard curve made with 10-fold serial dilutions of ATP and expressed as relative to the eATP concentration of the control (*i.e.* non exposed) sample. A positive hypotonic stress control, known to cause a massive ATP release, was performing by adding 1 mL of deionized sterile water to the Petri-Dish. The adenylate energy charge (AEC) was determined as previously described [30] and calculated with the following equation:

$$\text{AEC} = \left(\left[\text{ATP}\right] + 1/2\left[\text{ADP}\right]\right) / \left(\left[\text{ATP}\right] + \left[\text{ADP}\right] + \left[\text{AMP}\right]\right) \quad (1)$$

2.4. Clonogenic and Viability Assays

Clonogenic capacity was determined in Petri-dishes seeded with 5×10^3 cells in DMEM, 10% FBS in the presence (or absence) of 500 pM of the P2X receptor agonist BzATP. After 24 h, the cells were exposed to electromagnetic pulses and allowed to growth for 5 days. The colonies were counted after cell staining with crystal violet. The cell viability was evaluated in Petri-dishes seeded with 3×10^5 cells in DMEM, 10% FBS. The cells were exposed 24 h later to the electromagnetic pulses and the surviving cells counted 48 h later after trypan blue staining.

3. Results

3.1. Ultra Wide Band Electromagnetic Field Characteristics

The measured waveform of the transient electric field at the petri dish location and numerical simulation of electromagnetic field values are reported in **Figure 2**. The measured main peak signal amplitude reached 930 $kV \cdot m^{-1}$ with an apparent rise-time of 0.25 ns (measured from 10% to 90% of the amplitude value, **Figure 2(a)**). Very similar results were obtained from numerical simulations of the exposure, both for the rise-time and the peak amplitude (841 $kV \cdot m^{-1}$, **Figure 2(b)**). Furthermore, the spatial distribution of the EM field amplitude shows that this latter reached the maximum in the neighborhood where the Petri dish has been placed for the stimulations (**Figure 2(b)**).

3.2. Biological Responses of B16F10 Cells

The overall viability of B16F10 cells was not affected by the radiated electromagnetic pulses (**Table 1**). The percentage of living cells was high (>97%) in the control (unstimulated cells) and was not modified by the UWB radiation exposure, whatever the treatment configuration (1 set of 60,000 pulses or 2 sets of 30,000 pulses each). This result also indicated that the different manipulations of the Petri-dishes that were required for the UWB radiation exposure were not an issue for the B16F10 cells.

As expected, the eATP concentration was low in the culture medium of B16F10 cells and readily increased after the application of the radiated electromagnetic pulses (**Figure 3(a)**). The concentration more than doubled 2.5 min after the treatment to reach a maximum increase (about 5.5-fold) after 15 min, and started to decline after 30 min, where it was about 4-fold the level observed in the control condition. While this increase in ATP in the culture medium was strong and rapid, it was about 3-fold lower than the positive (hypotonic stress) control that caused an over 15-fold increase only 5 min after the treatment (**Figure 3(b)**), suggesting that the ATP

Table 1. Clonogenicity, cell survival and adenylate energy charge after exposure to UWB radiation. Clonogenicity was determined in control cells (sham), in the presence of bzATP and after exposure to UWB radiation (2 sets of 30,000 pulses or a single set of 60,000 pulses). Cell viability and Adenylate Energy Charge (AEC) was measured in control (sham) and after ex-posure to UWB radiation (2 sets of 30,000 pulses or a single set of 60,000 pulses). Each value is the mean of three independent experiments ± se.

A. Clonogenicity				
	Sham	Sham + bzATP	2 × 30,000 pulses	60,000 pulses
Colonies	100 ± 4.69	67.72 ± 4.70	77.20 ± 5.32	83.70 ± 5.63
B. Cell viability				
	Sham	2 × 30,000 pulses	60,000 pulses	
Alive	100 ± 3.03	101.70 ± 3.50	108.77 ± 3.51	
Dead	2.42 ± 0.35	3.16 ± 0.61	3.33 ± 0.17	
C. AEC				
	Sham	2 × 30,000 pulses		
	0.37 ± 0.067	0.58 ± 0.026		

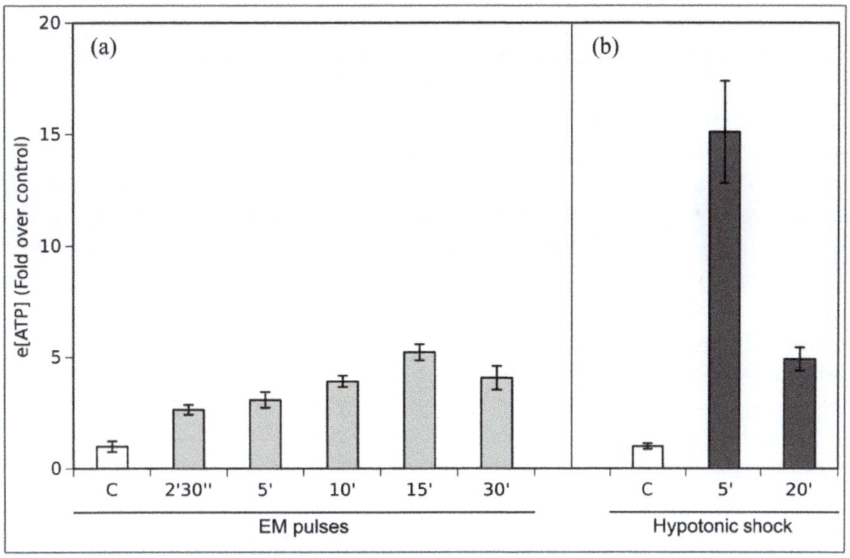

Figure 3. Extracellular ATP after UWB radiation exposure and hypotonic shock. (a) Extracellular ATP (eATP) concentration after UWB radiation exposure (2 × 30,000 pulses, 930 $kV \cdot m^{-1}$, 100 Hz pulse repetition frequency) expressed relatively to the control, unexposed sample taken as base 1. (b) ATP released after hypotonic shock. Each value is the mean of three independent experiments ± se.

externalization in response to UWB radiation is implying only part of the available cellular ATP. Concomitantly, intracellular ATP strongly decreased from 3.8 to 1.6 $pmole \cdot \mu g^{-1}$ protein 15 min after the application of electromagnetic pulses (**Figure 4**), confirming the intracellular origin of the eATP and suggesting that its increase mobilized over the half of the available ATP pool of the cells. Indeed, the adenylate energy charge, measured to a standard value of 0.73 in the control cells drop to the mean low value of 0.58 15 min after the treatment in exposed samples (**Table 1**), confirming a strong effect of radiating UWB radiation pulses toward the energetic status of the cells.

The clonogenic capacity of B16F10 cells was significantly affected by electromagnetic pulses (**Table 1**) that caused about 20% drop, whatever the treatment configuration (1 set of 60,000 pulses or 2 sets of 30,000 pulses

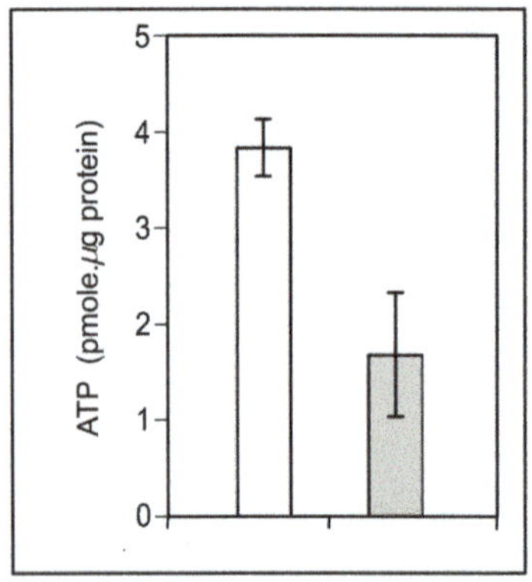

Figure 4. Intracellular ATP after UWB radiation exposure. Intracellular ATP (pmol ATP·µg^{-1} protein) after UWB radiation exposure (2 × 30,000 pulses, 930 kV·m^{-1}, 100 Hz pulse repetition frequency). Mean of three independent experiments ± se.

each). This result was similar to the clonogenic capacity decrease observed in the presence of 500 pM of the ATP agonist BzATP, suggesting that eATP interfere with the purinergic P2X7 receptors and thus is susceptible to act, outside of the cell, as a physiologically active molecule.

4. Discussion

The direct injection of UWB electromagnetic pulses using immersed electrodes in direct contact with the culture medium in small specialized designed cuvette is widely used to study their biological effects on living material [11]. However, exposure of living organisms in an open volume (where no direct contact of electrodes could be achieve) remains frequent both in civilian and military applications. The UWB pulse is in these cases delivered by specific antennas whom design is compatible with the high amplitude ultrashort complex UWB radiation pulses [31]. Such equipment is mainly related to medical devices in the civilian area and to certain kind of radar in the military domain. The Kentech amplifier and Koshelev antenna used in the present work were efficient to deliver a UWB radiation pulse with a waveform comparable in amplitude and duration to that observed through the use of direct (contact) electrodes (**Figure 2**).

It has long been demonstrated using immersed electrode that UWB pulses induce the formation of pores (electroporation) in the plasma membrane, and that their number, size, density and duration are proportional to the pulses amplitude and duration [17]-[19] [32]. These pores life is sufficient to allow molecules of diverse size to move across the cell membranes, thus giving the opportunity to highly polar molecules to rapidly and efficiently exit the cell because of the favorable concentration gradient. Among the whole set of cellular molecules that may exit the cell, ATP received a considerable attention these last years [33]. This molecule, present in low quantity in the cell, is continuously consumed and reformed. In addition to its well known function as an energy-related molecule, ATP is also acting as a signal molecule in both animal [25] and plant cell [34] when externalized after various kinds of environmental signals or injurious stimulus.

Here we observed that short exposure to radiated UWB radiation pulses evoke the rapid increase of ATP in the culture medium (**Figure 3(A)**), concomitantly with a decrease of both cellular ATP (**Figure 4**) and AEC (**Table 1**), demonstrating the cellular origin of ATP. This increase is maximum (5-fold over the control) after 15 min. Its amplitude is, however, much lower than that observed after a massive hypotonic shock (**Figure 3(B)**, 15-fold increase) showing that only about one third to the half of the cellular ATP readily exit the cell after UWB radiation. This observation is consistent with the internal decrease (about the half) of the ATP content after UWB exposure (**Figure 4**).

The kinetics of ATP externalization (maximal after 15 min) is consistent with the observation of Pakhomov *et al.* [21] [22] that reported long lasting (minutes) increased of membrane conductance of mammalian cells. The contribution of exocytosis to the ATP release, if any, remains to be proven even if short (10 min) 900 MHz phone-type radiation increase exocytosis on epidermal Merkel cells [35]. Membrane resealing is required to preserve cell viability [17] and appears to be ATP-dependant. Reference [36] brings evidence suggesting that RF modulation in the UWB pulse, as applied in our experiments, protect the cell from a lethal electroporation and therefore preserves their viability. However, it is not clear if electroporation is the unique way to externalize ATP since previous work have demonstrated that ATP release from cell may involve a mechano sensitive ATP channel [37] [38]. Indeed, cell treatment with gadolinium (Gd^{3+}, a potent inhibitor of stretch-activated channels) prevents ATP release after hypotonic stress [39]. Furthermore, [27] demonstrated that gadolinium blocked the membrane permeabilization induced by UWB treatment in cultured cells. Taken together, these observations suggest that similar ATP channels may account, at least partially, for the ATP externalization evoked after UWB surge through specific mechanosensitive channels. It may also explain that a substantial part of ATP actually remains in the cell while a pore-driven ATP exit may have cause a more massive exit. Further experiment would be necessary to decipher if either electroporation or mechanosensitive channels or both are involved in the externalization of ATP.

The eATP concentration started to diminish after 30 min (**Figure 3**), suggesting that it was either hydrolyzed or involved (consumed) into metabolic processes. Further experiments would be required to decipher if eATP is actually involved in physiological events surrounding cell death or recovery after UWB surge. Concomitantly to the ATP release, the energy charge of the cell dropped from 0.73 (control, unexposed cells) to 0.58 in the exposed cells (**Table 1**), highlighting a potential strong impact of the UWB radiation exposure on the energy metabolism. Similar observations have previously been reported in plants after high frequency (900 MHz), mobile phone-like exposure [30] and were proposed to be related to direct externalization of ATP [40].

Surprisingly, and despite deep impact on the energy metabolism of the cells, the UWB radiation pulse did not dramatically affect cells viability (**Table 1**). The UWB pulses used in the present work only slightly affect cell survival. Our observations are in accordance with those of reference [9] that reported that only non-fixed cultured cells are sensitive to UWB radiations. Thus, a substantial amount of ATP (**Figure 4**) and ATP synthesis capacity seems to be preserved in the exposed samples to allow cell survival and subsequent cell divisions. In contrast, UWB radiation pulses significantly inhibit clonogenicity by about 20% (**Table 1**). This effect could be mimicked by the addition of the ATP agonist bzATP in the culture medium (**Table 1**) indicating that the clonogenic inhibition was indeed directly related to the ATP release, therefore suggesting that eATP was a biologically active molecule. This observation is consistent with previous work [41] [42] that highlight an inhibitory effect of ATP on cell proliferation.

5. Conclusion

In summary, our results demonstrated that radiated UWB pulses can effectively evoke distant biological events (rapid, massive yet non-lethal release of ATP) similarly to that observed after direct, contact electrode-based application of UWB.

Acknowledgements

The authors would like to gratefully acknowledge the LRC BIOEM ("Laboratoire de Recherche Conventionné" —LRC No. 002-2011-CEA DAM/Institut Pascal) for its financial support and the Pr M. Gendraud (University Blaise Pascal) for advices on ATP metabolism.

References

[1] Giri, D.V. and Tesche, F.M. (2004) Classification of Intentional Electromagnetic Environments (IEME). *IEEE Transactions on Electromagnetic Compatibility*, **46**, 322-328. http://dx.doi.org/10.1109/TEMC.2004.831819

[2] Markov, M.S. (2007) Expanding Use of Pulsed Electromagnetic Field Therapie. *Electromagnetic Biology and Medicine*, **26**, 257-274. http://dx.doi.org/10.1080/15368370701580806

[3] Nuccitelli, R., Pliquett, U., Chen, X., Ford, W., Swanson, R.J., Beebe, S.J., Kolb, J.F. and Schoenbach, K.H. (2006) Nanosecond Pulsed Electric Fields Cause Melanomas to Self-Destruct. *Biochemical and Biophysical Research Communications*, **343**, 351-360. http://dx.doi.org/10.1016/j.bbrc.2006.02.181

[4] Breton, M. and Mir, L. (2012) Microsecond and Nanosecond Electric Pulses in Cancer Treatments. *Bioelectromagnetics*, **33**, 106-123. http://dx.doi.org/10.1002/bem.20692

[5] Camera, F., Paffi, A., Merla, C., Denzi, A., Apollonio, F., Marracino, P., d'Inzeo, G. and Liberti, M. (2012) Effects of Nanosecond Pulsed Electric Fields on the Activity of a Hodgkin and Huxley Neuron Model. *Annual International Conference of the IEEE Engineering in Medicine and Biology Society* (*EMBC*), San Diego, 28 August-1 September 2012, 2567-2570. http://dx.doi.org/10.1109/EMBC.2012.6346488

[6] Zito, D., Pepe, D., Mincica, M. and Zito, F. (2011) A 90nm CMOS SoC UWB Pulse Radar for Respiratory Rate Monitoring. *IEEE International Solid-State Circuits Conference Digest of Technical Papers* (*ISSCC*), San Francisco, 20-24 February 2011, 40-41.

[7] Deng, J., Schoenbach, K., Buescher, S., Hair, P., Fox, P. and Beebe, S. (2003) The Effects of Intense Submicrosecond Electrical Pulses on Cells. *Biophysical Journal*, **84**, 2709-2714. http://dx.doi.org/10.1016/S0006-3495(03)75076-0

[8] Huang, F., Fang, Z., Mast, J. and Chen, W. (2013) Comparison of Membrane Electroporation and Protein Denature in Response to Pulsed Electric Field with Different Durations. *Bioelectromagnetics*, **34**, 253-263. http://dx.doi.org/10.1002/bem.21773

[9] Stacey, M., Stickley, J., Fox, P., Statler, V., Schoenbach, K., Beebe, S.J. and Buescher, S. (2003) Differential Effects in Cells Exposed to Ultra-Short, High Intensity Electric Fields: Cell Survival, DNA Damage, and Cell Cycle Analysis. *Mutation Research*, **542**, 65-75. http://dx.doi.org/10.1016/j.mrgentox.2003.08.006

[10] Beebe, S.J., Fox, P.M., Rec, L.J., Willis, L.K. and Schoenbach, K.H. (2003) Nanosecond, High-Intensity Pulsed Electric Fields Induce Apoptosis in Human Cells. *FASEB Journal*, **17**, 1493-1495.

[11] Vernier, P.T., Sun, Y., Marcu, L., Salemi, S., Craft, C.M. and Gundersen, M.A. (2003) Calcium Bursts Induced by Nanosecond Electric Pulses. *Biochemical and Biophysical Research Communications*, **310**, 286-295. http://dx.doi.org/10.1016/j.bbrc.2003.08.140

[12] Sylvester, P.W., Shah, S.J., Haynie, D.T. and Briski, K.P. (2005) Effects of Ultra-Wideband Electromagnetic Pulses on Pre-Neoplastic Mammary Epithelial Cell Proliferation. *Cell Proliferation*, **38**, 153-163. http://dx.doi.org/10.1111/j.1365-2184.2005.00340.x

[13] Dorsey, W.C., Ford, B.D., Roane, L., Haynie, D.T. and Tchounwou, P.B. (2005) Induced Mitogenic Activity in AML-12 Mouse Hepatocytes Exposed to Low-Dose Ultra-Wideband Electromagnetic Radiation. *International Journal of Environmental Research and Public Health*, **2**, 24-30. http://dx.doi.org/10.3390/ijerph2005010024

[14] Chen, N., Schoenbach, K.H., Kolb, J.F., Swanson, R.J., Garner, A.L., Yang, J., Joshi, R.P. and Beebe, S.J. (2004) Leukemic Cell Intracellular Responses to Nanosecond Electric Fields. *Biochemical and Biophysical Research Communications*, **317**, 421-427. http://dx.doi.org/10.1016/j.bbrc.2004.03.063

[15] Xiao, D., Yao, C., Liu, H., Li, C., Cheng, J., Guo, F. and Tang, L. (2013) Irreversible Electroporation and Apoptosis in Human Liver Cancer Cells Induced by Nanosecond Electric Pulses. *Bioelectromagnetics*, **34**, 512-520.

[16] Kolb, J.F., Kono, S. and Schoenbach, K.H. (2006) Nanosecond Pulsed Electric Field Generators for the Study of Subcellular Effects. *Bioelectromagnetics*, **27**, 172-187. http://dx.doi.org/10.1002/bem.20185

[17] Rols, M.P., Delteil, C., Golzio, M. and Teissié, J. (1998) Control by ATP and ADP of Voltage-Induced Mammalian-Cell-Membrane Permeabilization, Gene Transfer and Resulting Expression. *European Journal of Biochemistry*, **254**, 382-388. http://dx.doi.org/10.1046/j.1432-1327.1998.2540382.x

[18] Joshi, R.P. and Schoenbach, K.H. (2000) Electroporation Dynamics in Biological Cells Subjected to Ultrafast Electrical Pulses: A Numerical Simulation Study. *Physical Review E*, **62**, 1025-1033. http://dx.doi.org/10.1103/PhysRevE.62.1025

[19] Vasilkoski, Z., Esser, A.T., Gowrishankar, T.R. and Weaver, J.C. (2006) Membrane Electroporation: The Absolute Rate Equation and Nanosecond Time Scale Pore Creation. *Physical Review E*, **74**, Article ID: 021904. http://dx.doi.org/10.1103/PhysRevE.74.021904

[20] Weaver, J.C. and Chizmadzhev, Y.A. (1996) Theory of Electroporation: A Review. *Bioelectrochemistry and Bioenergetics*, **41**, 135-160. http://dx.doi.org/10.1016/S0302-4598(96)05062-3

[21] Pakhomov, A.G., Kolb, J.F., White, J.A., Joshi, R.P., Xiao, S. and Schoenbach, K.H. (2007) Long-Lasting Plasma Membrane Permeabilization in Mammalian Cells by Nanosecond Pulsed Electric Field (nsPEF). *Bioelectromagnetics*, **28**, 655-663. http://dx.doi.org/10.1002/bem.20354

[22] Pakhomov, A.G., Shevin, R., White, J.A., Kolb, J.F., Pakhomova, O.N., Joshi, R.P. and Schoenbach, K.H. (2007) Membrane Permeabilization and Cell Damage by Ultrashort Electric Field Shocks. *Archives of Biochemistry and Biophysics*, **465**, 109-118. http://dx.doi.org/10.1016/j.abb.2007.05.003

[23] Pavlin, M., Kotnik, T., Miklavc, D., Kramar, P. and Lebar, A.M. (2008) Electroporation of Planar Lipid Bilayers and Membranes. In: Liu, L., Ed., *Advances in Planar Lipid Bilayers and Liposomes*, Vol. 6, Chapter 7, Elsevier B.V Academic Press, London, 165-226.

[24] Rols, M.P. and Teissié, J. (1990) Electropermeabilization of Mammalian Cells. Quantitative Analysis of the Phenomenon. *Biophysical Journal*, **58**, 1089-1098. http://dx.doi.org/10.1016/S0006-3495(90)82451-6

[25] Nowak, I. (2003) ATP as a Signaling Molecule: The Exocrine Focus. *Physiology*, **18**, 12-17.

[26] Sauer, H., Stanelle, R., Hescheler, J. and Wartenberg, M. (2002) The DC Electrical-Field-Induced Ca^{2+} Response and Growth Stimulation of Multicellular Tumor Spheroids Are Mediated by ATP Release and Purinergic Receptor Stimu-

lation. *Journal of Cell Science*, **115**, 3265-3273.

[27] André, F.M., Rassokhin, M.A., Bowman, A.M. and Pakhomov, A.G. (2010) Gadolinium Blocks Membrane Permeabilization Induced by Nanosecond Electric Pulses and Reduces Cell Death. *Bioelectrochemistry*, **79**, 95-100. http://dx.doi.org/10.1016/j.bioelechem.2009.12.007

[28] Koshelev, V.I., Buyanov, Y.I., Andreev, Y.A., Plisko, V.V. and Sukhushin, K.N. (2001) Ultrawideband Radiators of High-Power Pulses. *Proceedings of the IEEE Pulsed Power Plasma Science Conference*, **2**, 1661-1667.

[29] Cadilhon, B., Cassany, B., Modin, P., Diot, J.C., Bertrand, V. and Pécastaing, L. (2011) Ultra Wideband Antennas for High Pulsed Power Applications. In: Matin, M., Ed., *Ultra Wideband Communications*: *Novel Trends—Antennas and Propagation*, InTech, Rijeka, Croatia, 277-306.

[30] Roux, D., Vian, A., Girard, S., Bonnet, P., Paladian, F., Davies, E. and Ledoigt, G. (2008) High Frequency (900 MHz) Low Amplitude (5 Vm^{-1}) Electromagnetic Field: A Genuine Environmental Stimulus that Affects Transcription, Translation, Calcium and Energy Charge in Tomato. *Planta*, **227**, 883-891. http://dx.doi.org/10.1007/s00425-007-0664-2

[31] Godard, A., Desrumaux, L., Bertrand, V., Andrieu, J., Lalande, M., Jecko, B., Brishoual, M., Couderc, V. and Guillerey, R. (2010) A Transient UWB Antenna Array Used with Complex Impedance Surfaces. *International Journal of Antennas and Propagation*, **2010**, Article ID: 243145. http://dx.doi.org/10.1155/2010/243145

[32] Bennett, L.I., Mixon, D.G., Payne, A.B., Bowman, A., Sickendick, K., Wilmink, G.J., Roach, W.P. and Pakhomov, A.G. (2010) Plasma Membrane Permeabilization by Trains of Ultrashort Electric Pulses. *Bioelectrochemistry*, **79**, 114-121. http://dx.doi.org/10.1016/j.bioelechem.2010.01.001

[33] Vitiello, L., Gorini, S., Rosano, G. and la Sala, A. (2012) Immunoregulation through Extracellular Nucleotides. *Blood*, **120**, 511-518. http://dx.doi.org/10.1182/blood-2012-01-406496

[34] Tanaka, K., Gilroy, S., Jones, A.M. and Stacey, G. (2010) Extracellular ATP Signaling in Plants. *Trends Cell Biol.*, **20**, 601-608.

[35] Irmak, K.M., Oztas, E., Yagmurca, M., Fadillioglu, E. and Bakir, B. (2003) Effects of Electromagnetic Radiation from a Cellular Telephone on Epidermal Merkel Cells. *Journal of Cutaneous Pathology*, **30**, 135-138. http://dx.doi.org/10.1046/j.0303-6987.2003.00002.x

[36] Jordan, D.W., Gilgenbach, R.M., Uhler, M.D., Gates, L.H. and Lau, Y.Y. (2004) Effect of Pulsed, High-Power Radiofrequency Radiation on Electroporation of Mammalian Cells. *IEEE Transactions on Plasma Science*, **32**, 1573-1578.

[37] Braunstein, G.M., Roman, R.M., Clancy, J.P., Kudlow, B.A., Taylor, A.L., Shylonsky, V.G., Jovov, B., Peter, K., Jilling, T., Ismailov, I.I., Benos, D.J., Schwiebert, L.M., Fitz, J.G. and Schwiebert, E.M. (2001) Cystic Fibrosis Transmembrane Conductance Regulator Facilitates ATP Release by Stimulating a Separate ATP Release Channel for Autocrine Control of Cell Volume Regulation. *Journal of Biological Chemistry*, **276**, 6621-6630. http://dx.doi.org/10.1074/jbc.M005893200

[38] Bao, L., Locovei, S. and Dahl, G. (2004) Pannexin Membrane Channels Are Mechanosensitive Conduits for ATP. *FEBS Letters*, **572**, 65-68. http://dx.doi.org/10.1016/j.febslet.2004.07.009

[39] Boudreault, F. and Grygorczyk, R. (2002) Cell Swelling-Induced ATP Release and Gadolinium-Sensitive Channels. *American Journal of Physiology-Cell Physiology*, **282**, 219-226. http://dx.doi.org/10.1152/ajpcell.00317.2001

[40] Roux, D., Faure, C., Bonnet, P., Girard, S., Ledoigt, G., Davies, E., Gendraud, M., Paladian, F. and Vian, A. (2008) A Possible Role for Extra-Cellular ATP in Plant Responses to High Frequency, Low Amplitude Electromagnetic Field. *Plant Signaling & Behavior*, **3**, 383-385. http://dx.doi.org/10.4161/psb.3.6.5385

[41] Janssens, R. and Boeynaems, J.M. (2001) Effect of Extracellular Nucleotides and Nucleoside on Prostate Carcinoma Cells. *British Journal of Pharmacology*, **132**, 536-546. http://dx.doi.org/10.1038/sj.bjp.0703833

[42] Fang, W.G., Pirnia, F., Bang, Y.J., Myers, C.E. and Trepel, J.B. (1992) P2-Purinergic Receptor Agonists Inhibit the Growth of Androgen-Independent Prostate Carcinoma Cells. *Journal of Clinical Investigation*, **89**, 191-196. http://dx.doi.org/10.1172/JCI115562

5

Absorption of Microwaves in Low Intensity Eucalyptus Litter Fire

Douglas Letsholathebe, Kgakgamatso Mphale

Physics Department, University of Botswana, Gaborone, Botswana
Email: Mphalekm@mopipi.ub.bw

Abstract

A fuel bed was constructed where various vegetation species could be used as combustion fuel. The fuel bed was equipped with a thermocouple to measure fire temperature and a two-port automatic network analyser to measure microwave scattering parameters in flame medium. The parameters are then used to determine microwave propagation characteristics in fire. The measurements have implications on radio wave communication during wildfire suppression and in remote sensing. The attenuation data also provide an estimation of vegetation fire ionisation and conductivity. Eucalyptus litter fire with a maximum flame temperature of 976 K was set on the fuel bed and X-band microwaves (7.00 - 9.50 GHz) were caused to propagate through the flame. Attenuation of 0.35 - 0.90 dB was measured for microwaves in the frequency range. For the low intensity fire, conductivity was measured to range from 0.00021 - 0.00055 mho/m and electron density was to be the range of 1.83 - 2.24 × 10^{15} m^{-3}.

Keywords

Wildfire, Microwave Attenuation, Weakly Ionised Gas, Remote Sensing, Electric Conductivity

1. Introduction

Vegetation fires are slightly ionised gaseous medium. When the ionised medium is illuminated with microwaves, incumbent electrons are accelerated by electric field set by the incident electromagnetic energy. The electrons undergo repetitive collisions with relatively stationary heavier neutral incumbent flame particles. As they do, they transfer some of the incident electromagnetic energy into the fire medium. We consider a plane polarised electromagnetic wave which traverses a fire plume in x-direction. The electromagnetic energy dissipates such that its electric field strength ($E(x)$) varies as [1];

$$E(x) = E_0 \exp\left\{-\int_0^x \gamma(x)\,\mathrm{d}x\right\} \tag{1}$$

where

$$\gamma = \alpha_f + \beta_f \quad (2)$$

is propagation constant, α_f and β_f are attenuation and phase coefficients, respectively. E_0 is initial electric field strength. The absorption of energy by fire may result in a considerable signal intensity loss for electromagnetic waves propagating it. However, the absorption depends on temperature, type and concentration of metal impurities in the flame [2] [3].

Vegetation contains up to 3.4% of the element potassium on dry weight basis and a considerable quantity of other alkalis as nutrients [4]. During combustion, organic structure of plants thermally decomposes to release the alkalis and other inorganic particulates into the fire [5]. Thus, vegetation fires are essentially alkali seeded diffusion flames with temperatures up to 1300˚C [6]. High temperature flames seeded with alkali metal species show high microwave absorption, e.g., in [3] [7]. An explanation for the high absorption is that there is high electron concentration in the flames. Alkalis metals have low ionisation energies, e.g., potassium has ionisation energy of 4.34 eV, and are therefore easily ionised in high temperature environments. The absorption increases with the increase in concentration of alkali metal impurities [3] [7], thus high alkali content in plants could make forest fires detrimental to electromagnetic wave communication. In fact, cases of radio wave communication blackout at high frequency (HF) have been reported [8]. Numerical experiments also predict radio wave attenuation coefficients of up 25 dB/m in fire environments [9]. However, propagation experiments at super high (SHF) and extreme high (EHF) frequencies above vegetation fire flames, thus in the thermal plume region, have yielded insignificant attenuation [8] [10]. The negligible attenuation observed in [8] [10] could be attributable to cool flame conditions.

In the experiment, propagation measurements were carried out at the hottest region of low intensity eucalyptus litter fire (up to 37 $kW \cdot m^{-2}$). Attenuation measurements were carried out using a 2-port Automatic Network Analyser. Besides being a direct measure of microwave attenuation by fire, the measurements are important for wildfire-radio wave interaction model validation. The measurements were also used to estimate electrical conductivity of the low intensity vegetation fire and its average electron concentration.

2. Ionisation in the Fire

During vegetation combustion, volatile K and Na species diffuse and attach to oxygen containing carboxyl groups in charcoal matrix forming organo-alkali structures such as K-O-C [11]. The high combustion rates (up to 100 Ks^{-1}) combined with forced gas convection of light gases such as hydrogen radical (H^+) through the organic structure facilitate rapid evolution of alkali atoms from the thermally decomposing charcoal into the flame. This occurs when H^+ and other light volatiles react with the alkali containing carboxyl groups to displace alkali atoms from charcoal attachment sites [11]. The released alkali atoms (e.g., Na(g)) are excited by heat and consequently collide with other excited flame particles. This leads to thermal ionization according to the following equation [12]:

$$\mathrm{FS(g)} + \mathrm{Na}^*(\mathrm{g}) \Leftrightarrow \mathrm{Na}^+(\mathrm{g}) + \mathrm{FS(g)} + \mathrm{e}^-(\mathrm{g}) \quad (3)$$

where: FS(g) and Na*, are excited flame and sodium particles, respectively; A^+ and e^- are singly ionized alkali particle and an electron, respectively. At normal forest fire temperatures, Vodacek *et al.* [13] have estimated that about 10% - 20% of K in vegetation is ionized. The thermal ionization process is a dominant electron-producing mechanism in flames seeded with alkalis [12].

A considerable amount of ions in flames are also produced by chemi-ionization reactions [14]. The reaction is associated with the decomposition of metal oxides in fire to produce electrons according to the reaction [15]:

$$\mathrm{CO(g)} + \mathrm{MO(g)} \Leftrightarrow \mathrm{M}^+ + \mathrm{CO_2(g)} + \mathrm{e}^-(\mathrm{g}) \quad (4)$$

A considerable amount of chlorine exists in plants. It exists attached to potassium in cell vacuole [16]. In reactions where chlorine is in co-existence with alkali metals such as sodium, it may play a catalytic role in ionization of the latter by hydrogen atoms to produce electrons and metal ions. The process occurs according to the reaction equation [15]:

$$\mathrm{Na(g)} + \mathrm{H(g)} \Leftrightarrow \mathrm{Na}^+(\mathrm{g}) + \mathrm{H_2(g)} + \mathrm{e}^-(\mathrm{g}) \quad (5)$$

3. Microwave Absorption

Vegetation fires are highly collisional atmospheric pressure plasma [17]. When X-band microwaves illuminate fire, propagation coefficients, α_f and β_f, are related to momentum transfer electron-neutral collision frequency (φ_{eff}) and plume ionisation by the expressions [17]:

$$\alpha_f \cong \frac{\varphi_{eff}}{2c}\left[\frac{\omega_p^2}{\left(\omega^2+\varphi_{eff}^2\right)}\right] \tag{6}$$

and

$$\beta_f \cong \frac{\omega}{c}\left[1+\frac{\omega_p^4}{8\left(\omega^2+\varphi_{eff}^2\right)^2}\frac{\varphi_{eff}^2}{\omega^2}\right] \tag{7}$$

where ω, and c are propagation cyclic frequency and speed of light in vacuum, respectively. $\omega_p=\left(ne^2/m_e\varepsilon_0\right)^{1/2}$ is plasma frequency. ε_0 and m_e are dielectric permittivity and mass of electron, respectively.

Flame conductivity and ionisation are respectively given by the relations:

$$\sigma = 2c\varepsilon_0\alpha_f \tag{8}$$

and

$$n=\left\{\alpha_f\varphi_{eff}/5.292\times10^{-6}\right\} \tag{9}$$

Momentum transfer electron-neutral collision frequency is given by Uhm [18] as:

$$\varphi_{eff}=1.36\times10^{-12}\left(\frac{T_r}{T_g}\right)\left(T_e\right)^{0.5} \tag{10}$$

where T_r, T_g and T_e are ambient, flame and electron temperatures respectively. T_e for temperatures around 1000 K is 0.1 eV [19]. Ratio of transmitted microwave amplitude to that of the incident wave is propagation factor (T) and is given by the relation [20];

$$T=\mathrm{e}^{-i\gamma r} \tag{11}$$

where r is the length of the path traversed by the electromagnetic beam through the flame. Propagation factor could be measured using the ANA. It is determined by the measuring scattering parameters S_{11} and S_{21}. The scattering parameters are related to T by the following relations [20]:

$$S_{11}=\left[\frac{\Gamma\left(1-T^2\right)}{\left(1-\Gamma^2T^2\right)}\right] \tag{12}$$

and

$$S_{12}=\left[\frac{T\left(1-\Gamma^2\right)}{\left(1-\Gamma^2T^2\right)}\right] \tag{13}$$

where Γ is the reflection coefficient given by the expression, $\Gamma=\Omega\pm\sqrt{\Omega^2-1}$ where $\Omega=\left(\frac{S_{11}^2-S_{21}^2+1}{2S_{11}}\right)$. The sign in the expression for Γ is chosen such that $|\Gamma|<1$. Re-arranging (12) and (13), T can be written as:

$$T=\left[\frac{S_{11}^2+S_{21}^2-\Gamma}{1-\left(S_{11}+S_{21}\right)\Gamma}\right] \tag{14}$$

with T determined from S-parameters, the propagation constant γ can be worked out from (11), thus;

$$\gamma=\left[\ln\left(1/T\right)\right]/d \tag{15}$$

4. Experimental Setup

4.1. Combustion Fuel Bed and Automatic Network Analyser

The equipment used to measure the microwave absorption by fire consisted of a 0.03 m thick combustion tray, a Hewlett-Packard 8577C Automatic Network Analyser (ANA) with X-band horns and a computer. A layer of river sand, 0.04 m thick, was put on the tray to protect it from direct heat from combustion. The tray was 1.20 m long and 0.75 m wide (**Figure 1**). The microwaves were generated from ANA, transmitted and received by two X-band transmit-receive horn. The horns were connected to the ANA through the two-port S-parameter test set by coaxial cables. High quality mode transition adapters were used to make the connections between coaxial cables and the horn (see **Figure 1**).

Eucalyptus (poplar gum) litter of bulk density of 6.03 $kg\cdot m^{-3}$ was collected ten (10) days before the experiment and left to dry in a laboratory. This was done to maximise combustion efficiency during burning. The litter was laid over sand in the combustion tray and set alight.

4.2. Temperature Measurement

A hole was drilled at the bottom of the combustion tray to allow the insertion of a thermocouple for fire seat temperature measurement. The thermocouple (type K) wire had fibre glass shield which can withstand temperatures up to 723 K. To further insulate the fibreglass shield, the wire was wrapped with a thermal insulation material called Fiberfrax®. The thermocouple wire was cut from a 100 m double braided fiberglass insulated chromel-alumel (24-G/G) thermocouple wire 50 µm in diameter. The thermocouple wire was electro-fused at one end to make perfect junction and then tested with a hot air gun and a multimeter. The wire was then fixed to the bottom of the tray by means of a tape and the electro-fused junction was left protruded 0.01 m above river sand surface. The thermocouple was then wired to PICO® Tech TC-O8 data logger to read in the temperature into a computer throughout each experiment.

4.3. S-Parameter Measurements

The ANA was set to sweep from 7.00 to 9.50 GHz logging in 601 S-parameter data points. The analyser was calibrated using Transmit-Reflect-Line (TRL) as in Varadan *et al.* [21]. The S-parameters were then uploaded to the computer for analysis. The analyser took approximately 2 s to sample over one sweep, and then there was a latency of about 50 s before the next sweep could be initiated. Several sweeps and logging of S-parameters were

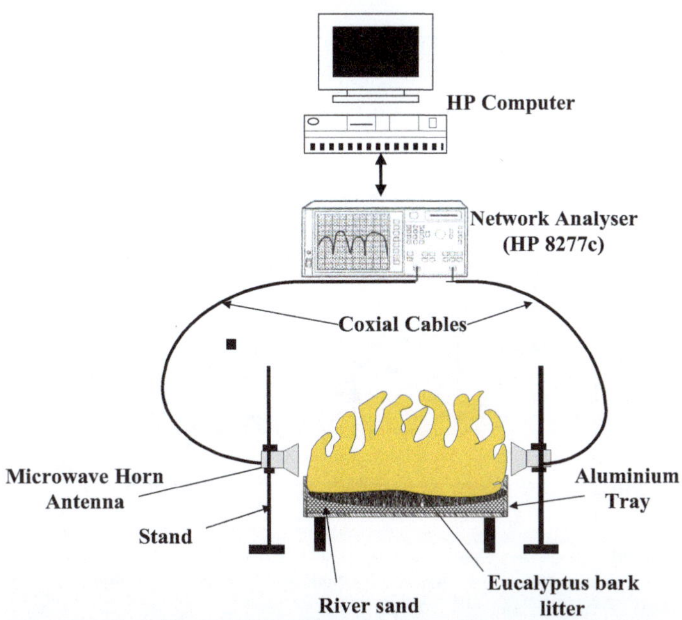

Figure 1. Automatic Network Analyser (ANA) set up for S-parameter measurements.

carried, e.g., in **Figure 2**, but those for which flames filled the entire propagation path (by visual inspection) were chosen for S-parameter analysis. Two such cases were **els2** and **els4** (see **Figure 2**) which corresponded to the times 178 and 429 s since ignition of the eucalyptus litter.

5. Experimental Results and Discussions

5.1. Fire Temperatures and S-Parameters

The eucalyptus litter was surface ignited. It took 42 s for the flame to cover the whole combustion tray. During this time, thermocouple measured temperature rose from an ambient of 302 to 894 K. It took 68 s for the fire to reach its maximum intensity at which flame temperature was 976 K. Logging of s-parameters was delayed until the fire assumed a steady rate of spread into the fuel layer. This was observed by a constant flame height. After reaching its maximum intensity, its temperature started to decrease steadily to 847 K after 97 s. At this point the first s-parameter measurement was made (**esp1** in **Figure 2**). The second s-parameter logging was carried out after 178 s (**esp2** in **Figure 2**). Temperature at **esp2** was observed to be 834 K. After 201 s, a slight wind deflected the flame such that temperature of the thermocouple was lower than the running average. The effect of the wind lasted over 48 s. Two other such instances occurred during the propagation measurement, one after 90 s and another after 323 s. They on average lasted for about 60 s. S-parameter measurements, **esp3**, **esp4** and **esp5** were made after 281, 429 and 540 s, respectively. Temperatures at the respective periods were, 827, 821 and 817 K. The uploading periods; **esp1**, **esp3** and **esp5**, were made when the flame covered nearly half of the combustion area and are therefore not used in the analysis. The flame extinguished 107 s after **esp5** and what remained were red hot charcoals.

5.2. Microwave Attenuation, Electron Density and Conductivity

Relative microwave absorption at times **esp2** and **esp4** for frequencies (7.00 - 9.50 GHz) are shown in **Figure 3** and **Figure 4**. The flames at **esp2** and **esp4** were of slightly different intensity as there was a temperature difference of 13 K. The litter flame at **esp2** caused a microwave absorption in the range of 0.90 - 0.42 dB while at **esp4** suffered a caused a signal loss of 0.72 - 0.35 dB for frequency range, respectively. It can be inferred from the measurements that absorption is a factor of rate of alkali emission from fuel. The rate of alkali emissions decreases with the increase in duration of combustion. Thus, the rate of alkali emissions follows mass loss rate curve in thermal decomposition of biomass. Alkali seeding at **esp2** is more than at **esp4** hence high absorption at **esp2** than **esp4**. Average electron density for **esp2** as determined from Equation (23) is 2.24×10^{15} m^{-3} while at

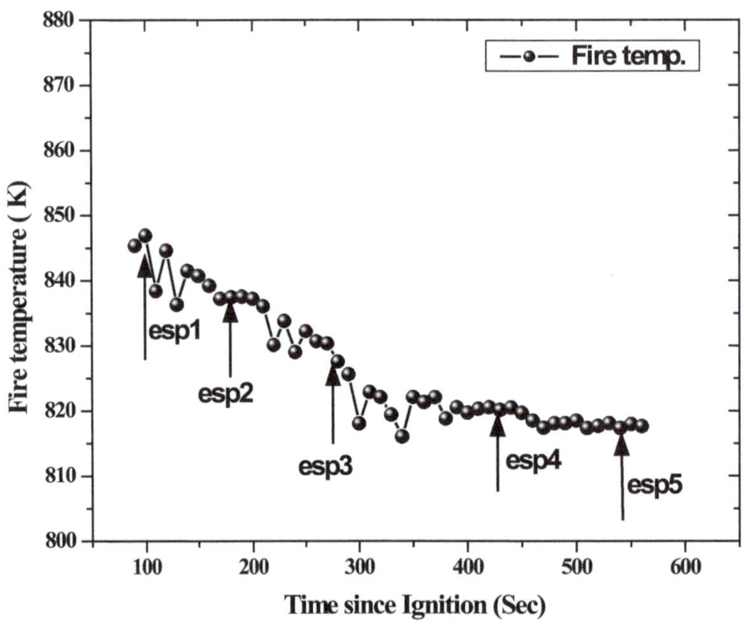

Figure 2. Eucalyptus litter fire temperature during the experiment.

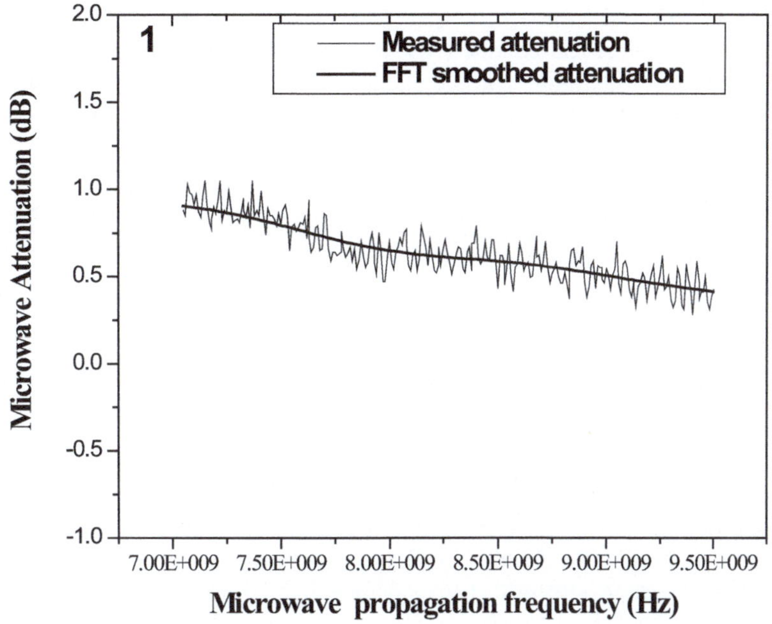

Figure 3. Variation of relative attenuation with propagation frequency at **esp2**.

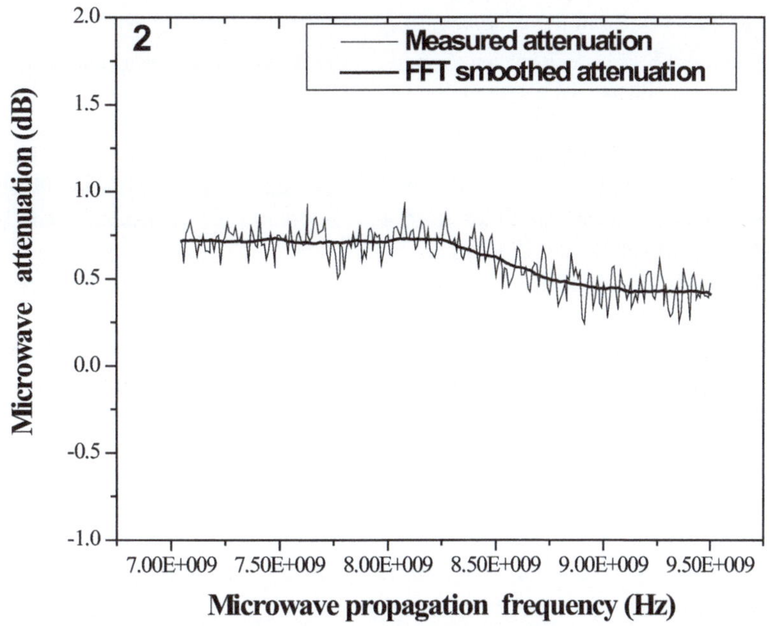

Figure 4. Variation of relative attenuation with propagation frequency at **esp4**.

esp3 is calculated to be 1.83×10^{15} m^{-3}. Electrical conductivity for the flame at **esp2** and **esp4** is calculated to have ranges $(2.57 - 5.50) \times 10^{-4}$ and $(2.11 - 4.40) \times 10^{-4}$ mho/m, respectively.

6. Conclusions

The electron concentration for the eucalyptus litter fire with maximum radiant heat up to about 37 kW·m^{-2} was in the range from $1.83 - 2.24 \times 10^{15}$m^{-3}. Comparatively, the electron densities were lower than those produced from non vegetation fuel flames with much higher temperatures e.g. Belcher *et al.* [3] and Koretzky *et al.* [22]. Electron concentration in the flame depends on the type of fuel used as well as its temperature because it is this combination that produces ionisable particles.

Electrical conductivity ranged from 2.11 - 5.50 × 10^{-4} mho/m for the frequency range considered in the experiment. Typical values of electrical conductivity of pure hydrocarbon flames are low than those measured in the experiment, e.g., in [23]. Maximum, alkali content in eucalyptus is 1.2%. It implies that forest fires that occur in higher alkali contaminated vegetation may seriously affect are Line-of-Sight (LOS) microwave communication systems.

Acknowledgements

We would like gratefully to acknowledge the Department of Electrical Engineering of James Cook University for providing the equipment for S-parameter measurement. The work is supported by the Staff Development Office of the University of Botswana. It is partly supported by Emergency Management Australia under project No. 60/2001.

References

[1] Santoru, J. and Gregorie, D.J. (1993) Electromagnetic Wave Absorption in Highly Collisional Plasma. *Journal of Applied Physics*, **74**, 3736-3743. http://dx.doi.org/10.1063/1.354520

[2] Schneider, J. and Hofmann, F.W. (1959) Absorption and Dispersion of Microwaves in Flames. *Physical Review*, **116**, 244-249. http://dx.doi.org/10.1103/physrev.116.244

[3] Belcher, H. and Sudden, T.M. (1950) Studies on the Ionisation Produced by Metallic Salts in Flames II. Reactions Governed by Ionic Equilibrium in Coal-Gas/Air Flames Containing Alkali Metal Salts. *Proceedings of the Royal Society of London Series A*, **202**, 17-39. http://dx.doi.org/10.1098/rspa.1950.0083

[4] Radojevic, M. (2003) Chemistry of Forest Fires and Regional Haze with Emphasis on Southeast Asia. *Pure and Applied Geophysics*, **12**, 157-187. http://dx.doi.org/10.1007/978-3-0348-7970-5_10

[5] Jensen, A.P., Frandsen, F.J., Dam-Johansen, K. and Sander, B. (2000) Experimental Investigation of the Transformation and Release to Gas Phase of Potassium and Chlorine during Straw Pyrolysis Conditions. *Energy and Fuels*, **11**, 1026-1032.

[6] Mphale, K.M., Heron, M. and Verma, T. (2007) Effect of Wildfire Induced Thermal Bubble on Radio Communications. *Progress in Electromagnetics Research* (*PIER*), **68**, 197-228. http://dx.doi.org/10.2528/pier06072202

[7] Boan, J. (2007) Radio Experiments with Fire. *IEEE Antennas Wireless Propagation Letters*, **6**, 411-414. http://dx.doi.org/10.1109/LAWP.2007.902809

[8] Williams, D.W., Adams, J.S., Batten, J.J., Whitty, G.F. and Richardson, G.T. (1970) Operation Euroka: An Australian Mass Fire Experiment. Report 386, Defense Standards Laboratory, Maribyrnor.

[9] Boan, J. (2006) Radio Communication in Fire Environments. *Proceedings of the Wars* 2006 *Conference*, Leura,

[10] Hata, M. and Shigeyuki, D. (1983) Propagation Tests for 23GHz and 40 GHz. *IEEE Journal on Selected Areas in Communications*, **1**, 658-673. http://dx.doi.org/10.1109/JSAC.1983.1145971

[11] Okuno, T., Sonoyama, N., Hayashi, J., Li, C., Sathe, C. and Chiba, T. (2005) Primary Release of Alkali and Alkaline Earth Metallic Species during Pyrolysis of Pulverized Biomass. *Energy and Fuels*, **19**, 2164-2171. http://dx.doi.org/10.1021/ef050002a

[12] Nesterko, N.A. and Taran, E.N. (1971) Ionization and Radiation of Alkali Metals in Acetylene—Air Flame Plasma, with Halogen Additions. *Journal of Applied Spectroscopy*, **14**, 242-244. http://dx.doi.org/10.1007/BF00613182

[13] Vodacek, A., Kremens, R.L., Fordham, S.C., Van Gorden, S.C., Luisi, D., Schott, J.R. and Latham, D.J. (2002) Remote Optical Detection of Biomass Burning Using Potassium Emission Signature. *International Journal of Remote Sensing*, **23**, 2721-2726. http://dx.doi.org/10.1080/01431160110109633

[14] Latham, D. (1999) Space Charge Generated by Wind Tunnel Fires. *Atmospheric Research*, **51**, 267-278. http://dx.doi.org/10.1016/S0169-8095(99)00012-5

[15] Alkemade, M.A. (1979) Fundamentals of Analytical Flame Spectroscopy. Hilger, Bristol.

[16] Bjorkmann, E. and Stromberg, B. (1997) Release of Chlorine from Biomass at Pyrolysis and Gasification Conditions. *Energy and Fuels*, **11**, 1026-1032. http://dx.doi.org/10.1021/ef970031o

[17] Mphale, K.M. and Heron, M. (2008) Nonintrusive Measurement of Ionization in Vegetation Fire Plasma. *European Physical Journal*: *Applied Physics*, **41**, 157-164. http://dx.doi.org/10.1051/epjap:2008013

[18] Uhm, H.S. (1999) Properties of Plasmas Generated by Electrical Breakdown in Flames. *Physics of Plasmas*, **6**, 4366-4374. http://dx.doi.org/10.1063/1.873701

[19] Sicha, M. (1979) Measurement of the Electron Energy Distribution Function in a Flame Plasma at Atmospheric Pres-

sure. *Czechoslovak Journal of Physics*, **29**, 640-645. http://dx.doi.org/10.1007/BF01600014

[20] Kadaba, P.K. (1984) Simultaneous Measurements of Complex Permittivity and Permeability in the Millimeter Region by a Frequency-Domain Technique. *IEEE Transactions on Instrumentation and Measurement*, **33**, 336-347. http://dx.doi.org/10.1109/TIM.1984.4315236

[21] Varadan, V.V., Jose, K.A. and Varadan, V.K. (2000) *In Situ* Microwave Characterization of Nonplanar Dielectric Objects. *IEEE Transactions on Microwave Theory and Techniques*, **48**, 388-394. http://dx.doi.org/10.1109/22.826837

[22] Koretzsky, E. and Kuo, S.P. (1998) Characterization of an Atmospheric Pressure Plasma Generated by a Plasma Torch Array. *Physics of Plasmas*, **5**, 3774-3780. http://dx.doi.org/10.1063/1.872741

[23] Adler, F.P. (1954) Measurement of Conductivity of a Jet Flame. *Journal of Applied Physics*, **25**, 903-908. http://dx.doi.org/10.1063/1.1721767

6

Feed Signal Influence and Potential Performances of a Compact Radiation Source Based on a Helical Antenna

Laurent Pecastaing[1], Marc Rivaletto[1], Antoine de Ferron[1], Romain Pecquois[2], René Vézinet[3], Jean-Christophe Diot[3], Stéphane Tortel[3]

[1]Laboratoire SIAME, Equipe Génie Electrique, Université de Pau et des Pays de l'Adour, Hélioparc Pau-Pyrénées, Pau, France

[2]Hi-Pulse, 300 rue des arts et métiers, Pont-de-Pany, France

[3]French Alternative Energies and Atomic Energy Commission (CEA), Gramat, France

Email: laurent.pecastaing@univ-pau.fr

Abstract

In the class of emerging high power electromagnetic sources, a complete pulsed power source, named MOUNA (French acronym of "Module Oscillant Utilisant une Nouvelle Architecture") has been developed. This device must transmit waveforms with a wide frequency band and a high figure-of-merit. To improve the overall performance of the MOUNA system while maintaining its compact size, two approaches are being explored in the paper: the replacement of the dipole antenna by a helical antenna and its feeding signal influence. Helical antenna is cylindrical shape and relatively compact. It offers relatively good gain factor and directivity. The waveform delivered to the antenna is directly related to the amplitude of the radiated electric field. Therefore, different waveforms (step pulse, Gaussian pulse, bipolar pulse and damped sinusoid) are compared to point out the feed signal influence on the radiated electric field. Switch oscillators appear to be considered as interesting resonant sources for driving an antenna. The novel radiating source consists of a primary power source, a resonant transformer, a coaxial transmission line damped oscillator (also termed as coaxial resonator), and a helical antenna. This high voltage pulsed source is very compact (volume of only 2500 cc without the antenna). Our study aims at designing the antenna (number of turns, size...) and a coaxial damped oscillator directly implemented at the output of the transformer. A CST-based simulation is proposed to predict the performances of this wideband source.

Keywords

Pulse Transformer, Helical Antenna, Coaxial Damped Oscillator, High Voltage, Peaking Switch, Electric Field

1. Introduction

High power microwave (HPM) sources have been developed over the past few decades for many applications, including defense for disrupting electronically-controlled system, medical for cell manipulation, and biological for food and water purification [1]. Defense requirements are the most prevalent. Ultra-Wide Band (UWB) source, one of the HPM sources, can be used for target identification, buried objects detection, radar system and directed energy weapon system [2]-[7]. Intentional electromagnetic interference describes a wide range of technologies that are designed to disrupt the performance of electronic systems by introduction of spurious signals in different frequency ranges. These systems must also typically be high power, since it is generally desired to have as large an effect possible at as long distances possible [8]. Even if the radiated energy of a single pulse UWB system is low, these systems can disrupt, disable, or destroy electronic target because of the repetitive operation. In principle, for affecting any target with such radiation, the frequency has to be matched to the frequency absorption spectrum of the target [7]. However, the main advantage of pulsed wideband radiation, with pulse durations in the nanosecond range, is that it is very aggressive for systems with wiring [9] [10].

There are a number of high power microwave efforts, and an equal number of antenna geometries being pursued. Example geometries generally include traditional TEM horns, traveling wave antenna [3] [4], the Impulse Radiating Antenna (IRA) [2], dipoles and bicones [9]-[11], among others [12]. However, most of these geometries are either large and not deployable, or radiate with undesirable radiation pattern. Furthermore, generating very large electric field strengths is difficult, since extremely high voltages are required. Insulating the high voltage increases the complexity of the overall source design, and typically results in larger volumes.

A previous work supported by the DGA aimed at developing an extremely compact source [11]. The electromagnetic pulse source named MOUNA is composed of a set of batteries, a DC/DC converter to charge four capacitors, four synchronized spark gap switches, a resonant transformer generating 600 kV/265 ns pulses, an oil peaking switch and a dipole antenna. This source is compact, with dimensions less than 1 m in length and 22 cm in diameter. These dimensions correspond to a 3800 cc volume. However, to radiate very high electric fields, the low gain of the dipole antenna is detrimental. The use of a directional antenna may improve performance really significantly focusing the radiation in a preferred direction. The main characteristics of the axial helical antenna (compactness, high gain on the axis, wideband spectrum and high impedance) make it an excellent candidate. Helical antennas are being studied for high power radiofrequency applications [12]. The helical geometry is very appealing for it is relatively small and conformal geometry, which makes it suitable for air platforms. Furthermore, the antenna is very directive, with typical beam widths of 20 - 40 degrees, and offering relatively good values of gain, from 8 - 15 dB.

Some sources already exist and use a Marx generator associated to a helical antenna [13]-[15]. Only one associates a 30 kV pulse transformer, a damped oscillator and a helical antenna [13]. Our structure uses also an ultra-compact high voltage resonant transformer which will allow an important volume gain versus other topologies (particularly Marx generators used more conventionally).

The waveform of the antenna pulse feeding is of importance for the overall source performance. Step pulse, bipolar pulse, Gaussian pulse and damped sinusoid are studied in the paper. One of the typical moderate band source is the damped sinusoidal system which employs a self breaking switch and antenna at each end of high voltage transmission line. The waveform delivered to the antenna will be characterized by one or more damped sinusoids. The pulse width (or number of cycles to some fraction of the peak) depends on the antenna impedance and other parasitic effects (non ideal switch, etc.). When the switch operates, the system generates a damped sine signal into an antenna. The system radiates N cycles of the same oscillation frequency wave with amplitude damp down. Radiated energy of damped sinusoid system might be more than UWB source, because duration of the former is longer than the one of the latter.

Including the transformer, the first stages principle is conserved to build the pulsed power generator of the new helical radiating system (**Figure 1**). The oil switch and the radial line of the previous version of the source are replaced by a multichannel annular gas switch [16] [17] associated to a coaxial quarter wave oscillator [18]-[20]. A resonant transformer is used to feed 500 kV high voltage pulse to the coaxial damped oscillator. This oscillator feeds a 140 Ω helical antenna.

Principle and/or realization of switched oscillators are described in many papers [18]-[20] where the line has a radial section and a coaxial section due to the use of only one switch. Our oscillator is only composed of a coaxial section switched by the multichannel annular gas switch. The design of the novel radiating source com-

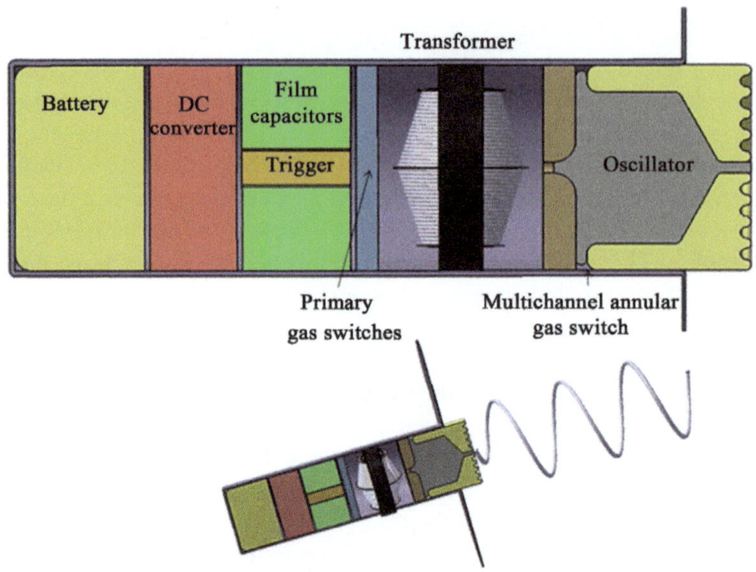

Figure 1. Arrangement of the new helical antenna radiating system.

posed of the MOUNA pulsed power source, the switch oscillator and the helical antenna is also described and the predicted overall performances of the electrical characteristics are detailed.

2. Helical Antenna Design

2.1. Antenna Design

The principle of a helical antenna was described for the first time by J.D. Kraus [12] (**Figure 2**). In these works, Kraus establishes relations between the dimensions of the antenna for which the field is a maximum in the direction of helix axis and is circularly polarized: the axial mode of radiation. If the helix circumference is comparable to the wavelength, the antenna operates as a directional antenna radiating a beam off the ends of the helix, along the antenna's axis. It radiates circularly polarized waves.

If the wavelength (λ) is the free-space wavelength of the radiating field, these relations follow:

$$C = \pi D \approx \lambda \tag{1}$$

$$\alpha = \tan^{-1} = \tan^{-1}\frac{S}{\pi D} \approx 12.5^{\circ} \tag{2}$$

$$D_{\text{ground}} \approx \frac{3\lambda}{4} \tag{3}$$

and the antenna input impedance R is given by:

$$R \approx 140 \times \left(\frac{C}{\lambda}\right) \approx 140\Omega \tag{4}$$

The design of a simple helical antenna follows from these equations, in which the circumference of the helix is one wavelength so that diameter D equals λ/π; the length of the helix is given by $N\cdot\lambda/4$, where N is the number of helix turns and $\lambda/4$ is the turn-to-turn spacing. The ground plane diameter (D) is approximately $3/4\lambda$. The pitch angle which controls how far the helix antenna grows in the z-direction per turn is given by α where S is the vertical spacing between turns for the antenna.

The MOUNA source radiates around 200 MHz, if we do not change this frequency, the helical antenna diameter will be around 48 cm. So, the MOUNA diameter (22 cm) will be increased drastically and its compactness will be altered. To keep MOUNA dimensions, the helical antenna diameter must be around 22 cm, in free-space the wavelength of the radiating field must be around 69 cm and its frequency around 434 MHz. That is why the axial mode helical antenna is designed to operate at a frequency around 434 MHz. So, the antenna mechanical characteristics are summarized in the **Table 1**.

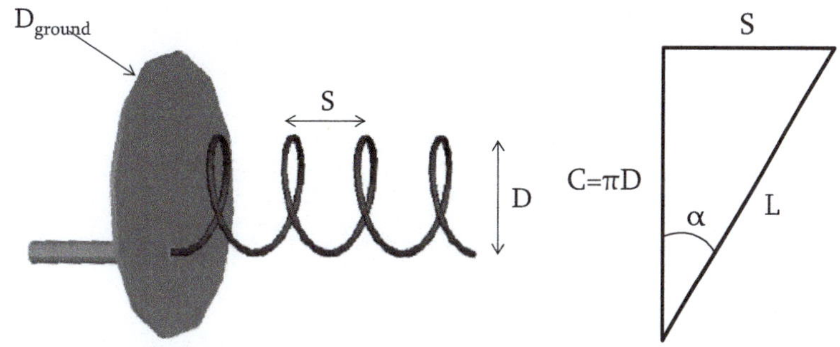

Figure 2. Geometric features of an axial helical antenna. *D*: Diameter of helix (m); *C*: Circumference of helix (m); *S*: Spacing between turns (m); D_{ground}: Diameter of the reflector (m); *α*: pitch angle (degrees); *L*: length of turn (m); *N*: number of turns.

Table 1. Antenna mechanichal design.

Diameter *D*	22.0 cm
Spacing *S*	15.3 cm
Reflector diameter D_{ground}	51.8 cm
Number of turns	3

So, a new damped oscillator must be designed. The MOUNA radial output line will be replaced by a coaxial quarter wave oscillator sized to oscillate at 434 MHz.

2.2. Feed Signal Influence

In [18] Giri *et al.* describe the problem of designing switched oscillators and their integrations into a helical antenna. This paper is following this track. But a few realizations, described in [13] [14], do not use a damped sinusoidal signal to feed the helical antenna. So, it was interesting to study the influence of the feed signal on the field intensity produced by the antenna. A CST Microwave Studio simulation of a 3 turn helical antenna model is computed.

For simulation, four different normalized voltages are applied on a 140 Ω adapted discrete port localized on the antenna input. The four signals are a step pulse (pink), a Gaussian pulse (green), a bipolar pulse (blue) and a 434 MHz damped sinusoid (red) (**Figure 3**).

The simulation provides the following results (**Figure 4**). These results clearly show a strong dependence of the shape of the radiated field with the shape of the supply signal. The antenna factor (*i.e.* rE/V) also varies greatly (**Table 2**). The highest antenna factor (0.82) is obtained when the antenna is fed by a 434 MHz damped sinusoid. Interesting results (0.6) are also noted in the case of a bipolar pulse.

3. Coaxial Damped Oscillator Design

Many papers [16]-[18] describe the principle and realization of a low impedance line damped oscillator to feed a high impedance antenna. In most cases, the quarter wave line is composed of a radial section followed by a coaxial section. Unlike other works where the transmission line oscillator consists of radial transmission line section and coaxial transmission line section, our oscillator design is simplified. The design of the coaxial section follows from simple coaxial transmission line theory.

According to a low impedance quarter wave transmission line oscillator, the transmission line is charged to a high voltage. A switch is located at one end of the transmission line and antenna is placed at the other end of the transmission line. By charging a transmission line to some high voltage and then shorting it by some fast-closing switch, a resonant waveform is applied to an antenna at the other end: a damped sinusoid oscillation is generated to the antenna. Indeed, when the oscillator is switched, the circuit is equivalent to a damped oscillator (high impedance of the transformer at its input and low impedance of the switch at its output) from which its energy is radiated by the helical antenna.

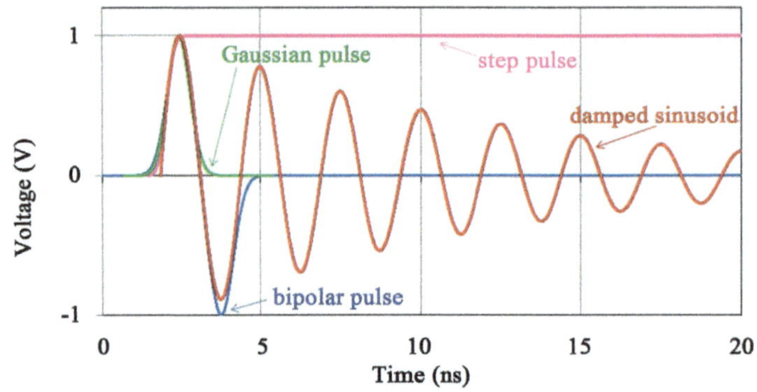

Figure 3. Normalized signals applied to the antenna port.

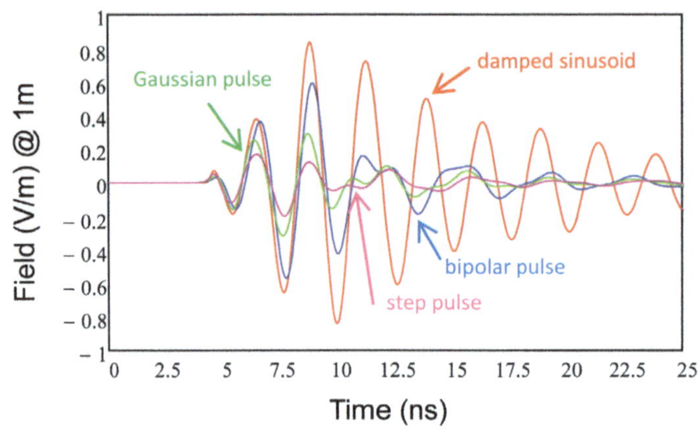

Figure 4. Radiated fields for the step pulse (pink), the Gaussian pulse (green), the bipolar pulse (blue) and the 434 MHz damped sinusoid (red).

Table 2. Variations of the antenna factor.

Supply signal	Antenna factor
Step pulse	0.2
Gaussian pulse	0.32
Bipolar pulse	0.6
Damped sinusoid	0.8

Due to the presence of two quite different sections, the design and the realization of this line oscillator is very accurate [15]. Design must be simplified if we can propose only a coaxial line oscillator with polyamide 6 dielectric (PA6). The volume of the oscillator is filled with this polymer. This choice allows the feeding of the line through the switch side. The need for a fast switch is clear. A real switch will have some inductance and resistance which will modify the resonance somewhat. The switch will need to be physically small compared to the transmission line length. To keep only one switch, this solution requires using a multichannel annular gas switch which will reduce the switch inductance. Its time of breakdown will be drastically reduced in high pressurized nitrogen atmosphere [21].

To keep the transformer performances (maximum output voltage: 600 kV, rise time: 265 ns, **Figure 5**), the capacitance of the coaxial line oscillator must remain close to 80 pF due to the resonance conditions of the transformer. This measurement (**Figure 8**) was provided by a Half-Megavolt capacitive probe developed in common by Université de Pau, Loughborough University (UK) and the North star company [22]. Its features are: attenuation, 1/10.000; input capacitance, 12.4 pF; maximum voltage, 620 kV; bandwidth, 8 Hz - 55 MHz; accuracy, <4% in a volume of only 1500 cc.

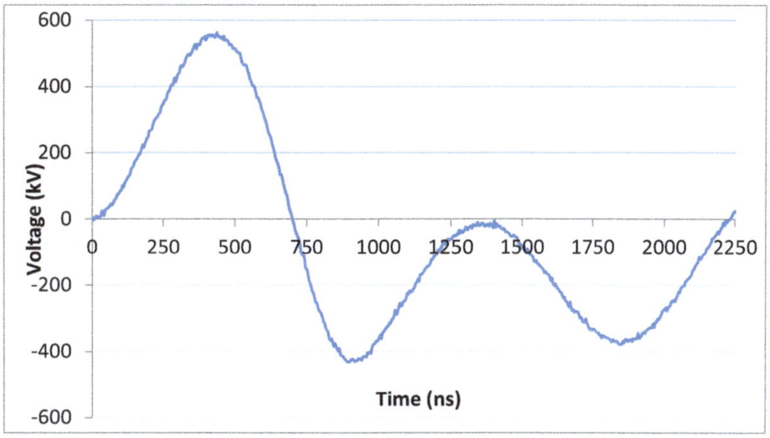

Figure 5. Typical output transformer voltage (input 9 kV, output 560 kV).

Insulation rules impose 18 cm for the outer diameter (D_{out}) maximum value. Indeed the geometry of the switch oscillator system must satisfy the static requirements for the self-breakdown switch operation in addition to the geometric constraint that define the resonance parameters of the system. Therefore, to oscillate at 434 MHz, the main characteristics of the line oscillator are summarized in the **Table 3**.

The length of the transmission line should be 9.11 cm because of the center frequency. The designed characteristic impedance and average switching voltage of the coaxial damped oscillator is 7.3 Ω and 500 kV respectively. The output of the high voltage transformer is fed to the transmission line coaxial damped oscillator which is designed to oscillate at center frequency of 434 MHz.

4. Coaxial Oscillator LTspice and CST Simulations

Prior to detailed design and manufacturing of the coaxial damped oscillator, the electrical parameters and performances of transmission line oscillator were simulated using electronic and electromagnetic simulation tools. The programs for the simulation activity are LTspice, and CST Microwave studio. The LTspice tool is used to simulate the oscillation frequency, and we assumed the uniform 7.3 Ω coaxial transmission line section is connected with the 140 Ω antenna.

Using predicted values above, an LTspice-based study of the line damped oscillator is realized (**Figure 6**).

(Rantenna is the impedance of the helical antenna considered as a constant for the electric simulation)

On **Figure 7**, the simulation results of the damped oscillator voltage are presented assuming that the breakdown time of the annular switch is about 500 ps. The simulation confirms calculations.

For the same conditions and dimensions, a CST Microwave Studio simulation of the damped oscillator is computed. For simulation, a 1 V normalized voltage is applied on an adapted discrete port localized on the multichannel gas annular gas switch (**Figure 8**).

The simulation provides the following results (**Figure 9**). The results are similar in the LTspice and CST simulations. In the results of the two types of simulation (**Figure 7** and **Figure 9**), the frequency is close to the expected 434 MHz.

5. Novel Multi Primary Winding Resonant High-Voltage Pulsed Transformer

As explained in Section 2, to keep the dimensions of the MOUNA source, the number of turns of the helical antenna is 3. The voltage signal present on **Figure 9** is applied to a 3 turn helical antenna simulated on CST. For simulation, this 1 V normalized voltage is applied on a 140 Ω adapted wave port localized on the antenna input.

Electric field distribution in far field of the system is simulated with CST Microwave Studio software. The electric fields of both vertical and horizontal directions are about 1 V/m at 1 m when 1 V power is applied to the system, the simulated electric fields distribution is shown in **Figure 10**. As observed the radiated electric field pulse is a damped sinusoid. The polarization is not perfectly circular, we suppose due to the low number of turns. In these conditions, the antenna factor is about 0.95.

The simulated beam pattern is provided in **Figure 11**. The center frequency spectrum of the radiated wave is

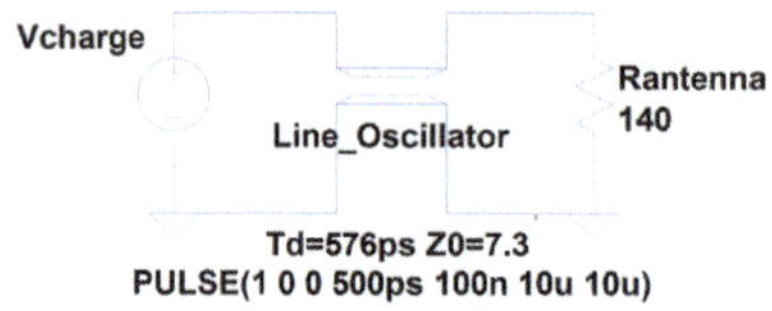

Figure 6. LTspice damped oscillator schematic.

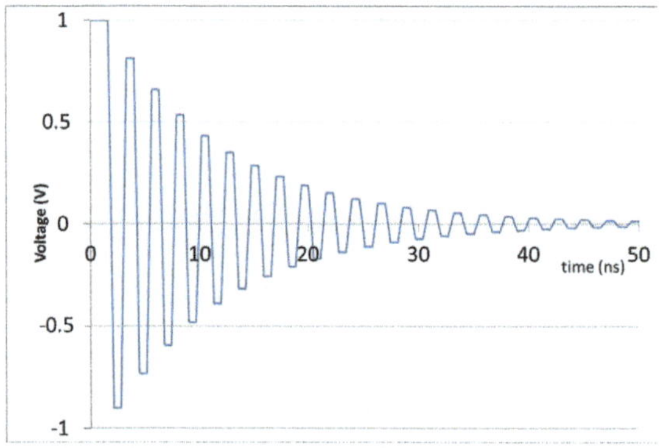

Figure 7. LTspice damped oscillator normalized voltage.

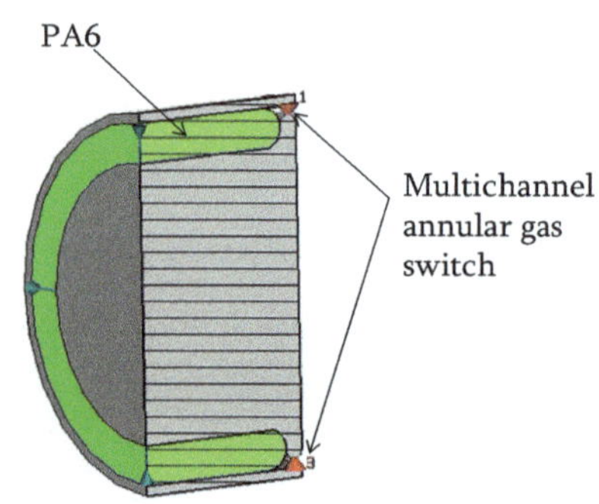

Figure 8. CST damped oscillator model.

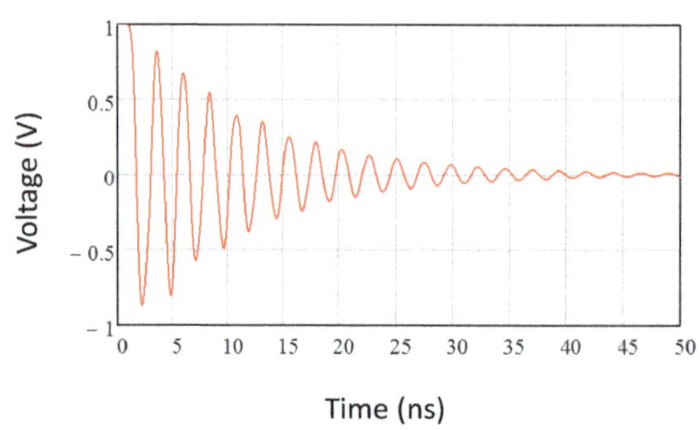

Figure 9. CST damped oscillator normalized voltage.

Table 3. Coaxial damped oscillator design.

Diameter D_{out} (outer diameter)	18.0 cm
Diameter D_{inn} (inner diameter)	14.3 cm
Length L	9.11 cm
Capacitance C_{line}	79.2 pF
Impedance Z_{line}	7.3 Ω
Propagation time T_{line}	576 ps

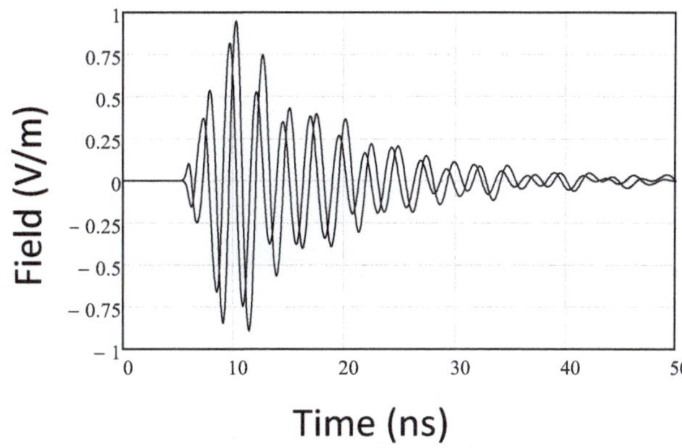

Figure 10. CST electric far fields in H and V directions measured at 10 m and then normalized. Red curve: electric far field in H direction; Blue curve: electric far field in the V direction.

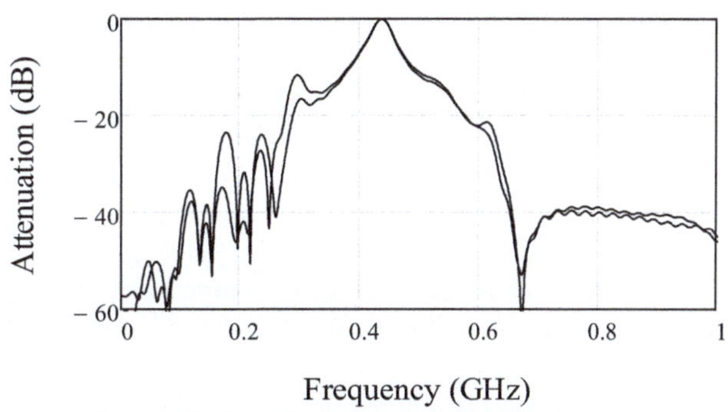

Figure 11. Corresponding spectrums of electric far fields in H and V directions measured. Red curve: electric far field in H direction; Blue curve: electric far field in the V direction.

close to 434 MHz which is coincide with the center frequency of the transmission line oscillator as seen in **Figure 9**. The spectral density distribution is between 277 MHz to 575 MHz (at −20 dB of the maximum). It was between 65 MHz to 350 MHz with a central frequency spectrum of 212 MHz for the MOUNA source [11]. The proposed changes lead to a slight increase of the bandwidth of the frequency spectrum, and this spectrum is pushed towards high frequencies. This is particularly important in terms of electric susceptibility for HPM systems.

One of the next steps in antenna study is the critical transition between the oscillator and the antenna. This transition (light grey in **Figure 12**) is the seat of the higher voltage and higher electric field. This point will require a specific study and a specific insulation. To improve radiation performances, Kraus [12] provides a very good example of a cupped helical antenna and this method could be applied in our case.

6. Conclusions

A wideband electromagnetic source, named MOUNA and based on a novel high-voltage resonant pulsed transformer, was successfully designed and manufactured. The sub-assembly formed by the novel resonant high-voltage pulsed transformer has been operated up to 600 kV with a rise time of 265 ns. However, to radiate very high electric fields, the low gain of the dipole antenna is detrimental. The use of a directional antenna may improve performances really significantly focusing the radiation in a preferred direction. In order to improve the overall electrical performances, two approaches were explored in the paper: the replacement of the dipole antenna by a helical antenna and its feeding signal influence. The theoretical simulation comparison between various feeding signals applied to a helical antenna points out a strong dependence of the shape of the radiated field with the shape of the feeding signal. The highest antenna factor is obtained with a damped sinusoid. Moreover the impulse feeding directly integrated in the antenna design improves both the overall dimensions and electromagnetic characteristics [23].

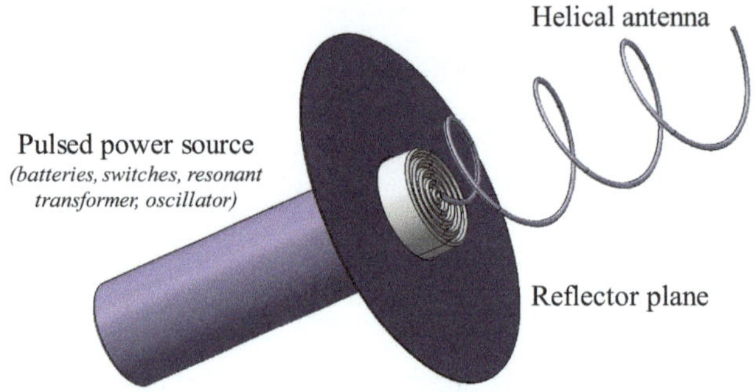

Figure 12. Schematic view of SPIRALE power source.

The concept of a switched oscillator as a highly capacitive quarter-wave transmission line terminated by an antenna load of high input impedance on one end and a self-breaking switch on the other end is proposed. After the switch closes, the stored energy is discharged into the high impedance load with a waveform that resembles a damped sinusoid. The predicted results concerning the design of the helical antenna, of a switch oscillator directly implemented at the output of the transformer have been presented. Simulation of the whole system is a useful way to predict electrical characteristics in the initial stage of manufacturing processes. This CST-based field simulation confirms the best expected performances. The antenna factor is 0.95 which allows a complete source with a very high figure of merit. In the present study the impedance of the closed switch has been assumed negligible. It is difficult to predict the rise time of the annular switch. However, it will have some inductance, lowering the resonant frequency, and some resistance, increasing the damping of the resonance. These effects need to be minimized. More accurate models can be developed to more precisely compute the performance. Probably the unknown losses in the circuit and particularly the performances of the switch will alter the calculated figure-of-merit.

The pulsed source with its new damped oscillator contains a volume of only 2500 cc. To achieve such compactness, all these parts are integrated within the antenna and immersed in oil. To keep such compactness, the helical antenna could be deployed in operational situation. The source performances could be increased by working on the reflector geometry or on the number of turns. Particularly a cupped ground plane could be implemented to improve the electrical performances.

Acknowledgements

These works are driven and financially supported by the Atomic Energy Commission (CEA), contract No. 4600233705.

References

[1] Giri, D.V. (2004) High Power Electromagnetic Radiators: Nonlethal Weapons and Other Applications. Harvard University Press, Cambridge.

[2] Morton, D., Banister, J., Levine, J., Naff, T., Smith, I., Sze, H., Warren, T., Giri, D., Mora, C., Pavlinko, J., *et al.* (2010) A 2 MV, <300 ps Risetime, 100 Hz Pulser for Generation of Microwaves. *Proceedings of the* 2010 *IEEE International Power Modulator and High Voltage Conference* (*IPMHVC*), Atlanta, 23-27 May 2010, 361-364. http://dx.doi.org/10.1109/ipmhvc.2010.5958368

[3] Koshelev, V.I., Plisko, V.V. and Sukhushin, K.N. (2010) Array Antenna for Directed Radiation of High-Power Ultra-wideband Pulses. In: Sabath, F., Giri, D.V., Rachidi, F. and Kaelin, A., Eds., *Ultra-Wideband, Short Pulse Electromagnetics* 9, Springer, New York, 259-267.

[4] Cadilhon, B., Pecastaing, L., Vauchamp, S., Andrieu, J., Bertrand, V. and Lalande, M. (2009) Improvement of an Ultra-Wideband Antenna for High-Power Transient Applications. *IET Microwaves Antennas and Propagation*, **3**, 1102-1109.

[5] Cadilhon, B., Cassany, B., Diot, J., Modin, P., Merle, E., Pecastaing, L., Rivaletto, M., Silvestre de Ferron, A. and Bertrand, V. (2011) Self-Contained, Hand-Portable, and Repetitive Ultrawideband Radiation Source. *IEEE Transactions on Plasma Science*, **39**, 1549-1559. http://dx.doi.org/10.1109/TPS.2011.2135867

[6] Cadilhon, B., Pecastaing, L., Reess, T., Silvestre de Ferron, A., Pignolet, P., Vauchamp, S., Andrieu, J. and Lalande, M.

(2010) High Pulsed Power Sources for Broadband Radiation. *IEEE Transactions on Plasma Science*, **38**, 2593-2603. http://dx.doi.org/10.1109/TPS.2010.2042732

[7] Delmote, P., Dubois, C., Andrieu, J., Beillard, B., Lalande, M., Bertrand, V., Jecko, B., Pecastaing, L., Gibert, A., Paillol, J., Domens, P., Guillerey, R., Monnier, F. and Legoff, M. (2003) UWB SAR System PULSAR: New Generator and Antenna Developments. *Proceedings of SPIE* 5077, *Passive Millimeter-Wave Imaging Technology VI and Radar Sensor Technology VII*, 25 August 2003, 223-234.

[8] Radasky, W.A., Baum, C.E. and Wik, M.W. (2004) Introduction to the Special Issue on High-Power Electromagnetics (HPEM) and Intentional Electromagnetic Interference (IEMI). *IEEE Transactions on Electromagnetic Compatibility*, **46**, 314-321. http://dx.doi.org/10.1109/TEMC.2004.831899

[9] Stark, R. (2011) Diehl HPEM—Source Development. The High Power Microwave Defense and Security Workshop, Saint-Louis.

[10] Hong, K.D. and Braidwood, S.W. (2002) Resonant Antenna-Source System for Generation of high-Power Wideband Pulses. *IEEE Transactions on Plasma Science*, **30**, 1705-1711. http://dx.doi.org/10.1109/TPS.2002.806637

[11] Pecquois, R., Pecastaing, L., Rivaletto, M., Silvestre De Ferron, A. and Vézinet, R. (2012) MOUNA: An Autonomous, Compact, High-Power, and Wideband Electromagnetic Source Based on a Novel Resonant Pulsed Transformer. *IEEE Transactions on Plasma Science*, **40**, 1407-1415. http://dx.doi.org/10.1109/TPS.2012.2189138

[12] Kraus, J. (2002) Antennas for All Applications. 3rd Edition, McGraw-Hill, New York.

[13] Carey, W., Wiebe, A., Schwindt, D., Altgilbers, L., Giesselmann, M., McHale, B. and Heinemann, K. (2005) Autonomous RF Radiation Package for Various Applications. *Proceedings of the* 2005 *IEEE Pulsed Power Conference*, Monterey, 13-17 June 2005, 218-221. http://dx.doi.org/10.1109/ppc.2005.300564

[14] Mayes, J.R., Mayes, M.G., Nunnally, W.C. and Hatfield, C.W. (2009) Helical Antennas for High Powered RF. *Proceedings of the* 2009 *Pulsed Power Conference*, Washington DC, 28 June-2 July 2009, 484-488. http://dx.doi.org/10.1109/ppc.2009.5386324

[15] Lee, J.M., Kwon, H.O., Hwang, S.M. and Ahn, J.W. (2011) Design of a Damped Sinusoidal Oscillator System. *Proceedings of the* 2011 *IEEE Pulsed Power Conference*, Chicago, 19-23 Jun 2011, 414-419. http://dx.doi.org/10.1109/ppc.2011.6191456

[16] Mosfegh, J., Shahabadi, M. and Rashed-Mohassel, J. (2011) Conditions of Maximum Efficiency for Wireless Power Transfer between Two Helical Wires. *IET Microwaves Antennas and Propagation*, **5**, 545-550.

[17] Sabath, F., Giri, D.V., Rachidi-Haeri, F. and Kaelin, A., Eds. (2010) Ultra-Wideband, Short Pulse Electromagnetics 9. Springer, New York, 315-324.

[18] Giri, D.V., Tesche, F.M., Abdalla, M.D., Skipper, M.C. and Nyffeler, M. (2010) Switched Oscillators and Their Integration Into Helical Antennas. *IEEE Transactions on Plasma Science*, **38**, 1411-1426. http://dx.doi.org/10.1109/TPS.2010.2047657

[19] Armanious, M. and Tyo, J.S. (2010) Field Management for a Self-Breakdown Switched Oscillator. *Proceedings of the IEEE International Symposium on Antennas and Propagation Society* (*APSURSI*), Toronto, 11-17 July 2010, 1-4. http://dx.doi.org/10.1109/aps.2010.5562335

[20] Baum, C.E. (2000) Switched Oscillators. Circuit and Electromagnetic System Design Note 45.

[21] Pecastaing, L., Reess, T., Paillol, J., Gibert, A. and Domens, P. (2001) Electrical Breakdown Studies of Pressurised Nitrogen in Non-Uniform Fields. *European Physical Journal Applied Physics*, **16**, 59-69. http://dx.doi.org/10.1051/epjap:2001194

[22] Pecquois, R., Pecastaing, L., De Ferron, A., Rivaletto, M., Pignolet, P., Novac, B.M., Smith, I.R. and Adler, R.J. (2012) Simple and Compact Capacitive Voltage Probe for Measuring Voltage Impulses up to 0.5 MV. *Review of Scientific Instruments*, **83**, Article ID: 035001. http://dx.doi.org/10.1063/1.3690906

[23] Bertrand, V., Negrier, R., Lalande, M., Andrieu, J., Couderc, V., Shalaby, B., Pecastaing, L. and De Ferron, A. (2014) Equivalent Model of Photoswitch: Application to the UWB Antenna Design Integrating Impulse Feeding. *Progress in Electromagnetic Research C*, **46**, 145-151. http://dx.doi.org/10.2528/PIERC13120301

7

Numerical Study of Highly Nonlinear Photonic Crystal Fiber with Tunable Zero Dispersion Wavelengths

Alejandro Barrientos García[1*], Igor A. Sukhoivanov[1], José Amparo Andrade Lucio[1], Oscar Gerardo Ibarra Manzano[1], Igor Guryev[2], Juan Carlos Hernández García[1,3], Gabriel Ramos Ortiz[4]

[1]Department of Electronics Engineering, Universidad de Guanajuato, Salamanca, Mexico
[2]Department of Multidisciplinary Studies, Universidad de Guanajuato, Yuriria, Mexico
[3]CONACYT, Cátedras CONACYT, Ciudad de Mexico, Mexico
[4]Centro de Investigaciones en Óptica, A. C., GPOM, León, Mexico
Email: *a.barrientosgarcia@ugto.mx

Abstract

Solid-core silica photonic crystal fiber is proposed borrowing the concept of golden ratio (1.618) and keeping it between pitch and air hole diameter Λ/d in a subset of six rings of air-holes with hexagonal arrangement. In the case when we have a pitch equal to one micron ($\Lambda = 1$ µm), we need air-holes diameters $d = 0.618$ µm in order to achieve two zero dispersion wavelength (ZDW) points at 725 nm and 1055 nm; this gives us the possibility to use the fiber for supercontinuum generation, by pumping close to that points, pulse compression or reshaping. We analyzed a series of fibers using this relation and showed the possibilities of tunable ZDW in a wide range of wavelengths from 725 nm to 2000 nm, with low losses and small effective area. In agreement with the ZDW point needed, the geometry of the structure can be modified to the point of having only three rings of air holes that surround the solid core with low losses and good confinement mode. The design proposed here is analyzed using the finite element method with perfectly matched layers, including the material dispersion directly into the model applying the Sellmeier's equation.

Keywords

Photonic Crystal Fibers, Supercontinuum Generation, Zero Dispersion Wavelength

*Corresponding author.

1. Introduction

Today the necessities for different applications in science and engineering of photonic crystal fibers (PCF) have boosted proposals on novel designs that satisfy specific conditions of linear and non-linear properties. Because of this, the control of dispersion, losses, mode confinement and conditions for generation of supercontinuum or pulse reshaping is a wide field of research. A series of techniques implying modification of some parameters of the fiber for various applications have been published. Since the invention of PCFs [1] [2], their designs and applications have been increased. However, some structures suffer from complex geometry. Traditionally optical fibers are used in various applications, for instance, to transport laser energy for material machining, as a gain medium in fiber lasers or optical telecommunication [3]. However, new possibilities of applications arise when nonlinear properties of PCFs are manifested. Specifically, nonlinearities in PCFs allow new light sources with very broad spectra, *i.e.*, supercontinuum generation (SCG) [4] [5]. The possibility to control group velocity dispersion (GVD) in PCFs, their small effective area and elevated nonlinearity [6] has allowed the observation of SCG in a much wider range of source parameters than has been possible with bulk media or conventional fibers [7]. It is known that in the case of normal GVD is present in the PCFs, the Raman scattering effect and self-phase modulation (SPM) dominate the spectrum broadening [8], and soliton dynamics govern spectral broadening when central wavelength of the propagating pulse is in anomalous GVD. Finite element and finite-difference methods as well as the multipole method and the plane wave expansion method [9] [10] were used to characterize the electromagnetic field distribution in PCFs. The accurate control of the dispersion characteristics is obligatory in variety of applications, such as pulse recompression [11] mode locked fiber lasers [12], nonlinear effects, including Raman scattering, four-wave mixing, SPM or soliton generation [13], and SCG [5] [7]. The flexible geometry of PCFs offers many unique and novel dispersion properties in comparison to conventional single mode fibers such as nearly zero ultraflattened dispersion [14]-[16], high negative chromatic dispersion [17], and anomalous dispersion in short-wavelength regime [18]. As is known, the chromatic dispersion coefficient of an optical fiber is a key quantity in various analysis and design issues in fiber transmission systems [19] [20].

According to the guidance mechanism, PCFs can be divided into two categories: first, total internal reflection (TIR), where air holes surround a high index solid core, and second, photonic band gap (PBG), where the light is confined to a central air hole [21]. In PCFs, the cladding region is usually formed by array of air holes in a regular triangular lattice with separation Λ [22]. Although there is a rich nomenclature of PCFs at this moment, we will restrict ourselves to index-guiding PCFs. In this kind of fibers, the core index is greater than the average index of the cladding because of the presence of air holes, and the fiber can guide the light by total internal reflection as a standard fiber does. That is, the guided mode has effective index n_{eff} that satisfies the condition

$$n_{co} > \left(n_{\text{eff}} = \frac{\beta}{k_0} \right) > n_{\text{FSM}} \tag{1}$$

where β is the mode propagation constant, n_{eff} is the core index, and n_{FSM} is the cladding effective index, which is termed also the effective index of the fundamental space-filling mode (FSM). In case of a PCF made of pure silica, n_{co} is the refractive index of silica. The losses in PCFs occur due to intrinsic material absorption, structural imperfection, Rayleigh scattering, and imperfect confinement. Fabrication-related losses can be reduced by carefully optimizing the fabrication process [23]. Confinement losses are an additional form of energy leakage that occurs in single-material fibers because of inherently leaky nature of the guided modes in these fibers [24]. This is because the core index is the same as the index of the outer cladding without air holes. Confinement losses can be reduced by increasing the number of rings of air holes which surround the solid core, and thus are determined by the geometry of the structure. It is important to know how many rings of air holes are required to reduce the confinement loss under the Rayleigh scattering limit. At the same time, geometry of the structure allows the chromatic dispersion to be easily controlled by varying the air hole diameter and the pitch of PCF. Here we show that some properties such as losses, ZDW and intermodal dispersion can be controlled keeping the ratio between fiber pitch and hole diameter of the microstructured cladding equal to the called number golden ratio, *i.e.*, 1.618.

Controllability of chromatic dispersion in PCFs is a very important problem for practical applications in optical communication systems, and nonlinear optics. The chromatic dispersion profile can be easily controlled by varying the values of any or all of the following parameters: air hole diameter (d), pitch (Λ), number of rings (Nr) and the number of missing holes (MNr). The aim of this work is to demonstrate how the dispersion profile may be modified by variation of the pitch and the hole diameter, keeping the golden ratio between them, *i.e.*, keeping a ratio between them equal to 1.618, providing a tunable ZWD in the range of 725 nm to 2000 nm.

2. Characteristics of the Fiber

Figure 1 shows the geometry of the solid core PCF proposed for SCG. The PCF consists of an arrangement of six rings with hexagonal structure and pitch size from 1 to 2 μm (Λ) and air hole diameter from 0.618 to 1.236 μm. This PCF has small effective area and tunable zero wavelength dispersion points in the range from 725 to 2000 nm. Although PCFs can be characterized by variety of methods including finite-difference time-domain method [25] we have used finite-element (FEM) approximation of the Helmholtz equation applying Sellmeier's equation to take the material dispersion into account when calculating effective mode index. The curve of total chromatic dispersion of the PCFs is obtained using following expression

$$D(\lambda) = -\frac{\lambda}{c}\frac{d^2\mathrm{Re}\{n_{\mathrm{eff}}\}}{\mathrm{d}\lambda^2} \tag{2}$$

where c is the speed of light in vacuum, n_{eff} is the effective mode index, and λ is the radiation wavelength.

This structure, proposed to tune the ZDW coinciding with the pump laser wavelengths to facilitate SCG or pulse reshaping applications, is inspired by the nature where the golden ratio can be found in perfect structures of the leaves, animals or the human body. Thus, we propose to use such a golden ratio in the design of a PCF and evaluate its impact on the optical properties. When this proportion is satisfied in the nature, the efficiency of the structures in macro- and micro-world is increased a lot, independently of the eye appeal. Talking about micro-world, we can found the golden ratio in the structure of some viruses, living cells, deoxyribonucleic acid, and bodies of honey bee, leaves or animals. In the macro-world, golden ratio can be found in the structure of galaxies. Due to this we probed to use the golden ratio for synthesis of photonic microstructures just like it happened in architecture to create some design of buildings, for example, keeping this relation. We found that this relation can improve the properties of the PCF; when we keep pitch-to-hole-diameter ratio equal to the golden ratio, it helps us to obtain simpler structures with better characteristics corresponding to the application of need.

In **Figure 2** and **Figure 3** we show the effective area and nonlinear coefficient of the designed PCF with the pitch Λ = 1, 1.5, 2 μm and air-holes diameter d = 0.618, 0.927, 1.236 μm. Using only six rings of air holes surrounding the solid core we have good confinement of the fundamental mode, small effective area, and high nonlinear coefficient. The structure keeps the golden ratio between pitch and diameter, and with this, we achieve

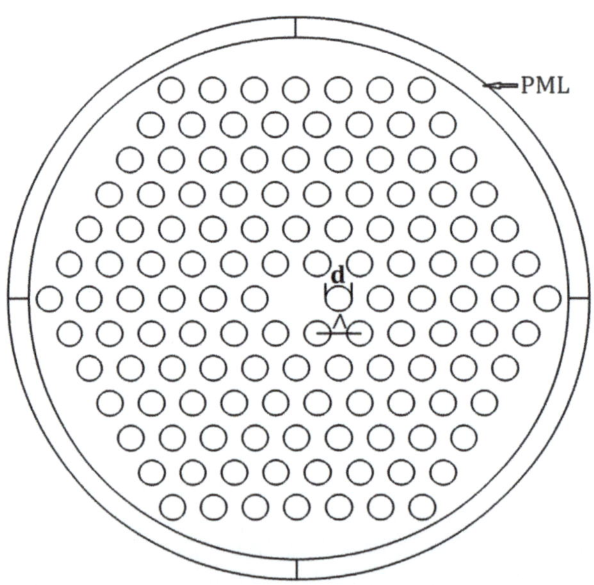

Figure 1. Cross-section of the PCF proposed where we keep the golden ratio between pitch and air hole diameter (Λ/*d*).

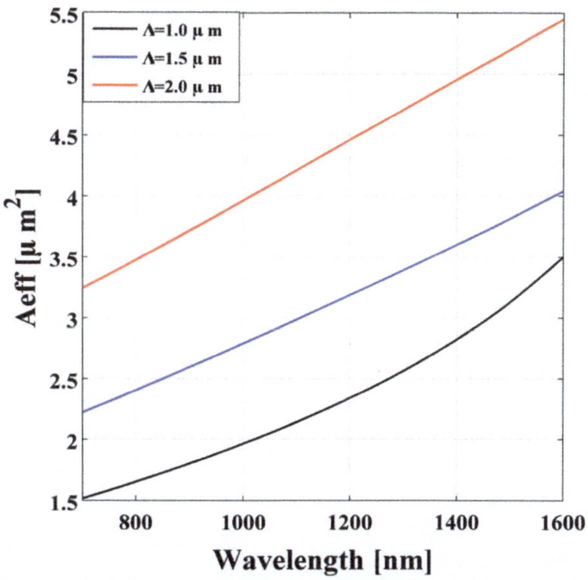

Figure 2. Effective area with Λ = 1, 1.5, 2 μm and d = 0.618, 0.927, 1.236 μm.

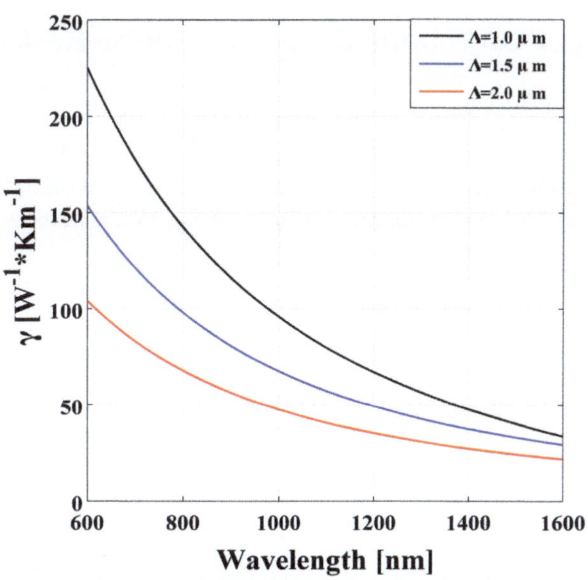

Figure 3. Nonlinear coefficient with Λ = 1, 1.5, 2 μm and d = 0.618, 0.927, 1.236 μm.

an improvement of the characteristics of PCF correspondingly to an application needed. The effective area and the nonlinear coefficient have been calculated applying the concepts of conventional fiber theory [26] with modifications necessary to include photonic-crystal cladding into account [26]-[28].

$$V_{\text{eff}} = \frac{2\pi}{\lambda} a_{\text{eff}} \sqrt{n_{co}^2 - n_{\text{FSM}}^2} \tag{3}$$

$$\frac{w}{a_{\text{eff}}} = 0.65 + \frac{1.619}{V_{\text{eff}}^{3/2}} + \frac{2.879}{V_{\text{eff}}^6} \tag{4}$$

$$A_{\text{eff}} \cong k_n \pi w^2 \tag{5}$$

$$\gamma(\lambda) = 2\pi n_2 / (\lambda A_{\text{eff}}) \tag{6}$$

where a_{eff} is the effective core radius, λ is the operating wavelength, n_{co} is the core index, n_{FSM} is the effective index of the fundamental space-filling mode, w is the half of mode field diameter (MFD) called also the effective modal spot size, A_{eff} is the effective mode area, k_n is the correction factor [29], γ is the nonlinear coefficient [30] and n_2 is the nonlinear index of silica. In Section 4 we show an application of the supercontinuum generation using this specific value of nonlinear coefficient shown in **Figure 3**.

3. Tunable Zero Dispersion Wavelength

In **Figure 4** we show the calculated total chromatic dispersion obtained using submicron air hole diameters in order to keep the golden ratio parameter. We can see that varying the pitch from 1 to 1.5 μm two different ZDW points are obtained which are tunable depending on the value of the pitch. Moreover, the curve becomes more gently sloped with increasing the value of pitch, giving the possibility to use this kind of fiber in different applications. In Section 4 we use this design in simulation of super continuum generation. The curves show a maximal chromatic dispersion from 38 to 80 ps/(nm∗km) and always we obtain two ZWD points which can be used for pumping by two different lasers to generate a more wide and flattened SC spectrum [31]-[33].

In **Figure 5** we show chromatic dispersion curves obtained when air hole diameters exceed 1 μm. Therefore, the pitch shall be from 1.618 to 2 μm in order to maintain the golden ratio proportion. The curves shows a maximal chromatic dispersion from 85 to 90 ps/(nm∗km), and, as in the previous case, we always have two ZWD points in the range of 725 - 2000 nm. We believe that fabrication of this kind of fiber is simple due to micrometric scale of the constituents.

4. Simulation of Super Continuum Generation, Pulse Reshaping and Compression

The PCFs proposed here were analyzed using commercial software COMSOL Multiphysics which uses the proven finite element method (FEM). This software runs the finite element analysis together with adaptive meshing and error control using a variety of numerical solvers. The RF Module of COMSOL provides the capabilities to design and is used to simulate electromagnetic wave propagation in PCFs. The mode analysis is made on a cross-section in the *xy*-plane of the fiber, by expressing the propagating wave in *z* direction in the following form

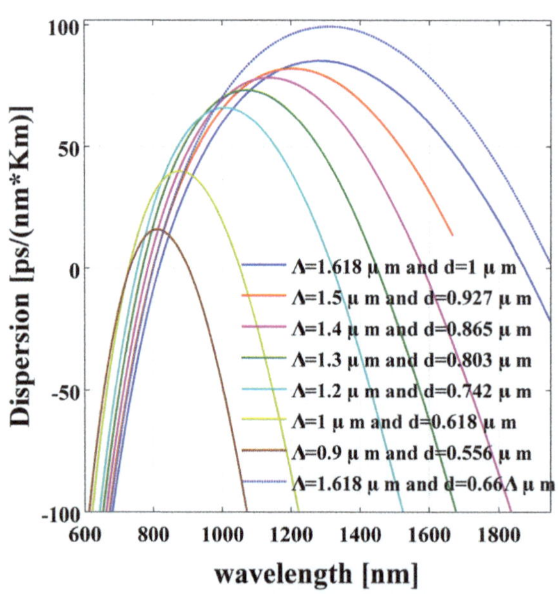

Figure 4. A family of the chromatic dispersion curves calculated by keeping the golden ratio. The ZDW points are moved from 725 to 1800 nm when pitch and air hole diameter are varied.

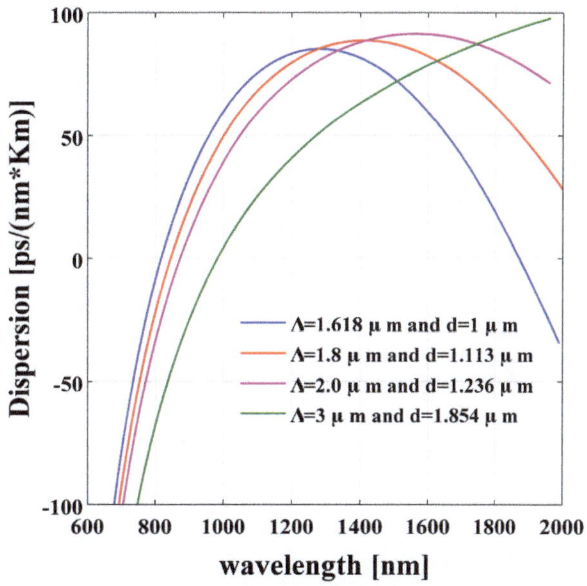

Figure 5. The behavior of chromatic dispersion with variation of geometry parameters, when diameter of holes is greater than 1 μm.

$$\boldsymbol{E}(x,y,z,t)=\boldsymbol{E}(x,y)\mathrm{e}^{j(\omega t-\beta z)} \tag{7}$$

and, thereupon, solving the homogeneous Helmholtz equation

$$\nabla\times(\nabla\times\boldsymbol{E})-k_0^2n^2\boldsymbol{E}=0 \tag{8}$$

for the eigenvalue $\lambda=-j\beta$ applying circular perfectly matched layer, with β being the propagation constant.

The pulse propagation in the PCFs is numerically investigated within the Generalized Nonlinear Schrödinger Equation (GNLSE) [34] [35]

$$\frac{\partial A}{\partial z}=-\frac{\alpha}{2}-\left(\sum_{n\geq 2}\beta_n\frac{i^{n-1}}{n!}\frac{\partial^n}{\partial T^n}\right)A+i\gamma\left(1+\frac{1}{\omega_0}\frac{\partial}{\partial T}\right)\left((1-f_R)A|A|^2+f_RA\int_0^\infty h_R(\tau)|A(z,T-\tau)|^2\,\mathrm{d}\tau\right) \tag{9}$$

where α is the attenuation constant, A is the complex field envelope, z is the distance, τ is the retarded time traveling at the envelope group velocity, γ is the nonlinear coefficient and β_n are the dispersion coefficients obtained by a Taylor series expansion of the propagation constant $\beta(\omega)$ around the center frequency. The GNLSE is solved by Fourth-Order Runge-Kutta in the Interaction Picture Method (RK4IP) [36] [37].

The Raman response function for silica fibers is described by [38]

$$R(t)=(1-f_r)\delta(t-\tau_e)+f_rh_R(t),\quad h_R(t)=\frac{\tau_1^2+\tau_2^2}{\tau_1\tau_2^2}\exp(-t/\tau_2)\sin(t/\tau_1) \tag{10}$$

To simulate the super continuum generation we consider the sech^2 waveform with the full width at half maximum (FWHM), T_{FWHM} = 50 fs, and 10 kW of peak power and 873 nm, propagating into a 5-mm length of a PCF with maximal chromatic dispersion at 873 nm of 39.78 ps/(nm∗km). The PCF designed has low losses, small effective area (A_{eff} = 1.76 μm^2 at 873 nm), and large nonlinear coefficient in the range of 800 to 1600 nm. Pulse evolution has been simulated using RK4IP method for solution of the generalized nonlinear Schrödinger equation The fiber parameters used in the calculation are given as $\gamma=122.5\ \mathrm{W^{-1}km^{-1}}$, $f_r=0.18$, $\lambda_c=873\ \mathrm{nm}$, $\beta_2=-16.09\ \mathrm{ps^2/km}$, $\beta_3=0.01686\ \mathrm{ps^3/km}$, $\beta_4=0.0001732\ \mathrm{ps^4/km}$ in which we achieve a more wide spectrum that reported using similar parameters.

Figure 6 shows the super continuum spectrum obtained by simulation of ultra short pulse propagation in the designed PCF, where we use the property of anomalous dispersion with high nonlinear coefficient of the PCF proposed, for creation a wide spectrum using femtosecond pump laser and shorter length of PCF. We can see a relatively smooth spectrum and wide SC spectrum of 2100 nm bandwidth at the level of −40-dB. In **Figure 7** we show temporal evolution of the pump pulse. We observe that soliton-like pump pulse has broken onto seven waveforms which can be interpreted as pulses; it gives the output pulse a form of the train of very short pulses. Taken this observation into account together with the fact that wavelength of the pump pulse is in anomalous dispersion region of the PCF designed, we conclude that **Figure 7** show us manifestation of the soliton fission, which is known as a primary nonlinear effect under these propagation conditions [7]. In the spectral domain the

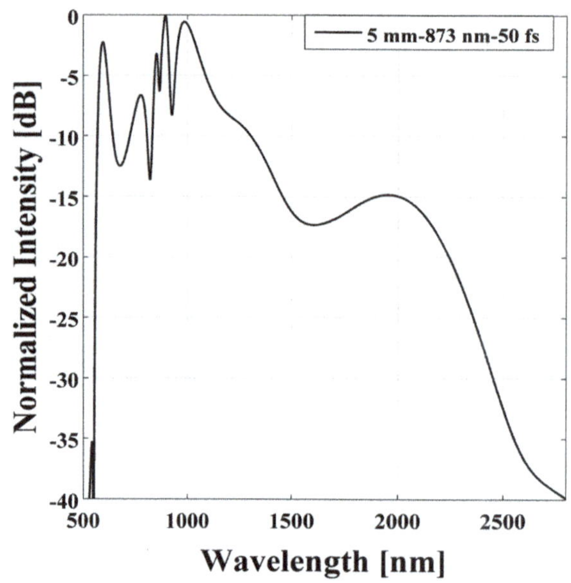

Figure 6. Output spectrum obtained at fiber length of 5 mm. The curve show SC spectrum with the 40-dB bandwidth of 2100 nm using a pump laser centered at 873 nm.

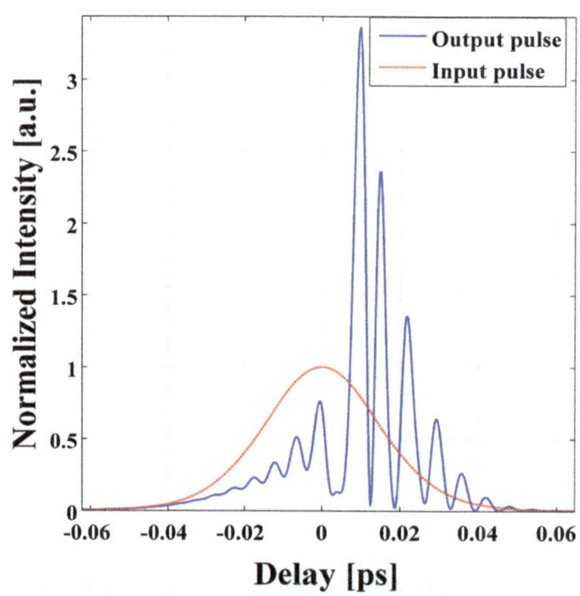

Figure 7. The red curve shows the initial pulse ($sech^2$ waveform) and the blue curve shows the waveform after propagation in 5 mm-long PCF.

soliton fission manifests itself as an extreme broadening of the spectrum of incident pulse. It is exactly what we observe in **Figure 6**. Thus, the main effect of spectral broadening in our case is the soliton fission.

The structure of the PCF designed has also been used for investigation of a possibility of pulse compression and transformation of the pulse shape supposing a sech^2 initial pulse envelope (IPE) waveform. Each of these applications requires a particular combination of dispersion and nonlinearity. Transformation of sech^2 pulse to another high-quality waveform in a fiber requires usually a normal dispersion at the central wavelength of the incident pulse [8] [39], as the primary phenomenon in this case is an interplay between the self-phase modulation and normal dispersion [8] [40]-[42]. Therefore, for simulation of pulse shape transformation we use following parameters and central wavelength of the pulse to be 600 nm: $T_{\text{FWHM}} = 140$ fs, peak power $P = 8$ kW, fiber length 13.5 mm, $\gamma = 102.5\ \text{W}^{-1}\text{km}^{-1}$, $f_r = 0.18$, $\beta_2 = 38\ \text{ps}^2/\text{km}$, $\beta_3 = 0.0385\ \text{ps}^3/\text{km}$, $\beta_4 = -0.00002147\ \text{ps}^4/\text{km}$. The wavelength of 600 nm has been chosen because of all dispersion curves we obtained using golden ratio, demonstrate normal dispersion at this wavelength; and this is exactly what we need to achieve pulse transformation with sufficiently high quality of the output pulse.

All of golden-ratio combinations of pitch and hole diameter demonstrate highly nonlinear dispersion profile; therefore, those fibers cannot be applied directly to pulse compression like it made in case of chirped mirrors, for example [43] [44]. However, these combinations provide wavelength ranges with anomalous group velocity dispersion where formation of higher-order solitons is possible. Thus, within these ranges one can compress the pulse using soliton-effect compression [45]. This has been implemented for the case of Λ = 1.618 μm and air-holes diameter $d = (2/3)\Lambda$ with five rings of holes in the cladding. In this case, zero dispersion wavelengths are 800 nm and 1956 nm, and maximum of the chromatic dispersion equals to 99.27 ps/(nm*km) @1303 nm, as shown in **Figure 4**. Other parameters are following: TFWHM = 250 fs @1303 nm, peak power P = 10 kW, $\gamma = 33\ \text{W}^{-1}\text{km}^{-1}$, $f_r = 0.18$, $\beta_2 = -90\ \text{ps}^2/\text{km}$, $\beta_3 = 0.1262\ \text{ps}^3/\text{km}$, $\beta_4 = -0.0001093\ \text{ps}^4/\text{km}$. The optimal length of the fiber when maximum compression is achieved, equals to 15 mm. The compression factor in this case equals to 100. **Figure 8** shows the incident pulses for cases of shape transformation and compression, compressed pulse, and reshaped pulse with a flat-top waveform. The shape of the compressed pulse demonstrates a pedestal, which is typical for pulses obtained via soliton-effect compression. Spectra of corresponding output pulses are shown in **Figure 9**. Spectrum of the flat-top pulse demonstrates signatures of the self-phase modulation. This nonlinear effect together with normal group velocity dispersion at 600 nm produces flat top pulses. A closer look to the reshaped pulse in **Figure 8** allows noting oscillations at the trailing edge of the pulse, which is manifestation of the optical wave breaking [46]. These results indicate a possibility to use the PCF proposed in variety of engineering problems of nonlinear optics.

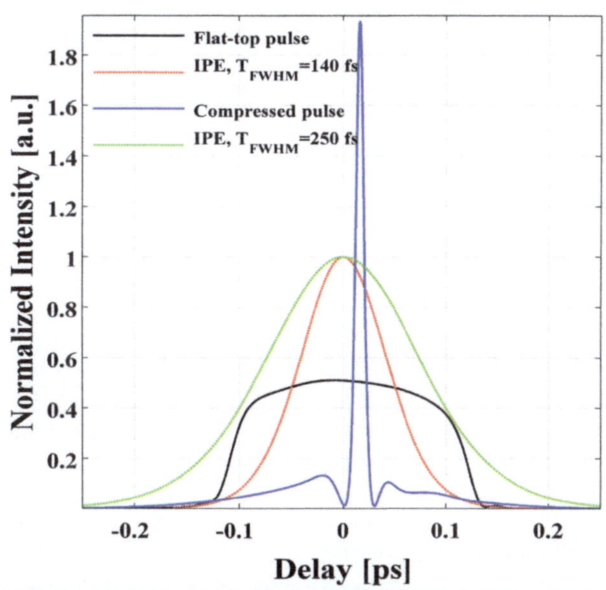

Figure 8. Initial pulse envelope (green and red curve), compressed pulse envelope (blue curve) and flat-top pulse (black curve).

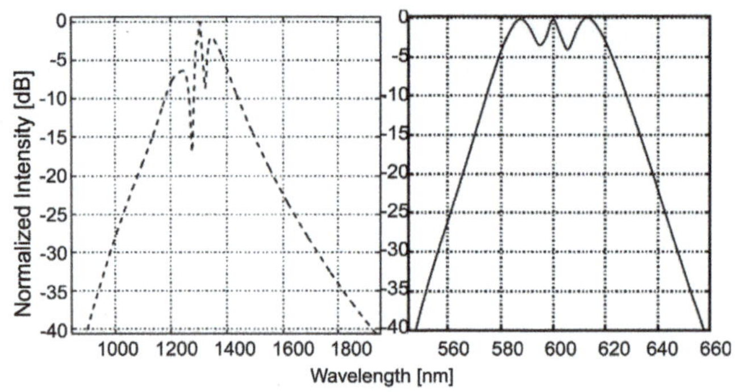

Figure 9. Spectrum for a compressed pulse and flat-top pulse. The dash line show the spectrum of compressed pulse and solid line show the spectrum of the flat-top pulse.

5. Conclusion

In this work, we proposed to evaluate how the optical properties of a PCF were influenced when the proportion between pitch and diameter of holes was equal to the golden ratio. In particular, when this ratio is maintained and the pitch and diameter holes are varied, the ZDW could be tunable between the range of 725 nm to 2000 nm. Additionally, with a low number of rings, we obtain a small loss confinement with small effective area and high nonlinear coefficient. Finally, we obtain a flattened chromatic dispersion curve between of two ZDW points. We present an example of applications of this kind of designs for supercontinuum generation, pulse recompression and pulse reshaping. A good quality of flattened and wide spectrum is achieved. The spectrum possesses 2100 nm bandwidth at 40 dB level. The compression factor achieved is 100 producing a compressed pulse of 2.5 fs from an initial pulse of 250 fs. It is believed that the design proposed could be fabricated easily with techniques available nowadays.

Acknowledgements

This work was supported in part by the National Council for the Science and technology (CONACYT) of Mexico, and by Universidad de Guanajuato (project DAIP-192/13 and DAIP-430/2014), FOMIX (CONACyT & CONCyTEG) under grant GTO-2012-C04-195229 and DAIP 458/2014. A. Barrientos-García and J. Carlos Hernandez-Garcia are grateful to CONACYT for a student research grant No. 329409/336847 and CONACYT project no. 3155.

References

[1] Knight, J.C. (2003) Photonic Crystal Fibres. *Nature*, **424**, 847-851. http://dx.doi.org/10.1038/nature01940

[2] Russell, P. (2003) Photonic Crystal Fibers. *Science*, **299**, 358-362. http://dx.doi.org/10.1126/science.1079280

[3] Böswetter, P., Baselt, T., Ebert, F., Basan, F. and Hartmann, P. (2011) Group-Velocity Dispersion in Multimode Photonic Crystal Fibers Measured Using Time-Domain White-Light Interferometry. *Proceedings of SPIE*, **7914**, Article ID: 791422. http://dx.doi.org/10.1117/12.873307

[4] Liao, M., Chaudhari, C., Qin, G., Yan, X., Suzuki, T. and Ohishi, Y. (2009) Tellurite Microstructure Fibers with Small Hexagonal Core for Supercontinuum Generation. *Optics Express*, **17**, 12174-12182. http://dx.doi.org/10.1364/OE.17.012174

[5] Gaeta, A.L. (2002) Nonlinear Propagation and Continuum Generation in Microstructured Optical Fibers. *Optics Letters*, **27**, 924-926. http://dx.doi.org/10.1364/OL.27.000924

[6] Reeves, W.H., Skryabin, D.V., Biancalana, F., Knight, J.C., Russell, P.St.J., Omenetto, F.G., Efimov, A. and Taylor, A.J. (2003) Transformation and Control of Ultra-Short Pulses in Dispersion-Engineered Photonic Crystal Fibres. *Nature*, **424**, 511-515. http://dx.doi.org/10.1038/nature01798

[7] Dudley, J.M., Genty, G. and Coen, S. (2006) Supercontinuum Generation in Photonic Crystal Fibers. *Reviews of Modern Physics*, **78**, 1135-1184. http://dx.doi.org/10.1103/RevModPhys.78.1135

[8] Sukhoivanov, I.A., Iakushev, S.O., Shulika, O.V., Díez, A. and Andrés, M. (2013) Femtosecond Parabolic Pulse Shaping in Normally Dispersive Optical Fibers. *Optical Express*, **21**, 17770-17785. http://dx.doi.org/10.1364/OE.21.017769

[9] Johnson, S.G. and Joannopoulos, J.D. (2001) Block-Iterative Frequency-Domain Methods for Maxwell's Equations in a Planewave Basis. *Optics Express*, **8**, 173-190. http://dx.doi.org/10.1364/OE.8.000173

[10] White, T.P., Kuhlmey, B.T., McPhedran, R.C., Maystre, D., Renversez, G., de Sterke, C.M. and Botten, L.C. (2002) Multipole Method for Microstructured Optical Fibers. I. Formulation. *Journal of the Optical Society of America B*, **19**, 2322-2330. http://dx.doi.org/10.1364/JOSAB.19.002322

[11] Laegsgaard, J. and Roberts, P.J. (2008) Dispersive Pulse Compression in Hollow-Core Photonic Band-Gap Fibers. *Optics Express*, **16**, 9628-9644. http://dx.doi.org/10.1364/OE.16.009628

[12] Lim, H. and Wise, F. (2004) Control of Dispersion in a Femtosecond Ytterbium Laser by Use of Hollow-Core Photonic Band-Gap Fiber. *Optics Express*, **12**, 2231-2235. http://dx.doi.org/10.1364/OPEX.12.002231

[13] Dudley, J.M. and Taylor, J.R. (2010) Supercontinuum Generation in Optical Fibres. Cambridge University Press, Cambridge, 32-96. http://dx.doi.org/10.1017/CBO9780511750465

[14] Ferrando, A., Silvestre, E., Miret, J.J., Monsoriu, J.A., Andres, M.V. and Russell, P.St.J. (1999) Designing a Photonic Crystal Fiber with Flattened Chromatic Dispersion. *Electronics Letters*, **35**, 325-326. http://dx.doi.org/10.1049/el:19990189

[15] Ferrando, A., Silvestre, E., Miret, J.J. and Andres, P. (2000) Nearly Zero Ultraflattened Dispersion in Photonic Crystal Fibers. *Optics Letters*, **25**, 790-792. http://dx.doi.org/10.1364/OL.25.000790

[16] Reeves, W.H., Knight, J.C., Russell, P.St.J. and Roberts, P.J. (2002) Demonstration of Ultra-Flattened Dispersion in Photonic Crystal Fibers. *Optics Express*, **10**, 609-613. http://dx.doi.org/10.1364/OE.10.000609

[17] Birks, T.A., Mogilevstev, D., Knight, J.C. and Russell, P.St.J. (1999) Dispersion Compression Using Single Material Fibers. *IEEE Photonics Technology Letters*, **11**, 674-676. http://dx.doi.org/10.1109/68.766781

[18] Knight, J.C., Arriaga, J., Birks, T.A., Ortigosa-Blanch, A., Wadsworth, W.J. and Russell, P.St.J. (2000) Anomalous Dispersion in Photonic Crystal Fiber. *IEEE Photonics Technology Letters*, **12**, 807-809. http://dx.doi.org/10.1109/68.853507

[19] Mogilevtsev, D., Birks, T.A. and Russell, P.S.J. (1998) Group-Velocity Dispersion in Photonic Crystal Fibers. *Optics Letters*, **23**, 1662-1664.

[20] Knight, J.C., Arriaga, J., Birks, T.A., Ortigosa-Blanch, A., Wadsworth, W.J. and Russell, P.S.J. (1998) Anomalous Dispersion in Photonic Crystal Fiber. *IEEE Photonics Technology Letters*, **12**, 817-819.

[21] Cheng, H., Crutchfield, W.Y., Doerg, M. and Greengard, L. (2004) Fast, Accurate Integral Equation Methods for the Analysis of Photonic Crystal Fibers I: Theory. *Optics Express*, **12**, 3791-3805. http://dx.doi.org/10.1364/OPEX.12.003791

[22] Saitoh, K., Koshiba, M., Hasegawa, T. and Sasaoka, E. (2003) Chromatic Dispersion Control in Photonic Crystal Fibers: Application to Ultra Flattened Dispersion. *Optics Express*, **11**, 843-852. http://dx.doi.org/10.1364/OE.11.000843

[23] Tajima, K., Zhou, J., Kurokawa, K. and Nakajima, K. (2011) Low Water Peak Photonic Crystal Fibers. *European Conference on Optical Communication* (*ECOC*), Rimini, September 2003, Th4.1.6.

[24] White, T.P., McPhedran, R.C., de Sterk, C.M., Botten, L.C. and Steel, M.J. (2001) Confinement Losses in Microstructured Optical Fibers. *Optics Letters*, **26**, 1660-1662. http://dx.doi.org/10.1364/OL.26.001660

[25] Shulika, O., Guryev, I., Gurieva, N., Sukhoivanov, I. and Andrade Lucio, J.A. (2013) Graphics Processing Unit-Accelerated Finite-Difference Time-Domain Method for Characterization of Photonic Crystal Fibers. *Optical Engineering*, **52**, Article ID: 126109. http://dx.doi.org/10.1117/1.OE.52.12.126109

[26] Marcuse, D. (1977) Loss Analysis of Single-Mode Fiber Splices. *Bell System Technical Journal*, **56**, 703-718. http://dx.doi.org/10.1002/j.1538-7305.1977.tb00534.x

[27] Koshiba, M. and Saitoh, K. (2004) Applicability of Classical Optical Fiber Theories to Holey Fibers. *Optics Letters*, **29**, 1739-1741. http://dx.doi.org/10.1364/OL.29.001739

[28] Mortensen, N.A., Folkenberg, J.R., Nielsen, M.D. and Hansen, K.P. (2003) Modal Cut-Off and the V Parameter in Photonic Crystal Fibers. *Optics Letters*, **28**, 1879-1881. http://dx.doi.org/10.1364/OL.28.001879

[29] Miyagi, K., Namihira, Y., Razzak, S.M.A., Kaijage, S.F. and Begum, F. (2010) Measurements of Mode Field Diameter and Effective Area of Photonic Crystal Fibers by Far-Field Scanning Technique. *Optical Review*, **17**, 388-392. http://dx.doi.org/10.1007/s10043-010-0072-x

[30] Liu, Z.L., Hou, L.T. and Wang, W. (2009) Tailoring Nonlinearity and Dispersion of Photonic Crystal Fibers Using Hybrid Cladding. *Brazilian Journal of Physics*, **39**, 50-54. http://dx.doi.org/10.1590/S0103-97332009000100009

[31] Frosz, M., Falk, P. and Bang, O. (2005) The Role of the Second Zero-Dispersion Wavelength in Generation of Supercontinua and Bright-Bright Soliton-Pairs across the Zero-Dispersion Wavelength. *Optics Express*, **13**, 6181-6192. http://dx.doi.org/10.1364/OPEX.13.006181

[32] Hilligsøe, K.M., Andersen, T., Paulsen, H., Nielsen, C., Mølmer, K., Keiding, S., Kristiansen, R., Hansen, K. and Larsen, J. (2004) Supercontinuum Generation in a Photonic Crystal Fiber with Two Zero Dispersion Wavelengths. *Optics Express*, **12**, 1045-1054. http://dx.doi.org/10.1364/OPEX.12.001045

[33] Tse, M.L.V., Horak, P., Poletti, F., Broderick, N.G., Price, J.H., Hayes, J.R. and Richardson, D.J. (2006) Supercontinuum Generation at 1.06μm in Holey Fibers with Dispersion Flattened Profiles. *Optics Express*, **14**, 4445-4451. http://dx.doi.org/10.1364/OE.14.004445

[34] Shulika, O.V., Sukhoivanov, I.A., Guryev, I.V., Andrade-Lucio, J.A., Ibarra-Manzano, O., Barrientos-Garcia, A. and Ramos-Ortiz, G. (2013) Numerical Study of Few-Cycle Pulses Generation from Supercontinuum in ANDi-PCF. *Acta Universitaria*, **23**, 40-44.

[35] Shulika, O.V., Sukhoivanov, I.A., Ramos-Ortiz, G., Guryev, I.V., Iakushev, S.O. and Andrade Lucio, J.A. (2013) Characterization of All-Normal Dispersion Microstructured Optical Fiber via Numerical Simulation of Passive Nonlinear Pulse Reshaping and Single-Pulse Flat-Top Supercontinuum. *Journal of Nanophotonics*, **8**, Article ID: 083890.

[36] Hult, J. (2007) A Fourth-Order Runge-Kutta in the Interaction Picture Method for Simulating Supercontinuum Generation in Optical Fibers. *Journal of Lightwave Technology*, **25**, 3770-3775. http://dx.doi.org/10.1109/JLT.2007.909373

[37] Hernandez-Garcia, J.C., Pottiez, O., Estudillo-Ayala, J.M. and Rojas-Laguna, R. (2012) Numerical Analysis of a Broadband Spectrum Generated in a Standard Fiber by Noise-Like Pulses from a Passively Mode-Locked Fiber Laser. *Optics Communications*, **285**, 1915-1919. http://dx.doi.org/10.1016/j.optcom.2011.12.069

[38] Blow, K.J. and Wood, D. (1989) Theoretical Description of Transient Stimulated Raman Scattering in Optical Fibers. *IEEE Journal of Quantum Electronics*, **25**, 2665-2673. http://dx.doi.org/10.1109/3.40655

[39] Iakushev, S.O., Shulika, O.V. and Sukhoivanov, I.A. (2012) Passive Nonlinear Reshaping towards Parabolic Pulses in the Steady-State Regime in Optical Fibers. *Optics Communications*, **285**, 4493-4499. http://dx.doi.org/10.1016/j.optcom.2012.06.024

[40] Agrawal, G. (2012) Nonlinear Fiber Optics. 5th Edition, Academic Press, Waltham, 648 p.

[41] Iakushev, S.O., Shulika, O.V., Sukhoivanov, I.A., Fesenko, V.I., Andrés, M.V. and Sayinc, H. (2014) Formation of Ultrashort Triangular Pulses in Optical Fibers. *Optics Express*, **22**, 29119-29134. http://dx.doi.org/10.1364/OE.22.029119

[42] Sukhoivanov, I.A., Iakushev, S.O., Shulika, O.V., Andrade Lucio, J.A., Díez, A. and Andrés, M. (2014) Supercontinuum Generation at 800 nm in All-Normal Dispersion Photonic Crystal Fiber. *Optics Express*, **22**, 30234-30250. http://dx.doi.org/10.1364/OE.22.030234

[43] Yakushev, S.O., Shulika, O.V., Petrov, S.I. and Sukhoivanov, I.A. (2008) Chirp Compression with Single Chirped Mirrors and Its Assembly. *Microelectronics Journal*, **39**, 690-695. http://dx.doi.org/10.1016/j.mejo.2007.07.056

[44] Iakushev, S.O., Shulika, O.V., Lysak, V.V. and Sukhoivanov, I.A. (2008) Air-Gap Silicon Nitride Chirped Mirror for Few-Cycle Pulse Compression. *Optoelectronics and Advanced Materials—Rapid Communications*, **2**, 686-688.

[45] Agrawal, G. (2008) Application of Nonlinear Fiber Optics. 2nd Edition, Academic Press, Waltham.

[46] Iakushev, S.O., Shulika, O.V. and Garcia, J.J.R. (2010) Optical Wave Breaking Cancelation and Formation of Quasi-Parabolic Ultrashort Pulses. 10*th International Conference on Laser and Fiber-Optical Networks Modeling*, Sevastopol, 12-14 September 2010, 176-179. http://dx.doi.org/10.1109/LFNM.2010.5624206

8

Performance Analysis on a Dual External Cavity Tunable Laser ECTL Source

Ahmed Fawzy[1], Osama M. El-Ghandour[2], Hesham F. A. Hamed[1]

[1]Department of Electrical Engineering, Minia University, Minia, Egypt
[2]Department of Electronics, Communications and Computer Engineering, Helwan University, Cairo, Egypt
Email: ah_fawzy86@hotmail.com

Abstract

This paper presents the diffraction effects on the performance of a dual external cavity tunable laser source, whose external cavities are constructed by micro electro mechanical systems (MEMS). One of the main problems in these structures is the optical diffraction as the emitting surface of the laser diode is usually quite limited in the transverse directions. The emitted beam diffracts rapidly in the air and only a small amount of light is coupled back to the source that usually limits the tuning range of the source. Device characteristics such as tuning range, wavelength shift and sensitivity are evaluated. New expression is used and multiple reflections inside external cavities are considered. The simulation results have shown that single external cavity has limited tuning range. It is shown that multiple reflections have significant effect in our model. To get a better engineering for the dual ECTL dimensions, diffraction effects must be taken into account.

Keywords

External Cavity Laser, Tunable Laser, MEMS, Optical Diffraction

1. Introduction

Tunable lasers have wide applications in communication, especially in wave length division multiplexing (WDM), sensing application and measurements [1]-[3]. There are three specifications that must be achieved in tunable laser to be applicable in modern applications. The specifications are the high tuning speed, miniaturized device dimensions and the wide tuning range [3]-[8]. These three specifications can be achieved using the Micro Electro Mechanical System (MEMS) technology [2] [3].

There are many features that make MEMS technology preferred in fabrication of micromechanical structures such as high accuracy, self aligned, high speed response in moving and enhanced mechanical stability [3] [9]

[10]. The MEMS technology can add a lot of features like the small size, low weight, high accuracy as well as fast speed in tuning. For this purpose, MEMS based tunable laser structures have been proposed and tested in the literature using different structures [2] [3]. These structures are mainly based on one external cavity semiconductor laser in which the external cavity is formed by a moving MEMS mirror but suffer from lower tuning range. To increase the tuning range, grating must be used in the structures such as Littrow or Littman but also these structures suffer from large size [2]-[4]. So in this paper dual external cavities are used to enhance tuning range rather than single external cavity and also to miniaturize size than Littrow and Littman [11] [12]. The configuration consists of two cavities: one reflective movable mirror element was placed close to the original cavity to form the first cavity; and the second mirror is a transparent mirror, it was placed far from original cavity to form the second cavity. One way to fabricate this moving mirror as well as its actuator is to use the Deep Reactive Ion Etching (DRIE) technology on SOI wafer [2] [9] [10]. The objective of this paper is to study the diffraction in the MEMS based on a dual external cavity tunable laser source and to prove that the tuning range is increased in case of the dual external cavity. Simulation of the optical field propagation in the structure is verified by accurate theoretical model [13]. In the model each external cavity can be represented by using effective reflection coefficient. The model is based on the plane wave expansion method. The calculation of tuning range through an effective phase perturbation technique is done by using effective reflection coefficient. In this technique phase oscillation condition for the laser oscillator is changed due to perturbation that come from external cavity phase. Based on that, the new resonance frequency (emission wavelength) is then calculated.

2. System Model

A MEMS dual external cavity tunable laser is shown in **Figure 1** with two external mirrors assumed to have finite dimensions limited by the technological capabilities of the DRIE technology. In our model, we assume that L is the primary cavity's length (containing the active medium), L_{ext1} is the length of the first external cavity, L_{ext2} is the length of the second external cavity, r_1 and r_2 are the reflectivities of the facets forming the original cavity, r_{ext1} is the reflectivity of the first external mirror and r_{ext2} is the reflectivity of the second external mirror. The external reflector can be combined with the facet of the primary cavity yielding an effective reflectivity r_{eff} [12] [14] that can be expressed as:

$$r_{\text{eff1}}(\nu) = r_1 - \frac{1-R_1}{r_1}\sum_{n=1}^{\infty}\left(-r_1 r_{\text{ext1}} e^{2i\theta_{r_1}}\right)^n \tag{1}$$

$$r_{\text{eff1}}(\nu) = \left|r_{\text{eff1}}\right| \exp\left(-j\varnothing r_1\right) \tag{2}$$

$$r_{\text{eff2}}(\nu) = r_2 - \frac{1-R_2}{r_2}\sum_{n=1}^{\infty}\left(-r_2 r_{\text{ext2}} e^{2i\theta_{r_2}}\right)^n \tag{3}$$

$$r_{\text{eff2}}(\nu) = \left|r_{\text{eff2}}\right| \exp\left(-j\varnothing r_2\right) \tag{4}$$

r_{eff1}, r_{eff2}, $\varnothing r_1$, $\varnothing r_2$ represent the amplitude and phase of effective reflectivity of the first external cavity and the second external cavity respectively, R_1 and R_2 are the power reflectivities of the two facets of the primary cavity and θ_r represents the phase term corresponding to each external cavity length at wavelength λ. In the dual external cavity tunable laser the external mirror of the second cavity is actually transparent, so in this

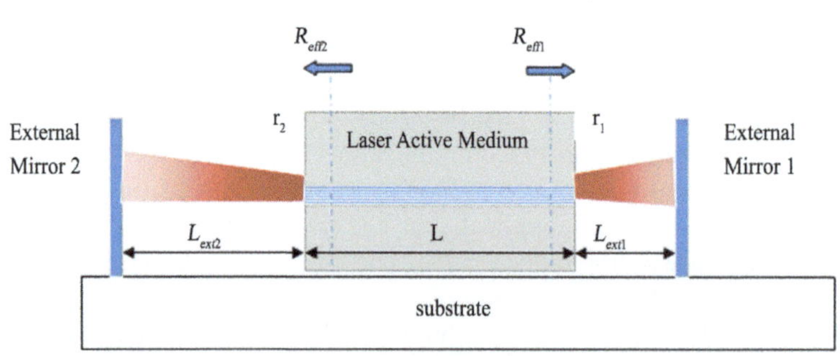

Figure 1. Schematic diagram of the studied dual ECTL source.

paper we make classification for the coupled feedback based on the first external mirror only. In case of weak feedback, $r_{\text{ext1}} \ll r_1$, so we can neglect multiple reflection and can substitute in Equation (1) by $n = 1$. In case of strong feedback, $r_{\text{ext1}} \gg r_1$, multiple reflections will affect the total amplitude and phase of effective reflectivity. From the previous work the best external cavity reflectivity is for the external cavity mirror to match the reflectivity of internal mirror of the primary cavity [12]. As the optical frequency is mainly determined from the phase condition in the round trip of the laser cavity, we calculate the change in the round trip phase due to the existence of the external cavities. When the phase takes values of multiples of 2π, the oscillation condition can be satisfied again and a laser longitudinal mode oscillates. This change in the phase can be calculated from Equation (3) [15]:

$$\Delta\varnothing_l = 2\pi\tau_l\left(\nu - \nu_{\text{th}}\right) + \varnothing_r - \alpha\left[\ln\left(1/|r_{\text{eff1}}||r_{\text{eff2}}|\right) - \ln\left(1/|r_1||r_2|\right)\right] \tag{5}$$

where τ_l is the round trip time delay of the optical beam inside the primary laser cavity, $\varnothing_r$ represents the total phase of effective reflectivity due to two cavities, ν is the new oscillation frequency and ν_{th} is the oscillation frequency before using an external cavities and α is the linewidth enhancement factor in the semiconductor material. From this relation, we can find a new formula for the frequency shift of the laser emission that result due to using dual cavities as follows:

$$\text{Freq shift} = \nu - \nu_{\text{th}} = \frac{\left[\alpha\left[\ln\dfrac{1}{|r_{\text{eff1}}||r_{\text{eff2}}|} - \ln\dfrac{1}{|r_1| * \left(|r_2|\right)}\right] - \varnothing_r\right]}{2\pi\tau_l} \tag{6}$$

This expression for the frequency shift is valid for both weak and strong feedback reflections. It can thus be used to study the effect of the diffraction on the frequency shift and hence on the resonance frequency of the laser. To calculate this frequency shift, it is required to evaluate both the amplitude and phase of the reflected beam after travelling in the external cavity. The suggested technique, the Plane Wave Expansion Method is based on the expansion of the optical field of finite spot size emitted from active cavity in terms of the proper propagation modes in the free space, the plane waves. The optical beam is thus decomposed into plane waves using the fast Fourier transform FFT algorithm. Propagation in the external cavity is represented for each plane wave component as a multiplication by a free space transfer function [16]. The total field composed of all the plan waves is then constructed again at the surface of the external reflector using the inverse Fourier transform IFFT to return to spatial domain. When the external mirror has finite dimensions, part of beam is reflected by mirror reflectivity and another part doesn't reflect due to finite mirror size and thus we have a truncated beam. Outside the mirror surface reflection coefficient is considered as zero. The part that reflected from mirror is modulated by the shape of mirror. The reflected beam is then transformed again to the spectral domain and propagated in the free space as before. When arriving back to the laser diode surface, the beam is returned back to the spatial domain using the IFFT and its coupling to the cavity is calculated using the overlap integral with the cavity mode [6].

For the infinite mirror, the tuning range of dual external cavity tunable laser is higher than single external cavity tunable laser as shown in **Figure 2**. In our model we assume that the laser output from primary cavity is an elliptical Gaussian beam with a spot size of about 1 μm in the y direction and 3.5 μm in the x direction. [16]-[18]. The field distribution at the output of the laser diode can thus be expressed as:

$$E(x,y) = E_m \exp\left(\left(-x^2/w_{0x}^2\right) + \left(-y^2/w_{0y}^2\right)\right) \tag{7}$$

where w_{0x} is the spot size in the x direction and w_{0y} is the spot size in the y direction, E_m is the field amplitude at the center of the beam and the beam phase front is assumed at the emitting edge of the laser diode surface. The wave length of the laser source is 1550 nm. Because of the spot diffraction and in the absence of any focusing element, the external cavity length needs to be very short to maintain feedback efficiency acceptable.

3. Results

For an external mirror with high reflection coefficient, the coupled power that come back to active cavity from external mirrors is affected by number of round trips through external cavities. For weak feedback with

$r_{ext1} \ll r_1$ in dual configuration, we assume only one round trip in the calculation after one round trip, the optical intensity may still be strong enough to perturb the operation of the laser. A model that accounts for the multiple reflections of the beam in the external cavity is thus developed. This situation is closer to the practical case as usually the mirror is coated by a metal to increase its reflection coefficient. The effect of multiple reflections on the laser performance when the external mirror reflection is moderate is calculated first and the obtained results are presented in **Figure 3**. On the other hand, when increasing the reflection coefficient, we observe a change in the tuning behavior as the tuning is not symmetric with respect to the positive and negative frequency shift as seen in **Figure 3**. Also multiple reflections show a change in tuning sensitivity. Tuning sensitivity is the change in $d\lambda/dx$ with respect to mirror position as can be seen in **Figure 4** [2] [19].

4. Conclusion

In this work we studied the diffraction effects on the single and dual external cavity tunable lasers and calculated the tuning range in both cases by using accurate developed model. The model shows that the dual external cavity has higher tuning range than single external cavity tunable laser. The model also accounts multiple reflection in the dual external cavity tunable laser. The model shows that multiple reflection has significant effect on the tun-

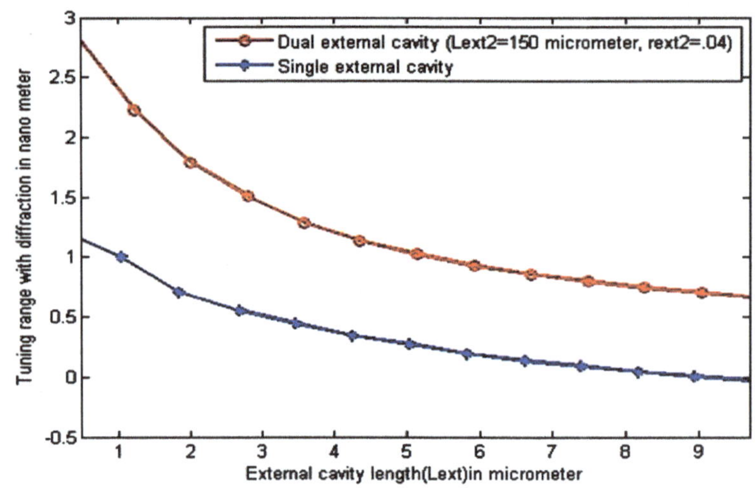

Figure 2. Tuning range of two different configurations as a function of the external mirror position.

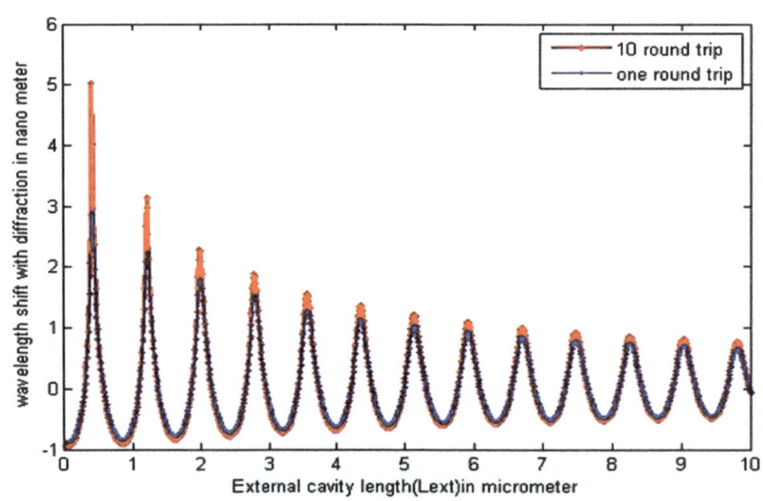

Figure 3. Wavelength shift of an dual external cavity tunable laser as a function of the external mirror position for moderate reflection coefficient first external mirror (0.32).

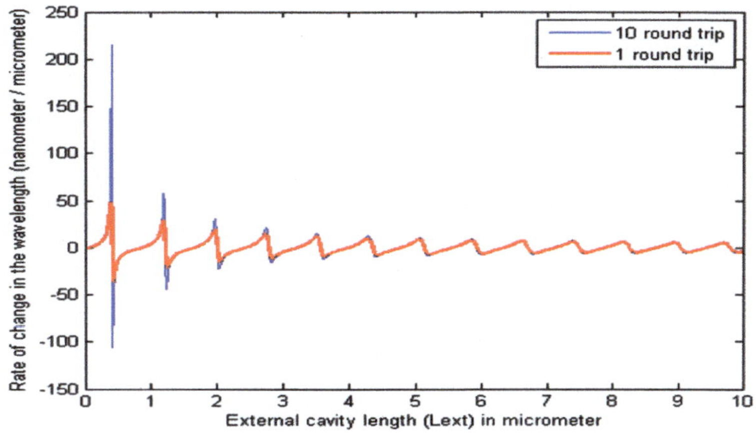

Figure 4. Comparison of wavelength variation predicted by the dual external cavity model in case of one round trip and multiple round trips.

ing range as well as the tuning sensitivity of the laser. By taking multiple reflections in our calculations, we simulate practical external cavity laser, so we can engineer accurate dimensions for a required tuning range.

References

[1] Coldren, L.A., Fish, G.A., Akulova, Y., Barton, J.S., Johansson, L. and Coldren, C.W. (2004) Tunable Semiconductor Lasers: A Tutorial. *Journal of Lightwave Technology*, **22**, 193-202. http://dx.doi.org/10.1109/JLT.2003.822207

[2] Liu, A.Q. (2004) Photonic MEMS Devices: Design, Fabrication and Control. CRC Press, Boca Raton.

[3] Liu, A.Q. and Zhang, X.M. (2004) A Review of MEMS External-Cavity Tunable Lasers. *Journal of Micromechanics and Microengineering*, **17**, R1-R13.

[4] Liu, A.Q., Zhang, X.M., Tang, D.Y. and Lu, C. (2004) Tunable Laser Using Micro Machined Grating with Continuous Wavelength Tuning. *Applied Physics Letters*, **85**, 3684-3686.

[5] Zhang, X.M., Liu, A.Q., Tang, D.Y. and Lu, C. (2004) Discrete Wavelength Tunable Laser Using Microelectromechanical Systems Technology. *Applied Physics Letters*, **84**, 329-331

[6] Solgaard, O. (2009) Photonic Microsystems: Micro and Nanotechnology Applied to Optical Devices and Systems. Springer, New York.

[7] Liu, A.Q., Zhang, X.M., Murukeshan, V.M. and Lam, Y.L. (2001) A Novel Integrated Micromachined Tunable Laser Using Polysilicon 3-D Mirror. *IEEE Photonics Technology Letters*, **13**, 427-429. http://dx.doi.org/10.1109/68.920739

[8] Zhu, W.M., Zhang, W., Cai, H., Tamil, J., Liu, B., Bourouina, T. and Liu, A.Q. (2009) A MEMS Digital Mirror for Tunable Laser Wavelength Selection. *International Conference on Solid-State Sensors, Actuators and Microsystems* (*TRANSDUCERS* 2009), Denver, 21-25 June 2009, 2206-2209.

[9] Sabry, Y.M., Khalil, D., Saadany, B. and Bourouina, T. (2013) Multi-Step Etching of Three-Dimensional Sub-Millimetre Curved Silicon Microstructures with In-Plane Principal Axis. *Journal of Microelectronic Engineering*, **114**, 78-84.

[10] Sabry, Y.M., Saadany, B., Khalil, D. and Bourouina, T. (2013) Silicon Micromirrors with Three-Dimensional Curvature Enabling Lensless Efficient Coupling of Free-Space Light. *Light*: *Science* & *Applications*, 2, e94.

[11] Zhu, X. and Cassidy, D.T. (1996) Liquid Detection with InGaAsP Semiconductor Lasers Having Multiple Short External Cavities. *Applied Optics*, **35**, 4689-4693. http://dx.doi.org/10.1364/AO.35.004689

[12] Abu-El-Magd, A.M. (2005) Double Tuning of A Dual External Cavity Semi Conductor Laser For Broad Wavelength Tuning With High Side Mode Suppression. Master Thesis, McMaster University, Hamilton.

[13] Fawzy, A., El-Sabban, S., Ismail, I. and Khalil, D. (2013) On the Modeling of an External Cavity Tunable Laser ECTL Source with Finite Mirror Dimensions. *Progress in Electromagnetics Research Symposium Proceedings*, Stockholm, 12-15 August 2013, 691-694.

[14] Voumard, C., Salathé, R. and Weber, H. (1977) Resonance Amplifier Model Describing Diode Lasers Coupled to Short External Resonators. *Applied Physics*, **12**, 369-378.

[15] Petermann, K. (1988) Laser Diode Modulation and Noise. Kluwer, London. http://dx.doi.org/10.1007/978-94-009-2907-4

[16] Saleh, B.E.A. and Teich, M.C. (2007) Fundamentals of Photonics. Wiley, Hoboken.

[17] Bandres, M.A. and Gutiérrez-Vega, J.C. (2008) Elliptical Beams. *Optics Express*, **16**, 21087-21092.

[18] Marincic, A. (2002) Huygens-Kirchhoff's Theory in Calculation of Elliptical Gaussian Beam Propagation through a Lens. Mikrotalasna revija.

[19] Heikkinen, V., Aikio, J.K., Alojoki, T., Hiltunen, J., Mattila, A.-J., Ollila, J. and Karioja, P. (2004) Single-Mode Tuning of a 1540-nm Diode Laser Using a Fabry-Pe/Spl Acute/Rot Interferometer. *IEEE Photonics Technology Letters*, **16**, 1164-1166. http://dx.doi.org/10.1109/LPT.2004.824619

Convective Schrödinger Equation: Insights on the Potential Energy's Role to Wave Particle Decay

Altair S. de Assis[1], Hector Torres-Silva[2], Göran T. Marklund[3]

[1]Universidade Federal Fluminense (IM-GMA), Niterói, Brasil
[2]Universidad de Tarapacá, Arica, Chile
[3]Kungliga Tekniska Högskolan—KTH, Stockholm, Sweden
Email: altair@vm.uff.br, htorres@uta.cl, goran.marklund@ee.kth.se

Abstract

In this paper, we coupled the Quantum Mechanics conventional Schrödinger's equation, for the particles, with the Maxwell's wave equation, in order to study the potential's role on the conversion of the electromagnetic field energy to mass and vice versa. We show that the dissipation ("conductivity") factor and the particle implicit proper frequency are both related to the potential energy. We have also derived a new expression for the Schrödinger's Equation considering the potential energy into this equation not as an ad hoc term, but also as an operator (Hermitian), which has the scalar potential energy as a natural eigenvalue of this operator.

Keywords

Schrödinger's Equation, Klein-Gordon's Equation, Maxwell's Wave Equation, Convection Displacement Current

1. Introduction

The conversion of electromagnetic energy to mass and vice versa is an old but still a very interesting problem to be studied in physics. One of the problems that still remain open is to explain in a more clear way the role of the potential energy on all those quantum wave-particle dissipation mechanisms addressed by quantum field theory—QFT/quantum mechanics. Here we study afresh this wave-particle dissipation problem using a simple alternative model, but considering the modification of the Schrödinger's Equation due to the effect of the vacuum zero point energy due to vacuum electromagnetic fluctuations, as described by Bo Lehnert [1].

2. Method

In order to study the quantum wave and particle Eigen modes excitation properties and the static potential role as a dissipation mechanism, we couple the Schrödinger's wave equation, which models particles as waves, to the Maxwell's wave equation for the electromagnetic field, since they are compatible mathematical structures. Here, we can view this alternative model as a plasma "wave-wave" interaction approach, where, instead of having only waves, we have particles and field interaction, using a linear wave model.

We start here with the non-relativistic Schrödinger's equation, which is given by the expression below:

$$\frac{\hbar^2}{2m}\nabla^2(\varphi)-V(\varphi)+j\hbar\frac{\partial(\varphi)}{\partial t}\equiv\left\{\nabla^2+j\left(\frac{2m}{\hbar}\right)\frac{\partial}{\partial t}-\frac{2m}{\hbar^2}V\right\}(\varphi)=0 \tag{1}$$

All the relevant terms present in Equation (1) are already defined in any basic quantum mechanics textbook ($j=\sqrt{-1}$) [2].

For the relativistic case, we have to use the Klein-Gordon equation given below:

$$\left\{\frac{1}{c^2}\left(j\hbar\frac{\partial}{\partial t}-V\right)^2-\left(\frac{\hbar}{j}\nabla\right)^2-m^2c^2\right\}(\varphi)=0$$

Expanding the squares:

$$\left\{\nabla^2+-\frac{1}{c^2}\frac{\partial^2}{\partial t^2}-j\frac{2V}{\hbar c^2}\frac{\partial}{\partial t}+\frac{1}{\hbar^2c^2}\left[V^2-m^2c^4\right]\right\}(\varphi)=0 \tag{2}$$

This equation is valid for zero spin particles, such as mesons, and photons with spin = 1.

The term $\frac{mc^2}{\hbar}$ (= ν_0) that can be obtained from Equation (2), should be considered as the particle intrinsic "Eigen frequency" (De Broglie frequency) via the dispersion relation:

$$\nu_0^2=\frac{m^2c^4}{\hbar^2c^2}=\left(\frac{mc}{\hbar}\right)^2c^2=k^2c^2.$$

Now, we will derive a fundamental equation relating the energy of an impinging electromagnetic wave to particles mass, which are created at a transition region defined below.

Let us now split the interaction space in three relevant regions, described as follows [3]-[5].

Region 1: Field region (Maxwell's region).

Region 2: Transition region (fuzzy wave-particle region where Maxwell's equation and Schrödinger's equation couple).

Region 3: Particle region (Schrödinger's region).

For the coupling to be consistent, we model the field region using the Klein-Gordon equation for non-massive photons, $m=0$.

Then, we obtain

$$\left\{-\nabla^2+\frac{1}{c^2}\frac{\partial^2}{\partial t^2}+j\frac{2V}{\hbar c^2}\frac{\partial}{\partial t}-\frac{1}{\hbar^2c^2}V^2\right\}(\varphi)=0 \tag{3}$$

The basic equation at the transition region is obtained by coupling region 1 to region 2, which is done just using the "momentum conservation" condition:

$$\nabla^2(\varphi)_{\text{region 1}}=\nabla^2(\varphi)_{\text{region 2}}.$$

Therefore,

$$\left\{\frac{\partial^2}{\partial t^2}+j\frac{2V}{\hbar}\frac{\partial}{\partial t}-\frac{V^2}{\hbar^2}\right\}(\varphi)=\left\{-j\left(\frac{2mc^2}{\hbar}\right)\frac{\partial}{\partial t}+\frac{2mc^2}{\hbar^2}V\right\}(\varphi) \tag{4}$$

Equation (4) is the basic equation to model the region—boundary layer—where particles and fields are not distinct—called here fuzzy region.

We have assumed that at the transition region a full energy and momentum conservation happens, so that the probability functions fully match.

To solve Equation (4), we consider the ansatz: $\phi = \phi_0 \exp(-j\omega t)$; We can then find easily the two proper frequencies:

$$\omega = \left(\frac{V}{\hbar} + \nu_0\right) \pm \sqrt{\left(\frac{V}{\hbar} + \nu_0\right)^2 - \frac{2V\nu_0}{\hbar} - \left(\frac{V}{\hbar}\right)^2} \quad (5)$$

Equation (5) can be re-written as,

$$\omega = \left(\frac{V}{\hbar} + \nu_0\right) \pm \nu_0 \quad (5')$$

The first root of (5') just states that it is possible to convert mass into energy and vice-versa.

For zero external potential ($V = 0$), we get the "resonance" (photon-particle) conditions:

For plus sign, $\omega = 2\nu_0 \left(\hbar\omega = 2mc^2\right)$ (two particles conversion) and for minus sign$\omega = 0$, no conversion, the trivial solution is obtained.

For non zero external potential, we can get another "resonance" (photon-particle) conditions:

If we consider the reflection of positively charged with spin = 0 particle from a barrier V, where this potential is zero to the left of the barrier and a constant V to the right, Equations (2) and (5) with $m = 0$ (massless boson) show that the mass of the particle to the right of the barrier V is $\left(m_f = V/c^2\right)$ having a mode conversion in this case.

However, it is necessary with a "dissipation" mechanism, which in our case related to V (no V, no dissipation, no energy-mass conversion).

We will show now that the factor $\frac{2V}{\hbar\mu c^2}$ plays exactly the role of a "dissipation term". It "permits" the vacuum to dissipate energy from photons. Where ν_0 is the "natural" proper frequency, as stated above. Of course, $\omega = 2\nu_0$ must be fulfilled, in order to guarantee energy conservation.

This shows explicitly why the particle production must be in the presence of a potential V, it makes it possible, somehow, that fields could be dissipated into the form of matter.

This result also indicates that the energy conservation at the transition region (mode conversion region) "requests" that the photon energy is at least twice the created particle rest energy—a well known old result. To consider the spin dynamics it is necessary to use the Dirac's relativistic quantum mechanics, but here we just want to see how V plays role in the field-mass conversion.

Now, for the sake of clarification, we will go further with the physical interpretation of the parameter $\frac{2V}{\hbar\mu c^2}$ and its role in quantum mechanics.

In order to gain more insight on this dissipation mechanism, we go back once more to Equation (2), assuming $m_0 \neq 0$, we have the dispersion relation:

$$\omega^2 + \frac{2V\omega}{\hbar} + \frac{\left(V^2 - m_0^2 c^4\right)}{\hbar^2} = 0.$$

We can rewrite this equation as,

$$\omega^2 + \mu\sigma c^2\omega - \frac{1}{4}\left[\nu_0^2 - \frac{\sigma^2}{\varepsilon^2}\right] = 0$$

Since

$$\frac{V^2}{\hbar^2} = \left(\frac{-j}{2}c^2\mu\sigma\right)^2 = \frac{-1}{4}c^4\mu^2\sigma^2,$$

$$\nu_0^2 = \left(\frac{m^2c^4}{\hbar^2}\right)^2, \text{ and } c^2 = \left(\frac{1}{\mu\varepsilon}\right).$$

Comparing now these results with the classical *RLC* circuit dispersion relation:

$$\omega^2 - j\frac{R}{L}\omega + \frac{1}{LC} = 0,$$

$$2\frac{V}{\hbar} = \mu\sigma c^2 = \frac{R}{L},$$

where L is the circuit inductance, C is the capacitance, and R the resistance.

Thus, the particle "natural" Eigen frequency can be written in this case as:

$$\omega_0^2 = \frac{1}{LC} = \frac{1}{4}\left[v_0^2 - \frac{\sigma^2}{\varepsilon^2}\right] = \frac{1}{4}\left[v_0^2 - \frac{R^2}{L^2}\right].$$

Therefore, for a pumped "circuit", the resonance condition is:

$$\omega = \omega_0 \text{, or } \hbar\omega \approx \frac{1}{2}mc^2\left(v_0 \gg \frac{\sigma^2}{\varepsilon^2}\right).$$

Here ω is the pumping (photon) energy/frequency.

The equivalent dissipation factor R/L for our case is $2\frac{V}{\hbar} = \mu\sigma c^2 = \frac{R}{L}$.

Concluding, we can consider the "mode conversion" regions as a resistance-impedance-capacitance-source circuit; the resonant region is driven by the impinging photons.

Also, due to the Equation (2), we have

$$\hbar^2\omega^2 + 2V\hbar\omega + V^2 - m^2c^4 = 0. \tag{2'}$$

$$\omega^2 = \frac{\left(V^2 - m_0^2c^4\right)}{\hbar^2}$$

And finally ($E = \hbar\omega$),

$$E^2 + 2VE + V^2 - m^2c^4 = 0,$$

or

$$E = -V \pm mc^2.$$

This is the relation used for the total energy to model the motion of particles in a static external potential. So we have the two possible particle energy modes, V can be positive or negative. It is possible to have zero energy mode taking the plus sign and $|V| = mc^2$, this shows how mass can become potential energy and vice-versa.

Adding the momentum contribution into (2'), one gets,

$$E^2 + 2VE + V^2 - p^2c^2 - m^2c^4 = 0,$$

So we get

$$E = -V \pm \sqrt{p^2c^2 + m^2c^4}$$

or

$$E = -V \pm mc^2\sqrt{1 + \frac{v^2}{c^2}\frac{1}{\sqrt{1 - \frac{v^2}{c^2}}}}$$

For nonrelativistic particle conversion, the usual relation is,

$$E \approx -V \pm m_{rest}c^2$$

For $V = 0$, we have the known conventional relation,

$$E^2 = p^2c^2 + m^2c^4.$$

L. Brillouin has also addressed the problem of relativity and potential energy using a different approach [6]. He considered an implicit mass content into the external potential energy.

Note that at high temperatures/energies *T*, it is possible to observe electron-positron and neutrino-antineutrino pairs, (*i.e.*, $T > 1$ MeV) [7] [8] and so the Eigen mode model presented here makes full sense. The potential V may be just a few eV, but similar results can be obtained with a different formalism [9].

3. Results

As a consequence of the above discussion, we will modify the original Schrödinger's equation just adding a constructed potential operator in a way that this equation will have a convective term, and we will call this novel (modified) Schrödinger's equation as Convective Schrödinger Equation.

By convective Schrödinger Equation, here, we just mean the conventional Schrödinger's Equation where we introduce a term that is related to the vacuum zero point energy (relating it to the potential) [1]. In doing so, we get a modified Schrödinger Equation, where the potential energy is not anymore introduced in an ad hoc manner into this equation, but it is considered as an Hermitian operator, such as the total energy and the kinetic energy, and where the eigenvalues related to it are exactly the scalar potential. Therefore, the Schrödinger Equation is now only left with one ad hoc term—the particle mass—that might be further treated as an operator also, by say, using the Higgs's Boson theory, for instance.

In order to move further with this study, it is important to say that both mass and potential, even if being observables quantities, are introduced ad hoc into Schrödinger's equation and not postulated as Hermitian operators in the same way that the total energy and the kinetic energy are. That is, there are no formal Hermitian operators associated to those physical quantities, which would aloud to obtain real eigenvalues related to both, such as the ones we have for energy and momentum. This more realistic model would permit the access to the mass particle content creation from a more fundamental entity (complete Schrödinger's function) and not just to access a probability density to find it, already massive, somewhere in a bounded or free state.

For the potential case, there are some discussions about its operator nature, but for mass this is still an open problem, in spite of Higgs's explanation for particle mass origin via Higgs Boson.

We leave the mass operator problem for a future work and address afresh the problem of the operator nature of the potential and try to find an expression that could be useful to possibly model that. We saw above that the scalar potential could be understood as a dissipation "inducing" factor for wave—particle interaction as pair production, but now we move further to understand the operator nature of this physical quantity.

Therefore, the question to be answered now here is: Can the potential in Schroedinger's equation be modeled as an Hermitian operator? If it is so, which form it would have? Can we get a measurable scalar potential as an eigenvalue of this operator? Has this question any physical meaning whatsoever?

There are people that consider the potential $V(x,t)$ to be an operator just due to its modification of the original wave function $\varphi(x,t)$ via $\varphi(x,t) \Rightarrow V(x,t)\varphi(x,t)$. Therefore, $\varphi(x,t)$ becomes $V(x,t)\varphi(x,t)$ via the "operation" $V(x,t)\varphi(x,t)$. But is that so?

There are also some/many other discussions about the operator nature of the potential, but we stop here, knowing that this is indeed a very interesting though controversy subject and therefore deserves further deeper discussion.

In order to proceed with this issue, let us consider Maxwell's wave equation for pure field, but with dissipation:

$$\nabla^2 F - \frac{1}{c^2}\frac{\partial^2 F}{\partial t^2} - \mu\sigma\frac{\partial F}{\partial t} = 0 \tag{6}$$

F is the wave field and σ is the material conductivity.

By analogy, using the Klein-Gordon equation for massless photons, we have,

$$\left\{\nabla^2 - \frac{1}{c^2}\frac{\partial^2}{\partial t^2} - j\frac{2V}{\hbar c^2}\frac{\partial}{\partial t} - \frac{1}{\hbar^2 c^2}V^2\right\}(\varphi) = 0 \tag{2'}$$

Examining Equations (6) and (2'), we can identify the wave function dissipation factor for a photon in vacuum, but treated as particle. The factor "is"

$$2j\frac{V}{\hbar c^2} = \mu\sigma ,$$

Which means that particles wave function can grow in the presence of electrostatic potentials and photons. This phenomenon is related to the ZPE—Zero Point Energy [2]-[4].

Therefore, the term $2\frac{V}{\hbar \mu c^2} = 2\frac{e^2 V}{\mu e^2 c\hbar c} = 2\alpha\frac{eV}{\mu c e^3}$ plays the role of vacuum "energy to mass dissipation", for photon. Where α is the fine structure constant.

If $V = 0$, there is no dissipation.

It is interesting to mention that if we consider the potential V as an observable, not included ad hoc at the Schrödinger's equation, as it is in the conventional quantum mechanics, a "potential operator" could be understood as:

$$\hat{V} = \left(\frac{\mu\hbar c^2}{2j}\right)\hat{\sigma} .$$

The important point is now to known the physical meaning of the operator $\hat{\sigma}$.

We consider at this point the "convection" displacement current (using a revised electromagnetic theory), that is valid for a point charged particle, which moves with velocity v. The "operator condutivity" can be writen as [3] [4]:

$$\hat{\sigma} \equiv \varepsilon_0\left(\frac{\partial}{\partial t} + \boldsymbol{v}\cdot\nabla\right) .$$

Therefore, it might well be that the operator potential could be written as:

$$\hat{V} \equiv j\hbar\left(\frac{\partial}{2\partial t} - \frac{1}{2}\boldsymbol{v}\cdot\nabla\right) \tag{7}$$

The second term in (7) is related to the vacuum longitudinal Eigen modes. They do not appears (exist) in the conventional Maxwell's electromagnetic theory. This factor is considered a convection term related to a convection displacement current, as already mentioned it describes the contribution to the displacement current of a single point-shaped charged particle moving with velocity v [3]-[5].

We apply now this operator (7) on the particle standing wave function,

$$\varphi = \varphi(\boldsymbol{x})\exp\left(\left(-i\frac{E}{\hbar}\right)t\right),$$

The above solution $\varphi(x,t)$, with $\varphi(\boldsymbol{x})$ obtained satisfying the boundary conditions, describes standing waves of the time-dependent equation, which are the states with definite energy, those standing waves are called energy ("frequency") Eigen state.

We have then, of course, for a particle (electron for instance) in the presence of a potential V.

$$\begin{aligned}\hat{V}(\varphi) &= \hat{V}\left(\varphi(\boldsymbol{x})\exp\left(\left(-j\frac{E}{\hbar}\right)t\right)\right)\\ &= \left\{j\hbar\left(\frac{\partial}{\partial t} - \frac{1}{2}\boldsymbol{v}\cdot\nabla\right)\right\}\left(\varphi(\boldsymbol{x})\exp\left(\left(-j\frac{E}{\hbar}\right)t\right)\right)\\ &= \left\{E - -\frac{1}{2}\boldsymbol{v}\cdot\nabla\right\}\left(\varphi(x)\exp\left(\left(-j\frac{E}{\hbar}\right)t\right)\right)\end{aligned} \tag{7'}$$

Considering here the wave function of a free particle, but under the influence of a potential V, we have:

$$\varphi(\boldsymbol{x},t) = A\exp\left[j(\boldsymbol{p}\cdot\boldsymbol{x} - Et)/\hbar\right].$$

The potential "double-eigenvalue" is obtained via,

$$\hat{V}\varphi(\boldsymbol{x},t) = \hat{V}A\exp\left[j(\boldsymbol{p}\cdot\boldsymbol{x} - Et)/\hbar\right]$$
$$= \left[E - \frac{1}{2}jpv\right]\varphi(\boldsymbol{x},t) = \left[E - \frac{1}{2}mv^2\right]\varphi(\boldsymbol{x},t).$$

Thus: $\hat{V}\phi(\boldsymbol{x},t) = V\phi(\boldsymbol{x},t)$.

It means that E is the total energy eigenvalue via the time operator and $\frac{1}{2}mv^2$ is the kinetic energy eigenvalue via the convection term. So, V is type "bilinear" operator.

So, from (7'), we can write a new type of Schrödinger's equation, "convection" Schrödinger equation—which contains in its structure the energy equipartion (factor 1/2) for particle and anti-particle:

$$-\frac{\hbar^2}{2m}\nabla^2\varphi + j\hbar\left(\frac{\partial}{\partial t} - \frac{1}{2}\boldsymbol{v}.\nabla\right)\varphi = j\hbar\frac{\partial\varphi}{\partial t}.$$

Now, we get a real formal "full" energy operator Schrödinger's equation, self-consistently: $E = T + V$ e $V = E - T$, eigenvalues of $\hat{V}$ and $\hat{E}$.

4. Conclusions

The results presented in this paper show that the potential, combined with the magnetic permeability/electric permittivity (implicit vacuum entities) and the Planck's constant, plays the role of dissipation factor for the conversion of electromagnetic fields to particles and vice-versa.

Also, considering the contribution of the time convective derivative term (convective displacement current) [3] [4], for a motion particle, we have derived a novel modified Schrödinger's equation, which we have called here the "Convective" Schrödinger Equation. This equation takes into consideration, in a formal way, the potential as an operator related to the observable potential energy, and not adding it in a ad hoc manner as done in the usual Schrödinger's Equation.

In the past, the study of relativistic particles has been in the exclusive domain of high-energy and particle physics. In grapheme theory, nonetheless, the linear electronic band dispersion near the Dirac points gave rise to charge carriers (electrons or holes) that propagate as if they were massless fermions with speeds of the order of 10^6 m/s rather than the speed of light. Hence, charge carriers in this structure had to be described by the massless Dirac equation or the Schrödinger's equation when the compaction of graphene device generates mass proportional to conductivity. The physics of relativistic electrons is thus now experimentally accessible in graphene based solid-state devices, whose behavior differs drastically from that of similar devices fabricated with usual semiconductors. Consequently, new unexpected phenomena have been observed while other phenomena that were well-understood in common semiconductors such as the quantum Hall effect and weak-localization, exhibited surprising behavior in graphene. Thus, the graphene devices enabled the study of relativistic dynamics in controllable nano-electronic circuits (relativistic electrons on-a-chip). The Convective Schrödinger Equation (convection displacement current) allows basic understanding of electronic processes in graphene devices, which have conductivity in presence of electromagnetic fields (see Equations (10)-(12) of ref. [8]). It also allowed for the observation of some subtle effects, previously accessible only to high-energy physics, such as Klein tunneling and vacuum breakdown [10]-[12].

A further and deeper discussion on novel insights of field, matter (particles and anti-particles) and vacuum dynamics can be found in references [1] [3]-[5].

Acknowledgements

The authors would like to thank Bo Lehnert for general discussions and suggestions considered in this work.

References

[1] Lehnert, B. (2014) Some Consequences of Zero Point Energy. *Journal of Electromagnetic Analysis and Applications*, **6**, 319-327.
Lehnert, B. (2015) Zero Point Energy Effects on Quantum Electrodynamics. *Journal of Modern Physics*, **6**, 448-452.

[2] Cohen-Tannoudji, C., Diu, B. and Laloe, F. (1977) Quantum Mechanics, Volumes 1 and 2. Hermann and John Wiley

& Sons, New York.

[3] Lehnert, B. (1990) Complementary Aspects on Matter-Antimatter Boundary Layers. TRITA-EPP-90-04, Royal Institute of Technology, Stockholm, Sweden, and References Therein.

[4] Lehnert, B. (2013) Revised Quantum Electrodynamics. NOVA Publishers, New York.

[5] Lehnert, B. (2008) A Revised Electromagnetic Theory with Fundamental Applications, Swedish Physics Archive.

[6] Brillouin, L. (1965) The Actual Mass of Potential Energy, A Correction to Classical Relativity. *Proceedings of the National Academy of Sciences*, **53**, 475-482. http://dx.doi.org/10.1073/pnas.53.3.475

[7] Torres-Silva, H. (2013) Physical Interpretation of the Dirac Neutrino with Electromagnetic Mass. *Journal of Electromagnetic Analysis and Applications*, **5**, 294-301. http://dx.doi.org/10.4236/jemaa.2013.57046

[8] Torres-Silva, H. and Cabezas, D.T. (2012) Chiral Current in a Graphene Battery. *Journal of Electromagnetic Analysis and Applications*, **4**, 426-431. http://dx.doi.org/10.4236/jemaa.2012.410059

[9] Arbab, A. (2011) A New Wave Equation of the Electron. *Journal of Modern Physics*, **2**, 1012-1016. http://dx.doi.org/10.4236/jmp.2011.29121

[10] Novoselov, K.S., Geim, A.K., Morozov, S.V., Jiang, D., Zhang, Y., Dubonos, S.V., Grigorieva, I.V. and Firsov, A.A. (2004) Electric Field Effect in Atomically Thin Carbon Films. *Science*, **306**, 666. http://dx.doi.org/10.1126/science.1102896

[11] Zhang, Y., Small, J.P., Amori, M.E.S. and Kim, P. (2005) Electric Field Modulation of Galvanomagnetic Properties of Mesoscopic Graphite. *Physical Review Letters*, **94**, Article ID: 176803. http://dx.doi.org/10.1103/PhysRevLett.94.176803

[12] Castro, A.H., Guinea, F., Peres, N.M.R., Novoselov, K.S. and Geim, A.K. (2009) The Electronic Properties of Graphene. *Reviews of Modern Physics*, **81**, 109. http://dx.doi.org/10.1103/RevModPhys.81.109

10

A Device for Measuring 3-Phase Alternating Current at a Distance

Seppo Mäkinen

School of Technology, Vaasa University of Applied Sciences, Vaasa, Finland
Email: seppo.makinen@vamk.fi

Abstract

We have made a preliminary study on a possibility to build a device, which could measure three-phase alternating current at a distance. This report gives a brief look at the theoretical background related with such a device and derives several mathematical results associated with the theory and the structure of the device. Our results suggest that such a device can be built, assuming the distances between the three current-carrying wires and the two sensors of the device are known accurately enough. There are ways to measure these distances, but this article does not concentrate on them.

Keywords

Electromagnetic Induction, 3-Phase Alternating Current, Ammeter

1. Introduction

For energy companies, it is important to know if the transmission lines are working as they should or not. Further, they need this information all the time. They also want to know about the quality of the electric current flowing in the lines. Naturally, they have meters to find out the characteristics of the alternating current in their interest, but installation of a meter to a transmission line requires cutting off the current in the line. This is time-consuming and requires more work than the mere installation.

In our recent paper, we discussed the use of a toroidal coil in measuring alternating current flowing in one wire [1]. We wanted to increase the challenge, and decided to consider three-phase alternating current flowing in three wires. In addition, we wanted to consider a meter, which could measure all three currents as functions of time—and so that such a meter could be installed to the transmission line safely without cutting off the current in the line. We decided to concentrate on the typical 20 kV transmission line, which is shown in **Figure 1**.

Since the geometry shown in **Figure 1** is not the only possible one, we need to build our model so that the

Figure 1. A typical 20 kV transmission line with three current-conducting wires.

locations of the three wires are not restricted in any way. This means that we need to introduce a system of coordinates fixed to the device, relative to which the locations of the wires are specified. An illustration of the device is added to **Figure 1**.

2. The Numerical Model

Let us consider three straight and parallel conductors, which carry sinusoidally oscillating currents. We assume that the positions of the wires are not known, and that the phase differences between the three currents are unknown, as well. Further, we assume that the currents may have different amplitudes, and that we do not know these amplitudes beforehand. Thus, the only thing we do know about the case is that the currents are sinusoidal in shape, and that they oscillate at a given frequency. In addition, our analysis is built so that the number of currents is known to be three.

Our task is to find the three currents as functions of time. This means that we have altogether 9 unknowns (if the phase angles φ_1, φ_2 and φ_3 are ignored in the analysis): the x and the y coordinates of each wire, and the amplitudes of the three currents. For an unequivocal solution, we would thus need at least 9 equations in terms of these 9 variables. Solving such a set of equations seems too time-consuming, and we see that the only practical way to find the currents is to start with more initial information about the system.

In our opinion, the easiest way to obtain more information about the system is to locate the three wires by measuring the distances between these three and some known reference points. With these lengths, one can easily calculate the x and the y coordinates of each wire relative to one of the reference points, and relative to some chosen directions for the x and the y axes.

It appears that only two such points are needed. One point serves as the origin of the Cartesian system of coordinates, and the other shows the direction of, e.g., the x axis. Hence, it would be wise to place two sets of induction coils (or Hall sensors) at two points in the vicinity of the wires, and locate the wires relative to these two.

Using a laser-beam-based device, one can measure the distances between the two reference points—at which the induction coils (or Hall sensors) are placed—and the three wires to a reasonable accuracy. In the following analysis we assume that these lengths have been measured on the installation of the induction coils, and that these lengths are determined accurately enough. We also assume that the distance between the two reference points is known precisely enough.

Relative to the chosen system of coordinates, the measured lengths and the coordinates of the wires are defined according to **Figure 2**. In this figure, we have also illustrated the magnetic field vectors corresponding to the field generated by the current $I_1(t)$ at points P1 and P2, $\boldsymbol{B}_{11}$ and $\boldsymbol{B}_{12}$, respectively. Similarly, we have illustrated the magnetic field generated by the second current, $I_2(t)$, at the two points, $\boldsymbol{B}_{21}$ and $\boldsymbol{B}_{22}$, and the field produced by the third current, $I_3(t)$, at points P1 and P2, $\boldsymbol{B}_{31}$ and $\boldsymbol{B}_{32}$, respectively. The vectors are drawn by

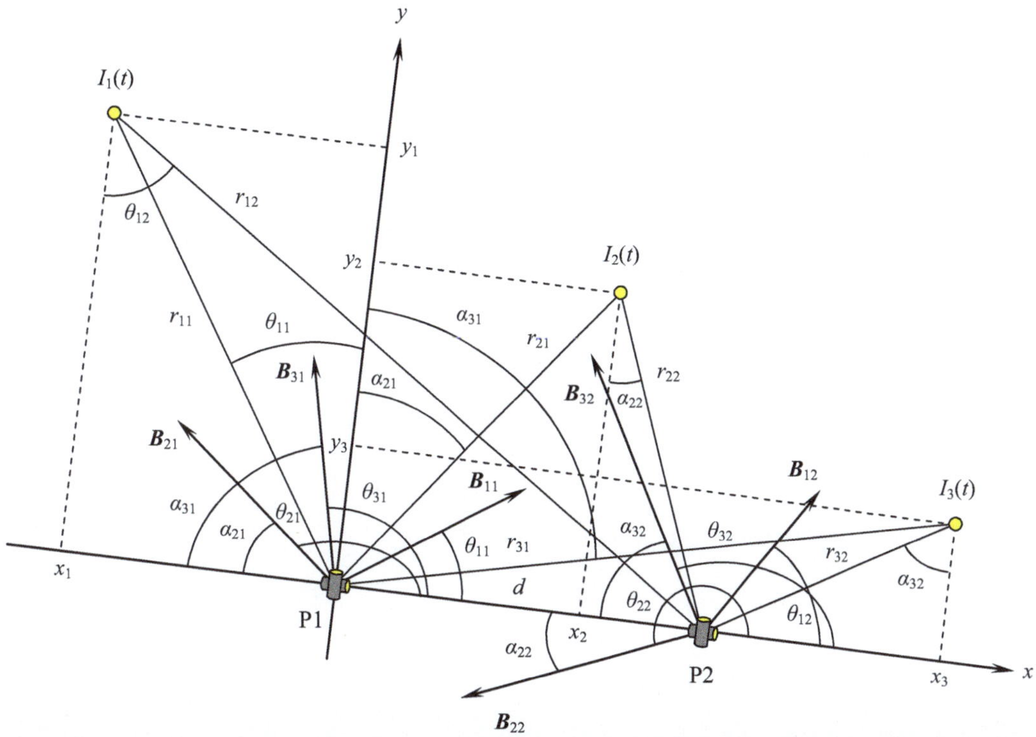

Figure 2. The two sets of induction coils measure the induced emfs generated by the three sinusoidally oscillating currents.

assuming, as an example which does not restrict the generality of our analysis, that $t = 0$ and that the currents are given by:

$$\begin{cases} I_1(t) = \hat{I}_1 \cos(\omega t - \varphi_1) \\ I_2(t) = \hat{I}_2 \cos(\omega t - \varphi_2); \\ I_3(t) = \hat{I}_3 \cos(\omega t - \varphi_3) \end{cases} \begin{cases} \varphi_1 = 0^\circ \\ \varphi_2 = 120^\circ \\ \varphi_3 = 240^\circ \end{cases}$$

The figure was drawn somewhat slanted to remind the reader about the fact that the x axis does not necessarily have to be horizontal; it only needs to point along a straight line passing through the two reference points. In our analysis, we assume that all the wires are parallel to each other, and that they point along the z axis.

For a case of this kind, we need to use two induction coils at both reference points. One of these is directed along the x axis, while the other points along the y axis.

If the perpendicular distances r_{11}, r_{12}, r_{21}, r_{22}, r_{31} and r_{32} are measured during installation of the induction coils, for instance by using laser beam, one can determine the coordinates of the three wires by applying the Law of Cosines, see **Figure 3**:

$$\begin{cases} r_{11}^2 = r_{12}^2 + d^2 - 2r_{12}d\cos\beta_1 \\ r_{21}^2 = r_{22}^2 + d^2 - 2r_{22}d\cos\beta_2 \\ r_{31}^2 = r_{32}^2 + d^2 - 2r_{32}d\cos\beta_3 \end{cases}$$

These yield:

$$\beta_1 = \cos^{-1}\left(\frac{r_{12}^2 + d^2 - r_{11}^2}{2r_{12}d}\right); \begin{cases} x_1 = d - r_{12}\cos\beta_1 \\ y_1 = r_{12}\sin\beta_1 \end{cases}$$

$$\beta_2 = \cos^{-1}\left(\frac{r_{22}^2 + d^2 - r_{21}^2}{2r_{22}d}\right); \begin{cases} x_2 = d - r_{22}\cos\beta_2 \\ y_2 = r_{22}\sin\beta_2 \end{cases}$$

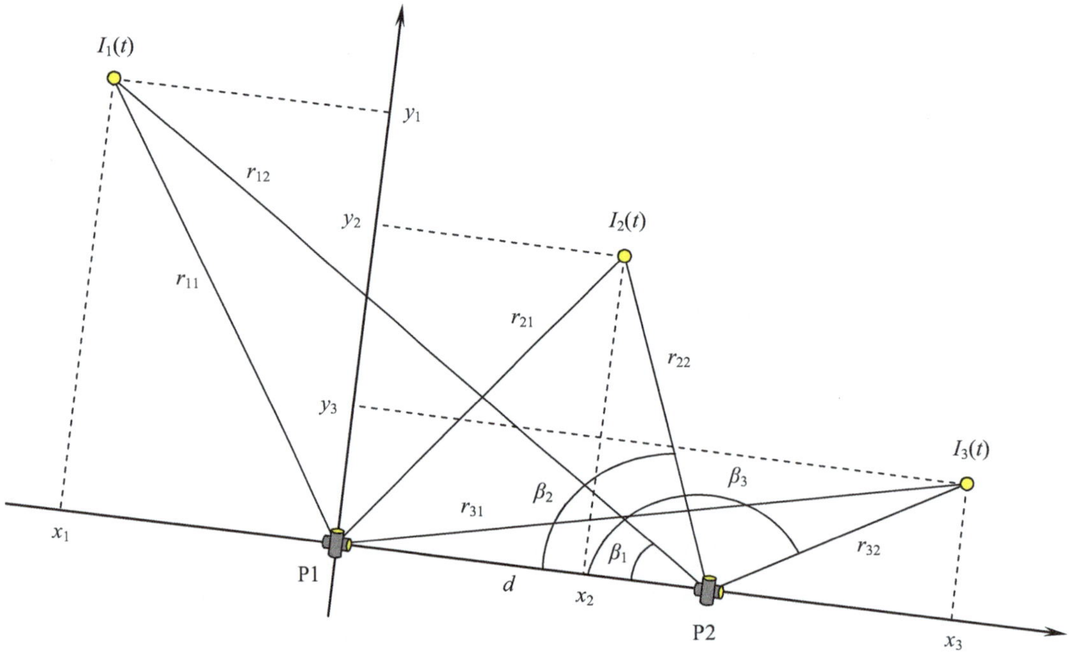

Figure 3. A simplified illustration of the geometry.

$$\beta_3 = \cos^{-1}\left(\frac{r_{32}^2 + d^2 - r_{31}^2}{2r_{32}d}\right); \begin{cases} x_3 = d - r_{32}\cos\beta_3 \\ y_3 = r_{32}\sin\beta_3 \end{cases}$$

The emf induced in the x-directed coil at point P1 can be given in terms of the three currents, the calculated wire coordinates and the angles defined in **Figure 2**:

$$\begin{aligned}\varepsilon_{1x}(t) &= -N_{1x}\frac{\mathrm{d}}{\mathrm{d}t}\left\{\boldsymbol{A}_{1x}\cdot\boldsymbol{B}_{11}(t) + \boldsymbol{A}_{1x}\cdot\boldsymbol{B}_{21}(t) + \boldsymbol{A}_{1x}\cdot\boldsymbol{B}_{31}(t)\right\} \\ &= -N_{1x}\frac{\mathrm{d}}{\mathrm{d}t}\left\{\pi r_{1x}^2\left[B_{11}(t)\cos\theta_{11} + B_{21}(t)\cos\theta_{21} + B_{31}(t)\cos\theta_{31}\right]\right\} \\ &= -N_{1x}\pi r_{1x}^2\frac{\mathrm{d}}{\mathrm{d}t}\left\{\frac{\mu_0 I_1(t)}{2\pi r_{11}}\cos\theta_{11} + \frac{\mu_0 I_2(t)}{2\pi r_{21}}\cos\theta_{21} + \frac{\mu_0 I_3(t)}{2\pi r_{31}}\mathrm{co}\,\theta_{31}\right\}\end{aligned}$$

In this expression, N_{1x} is the number of turns of wire in the induction coil oriented along the x axis and placed at P1; r_{1x} is the radius of this coil. According to **Figure 2**, the cosines of the angles in this result can be given as:

$$\cos\theta_{11} = \frac{y_1}{r_{11}}; \cos\theta_{21} = -\frac{y_2}{r_{21}}; \cos\theta_{31} = -\frac{y_3}{r_{31}}$$

Therefore, we can give the emf induced in the x-directed coil at P1 as:

$$\begin{aligned}\varepsilon_{1x}(t) &= -\frac{N_{1x}r_{1x}^2\mu_0}{2}\frac{\mathrm{d}}{\mathrm{d}t}\left\{\frac{\hat{I}_1\cos(\omega t - \varphi_1)}{r_{11}}\frac{y_1}{r_{11}} + \frac{\hat{I}_2\cos(\omega t - \varphi_2)}{r_{21}}\left(-\frac{y_2}{r_{21}}\right) + \frac{\hat{I}_3\cos(\omega t - \varphi_3)}{r_{31}}\left(-\frac{y_3}{r_{31}}\right)\right\} \\ &= \frac{N_{1x}r_{1x}^2\mu_0\omega}{2}\left\{\frac{y_1}{r_{11}^2}\hat{I}_1\sin(\omega t - \varphi_1) - \frac{y_2}{r_{21}^2}\hat{I}_2\sin(\omega t - \varphi_2) - \frac{y_3}{r_{31}^2}\hat{I}_3\sin(\omega t - \varphi_3)\right\}\end{aligned}$$

Similarly, we can show that the emf induced in the x-directed coil at P2 is given by:

$$\varepsilon_{2x}(t) = \frac{N_{2x}r_{2x}^2\mu_0\omega}{2}\left\{\frac{y_1}{r_{12}^2}\hat{I}_1\sin(\omega t - \varphi_1) - \frac{y_2}{r_{22}^2}\hat{I}_2\sin(\omega t - \varphi_2) - \frac{y_3}{r_{32}^2}\hat{I}_3\sin(\omega t - \varphi_3)\right\}$$

Here, N_{2x} is the number of turns of wire in the second induction coil oriented along the x axis and placed at P2;

r_{2x} is the radius of this coil.

We can express these results more simply if we rewrite them in terms of factors $k_1, k_2, \cdots, k_6$ as follows:

$$\varepsilon_{1x}(t) = k_1\hat{I}_1 \sin(\omega t - \varphi_1) + k_2\hat{I}_2 \sin(\omega t - \varphi_2) + k_3\hat{I}_3 \sin(\omega t - \varphi_3); \begin{cases} k_1 = \dfrac{N_{1x}r_{1x}^2\mu_0\omega}{2}\dfrac{y_1}{r_{11}^2} \\ k_2 = -\dfrac{N_{1x}r_{1x}^2\mu_0\omega}{2}\dfrac{y_2}{r_{21}^2} \\ k_3 = -\dfrac{N_{1x}r_{1x}^2\mu_0\omega}{2}\dfrac{y_3}{r_{31}^2} \end{cases}$$

$$\varepsilon_{2x}(t) = k_4\hat{I}_1 \sin(\omega t - \varphi_1) + k_5\hat{I}_2 \sin(\omega t - \varphi_2) + k_6\hat{I}_3 \sin(\omega t - \varphi_3); \begin{cases} k_4 = \dfrac{N_{2x}r_{2x}^2\mu_0\omega}{2}\dfrac{y_1}{r_{12}^2} \\ k_5 = -\dfrac{N_{2x}r_{2x}^2\mu_0\omega}{2}\dfrac{y_2}{r_{22}^2} \\ k_6 = -\dfrac{N_{2x}r_{2x}^2\mu_0\omega}{2}\dfrac{y_3}{r_{32}^2} \end{cases}$$

The difference in voltage across the two x-directed coils can be given as:

$$\Delta\varepsilon_x(t) = \varepsilon_{2x}(t) - \varepsilon_{1x}(t) = \frac{\mu_0\omega y_1\hat{I}_1 \sin(\omega t - \varphi_1)}{2}\left\{\frac{N_{2x}r_{2x}^2}{r_{12}^2} - \frac{N_{1x}r_{1x}^2}{r_{11}^2}\right\}$$

$$+\frac{\mu_0\omega y_2\hat{I}_2 \sin(\omega t - \varphi_1)}{2}\left\{-\frac{N_{2x}r_{2x}^2}{r_{22}^2} + \frac{N_{1x}r_{1x}^2}{r_{21}^2}\right\} + \frac{\mu_0\omega y_3\hat{I}_3 \sin(\omega t - \varphi_1)}{2}\left\{-\frac{N_{2x}r_{2x}^2}{r_{32}^2} + \frac{N_{1x}r_{1x}^2}{r_{31}^2}\right\}$$

If we call the factors that multiply the current amplitudes as k_7, k_8 and k_9, this result can be given in a shorter form as:

$$\Delta\varepsilon_x(t) = k_7\hat{I}_1 \sin(\omega t - \varphi_1) + k_8\hat{I}_2 \sin(\omega t - \varphi_2) + k_9\hat{I}_3 \sin(\omega t - \varphi_3); \begin{cases} k_7 = \dfrac{\mu_0\omega y_1}{2}\left(\dfrac{N_{2x}r_{2x}^2}{r_{12}^2} - \dfrac{N_{1x}r_{1x}^2}{r_{11}^2}\right) \\ k_8 = \dfrac{\mu_0\omega y_2}{2}\left(\dfrac{N_{1x}r_{1x}^2}{r_{21}^2} - \dfrac{N_{2x}r_{2x}^2}{r_{22}^2}\right) \\ k_9 = \dfrac{\mu_0\omega y_3}{2}\left(\dfrac{N_{1x}r_{1x}^2}{r_{31}^2} - \dfrac{N_{2x}r_{2x}^2}{r_{32}^2}\right) \end{cases}$$

Note that the current terms in this result are the actual currents flowing in the three wires, except for the phase difference of $\frac{\pi}{2}$ radians. This means that we can find out the time-dependent currents $I_1(t)$, $I_2(t)$ and $I_3(t)$ —with the phase difference of $\frac{\pi}{2}$—by solving the following matrix equation:

$$\begin{pmatrix} k_1 & k_2 & k_3 \\ k_4 & k_5 & k_6 \\ k_7 & k_8 & k_9 \end{pmatrix}\begin{pmatrix} I_1(t)' \\ I_2(t)' \\ I_3(t)' \end{pmatrix} = \begin{pmatrix} \varepsilon_{1x}(t) \\ \varepsilon_{2x}(t) \\ \Delta\varepsilon_x(t) \end{pmatrix}$$

Here, the primed symbols refer to the currents, which differ from the real ones by the phase difference of $\frac{\pi}{2}$ rad.

Following a similar approach, one can derive the following result for the y-directed coil at P1:

$$\varepsilon_{1y}(t) = -N_{1y}\frac{\mathrm{d}}{\mathrm{d}t}\left\{\boldsymbol{A}_{1y}\cdot\boldsymbol{B}_{11}(t)+\boldsymbol{A}_{1y}\cdot\boldsymbol{B}_{21}(t)+\boldsymbol{A}_{1y}\cdot\boldsymbol{B}_{31}(t)\right\}$$
$$= -N_{1y}\frac{\mathrm{d}}{\mathrm{d}t}\left\{\pi r_{1y}^2\left[B_{11}(t)\cos\left(\frac{\pi}{2}-\theta_{11}\right)+B_{21}(t)\cos\left(\theta_{21}-\frac{\pi}{2}\right)+B_{31}(t)\cos\left(\theta_{31}-\frac{\pi}{2}\right)\right]\right\}$$

In this expression, N_{1y} is the number of turns of wire in the induction coil oriented along the y axis and placed at P1; r_{1y} is the radius of this coil. According to **Figure 2** and the fundamental rules of trigonometry, the trigonometric terms in this result can be given as:

$$\cos\left(\frac{\pi}{2}-\theta_{11}\right)=\cos\left(\theta_{11}-\frac{\pi}{2}\right)=\sin\theta_{11}=\frac{x_1}{r_{11}}$$
$$\cos\left(\theta_{21}-\frac{\pi}{2}\right)=\sin\theta_{21}=-\sin\left(2\pi-\theta_{21}\right)=-\sin\alpha_{21}=-\frac{x_2}{r_{21}}$$
$$\cos\left(\theta_{31}-\frac{\pi}{2}\right)=\sin\theta_{31}=-\sin\left(2\pi-\theta_{31}\right)=-\sin\alpha_{31}=-\frac{x_3}{r_{31}}$$

These results yield:

$$\varepsilon_{1y}(t) = -N_{1y}\pi r_{1y}^2\frac{\mathrm{d}}{\mathrm{d}t}\left\{\frac{\mu_0 I_1(t)}{2\pi r_{11}}\frac{x_1}{r_{11}}-\frac{\mu_0 I_2(t)}{2\pi r_{21}}\frac{x_2}{r_{21}}-\frac{\mu_0 I_3(t)}{2\pi r_{31}}\frac{x_3}{r_{31}}\right\}$$
$$= -\frac{N_{1y}r_{1y}^2\mu_0}{2}\frac{\mathrm{d}}{\mathrm{d}t}\left\{\frac{x_1}{r_{11}^2}\hat{I}_1\cos(\omega t-\varphi_1)-\frac{x_2}{r_{21}^2}\hat{I}_2\cos(\omega t-\varphi_2)-\frac{x_3}{r_{31}^2}\hat{I}_3\cos(\omega t-\varphi_3)\right\}$$
$$= \frac{N_{1y}r_{1y}^2\mu_0\omega}{2}\left\{\frac{x_1}{r_{11}^2}\hat{I}_1\sin(\omega t-\varphi_1)-\frac{x_2}{r_{21}^2}\hat{I}_2\sin(\omega t-\varphi_2)-\frac{x_3}{r_{31}^2}\hat{I}_3\sin(\omega t-\varphi_3)\right\}$$

Again, we can express this result more simply if we rewrite it in terms of factors p_1, p_2 and p_3 as follows:

$$\varepsilon_{1y}(t) = p_1\hat{I}_1\sin(\omega t-\varphi_1)+p_2\hat{I}_2\sin(\omega t-\varphi_2)+p_3\hat{I}_3\sin(\omega t-\varphi_3);\begin{cases} p_1 = \dfrac{N_{1y}r_{1y}^2\mu_0\omega}{2}\dfrac{x_1}{r_{11}^2} \\ p_2 = -\dfrac{N_{1y}r_{1y}^2\mu_0\omega}{2}\dfrac{x_2}{r_{21}^2} \\ p_3 = -\dfrac{N_{1y}r_{1y}^2\mu_0\omega}{2}\dfrac{x_3}{r_{31}^2}\end{cases}$$

Similarly, we can show that the emf induced in the y-directed coil at P2 is given by:

$$\varepsilon_{2y}(t) = -N_{2y}\frac{\mathrm{d}}{\mathrm{d}t}\left\{\boldsymbol{A}_{2y}\cdot\boldsymbol{B}_{12}(t)+\boldsymbol{A}_{2y}\cdot\boldsymbol{B}_{22}(t)+\boldsymbol{A}_{2y}\cdot\boldsymbol{B}_{32}(t)\right\}$$
$$= -N_{2y}\pi r_{2y}^2\frac{\mathrm{d}}{\mathrm{d}t}\left\{\frac{\mu_0 I_1(t)}{2\pi r_{12}}\cos\left(\frac{\pi}{2}-\theta_{12}\right)+\frac{\mu_0 I_2(t)}{2\pi r_{22}}\cos\left(\theta_{22}-\frac{\pi}{2}\right)+\frac{\mu_0 I_3(t)}{2\pi r_{32}}\cos\left(\theta_{32}-\frac{\pi}{2}\right)\right\}$$

Here, N_{2y} is the number of turns of wire in the induction coil oriented along the y axis and placed at P2; r_{2y} is the radius of this coil. According to laws of trigonometry and **Figure 2**, the cosines of the angles in this result can be given as:

$$\cos\left(\frac{\pi}{2}-\theta_{12}\right)=\sin\theta_{12}=\frac{d-x_1}{r_{12}}$$
$$\cos\left(\theta_{22}-\frac{\pi}{2}\right)=\cos\left(\frac{\pi}{2}-\theta_{22}\right)=\sin\theta_{22}=-\sin\left(\theta_{22}-\pi\right)=-\sin\alpha_{22}=-\frac{d-x_2}{r_{22}}$$
$$\cos\left(\theta_{32}-\frac{\pi}{2}\right)=\cos\left(\frac{\pi}{2}-\theta_{32}\right)=\sin\theta_{32}=\sin\left(\pi-\theta_{32}\right)=\sin\alpha_{32}=\frac{x_3-d}{r_{32}}$$

Thus, we can give the emf induced in the y-directed coil at P2 as:

$$\varepsilon_{2y}(t)=-\frac{N_{2y}r_{2y}^2\mu_0}{2}\frac{\mathrm{d}}{\mathrm{d}t}\left\{\frac{\hat{I}_1\cos(\omega t-\varphi_1)}{r_{12}}\frac{d-x_1}{r_{12}}+\frac{\hat{I}_2\cos(\omega t-\varphi_2)}{r_{22}}\left(-\frac{d-x_2}{r_{22}}\right)+\frac{\hat{I}_3\cos(\omega t-\varphi_3)}{r_{32}}\frac{x_3-d}{r_{32}}\right\}$$

$$=-\frac{N_{2y}r_{2y}^2\mu_0\omega}{2}\left\{-\frac{d-x_1}{r_{12}^2}\hat{I}_1\sin(\omega t-\varphi_1)+\frac{d-x_2}{r_{22}^2}\hat{I}_2\sin(\omega t-\varphi_2)+\frac{d-x_3}{r_{32}^2}\hat{I}_3\sin(\omega t-\varphi_3)\right\}$$

$$=\frac{N_{2y}r_{2y}^2\mu_0\omega}{2}\left\{\frac{d-x_1}{r_{12}^2}\hat{I}_1\sin(\omega t-\varphi_1)-\frac{d-x_2}{r_{22}^2}\hat{I}_2\sin(\omega t-\varphi_2)-\frac{d-x_3}{r_{32}^2}\hat{I}_3\sin(\omega t-\varphi_3)\right\}$$

Let us rewrite this result in terms of factors p_4, p_5 and p_6:

$$\varepsilon_{2y}(t)=p_4\hat{I}_1\sin(\omega t-\varphi_1)+p_5\hat{I}_2\sin(\omega t-\varphi_2)+p_6\hat{I}_3\sin(\omega t-\varphi_3);\begin{cases}p_4=\frac{N_{2y}r_{2y}^2\mu_0\omega}{2}\frac{d-x_1}{r_{12}^2}\\ p_5=-\frac{N_{2y}r_{2y}^2\mu_0\omega}{2}\frac{d-x_2}{r_{22}^2}\\ p_6=-\frac{N_{2y}r_{2y}^2\mu_0\omega}{2}\frac{d-x_3}{r_{32}^2}\end{cases}$$

The difference in voltage across the two y-directed coils can be given as:

$$\Delta\varepsilon_y(t)=\varepsilon_{2y}(t)-\varepsilon_{1y}(t)=\frac{\mu_0\omega\hat{I}_1\sin(\omega t-\varphi_1)}{2}\left\{N_{2y}r_{2y}^2\frac{d-x_1}{r_{12}^2}-N_{1y}r_{1y}^2\frac{x_1}{r_{11}^2}\right\}$$

$$+\frac{\mu_0\omega\hat{I}_2\sin(\omega t-\varphi_2)}{2}\left\{N_{2y}r_{2y}^2\frac{x_2-d}{r_{22}^2}+N_{1y}r_{1y}^2\frac{x_2}{r_{21}^2}\right\}+\frac{\mu_0\omega\hat{I}_3\sin(\omega t-\varphi_3)}{2}\left\{N_{2y}r_{2y}^2\frac{x_3-d}{r_{32}^2}+N_{1y}r_{1y}^2\frac{x_3}{r_{31}^2}\right\}$$

This result can be given in a shorter form as:

$$\Delta\varepsilon_y(t)=p_7\hat{I}_1\sin(\omega t-\varphi_1)+p_8\hat{I}_2\sin(\omega t-\varphi_2)+p_9\hat{I}_3\sin(\omega t-\varphi_3);\begin{cases}p_7=\frac{\mu_0\omega}{2}\left(N_{2y}r_{2y}^2\frac{d-x_1}{r_{12}^2}-N_{1y}r_{1y}^2\frac{x_1}{r_{11}^2}\right)\\ p_8=\frac{\mu_0\omega}{2}\left(N_{2y}r_{2y}^2\frac{x_2-d}{r_{22}^2}+N_{1y}r_{1y}^2\frac{x_2}{r_{21}^2}\right)\\ p_9=\frac{\mu_0\omega}{2}\left(N_{2y}r_{2y}^2\frac{x_3-d}{r_{32}^2}+N_{1y}r_{1y}^2\frac{x_3}{r_{31}^2}\right)\end{cases}$$

Note that the current terms in this result are the actual currents flowing in the three wires, except for the phase difference of $\frac{\pi}{2}$ radians.

We can find out the time-dependent currents $I_1(t)$, $I_2(t)$ and $I_3(t)$—with the phase difference of $\frac{\pi}{2}$—by solving the following matrix equation:

$$\begin{pmatrix}p_1 & p_2 & p_3\\ p_4 & p_5 & p_6\\ p_7 & p_8 & p_9\end{pmatrix}\begin{pmatrix}I_1(t)'\\ I_2(t)'\\ I_3(t)'\end{pmatrix}=\begin{pmatrix}\varepsilon_{1y}(t)\\ \varepsilon_{2y}(t)\\ \Delta\varepsilon_y(t)\end{pmatrix}$$

Here, the primed symbols refer to the currents, which differ from the real ones by the phase difference of $\frac{\pi}{2}$ rad.

3. Applying the Results

The device described in this article could be used so that both sensors, at P1 and at P2, measure the induced voltage at least once in every millisecond. One measurement sequence could last for about 20 periods of current oscillations, and such sequences could be measured once in every 10 minutes, for example. The collected data could then be transmitted wirelessly to a remote receiver. The received data could be analyzed in the office using software based on the matrix equations derived in this article.

The software could calculate the currents $I_1(t)'$, $I_2(t)'$ and $I_3(t)'$ with both of the given matrix equations. After this, one could compare the results with each other, and if they were close to each other within reasonable accuracy, the final results could be found by calculating the average values of the two. Finally, the actual currents, $I_1(t)$, $I_2(t)$ and $I_3(t)$ would be obtained by shifting the sinusoidal functions of time by $\frac{\pi}{2}$ rad.

The device should be equipped with two additional induction coils along the device's z axis. On installation, one should make sure that the voltage in these coils is zero. This ensures that the two sets of perpendicular coils at P1 and at P2 are in the xy plane perpendicular to the three current-carrying wires.

4. Conclusion

Our analysis indicates that it is possible to build a device that can determine three sinusoidally oscillating currents as functions of time from a reasonable distance. The device can be installed safely to the measuring site without turning the power off in the transmission lines. The next logical step in our process would be to build a device explained in this article, and to make test runs with it and real transmission lines.

References

[1] Mäkinen, S. (2014) Toroidal Coil in Measuring Alternating Current at a Distance. *Journal of Electromagnetic Analysis and Applications*, **6**, 367-371. http://dx.doi.org/10.4236/jemaa.2014.612037

11

Real Time MR Thermometry Using Tm-DOTMA

Bu S. Park[1], Martin J. Lizak[2], Leonardo M. Angelone[1], Sunder S. Rajan[1*]

[1]Division of Biomedical Physics (DBP)/OSEL/CDRH, Food and Drug Administration (FDA), Silver Spring, MD, USA
[2]Mouse Imaging Facility, National Institute of Neurological Disorders and Stroke (NINDS), National Institutes of Health (NIH), Bethesda, MD, USA
Email: *sunder.rajan@fda.hhs.gov

Abstract

We present results of real-time and sensitive MR Thermometry (MRT) using a paramagnetic lanthanide complex thulium 1,4,7,10-tetraazacyclododecane-1,4,7,10-tetramethyl-1,4,7,10-tetraa-cetate (Tm-DOTMA) to study radio frequency (RF) heating induced by a copper wire and a titanium bone screw in an agarose gel phantom. The temperature dependent chemical shift coefficient (TDCSC) of the methyl resonance was found to be 0.7 ± 0.03 ppm/°C in agarose gel. The methyl protons of Tm-DOTMA were imaged using 2D chemical shift imaging (CSI) and 3D phase mapping methods (PMM), approximately 7 sec long, and compared with conventional water proton resonance frequency (PRF) method. Two RF-induced heating approaches were tested: 1) using a pre-scan before the MRT; or 2) using the heating caused by the imaging pulse during continuous imaging. Both approaches allowed detection of temperature changes which are less than 1°C and continuously mapping temperature changes around the copper wire. Using a heating pre-scan, the Tm-DOTMA 2D-CSI allowed better qualitative visualization of the temperature changes around the titanium screw compared with water phase shift thermometry. Numerical electromagnetic field simulations were also conducted for the evaluation of orientation dependency using the copper wire in 4.7 T (200 MHz). Thermometry approach using Tm-DOTMA can detect smaller temperature changes with decreased scanning time resulting in real-time and sensitive temperature mapping.

Keywords

Tm-DOTMA, CSI, PRF, Real-Time, FDTD, RF, Temperature Mapping

1. Introduction

*Corresponding author.

Magnetic resonance imaging (MRI) is widely used in clinical diagnosis because of benefits such as excellent soft tissue contrast and the use of non-ionizing radiation. Many patients—including patients with metallic implanted medical devices—will undergo MRI during the course of their life. While MRI is considered to be a safe technology, there are possible risks, such as tissue heating caused by the radio frequency (RF) electromagnetic energy used to elicit the MRI signal [1]-[10]. This issue may be exacerbated in patients with metallic implants, because of the RF currents induced along the implant and the related electric field enhancement at the interface between implant and tissue [11]. For this reason, many patients with medical implants cannot undergo MRI. In order to facilitate the development of medical implants that are safe for MRI use, there is a need to design real time and robust methods to test the extent of RF-induced heating of tissue generated during an MRI scan. Currently, heating tests are performed using point temperature measurements made with fluoroptic probes. The measurements only provide the temperature of a small volume (1 mm^3) near the probe, and are limited to the total number of probes available. Because of this, these measurements are inherently difficult due to the need for precise probe positioning related to the highly geometry-dependent and spatially inhomogeneous RF-induced heating [12].

Over the past years, researchers have focused on using MR Thermometry because of the inherent ability of providing a volumetric assessment of temperature that overcomes the limitations of the fluoroptic probes [13]-[21].

The most commonly used method in MR Thermometry (MRT) is the water proton resonance frequency (PRF) method, which is based on the phase shifts caused by the temperature dependence of the chemical shift of the water proton (^1H) resonance [13]-[15]. However, MRT of the water proton is characterized by weak temperature sensitivities (~ 0.01 ppm/°C), high sensitivity to main magnetic field (B_o) inhomogeneity, and susceptibility effects [16] [17] due to the relatively long echo times used. Therefore, the PRF method is not well suited for imaging around medical implants. Recently, the dependence of the water spin-lattice relaxation time (T_1) on temperature has also been used in MR Thermometry [8]. However, because the T_1 dependence is weak (*i.e.*, 10 - 35 ms/°C), this approach is less suited for measuring small temperature changes. Another difficulty in using the water resonance is the need for interleaved off-resonance RF pulses for heating, which is difficult to implement on many scanners due to software constraints.

Recent studies have proposed the NMR spectroscopic use of paramagnetic lanthanide complexes, namely thulium 1,4,7,10-tetraazacyclododecane-1,4,7,10-tetramethyl-1,4,7,10-tetraacetate (Tm-DOTMA) and thulium 1,4,7,10-tetraazacyclododecane-1,4,7,10-tetrakis (methylene phosphonate) (Tm-DOTP) for measuring temperature changes [16] [17]. These compounds are characterized by a high sensitivity of the chemical shift to temperature changes (methyl resonance: 0.57 - 0.7 ppm/°C of Tm-DOTMA and ~1 ppm/°C of Tm-DOTP [17]-[22]). Compared to Tm-DOTP, Tm-DOTMA has lower temperature sensitivity, but much higher (about five-fold) signal, longer T_2 and it is less sensitive to other parameters such as pH and concentrations of Ca^{2+} [18]-[22]. Additionally, these lanthanide complexes have short T_1 and spin-spin relaxation time (T_2) values (e.g., T_1 and T_2 values of CH_3 proton in Tm-DOTMA at 9.4 T are 5.3 ms and 4.1 ms [21], respectively), which could allow the use of shorter echo time (TE) and repetition time (TR). In addition, high flip angle and short TR were used in the *in vitro* experiments to increase the amount of RF heating to be detected with MRT.

Despite some of the positive attributes described above, the utility of Tm-DOTMA for real time MR thermometry imaging has not been previously reported. In this study, we have characterized the behavior of Tm-DOTMA in agarose gel and explored the use of Tm-DOTMA for real time MRT using both chemical shift imaging (CSI) and phase mapping method (PMM) [13]-[15] of PRF shift to map real time temperature changes in a phantom containing a metallic object, namely either a copper wire or a titanium screw. In addition, we have compared the results with those using conventional water PRF methods.

2. Materials and Methods

MR Thermometry experiments using water proton and paramagnetic lanthanide complex, Tm-DOTMA (Macrocyclics, Dallas, TX), were performed on a 4.7 T (200 MHz) Avance system (Bruker Biospin, Billerica, MA, USA) with a shielded high-pass birdcage RF coil.

A. TDCSC measurement and CSI implementation

Temperature dependent chemical shift coefficient (TDCSC) measurements of Tm-DOTMA were performed in agarose gel (Sigma-Aldrich, St. Louis, MO) and in distilled water. Three agarose gel samples (20 mM, 10 mM, and 4 mM Tm-DOTMA) as well as a 7.15 mM liquid solution were tested. The gel and liquid samples were prepared in a plastic tube that snugly fit inside a glass water jacket, which in turn was placed inside a saddle RF coil (diameter: 18 mm and length: 8 mm) as shown in **Figure 1(a)**. Temperature control was achieved by circulating warm water through the jacket using a conventional hyper-hypothermia system (CSZ medical Inc., Cincinnati, OH, USA). The temperature was monitored using a multichannel thermometer system with two fluoroptic fiber probes (Luxtron 3100, Santa Clara, CA, USA), of which one was located at the center of the sample tube and the second one outside of the sample but inside the jacket to ensure temperature equilibration (**Figure 1(a)**). The system allowed obtaining a temperature over the whole tube spatially homogenous and constant over time.

Using the above setup, both 1D spectroscopy and 2D-CSI were performed to calibrate the chemical shift changes as a function of temperature and to validate the implementation of the 2D-CSI pulse sequence. 1D spectroscopy for the TDCSC measurements was performed using the following parameters: TR = 220 ms, number of averages (NA) = 32, 1024 data points, flip angle = 90° and bandwidth = 5 KHz, FID Acquisition time of 0.2 sec. The acquired data was processed using interactive left shifting, Fourier transform, zero filling, filtering, and phase correction. Analysis of the results was performed in Matlab (The MathWorks, Inc., Natick, MA). The 2D-CSI imaging sequence was implemented with the following parameters: TR = 55 ms, NA = 4, FOV = 25.6 × 25.6 mm, matrix size = 32 × 32, slice gap = 2 mm, and slice thickness = 5 mm. The temperature change (ΔT) was calculated using the relationship between measured frequency shift of each voxel and the measured TDCSC of Tm-DOTMA.

B. Setup for real-time thermometry

A rectangular 1.0% agarose gel phantom (length L_R = 116 mm, width W_R = 55 mm and height = 20 mm, as shown in **Figure 1(d)**) containing 15 mM Tm-DOTMA was used. MRT was performed with either an copper wire (84 mm long) or a titanium bone screw (35 mm long and with 3 mm diameter), placed inside the gel. The copper wire was insulated, except for one of the two tips where a 1 cm-section of the insulation was removed. The implant (copper wire/titanium screw) was oriented along the z-axis and placed by the edge on one side of the phantom. A fiber optic probe was positioned near the tip of the implant and the second one was positioned on the contra-lateral side of the rectangular phantom (at 45 mm from the tip) to measure a reference temperature. Measurements were performed with a linear transmit/receive birdcage coil (diameter: ~ 72 mm, length: ~110 mm).

The temperature increase was achieved using two strategies. The first consisted of using pre-scans with high RF power (450° flip angle) followed by ~7 second imaging; progressively higher temperatures were achieved by

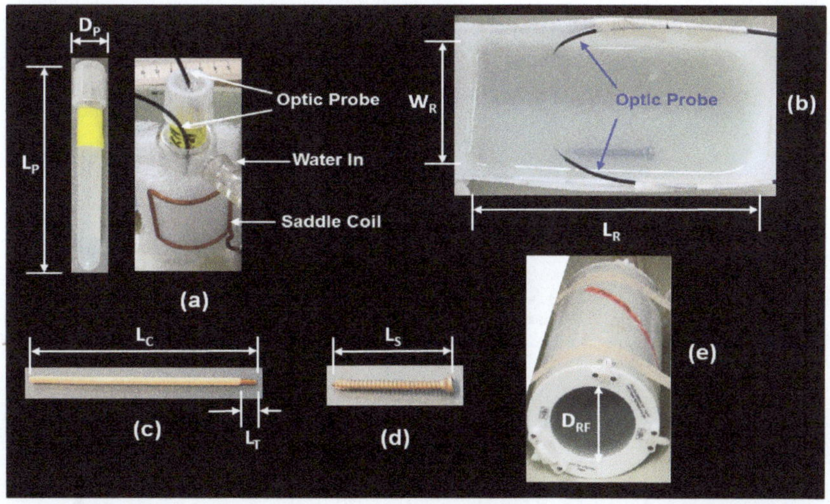

Figure 1. Experimental setup for measurement of temperature dependent chemical shift coefficient (TDCSC) (a) and fast thermometry (b) using a copper wire (c), a titanium screw ((b) and (d)) and a birdcage coil (e). The temperature increase was induced using flowing water (a) or RF heating (b). Four different solutions, *i.e.*, three agarose gel samples (20 mM, 10 mM, and 4 mM Tm-DOTMA) as well as a 7.15 mM liquid solution, with same dimension of (a) were used in the TDCSC measurement. Two fluoroptic temperature probes were used to record the temperature changes near the implant and at a reference location. Physical dimensions: L_P = 70 mm, D_P = 10 mm, W_R = 55 mm, L_R = 116 mm, L_C = 84 mm, L_T = 10 mm, L_S = 35 mm, and D_{RF} = 72 mm.

increasing the number of pre-scans from 0 to 800. The second strategy was based on the heating generated from the RF pulses used for continuous dynamic imaging. With the two heating strategies described above, fast imaging of the methyl peak of Tm-DOTMA was performed using both 2D-CSI and 3D-PMM of PRF shift with the following parameters:

2D-CSI: 2D slice selective pulse (1 ms), TR = 25 ms, flip angle = 90°, phase encode gradient time = 0.86 ms, FID acquisition time =12 ms, NA=1, variable FOV = (120 to 80) × (80 to 30) mm, matrix size = 16 × 16, variable spatial resolution = (7.5 to 5.0) × (5 to 1.9) mm/pixel, and spectral width = 10 kHz (~49.8 ppm).

3D-PMM: TR/TE = 14/(2.2 to 2.7) ms, a frequency selective pulse with flip angle = 90°, NA = 2, FOV = 150 × 80 × 30 mm, matrix size = (64 to 128) (Readout) × 32 × 8, spatial resolution = (1.17 to 2.34) × 2.50 × 3.75 mm/pixel, and scanning time = 7.2 sec for each scan. Centric phase encoding was used for both 2D and 3D scans to minimize the effects of phase cancellation during the scan.

Water proton MRT was performed using a conventional 2D GRE sequence with the following parameters: TR/TE = 100/10 ms, flip angle = 30°, NA = 1, FOV = 120 × 80 mm, matrix size = 120 × 80, spatial resolution = 1 × 1 mm/pixel, slice thickness = 2 mm, and scanning time = 8.0 sec. Because the RF pulse power was not high enough, only the pre-scan heating approach was used with water proton MRT.

MRT phase maps were generated off-line from the phase difference images ($\Delta\phi$) using MATLAB scripts and phase unwrapping. The later was performed using edge detection and region-merging algorithm, based on a method by Mark [23]. The pixel wise ΔT computed with respect to the initial image was calculated as [14]:

$$\Delta T = -\frac{\Delta\phi}{\alpha\gamma B_0 TE} \tag{1}$$

where $\Delta\phi$ is the phase difference computed from each image at various temperatures and the initial image acquired before the first heating period, B_0 is the static main magnetic field, γ is the gyromagnetic ratio, and α is the TDCSC of each proton (*i.e.*, Tm-DOTMA or water).

C. Numerical electromagnetic field simulations

Numerical simulations were conducted for the evaluation of orientation dependency using the copper wire in 4.7 T (200 MHz). The geometry of the RF coil, RF shield, sample and copper wire with different orientations used in the simulation is shown in **Figure 2**. The modeled RF coil (ID: 72 mm, Length: 110 mm), RF shield (ID: 120 mm, Length: 110 mm) and sample (Height: 32 mm, Width: 55 mm, Length: 116 mm) geometries closely matched those used in experiment. Four different orientations of the copper wire, *i.e.*, 0, 30, 60 and 90°, were used in the simulation. A total of 12 radially oriented voltage sources located along the end rings and having 1 V peak-to-peak amplitude were used for ideal HP birdcage coil excitation. A 2 × 2 × 2 mm^3 isotropic resolution

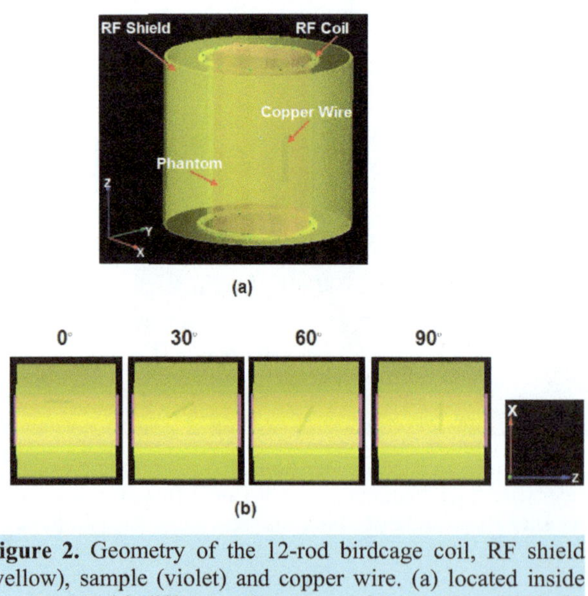

Figure 2. Geometry of the 12-rod birdcage coil, RF shield (yellow), sample (violet) and copper wire. (a) located inside the sample with different orientations of 0, 30, 60 and 90°; (b) used for numerical simulations.

was used for the numerical simulations. All simulation results were normalized to obtain rotating RF magnetic field ($|\mathbf{B}_1^+|$) = 4 μT at the center of the coil.

3. Results

The measured frequency shift relative to temperature using 1D spectroscopy and four different Tm-DOTMA samples (20 mM, 10 mM and 4 mM agarose gel and a 7.16 mM aqueous solution) (**Figure 1(a)**) were found to be 137 ± 3, 140 ± 6, 139 ± 5 and 141 ± 3 Hz/°C, respectively. These values correspond to a TDCSC value of 0.685 ± 0.015, 0.7 ± 0.03, 0.695 ± 0.025, and 0.705 ± 0.015 ppm/°C, respectively. In addition, the line widths of the spectra were found to be typically less than 120 Hz.

Figure 3(a) shows CSI color maps for a single slice, for the methyl peak of 20 mM Tm-DOTMA agarose sample at four different temperatures. The peak frequency was converted to a color scale when generating the images. The corresponding spectra for the four temperatures are shown in **Figure 3(b)**. When the temperature measured by fiber optic probes increased from 16.6°C (A) to 24.7°C (D), the frequency changed from 273 Hz to 1386 Hz. The corresponding TDCSC value from 2D CSI was 0.685 ± 0.005 ppm/°C, equals to 137 ± 1 Hz/°C, in good agreement with the result of the 1D spectroscopy.

Figure 4 shows examples of thermal maps depicting temperature changes (ΔT) that were obtained by application of 800 high power (450° flip) pre-scans, using the 2D CSI (a) and a PMM (b) of PRF technique, in a 15 mM Tm-DOTMA agarose-gel phantom with the copper wire. The pre-scan caused a temperature increase of approximately 1°C near the tip as indicated by the fluoroptic probe. Both imaging techniques (2D-CSI and 3D-PMM of PRF shift) clearly revealed the temperature change using the Tm-DOTMA resonance, even though the scan time was 7 to 8 sec. Conversely, the corresponding water 2D-PMM of PRF shift scan (8 sec scan time) revealed in comparison with only minimal changes. In addition, the coronal Tm-DOTMA image showed tip heating at both ends of the wire, with a smaller effect on the side that was fully insulated. This weaker heating effect was not detected with the water PRF scan.

Figure 5 shows the real time temperature changes induced by the RF pulse used for imaging in the rectangular phantom. Sample images obtained using 2D CSI (a) and 3D- PMM of PRF shift (b) are shown. Each thermal map was computed by determining the frequency/phase shifts relative to the first image (reference). The temperatures measured by the fluoroptic probes and by the images acquired with both 2D CSI and 3D-PMM are shown in the lower panel.

Figure 6 shows an example of the thermal maps obtained using the gel phantom (**Figure 1(b)**) with the titanium screw instead of the copper wire. RF heating was induced by 800 pre-scan RF pulses. Thermal maps were

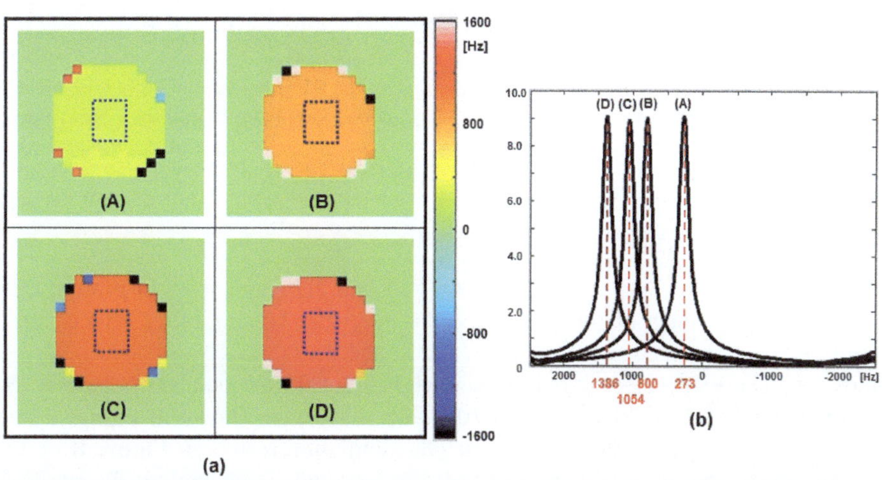

Figure 3. (a) Frequency shift map along an axial slice acquired using 2D-CSI with 20 mM agarose gel Tm-DOTMA phantom. The dashed rectangular box in (A) represents the region of interest (ROI) used for the calculation of an average frequency shift. The temperature measured using fiber optic probes located inside the phantom was 16.6 °C (A), 20.5 °C (B), 22.3°C (C), and 24.7°C (D). (b) Corresponding averaged spectra within the ROI. The pulse sequence parameters were: repetition time (TR) = 55 msec, FOV = 25.6 × 25.6 mm, matrix = 32 × 32, number of slice = 1, and scan time = 3 min and 45 sec. The calculated TDCSC was 137 ± 1 Hz/°C, in line with the previous results of 1D spectroscopy.

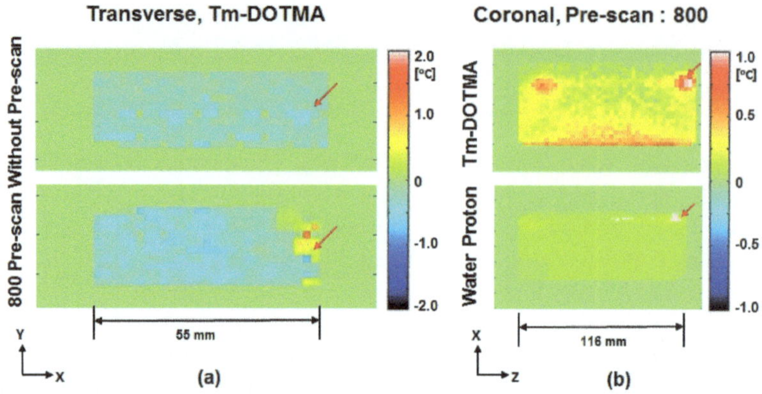

Figure 4. Temperature change maps (ΔT) obtained using the 2D CSI (a) and phase mapping method (b) of PRF technique in a 15 mM Tm-DOTMA agarose gel phantom containing the copper wire. The temperature change was induced by application of 800 high-power pre-scans. With the 2D CSI method tip heating can be observed in the transverse slice position near the tip of the wire. The coronal 3D-PRF maps of the methyl resonance of Tm-DOTMA ((b)-top) showed hot spots at both ends of the copper wire. The corresponding water proton 2D-PRF map ((b)-bottom) showed a smaller phase shift consistent with the lower sensitivity. The position of the probe is indicated by the red arrow.

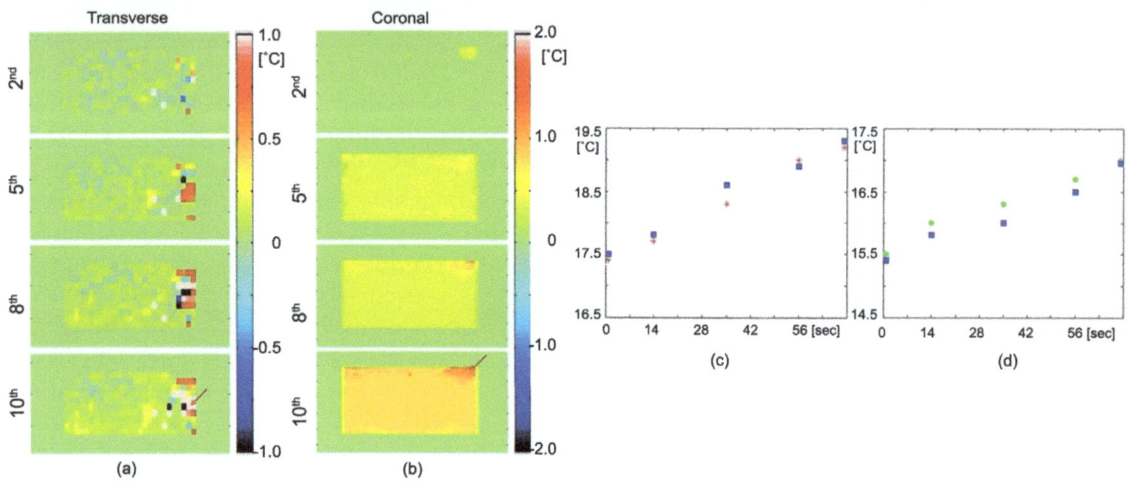

Figure 5. Real time temperature changes due to heating caused by the RF pulse used for imaging are shown using 2D CSI (a) and 3D PMM of PRF shift (b) with the 15 mM Tm-DOTMA agarose gel phantom and the copper wire. Each ΔT image was calculated using a frequency shift relative to the first image (reference). (c) Temperatures measured by the optic probes (blue rectangles in (c)) and by the acquired images (red asterisk) with the 2D-CSI. (d) Temperatures measured by the optic probes (blue rectangles) and by the acquired images (green asterisk) with the 3D PMM. The position of the probe is indicated by the red arrow.

generated using PRF technique, 3D scan for Tm-DOTMA and 2D scan for water proton. Despite the considerable signal loss seen at both ends of the screw due the magnetic susceptibility effects, the Tm-DOTMA map showed a clearer heating pattern compared to the water image.

2D simulation results of $|\mathbf{B_1}^+|$, electric field intensity($\|\mathbf{E}\|$), local specific absorption rate (SAR) defined as absorbed power divided by mass density [14] and 10 g-avg. SAR were shown in **Figure 7** and 1D profile of the $\|\mathbf{E}\|$ at the tip of the copper wire is shown in **Figure 8**. Simulation results in **Figure 8** show that the maximum $\|\mathbf{E}\|$ in the 1D profile is increased over 3 times (80 vs. 270 V/m) when the angle is decreased from 90° to 0°.

4. Discussion

In this study we have explored the use of Tm-DOTMA to obtain near real time thermal maps in a phantom containing metal objects. Because the methyl resonance of Tm-DOTMA has much higher temperature sensitivity and substantially smaller T_1 and T_2 values than water proton [16]-[21], it has the potential to decrease scan

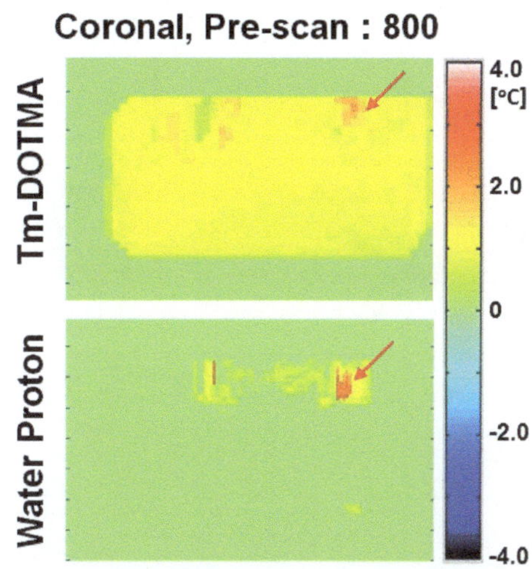

Figure 6. Thermal maps obtained using a 15 mM Tm-DOTMA agarose-gel phantom containing the titanium screw. The phantom was heated with the pre-scan method (*i.e.*, 800 RF pulses) (see also **Figure 4**). The top panel shows the 3D-PRF shift image from methyl resonance of Tm-DOTMA. The bottom panel shows the corresponding water PRF shift image. Considerable signal loss is seen at the tips due the magnetic susceptibility effects in both images. The scanning time without a pre-scan was 7.2 sec (Tm-DOTMA) and 8.0 sec (water proton). The MRT performed using Tm-DOTMA showed better sensitivity than that of a water proton. The position of the probe is indicated by the red arrow.

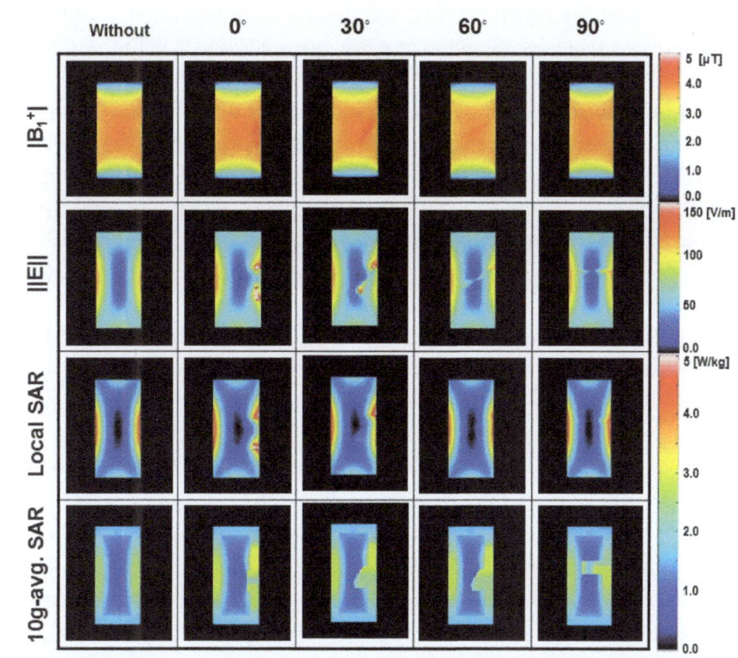

Figure 7. Numerical simulation results of $|\mathbf{B_1}^+|$ (first row), $||\mathbf{E}||$ (second row), Local SAR (third row), and 10 g-avg. SAR (fourth row) calculated without (first column) and with the copper wire having different orientations of 0(second column), 30 (third column), 60 (fourth column) and 90˚ (fifth column). All simulation results were normalized to obtain $|\mathbf{B_1}^+| = 4\ \mu T$ at the center of the coil.

times by shortening TR and TE to about 2.2 (**Figure 5(b)**) or 2.7 ms (**Figure 4(c)** and **Figure 6**) compared to the range of 10 - 20 ms with water proton used in previous studies [14] [15].

Gensler *et al.* [8] recently proposed a method of fast T_1-based thermometry to detect the RF-induced heating of medical devices using off-resonance heating pulses. This method has two key limitations: 1) the quantifications at high temperature levels are not accurate because of the nonlinear and weak temperature dependence of T_1; and 2) the method requires interleaved RF pulses for heating. Compared to the T_1-based method, the Tm-

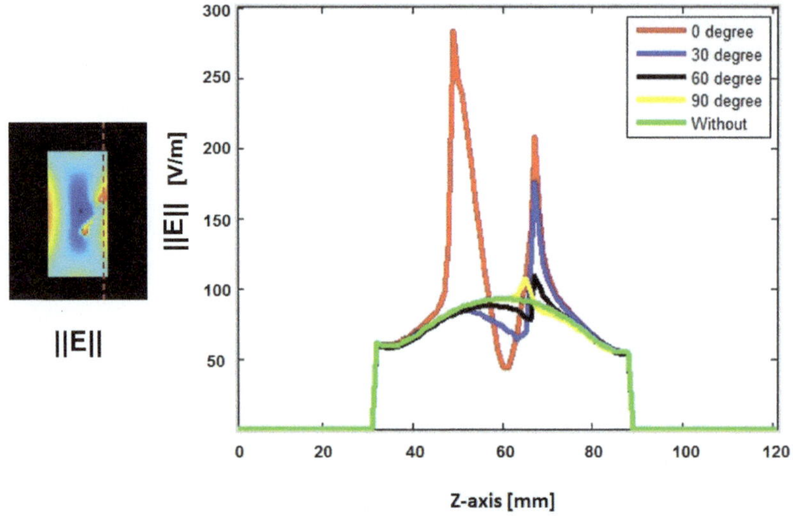

Figure 8. 1D profile of electric field intensity (||**E**||) without (green line) and with the copper wire having different orientations of 0 (red line), 30 (blue line), 60 (black line) and 90° (yellow line). Dashed red lines in the left image indicate the line selected for the graph. Values were normalized to obtain $|\mathbf{B_1}^+| = 4$ µT at the center of the coil.

DOTMA method is based on the PRF shift and offers several advantages, including linear and sensitive dependence on temperature, potential for short TR/TE, and absence of dependence on tissue type [8] [24]. Additionally, the utility of T_1 approach used by Gensler has not been studied for higher fields (>1.5 T) where the T_1 is significantly longer.

The experiments used to acquire frequency shift maps of the four different Tm-DOTMA solutions (**Figure 3**) were performed by varying the temperature using a temperature controlled flow system which allowed excellent temperature control. Additionally, using a small cylindrical tube yielded uniform temperature throughout the sample by equilibration. The calculated TDCSC using the acquired frequency shift map was about 0.7 ± 0.03 ppm/°C, which was consistent with previous research reporting 0.7 ± 0.1 ppm/°C at 4.0 T using samples containing 4 mM of the agent, 3 mM of 3-(trimethylsilyl) propionic-2,2,3,3-d4 acid (TSP), and 10% of D_2O with different cation (e.g., K^+, Na^+, Ca^{2+}) concentrations [20].

The experiments performed with the heating induced by pre-scan (**Figure 4** and **Figure 6**) were limited by the maximum RF power tolerated by the coil, which allowed a 450° RF pulse. The use of Tm-DOTMA with either CSI or phase mapping allowed detecting a temperature change of less than 1°C in near real time (7-8 sec). Additionally, Tm-DOTMA allowed capturing more details than the water PRF, such as the difference in temperature between the two tips—insulated vs. non-insulated—of the copper wire **Figure 4(b)**.

The 2D CSI map (**Figure 4(a)**) agreed the difference qualitatively, but contained a few artifacts. The negative temperature change pixels may be due to intra-voxel phase cancelation, or due to inhomogeneous broadening from thermal gradients. Further investigation is required to understand the underlying mechanism. One of the drawbacks of the CSI method is that the TR time is limited by the long FID read time (~20 ms) necessary for the spectral resolution to detect small changes in chemical shift. In the 2D CSI scans, water suppression pulse sequences, such as CHESS [25] and VAPOR [26], were not used in our experiment because of the relatively big frequency difference between water and methyl peak of Tm-DOTMA [20], *i.e.*, 100 ppm. Instead the water peak was filtered out by using a relatively narrow receive bandwidth. For the Tm-DOTMA PRF scan, the water signal was eliminated by using a frequency selective excitation of the methyl resonance in conjunction with phase encoding in two dimensions. Further temporal improvements could be potentially achieved using a spectral-spatial excitation with 2D imaging, which was not available on our scanner.

The experiment performed by heating the gel with the RF pulse used for imaging (**Figure 5**)—instead of the prescans—exploited one of the key strengths of Tm-DOTMA, namely the ability of using minimum TR and high flip angle due to the short T_1 of the methyl resonance of Tm-DOTMA (~5.3ms at 9.4 T [21]). The TR/TE used here is MRI scanner limited and can be shortened by a factor of two or three in modern clinical scanners. Shortening the TE further also has the potential to minimize susceptibility effects. The measured temperature changes from these near real time acquired images (red and green asterisk in **Figure 5(c)** and **Figure 5(d)**) matches well with that of the optic probes (blue rectangle in **Figure 5(c)** and **Figure 5(d)**).

In the experiments performed with the titanium bone screw (**Figure 6**), signal dropout was seen for both water proton PRF scan and Tm-DOTMA, however, the heating pattern was well visualized for 2D CSI of Tm-DOTMA but not for the water proton PRF.

In the context of testing for RF-induced heating due to the presence of metallic implants, the thermometry scan has to be fast compared to temperature changes due to heating. However, a unique challenge emerges for monitoring real time changes in temperature caused by the RF pulses of the imaging pulse sequence. If the heating is too rapid, or if the imaging scan is too slow, this can cause chemical shift related artifacts. Ideally, the heating from the RF pulses should be fast compared to heat dissipation due to conduction, and imaging scan need to be sensitive and fast enough to minimize artifacts. In addition, the estimates of local SAR depend on the initial slope of temperature increase. Typically, the slopes are under a few degrees/min [27]-[29]. Hence the need for a sub-minute time resolution for real time thermometry.

The source of the heating is the electric field as well as the physical properties (e.g., conductivity, relative permittivity, mass density, heat capacity) of the copper wire and the surrounding phantom [13]-[15]. The electric field is composed of conservative electric field (primarily originating with the scalar electric potential in the coil), and the magnetically-induced electric fields (caused by the time-varying magnetic field) [14] [27]. The two components may affect unevenly to the heating. However, it is difficult to separate these two components in our study because of a significant wavelength effect (200 MHz, 150 cm wavelength in air, ~19 cm wavelength in the phantom, RF coil diameter = 10 cm) [27].

Our approach using Tm-DOTMA presents some limitations. Tm-DOTMA is not approved for human use, but can be used in phantom or animal testing [5]-[10]. Tests in gel-phantom have been used to evaluate RF-induced heating of metallic implants in high field MRI. In this study, we focused on the possibility of medical device evaluation using a phantom, not a human volunteer.MR images acquired in this study with Tm-DOTMA showed substantially lower SNR compared to those of a water proton images because of the significant difference in concentration (*i.e.*, Molar level concentration for the Tm-DOTMA vs. Molar level for the water proton). However, this limitation can be partially overcome if a higher concentration Tm-DOTMA is used for phantom experiments.

5. Conclusion

We have implemented real-time and sensitive MR Thermometry using CSI and phase mapping of PRF shift with a Tm-DOTMA phantom to image small temperature changes caused by RF-induced heating near two types of implants, namely a copper wire and a titanium bone screw. We have compared the results with those using the conventional water PRF method to show that the Tm-DOTMA based approach is more sensitive and able to image changes that are less than 1˚C.

Acknowledgements

The authors thank Howard Bassen and Daniel M. Krainak for technical discussion. This work was supported by the FDA Center of Devices and Radiological Health Critical Path Program.

Disclaimer: The mention of commercial products, their sources, or their use in connection with material reported herein is not to be construed as either an actual or implied endorsement of such products by the US Government Department of Health and Human Services.

References

[1] US Department of Health and Human Services, Food and Drug Administration (2003) Criteria for Significant Risk Investigations of Magnetic Resonance Diagnostic Devices—Guidance for Industry and FDA Staff. FDA, Silver Spring.

[2] IEC (2010) International Standard, Medical Equipment Part 2-33: Particular Requirements for the Safety of Magnetic Resonance Equipment for Medical Diagnosis. 3rd Edition, International Electrotechnical Commission, Geneva, No. 60601-2-33.

[3] Mattei, E., Calcagnini, G., Censi, F., Triventi, M. and Bartolini, P. (2010) Numerical Model for Estimating RF-Induced Heating on a Pacemaker Implant During MRI: Experimental Validation. *IEEE Transactions on Biomedical Engineering*, **57**, 2045-2052.

http://dx.doi.org/10.1109/TBME.2010.2043102

[4] ASTM (2008) Standard Practice for Marking Medical Devices and Other Items for Safety in the Magnetic Resonance Environment. ASTM International, West Conshohocken, ASTM F2503-13. http://www.astm.org/Standards/F2503.htm

[5] Neufeld, E., Kuhn, S., Szekelyand, G. and Kuster, N. (2009) Measurement, Simulation and Uncertainty Assessment of Implant Heating during MRI. *Physics in Medicine and Biology*, **54**, 4151-4169. http://dx.doi.org/10.1088/0031-9155/54/13/012

[6] Liu, Y., Chen, J., Shellock, F.G. and Kainz, W. (2013) Computational and Experimental Studies of an Orthopedic Implant: MRI-Related Heating at 1.5-T/64-MHz and 3-T/128-MHz. *Journal of Magnetic Resonance Imaging*, **37**, 491-497. http://dx.doi.org/10.1002/jmri.23764

[7] Detti, V., Grenier, D., Perrin, E. and Beuf, O. (2011) Assessment of Radiofrequency Self-Heating Around a Metallic Wire with MR T1-Based Thermometry. *Magnetic Resonance in Medicine*, **66**, 448-455. http://dx.doi.org/10.1002/mrm.22834

[8] Gensler, D., Fidler, F., Ehses, P., Warmuth, M., Reiter, T., Düring, M., Ritter, O., Ladd, M.E., Quick, H.H., Jakob, P.M., Bauer, W.R. and Nordbeck, P. (2012) MR Safety: Fast T_1 Thermometry of the RF-Induced Heating of Medical Devices. *Magnetic Resonance in Medicine*, **68**, 1593-1599. http://dx.doi.org/10.1002/mrm.24171

[9] Shrivastava, D., Hanson, T., Kulesa, J., Tian, J.F., Adriany, G. and Vaughan, J.T. (2011) Radiofrequency Heating in Porcine Models with a ''Large'' 32 cm Internal Diameter, 7 T (296 MHz) Head Coil. *Magnetic Resonance in Medicine*, **66**, 255-263. http://dx.doi.org/10.1002/mrm.22790

[10] Hofstetter, L.W., Yeo, D.T., Dixon, W.T., Kempf, J.G., Davis, C.E. and Foo, T.K. (2012) Fat-Referenced MR Thermometry in the Breast and Prostate Using IDEAL. *Journal of Magnetic Resonance Imaging*, **36**, 722-732. http://dx.doi.org/10.1002/jmri.23692

[11] John, C.C. and Guy, A.W. (1974) Nonionizing Electromagnetic Wave Effects in Biological Materials and Systems, *Proceedings of the IEEE*, **60**, 692-718. http://dx.doi.org/10.1109/PROC.1972.8728

[12] Mattei, E., Triventi, M., Calcagnini, G., Censi, F., Kainz, W., Bassenand, H.I. and Bartolini, P. (2007) Temperature and SAR Measurement Errors in the Evaluation of Metallic Linear Structures Heating during MRI Using Fluoroptic® Probes. *Physics in Medicine and Biology*, **52**, 1633-1646. http://dx.doi.org/10.1088/0031-9155/52/6/006

[13] Shapiro, E.M., Borthakur, A., Shapiro, M.J. Reddy, R. and Leigh, J.S. (2002) Fast MRI of RF Heating via Phase Difference Mapping. *Magnetic Resonance in Medicine*, **47**, 492-498. http://dx.doi.org/10.1002/mrm.10067

[14] Park, B.S., Neuberger, T., Webb, A.G., Bigler, D.C. and Collins, C.M. (2010) Faraday Shields within a Solenoidal Coil to Reduce Sample Heating: Numerical Comparison of Designs and Experimental Verification. *Journal of Magnetic Resonance*, **202**, 72-77. http://dx.doi.org/10.1016/j.jmr.2009.09.023

[15] Oh, S., Webb, A.G., Neuberger, T., Park, B.S. and Collins, C.M. (2010) Experimental and Numerical Assessment of MRI-Induced Temperature Change and SAR Distributions in Phantoms and *in Vivo*. *Magnetic Resonance in Medicine*, **63**, 218-223.

[16] James, J.R., Gao, Y., Miller, M.A., Babsky, A. and Bansal, N. (2009) Absolute Temperature MR Imaging with Thulium 1,4,7,10-Tetraazacyclododecane-1,4,7,10-Tetramethyl-1,4,7,10-Tetraacetic Acid ($TmDOTMA^-$). *Magnetic Resonance in Medicine*, **62**, 550-556. http://dx.doi.org/10.1002/mrm.22039

[17] Coman, D., Trubela, H.K., Rycyna, R.E. and Hyder, F. (2009) Brain Temperature and pH Measured by ^1H Chemical Shift Imaging of a Thulium Agent. *NMR in Biomedicine*, **22**, 229-239. http://dx.doi.org/10.1002/nbm.1312

[18] Aime, S., Botta, M., Fasano, M., Terreno, E., Kinchesh, P., Calabi, L. and Paleari, L. (1996) A New Ytterbium Chelate as Contrast Agent in Chemical Shift Imaging and Temperature Sensitive Probe for MR Spectroscopy. *Magnetic Resonance in Medicine*, **35**, 648-651. http://dx.doi.org/10.1002/mrm.1910350504

[19] Zuo, C.S., Bowers, J.L., Metz, K.R., Nosaka, T., Sherry, A.D. and Clouse, M.E. (1996) $TmDOTP^{5-}$: A Substance for NMR Temperature Measurements *in Vivo*. *Magnetic Resonance in Medicine*, **36**, 955-959. http://dx.doi.org/10.1002/mrm.1910360619

[20] Coman, D., Trubela, H.K. and Hyder, F. (2010) Brain Temperature by Biosensor Imaging of Redundant Deviation in Shifts (BIRDS): Comparison between $TmDOTP^{5-}$ and $TmDOTMA^-$. *NMR in Biomedicine*, **23**, 277-285.

[21] Hekmatyar, S.K., Paige, H., Sait, K.P., Andriy, B. and Navin, B. (2005) Noninvasive MR Thermometry Using Paramagnetic Lanthanide Complexes of 1,4,7,10-Tetraazacyclodoecane-α, α', α'', α'''-tetramethyl-1,4,7,10-tetraacetic acid ($DOTMA^{4-}$). *Magnetic Resonance in Medicine*, **53**, 294-303. http://dx.doi.org/10.1002/mrm.20345

[22] Graedel, N.N., Polimeni, J.R., Guerin, B., Gagoski, B. and Wald. L.L. (2015) An Anatomically Realistic Temperature Phantom for Radiofrequency Heating Measurements. *Magnetic Resonance in Medicine*, **73**, 442-450. http://dx.doi.org/10.1002/mrm.25123

[23] Jenkinson, M. (2003) Fast, Automated, *N*-Dimensional Phase-Unwrapping Algorithm. *Magnetic Resonance in Medicine*, **49**, 193-197. http://dx.doi.org/10.1002/mrm.10354

[24] Rieke, V. and Pauly, K.B. (2008) MR Thermometry. *Journal of Magnetic Resonance Imaging*, **27**, 376-390. http://dx.doi.org/10.1002/jmri.21265

[25] Haase, A., Frahm, J., Hanicke, W. and Matthaei, D. (1985) ^{1}H NMR Chemical Shift Selective (CHESS) Imaging. *Physics in Medicine and Biology*, **30**, 341-344. http://dx.doi.org/10.1088/0031-9155/30/4/008

[26] Tkáč, I., Starčuk, Z., Choi, I.Y. and Gruetter, R. (1999) *In Vivo* ^{1}H NMR Spectroscopy of Rat Brain at 1 ms Echo Time. *Magnetic Resonance in Medicine*, **41**, 649-656. http://dx.doi.org/10.1002/(SICI)1522-2594(199904)41:4<649::AID-MRM2>3.0.CO;2-G

[27] Park, B.S., Webb, A.G. and Collins, C.M. (2009) A Method to Separate Conservative and Magnetically-Induced Electric Fields in Calculations for MR Microscopy. *Journal of Magnetic Resonance*, **199**, 233-237. http://dx.doi.org/10.1016/j.jmr.2009.05.007

[28] Institute for Magnetic Resonance Safety, Education, and Research (IMRSER). http://www.imrser.org/

[29] Muranaka, H., Horiguchi, T., Ueda, Y. and Tanki, N. (2011) Evaluation of RF Heating due to Various Implants during MR Procedures. *Magnetic Resonance in Medical Sciences*, **10**, 11-19. http://dx.doi.org/10.2463/mrms.10.11

12

EMF of Hot Charge Carriers Arising at the p-n-Junction under the Influence of the Microwave Field and Light

Gafur Gulyamov, Muhammadjon Gulomkodirovich Dadamirzaev*, Nosir Yusupjanovich Sharibayev, Ne'matjon Zokirov

Namangan Engineering Pedagogical Institute, Namangan, Uzbekistan

Email: *dadamirzaev70@umail.uz

Abstract

It is shown that the increase in the current of an asymmetric p-n-junction, caused by perturbation of potential barrier height and increasing recombination current in a strong microwave field, is suppressed by light generated photo carriers, leading to the displacement of current-voltage characteristics of p-n-junction into the direction of smaller current values.

Keywords

Hot Electrons, The Microwave Field, p-n-Junction, Light, Photocurrent, Generation and Recombination Currents

1. Introduction

It is known that when in a non-uniform semiconductor temperature gradient occurs between the edges of the semiconductor thermo EMF will arise. Inhomogeneous concentration of charge carriers in semiconductor can be created artificially, for example by means of lighting. If to heat up one edge of the illuminated region and another left at former temperature, then at the end, the semiconductor EMF arises [1].

Tautz [2] for the first time has considered this effect, and called it thermal photoelectric effect. Tautz's experimental measurements have shown that under normal temperature gradient this effect is small, although it can be observed.

However, if to heat up only the charge carriers, by such as a microwave field, it is possible to reach higher temperatures, and hence high values of voltage. This effect was observed by Repshas and Ashmontas [3], but

*Corresponding author.

they got very small voltage values, which they called photo-gradient EMF of hot carriers. This name was used in some cases to refer the bulk photo-emf caused by nonuniform distribution of impurities in semiconductors [3].

A. I. Veynger *et al.* [4] [5] theoretically and experimentally examined the current and the EMF of hot carriers in germanium n- and p-type at temperatures of lattice 300 K and 77 K.

It is shown that in weak fields the effect is proportional to the square of the field warming up, and strongly heated field. It was experimentally observed that in strong fields there were saturation and even a decrease of this effect. Furthermore, it was found that the effect was very sensitive to the form of the momentum relaxation time depending on the energy of the carriers and the potential on the surface. Also the variations of the photo-EMF were investigated at the germanium p-n-junction during heating of the carriers. It is shown that the photo-EMF increases with the temperature of the minority carriers. Measurement of photo-EMF on hot carriers is a simple and accurate method of determining the temperature of the hot minority carriers.

The experimental scheme is shown in **Figure 1** [4]. The sample was placed in a waveguide and illuminated through a slit in the broad wall of the waveguide. Those portions of the sample that should not have been exposed to light (in particular, contacts), were covered with black paper. This figure also schematically illustrates how the region located along the action of a strong microwave field, and light. As the sample contacts were out of the action of light and the microwave field, the EMF occurring in the semiconductor is determined only by the change in temperature and carrier concentration in the sample. Also changes in photoelectric characteristics of the silicon p-n-junctions have been investigated in a strong microwave field [1]. It is shown that the barrier height of p-n-junction in a strong microwave field and reducing the barrier during the lighting of p-n-junction are proportional to the height of the initial barrier, if the latter is reduced by direct displacement. Near zero bias and at reverse bias, this proportionality will be disturbed due to the influence of the reverse current through the p-n-junction.

A mechanism is proposed to increase the barrier of p-n-junction based on the account of generation-recombination processes in space-charge layer. During the experiment, CVC of unilluminated and illuminated p-n-junction were pre-measured without the influence of the microwave field. For one of the samples they are shown in **Figure 2** (curves 1 and 2, respectively). Under the influence of the microwave field in the sample an additional impulse EMF arises and the additional pulse current passes through the load. Therefore, the CVC of silicon p-n-junction under the influence of the microwave field were built without illumination (**Figure 2**, curve 3) and under illumination (**Figure 2**, curve 4) [1].

However, in the above-mentioned works current and EMF of asymmetric p-n-junction are not theoretically investigated at simultaneous impact of the microwave field and light.

The aim of this work is a theoretical study of currents and EMF of hot carriers in the asymmetric p-n-junction.

2. Theoretical Study of Currents and EMF of Hot Charge Carrier's Asymmetric p-n-Junction under the Influence of the Microwave Field and Light

The average value of the total current passing through the diode consists of the electron and hole currents and defined by the following equation [6] [7]:

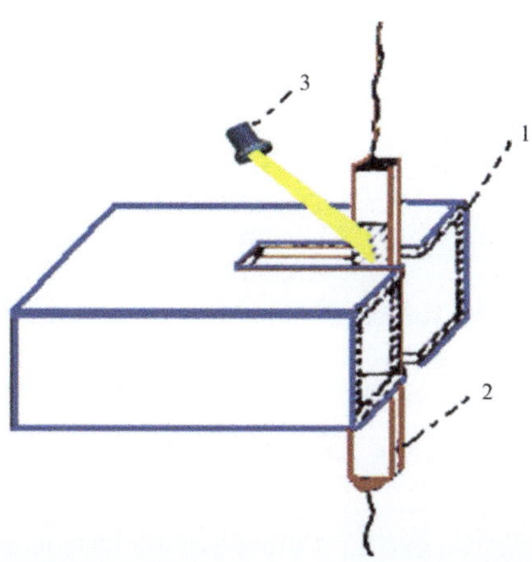

Figure 1. Illuminating scheme of the sample in the waveguide: 1—waveguide, 2—sample, 3—light [4].

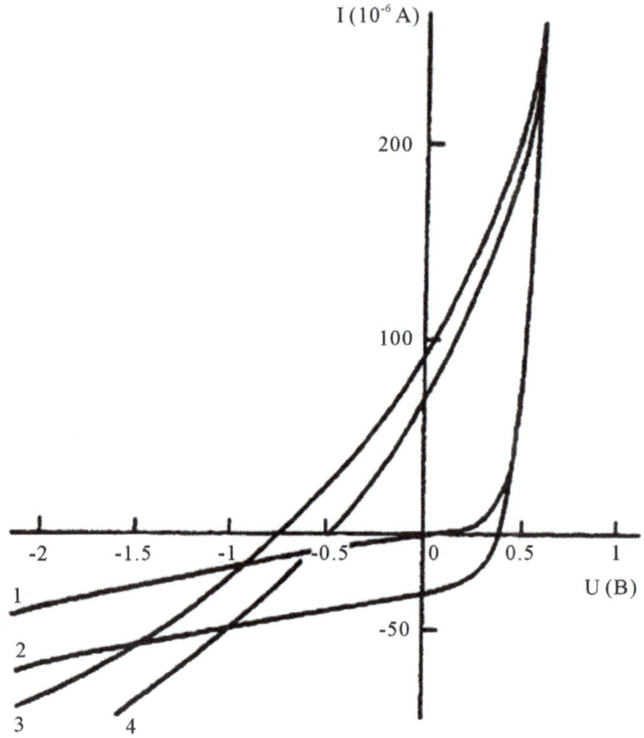

Figure 2. CVC p-n-junction under various conditions: 1—without lighting and the microwave field; 2—illuminated without the microwave field; 3—without light in the microwave exposure; 4—illuminated and microwave exposure [1].

$$\bar{I} = I_{se}\left\{\left(\frac{T_e}{T}\right)^{\frac{1}{2}} \exp\left(\frac{e\varphi_0}{kT} - \frac{e(\varphi_0 - U)}{kT_e}\right)\int_0^{2\pi} \exp\left(-\frac{eU_B\cos(\omega t)}{kT_e}\right)\frac{d(\omega t)}{2\pi} - 1\right\}$$
$$+ I_{sh}\left\{\left(\frac{T_h}{T}\right)^{\frac{1}{2}} \exp\left(\frac{e\varphi_0}{kT} - \frac{e(\varphi_0 - U)}{kT_h}\right)\int_0^{2\pi} \exp\left(-\frac{eU_B\cos(\omega t)}{kT_h}\right)\frac{d(\omega t)}{2\pi} - 1\right\}. \quad (1)$$

where, $j_{se} = \frac{eD_e n_p}{L_e}$; $j_{sh} = \frac{eD_h p_n}{L_h}$ —saturation currents for electrons and holes; φ_0—the height of the potential barrier in the absence of an electromagnetic wave; $\varphi = \varphi_0 - U$; U—a voltage across the diode; $U_B = -\int_0^d E_B dx$ — AC voltage the incident wave created by the barrier diode; T is the temperature of the lattice; k—Boltzmann constant; T_e, and T_h—temperature electrons and holes; E_b—electric field of the wave; e—is the charge of an electron; D_e and D_h—diffusion coefficients of electrons and holes, L_e and L_h—their diffusion length; n_p and p_n—the concentration of minority carriers.

If you do not take into account in the formula (1) heating the charge carriers and the indignation of the potential barrier height ($T_e = T_h = T$; $U_B = 0$), we have

$$I = I_s\left[\exp\left(\frac{eU}{kT}\right) - 1\right] \quad (2)$$

This formula is the same as the classic current-voltage characteristic at the p-n-junction as shown in [8].

After creating the first copper-oxide rectifiers it has been found that when covering the surface of the cuprous light of the appropriate wavelength in the rectifier circuit current starts to flow in the absence of external EMF

On this basis we created special devices—photocells barrier layer for converting light into electrical signals.

Copper-ferrous solar cells are not used for energy purposes, as their efficiency is less than a percent. Just after the manufacture of photovoltaic cells for use steel germanium and particularly silicon, it became possible to use solar cells as efficient power converters of sunlight into electrical energy [8].

As you know, the photocurrent arises only when illuminated by such light, the photon energy is sufficient for the formation of electron-hole pairs, this exceeds the band gap. Light of this wavelength is strongly absorbed, so the pairs are formed near the boundary.

According to (34.16) in [8], the total current of the photodiode can be written as (where $I_c \neq 0$; $T_e = T_h = T$; $U_B = 0$)

$$I = I_s\left[\exp\left(\frac{eU}{kT}\right)-1\right]-eI_0\left(1-\beta_0\right), \tag{3}$$

where, $I_c = eI_0\left(1-\beta_0\right)$, $\beta_0 = \beta_s + \beta_v = \dfrac{\dfrac{d}{L_s}+\dfrac{1}{2}\left(\dfrac{d}{L_n}\right)^2}{1+\dfrac{d}{L_s}}$—the total loss on the bulk and surface recombination,

$\beta_c = \dfrac{\dfrac{d}{L_s}}{1+\dfrac{d}{L_s}}$—relative to the surface recombination losses, $\beta_v = \dfrac{1}{2}\dfrac{\left(\dfrac{d}{L_n}\right)^2}{1+\dfrac{d}{L_s}}$—the relative loss of volume recombination, $L_s = \dfrac{D_n}{s}$, d—the thickness of the illuminated region.

The first term—the so-called dark current, second term—a light current. Equation (3) expresses the additivity rule, according to which the total current is the sum of the photodiode dark current is determined by the applied voltage and is independent of lighting and light current, equal to the number of pairs generated net loss of recombination that does not depend on the applied voltage. In the formula (3) formula expresses generally additive, according to which the total current of the photodiode is equal to the sum of the dark current is determined by the applied voltage and is independent of lighting and light current, equal to the number of generated pairs of net losses on recombination, which is independent of the applied voltage.

According to (1), at low powers the microwave heating can be ignored electrons and holes (when $I_c = 0$; $T_e = T_h = T$; $U_B \neq 0$), then the current through the diode occurs only as a result, the height of the potential barrier, and it is defined by the formula:

$$\bar{I}\left(U_B\right) = I_s\left[\exp\left(\frac{eU}{kT}\right)\int_0^{2\pi}\left(\exp\left(-\frac{eU_B\cos(\omega t)}{kT}\right)\right)\frac{\mathrm{d}(\omega t)}{2\pi}-1\right], \tag{4}$$

where $I_s = e\left(\dfrac{D_h p_n}{L_h}+\dfrac{D_e n_p}{L_e}\right)$.

Using (4) define Thermo photovoltaic current-voltage characteristic at low microwave powers of the wave (when $I_c \neq 0$; $T_e = T_h = T$; $U_B \neq 0$):

$$\bar{I}\left(U_B, I_c\right) = I_s\left[\exp\left(\frac{eU}{kT}\right)\int_0^{2\pi}\left(\exp\left(-\frac{eU_B\cos(\omega t)}{kT}\right)\right)\frac{\mathrm{d}(\omega t)}{2\pi}-1\right]-I_c \tag{5}$$

This shows that you are influenced outrage potential barrier height and light. High power microwave wave (when $I_c = 0$; $T_e \neq T_h > T$; $U_B \neq 0$) CVC diode is determined by (1).

Honey CVC p-n-junctions at high power microwave wave (when $I_c \neq 0$; $T_e \neq T_h > T$; $U_B \neq 0$) is determined using (1) as follows:

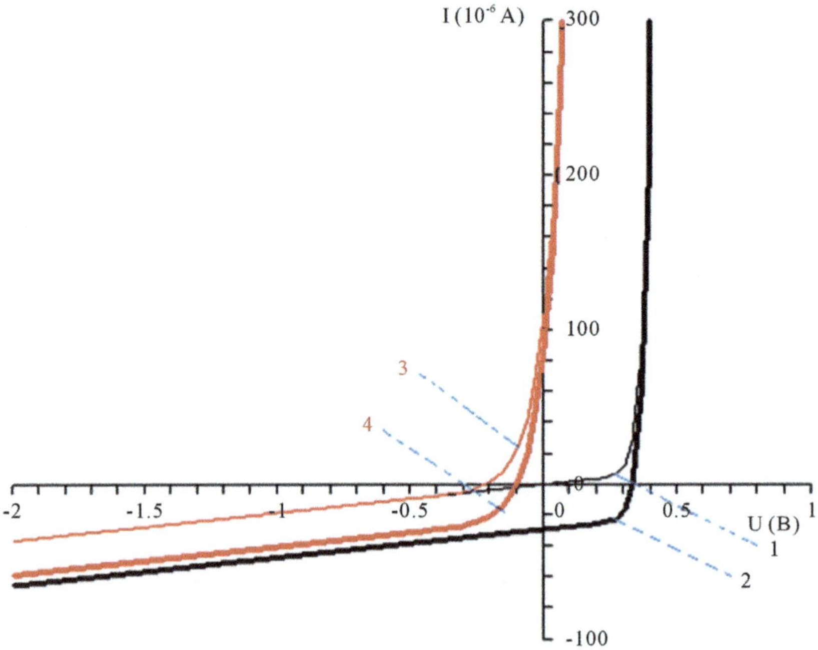

Figure 3. CVC p-n-junction in the various capacities of the microwave and lighting fittings: 1—without lights and without the microwave field, 2—illuminated and without the microwave field; 3—no lighting and high power microwave energy, 4—at the light and at high power microwave energy.

$$\begin{aligned}\bar{I} = I_{se}&\left\{\left(\frac{T_e}{T}\right)^{\frac{1}{2}}\exp\left(\frac{e\varphi_0}{kT}-\frac{e(\varphi_0-U)}{kT_e}\right)\int_0^{2\pi}\exp\left(-\frac{eU_B\cos(\omega t)}{kT_e}\right)\frac{\mathrm{d}(\omega t)}{2\pi}-1\right\}\\ &+I_{sh}\left\{\left(\frac{T_h}{T}\right)^{\frac{1}{2}}\exp\left(\frac{e\varphi_0}{kT}-\frac{e(\varphi_0-U)}{kT_h}\right)\int_0^{2\pi}\exp\left(-\frac{eU_B\cos(\omega t)}{kT_h}\right)\frac{\mathrm{d}(\omega t)}{2\pi}-1\right\}-I_c\end{aligned} \quad (6)$$

From these expressions can build CVC p-n-junctions at different conditions. **Figure 3** given CVC p-n-junction with the simultaneous action of the microwave and light.

Theoretical graphs (**Figure 3**) and experimental results (**Figure 2**) can be compared qualitatively interpretation. Since the concentration distribution of electrons and holes in the p-n-junction [4] is not known, a quantitative comparison of these graphs do not give reliable information about the passage of current through the diode.

3. Conclusion

The theory is developed for the photo-EMF of hot carriers in the p-n-junction in a strong microwave field. An analytical expression is obtained for CVC of illuminated p-n-junction in a strong microwave field. The strong increase of currents of p-n-junction conditioned by the fact that the heating perturbs the height of the potential barrier and increases the recombination currents in the p-n-junction. While lighting increases the generation current in the diode. Since generation and recombination currents are directed oppositely, the illumination of the sample moves the current-voltage characteristics of p-n-junction in the direction of smaller current values. Lighting reduces the overall current flowing through the diode in a strong microwave field. The theoretical results are in qualitative agreement with the experimental results of A. I. Veynger [1].

References

[1] Ablyazimova, N.A., Veynger, A.I. and Food, V.S. (1992) The Impact of a Strong Microwave Field in the Photovoltaic Properties of Silicon p-n-Junctions. *Physics and Engineering of St. Petersburg Semiconductors*, **26**, 1041-1047.

[2] Tautz, J. (1962) Photo and Thermoelectric Phenomena in Semiconductors. Moscow, IL, 192s.

[3] Ashmontas, S.P. (1984) Monograph Series. Electrons in Semiconductors. No. 5. Elektrogradientnye Phenomena in Semiconductors. In: Pozhela, Yu., Ed., Mokslas, Vilnius, 183.

[4] Veynger, A.I., Kramer, N.I., Paritssky, L.G. and Abdinov, A.S. (1972) Thermophotovoltaic Effect on Heating Carriers in Germany. *Semiconductors*, **6**, 353-359.

[5] Akopyan, E.A., Veynger, A.I. and Paritssky, L.G. (1974) Photovoltage Hot Carriers at the pn-Junction. *Semiconductors*, **8**, 754-759.

[6] Dadamirzaev, M.G. (2011) Heating of Charge Carriers and Rectification of Current in Asymmetrical p-n Junction in a Microwave Field. *Semiconductors*, **45**, 288-291. http://dx.doi.org/10.1134/S1063782611030092

[7] Dadamirzaev, M.G. (2015) Influence of Deformation on CVC p-n-Junction in a Strong Microwave Field. *Journal of Modern Physics-USA*, **6**, 176-180. http://dx.doi.org/10.4236/jmp.2015.62023

[8] Pikus, G.E. (1965) Fundamentals of the Theory of Semiconductor Devices. Science, Moscow, 448.

13

A New Approach Based on MoM-GEC Method to Mutual Coupling Analysis of Symmetric Twin Waveguides for Antenna Applications

Abdessalem Kaddouri, Mourad Aidi, Taoufik Aguili

Communication System Laboratory Sys'Com, National Engineering School of Tunis, University Tunis El Manar, Tunis, Tunisia

Email: kaddouri.abdessalam@sotetel.tn

Abstract

In previous modeling works, the waveguide radiation in the free space is modeled using an infinite flange in aperture plan. In this paper, we propose a new formulation to analyze the radiation of twin rectangular waveguides in free space. Our formulation consists firstly in simulating the free space as rectangular waveguide and seeking the appropriate dimensions that do not affect scattering parameters. In the second step, we use the symmetry principle to reduce the coupling problem to single guide radiating in the free space. Moreover, for more simplification, we consider a concentric discontinuity to solve this latter. This approach is based on moments method combined to the generalized equivalent circuit method (MoM-GEC) in order to reduce the number of unknown problems and alleviate their computation. Obtained numerical results are presented and discussed. A good agreement with literature is shown and the boundary conditions are verified.

Keywords

Waveguide, Coupling, MoM-GEC, Propagation, Scattering, Electric Field

1. Introduction

Open-ended rectangular waveguide is the elementary component for the millimeter wave's propagation. Several kinds of design have been proposed (single, coupled, networked...) to be used to achieve several applications such as satellite communication, radar, medical systems, and geophysical applications [1]-[3]. The variational method is the first approach used to solve the electromagnetic problem concerning these areas [4] [5]. The boundary conditions require a relationship between the tangential electromagnetic fields, which leads to an integral equation [6]. This integral equation is solved using the moment method to investigate the single or coupled waveguides behaviors.

In [7], the aperture field is modeled using the LSE modes and is based on point-matching approach. This formulation is extended in [8] by including the cross-polarized component and a dominant mode approximation is employed. Another approach consists in formulating the waveguide coupling with an iris [9] by employing orthogonal functions associated to the LSE modes. In fact, a general asymptotic formula to study the coupling between modes in separate waveguides has been applied to investigate the response of a uniform [10], and non-uniform rectangular waveguide array [11].

Some other authors have used the generalized network formulation, which is based essentially on the equivalence principle and MoM method. This formulation was applied to finite phased arrays [12] and reactively loaded waveguide arrays [13].

There are many other methods which have been applied to solve the flanged waveguide problem. The correlation matrix method which is based on the energy conservation low has been applied in [14], while Teodoridis *et al.* [15] have used the characteristic modes and the problem is solved by investigating the eigenvalues. In [16], Boudrant *et al.* have obtained an integral equation by using the operator transverse method. This integral equation was solved using the Galerkin's method.

In [17], a rigorous formulation based on integral equation is proposed and solved by MoM method in order to analyse the radiation of rectangular waveguides array. Another approach consists to consider this last as a flanged parallel-plate waveguides array [18]. Serizawa *et al.* [19] used an analytical method based on Kobayashi potential method (KP) and derived the exact expressions of the fields to obtain reference data of physical quantities.

Mongiado *et al.* [20] proposed a singular integral equation approach including the edge condition on the apertures from the start, in order to analyze single and twin radiating waveguide

In this paper, a new rigorous approach is proposed to model two opened waveguides radiating in free space. The concept consists, firstly, in using the symmetry principle to reduce structure to half, and simulate properly the free space by another rectangular waveguide. Secondly, it consists in studying electromagnetic coupling between these waveguides. For validity purpose, obtained results are compared to those in previous works, and good agreement is shown.

2. MoM-GEC Formalism

Basically, in electromagnetic study, problems are well described by Maxwell's equations. But in order to alleviate their complexity, we always need to choose the appropriate method that can approximate environment and boundary conditions to compute it within a reasonable time [11] [21] [22]. In our case, we combine two methods. The first method is the generalized equivalent circuits (GEC) [22]. It can present integral equations by equivalent circuit that describes faithfully all the problem. Electromagnetic phenomenon in discontinuity, which is the unknown problem, is presented by a virtual source. The environments are brought back to discontinuity plan as admittance or impedance operator. Excitation is illustrated as a localized or a modal source, in discontinuity too [23]. The second is the method of moment (MoM) [21]. It associates the known test functions to a virtual source, injects a modal basis in environment operator, and projects test function on a modal basis to resolve the problem. The good choices of test function lead to accelerate the convergence, and reduce greatly the computation time [24].

3. Formulation

In this paper, we propose a modular formulation, in other terms, we simplify the problem step by step.

As shown in **Figure 1**, the study structure is composed by two opened rectangular waveguides, called network guides (NG), radiate in free space. This latter is considered as a third waveguide, called space guide (SG); with large dimensions.

In the second time, we use the symmetry principle to reduce the number of NGs to one. Then, odd and even input impedance are determined respectively when we apply structures presented in **Figure 2(a)** and **Figure 2(b)**. Thus, the input impedance of initial structure is easily deduced as it will be shown later.

Finally, we propose to study the SG influence, and we will make sure that considered dimensions do not affect the results.

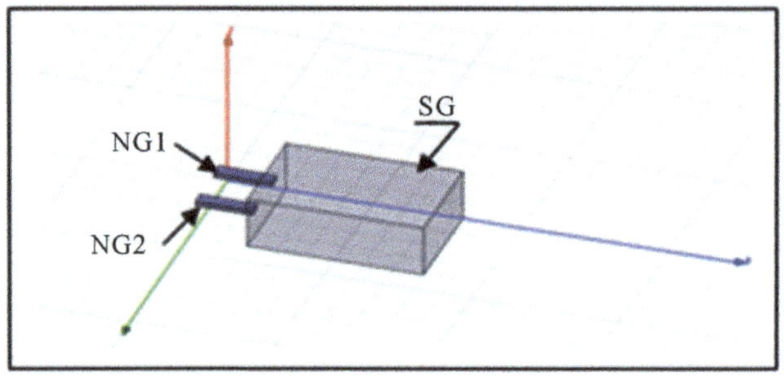

Figure 1. Studied structure.

Figure 2. Symmetry principle.

So, now the problem is reduced to study NG1 radiates in SG. For this, we present the new considered structure in **Figure 3(a)**; and its relative equivalent circuit model in **Figure 3(b)**.

The excitation is brought back to discontinuity plan as a modal source of current, its value is the current density of the fundamental mode, and its internal admittance is $\hat{Y}_1$. This latter represents the evanescent modes contribution of NG1; and its formal writing is:

$$\hat{Y}_1 = \sum \left| f_{pq} \right\rangle y_{pq} \left\langle f_{pq} \right| \tag{1}$$

where f_{pq} is the NG1 modal basis without the fundamental f_0, and y_{pq} is the mode admittance of each f_{pq}. The voltage at terminals of this source is e_1, who is its dual greatness.

E_e is the virtual voltage source defined in discontinuity plan, and j_1 is the current flowing it. E_e is the problem unknown, and j_1 is expressed as a serial of test functions g_p weighted by unknown modal amplitudes x_p:

$$E_e = \sum_p x_p g_p. \tag{2}$$

The modes contribution of SG is expressed in the discontinuity plane by admittance operator $\hat{Z}^{-1}$.

$$\hat{Z}^{-1} = \sum \left| F_{mn} \right\rangle y_{mn} \left\langle F_{mn} \right| \tag{3}$$

where F_{mn} is the SG modal basis and y_{mn} is the relative mode admittance.

Based on the equivalent circuit shown in **Figure 3(b)**, the integral equations associated with the problem can be easily derived by applying Kirchhoff laws generalized:

$$\begin{cases} J = \hat{Z}^{-1} E_e \\ J = -\hat{Y}_1 E_e + I_0 f_0 + j_1. \end{cases} \tag{4}$$

Using the equations system (4), we can write:

$$\hat{Z}^{-1} E_e = -\hat{Y}_1 E_e + I_0 f_0 + j_1. \tag{5}$$

Then, we can deduce the current j_1:

$$j_1 = \left(\hat{Y}_1 + \hat{Z}^{-1} \right) E_e - I_0 f_0. \tag{6}$$

Taking into account the expression $e_1 = E_e$, we can find the relation between source variables and their duals:

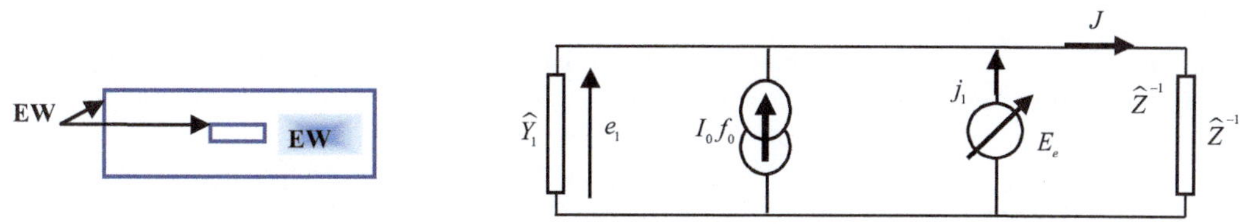

(a) (b)

Figure 3. Simplified structure and the relative equivalent circuit.

$$\begin{pmatrix} e_1 \\ j_1 \end{pmatrix} = \begin{pmatrix} 1 & 0 \\ \hat{Z}^{-1} + \hat{Y}_1 & -1 \end{pmatrix} \begin{pmatrix} E_e \\ I_0 f_0 \end{pmatrix}. \tag{7}$$

Applying the Galerkin's method:

$$\begin{cases} V_0 = \langle f_0 | E_e \rangle = \sum_q x_q \langle f_0 | g_q \rangle = A^t X \\ -I_0 \langle g_p | f_0 \rangle + \sum_q \left\langle g_p \middle| \left(\hat{Y}_1 + \hat{Z}^{-1} \right) g_q \right\rangle x_q = 0. \end{cases} \tag{8}$$

This equations system can be rewritten as:

$$\begin{cases} A^t X = V_0 \\ -A I_0 + [y] X = 0 \end{cases} \tag{9}$$

where A is the excitation vector, and $[y]$ is the admittance matrix:

$$A = \langle f_0 | g_p \rangle \tag{10}$$

$$[y] = \begin{bmatrix} \left\langle g_{10} \middle| \left(\hat{Y}_1 + \hat{Z}^{-1} \right) g_{10} \right\rangle & \cdots & \left\langle g_{10} \middle| \left(\hat{Y}_1 + \hat{Z}^{-1} \right) g_{pq} \right\rangle \\ \vdots & \ddots & \vdots \\ \left\langle g_{pq} \middle| \left(\hat{Y}_1 + \hat{Z}^{-1} \right) g_{10} \right\rangle & \cdots & \left\langle g_{pq} \middle| \left(\hat{Y}_1 + \hat{Z}^{-1} \right) g_{pq} \right\rangle \end{bmatrix} \tag{11}$$

By separating the admittance operators, we can rewrite $[y]$ matrix as:

$$[y] = \begin{bmatrix} 0 & 0 & \cdots & 0 \\ 0 & y_{22} & 0 & \vdots \\ \vdots & 0 & \ddots & 0 \\ 0 & \cdots & 0 & y_{pq} \end{bmatrix} \begin{bmatrix} \left\langle g_{10} \middle| \hat{Z}^{-1} g_{10} \right\rangle & \cdots & \left\langle g_{10} \middle| \hat{Z}^{-1} g_{pq} \right\rangle \\ \vdots & \ddots & \vdots \\ \left\langle g_{pq} \middle| \hat{Z}^{-1} g_{10} \right\rangle & \cdots & \left\langle g_{pq} \middle| \hat{Z}^{-1} g_{pq} \right\rangle \end{bmatrix} \tag{12}$$

where the first term presents the projection of test functions on the evanescent modes of the same guide, and the second term their projection on the load operator.

Finally, we deduce the unknown problem:

$$X = [y]^{-1} A I_0. \tag{13}$$

Taking into account relations (8) and (12), we deduce the input impedance Z:

$$Z = A^t [y]^{-1} A. \tag{14}$$

And scattering matrix is given by:

$$S = \left(\frac{Z}{Z_0} - 1 \right) \left(\frac{Z}{Z_0} + 1 \right)^{-1} \tag{15}$$

where Z_0 is impedance of free space.

To pass to total structure, we must replace $\boldsymbol{Z}$, in Equation (13), by Z_{in}^{even} or Z_{in}^{odd} depending on boundary conditions described in **Figure 2**.

So, the total input impedance matrix $\boldsymbol{Z}^{\mathrm{T}}$ is given by:

$$\boldsymbol{Z}^{\mathrm{T}} = \frac{1}{2}\begin{pmatrix} Z_{in}^{even} + Z_{in}^{odd} & Z_{in}^{even} - Z_{in}^{odd} \\ Z_{in}^{even} - Z_{in}^{odd} & Z_{in}^{even} + Z_{in}^{odd} \end{pmatrix} \tag{16}$$

We note that z is the reduced matrix, and $\boldsymbol{I}_2$ is the identity matrix.

$$z = \boldsymbol{Z}^{\mathrm{T}} / Z_0 . \tag{17}$$

The total scattering matrix $\boldsymbol{S}^{\mathrm{T}}$ is expressed as:

$$\boldsymbol{S}^{\mathrm{T}} = \left(z - \boldsymbol{I}_2\right)\left(z + \boldsymbol{I}_2\right)^{-1} . \tag{18}$$

Here, the formulation gives us $\boldsymbol{S}^{\mathrm{T}}$ parameters between NG1 and NG2; and SG has no more influence.

4. Numerical Results

In this section, we propose first a quantitative discussion about the radiation of a real opened waveguide (WR90) in the free space, in order to determine the appropriate model that simulates the free space. Second, we study the electromagnetic coupling between NG1 and NG2.

To set certain parameters of the problem, a convergence study is strongly recurred. **Figure 4** presents the $\boldsymbol{S}11$ norm as a function of a mode number for different test function number.

As shown in **Figure 4**, the convergence is obtained for 15 test functions and 1600 mode functions. To ensure the convergence, in the following, we always use these last values.

First main target of this study is to determine the appropriate SG dimensions to simulate the free space. For this, we present in **Figure 5** the S parameters as function of $r = D/d_1$. Where D and d_1 are respectively dimensions of SG and NG1.

We can consider that from $D = 7d_1 - 4.6\lambda$ there is no influence of SG on results. For a validation propose, we present in **Table 1**, the obtained reflection coefficient S11 and that obtained in previous works for operating frequency $f = 9.33$ GHz. It is found that a good agreement is shown.

The second main target of this study is to determine coupling effects between two NGs. For this we explore $\left|S_{21}^{T}\right|$ parameter, and validate it with previous work [7] (**Figure 6**). We notice that a good agreement is shown around the operating frequency $f = 9.33$ GHz.

We present in **Figure 7**, the S_{21}^{T} parameter as function of the coupling distance d. And we can notice that for $d = 2\lambda$, the coupling coefficient reaches a value of $\left|S_{21}^{T}\right| = 0.157$. This coupling value corresponds to 2% of the total energy. So, from this distance, we can consider the two waveguides as isolated.

At $d = 2\lambda$, the scattering matrix $\boldsymbol{S}^{\mathrm{T}}$ worth:

$$\boldsymbol{S}^{\mathrm{T}} = \begin{pmatrix} 0.34 & 0.16 \\ 0.16 & 0.34 \end{pmatrix} \tag{19}$$

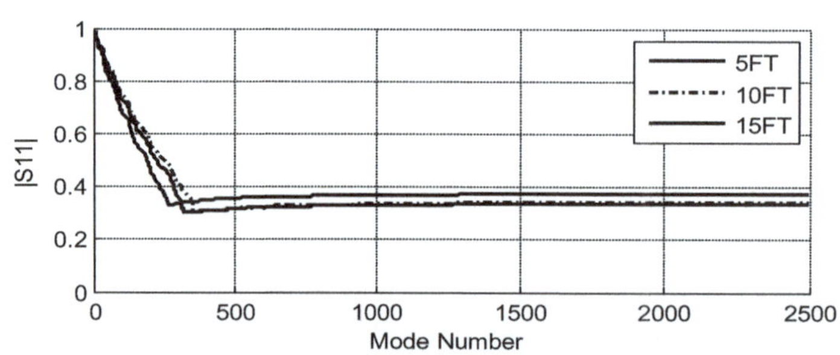

Figure 4. $|S_{11}|$ based on mode number and at several test functions number ratio for operating frequency $f = 9.33$ GHz and for NG1 is a WR90 waveguide.

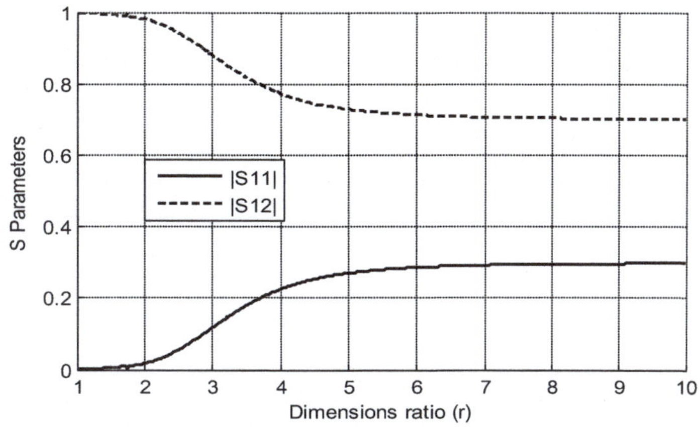

Figure 5. ***S*** parameters based on dimensions ratio for operating frequency $f = 9.33$ GHz and for NG1 is a WR90 waveguide.

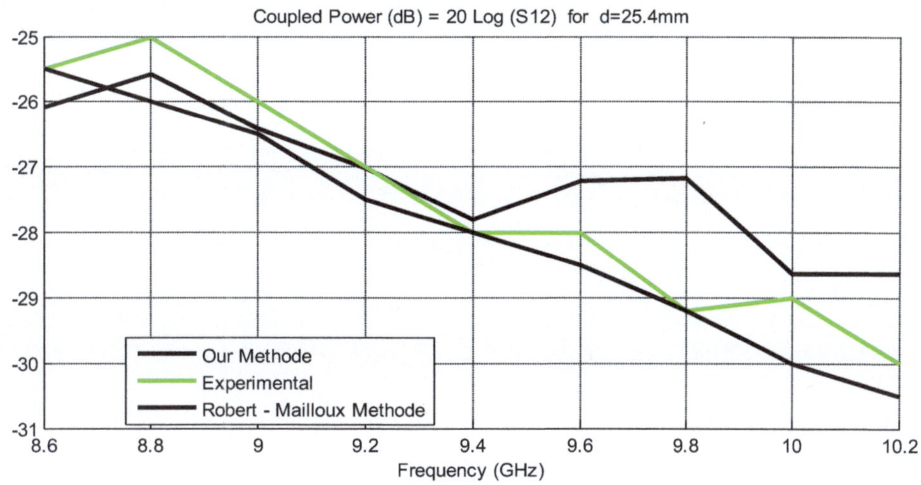

Figure 6. $\left|S_{21}^{T}\right|$ as function of frequency for $r = 10$, coupling distance $d = 25.4$ mm, and NGs are WR90 waveguides.

Table 1. Comparison of reflection coefficient with published results.

Method	Re (S11)	Im (*S*11)	|*S*11|	E_r
Present method	0.0768	−0.2662	0.2771	
Serizawa, Hongo [19]	0.0713	−0.2344	0.2450	13%
Bird [10]	0.0706	−0.2352	0.2456	12%
Macphie [14]	0.0632	−0.2403	0.2484	11%
Baudrand [16]	0.0655	−0.2709	0.2787	0.5%
Mongiardo [20]	0.0803	−0.2365	0.2498	10%

We can notice that $\boldsymbol{S}^{\mathrm{T}}$ does not contain the entire energy.

$$\left(S_{11}^{T}\right)^{2} + \left(S_{21}^{T}\right)^{2} = 15\% \tag{20}$$

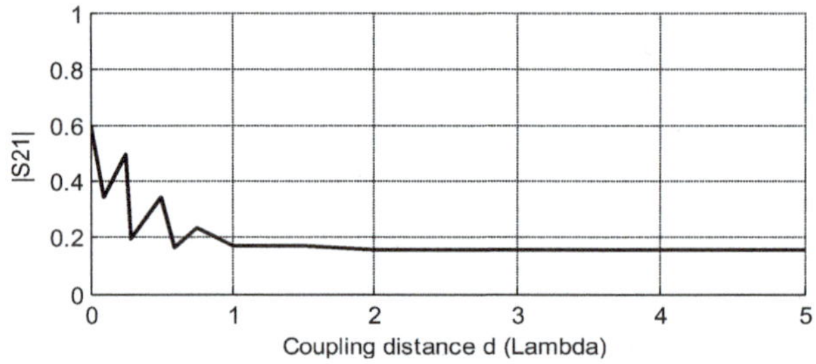

Figure 7. $|S_{21}^{T}|$ parameter as function of distance d, f = 9.33 GHz, r = 10, and NGs are WR90 waveguides.

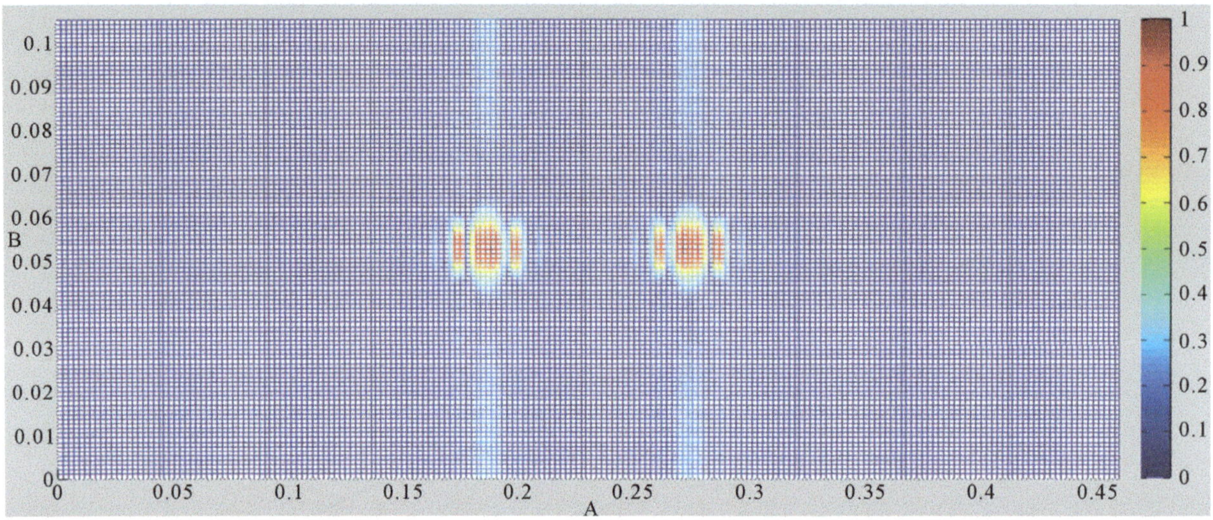

Figure 8. Normalized electric field in discontinuity plan, r = 10, $d = 2\lambda$, f = 9.33 GHz.

If one considers the load operator $\hat{Z}^{-1}$ which is in fact the free space, we can explain the difference by the radiating energy in the free space. This radiation is given by:

$$R = 1 - \left(S_{11}^{T}\right)^{2} - \left(S_{21}^{T}\right)^{2}. \tag{21}$$

Moreover, we present the normalized electric field in the discontinuity plan, with $d = 2\lambda$, r = 10, and we can verify that it is maximum in the radiating apertures and vanishes on the metal surface. This shows (**Figure 8**) that the electric field verifies the boundary conditions.

5. Conclusion

In this study, we apply a rigorous formulation based on MoM-GEC method to explore a radiation of twin open ended waveguides in the free space. The main idea is composed by two steps. First, we consider two waveguides with a simple concentric discontinuity in the propagation axis. One represents the radiating waveguide, and the other is the load waveguide; and we seek how this latter can properly simulate the free space. So, we determine its appropriate dimensions so that it no longer affects the scattering parameters. Second, we studied the electromagnetic coupling between these two open ended waveguides and we present the electric field in the discontinuity plan. For a validation purpose, obtained results are compared to previous published works and boundary conditions are verified and a good agreement is shown. From the numerical results discussion, we can deduce that the load waveguide emulates the free space and no mismatch problem is shown from r = 7 and that the electromagnetic coupling for twin waveguides can be considered neglected from a coupling distance $d = 2\lambda$.

References

[1] Hirokawa, J. (2012) Plate-Laminated Waveguide Slot Array Antennas and Its Polarization Conversion Layers. *Auto-*

matika—Journal for Control, Measurement, Electronics, Computing and Communications, **53**. http://dx.doi.org/10.7305/automatika.53-1.143

[2] Douvalis, V., Hao, Y. and Parini, C.G. (2006) A Monolithic Active Conical Horn Antenna Array for Millimeter and Submillimeter Wave Applications. *IEEE Transactions on Antennas and Propagation*, **54**, 1393-1398. http://dx.doi.org/10.1109/TAP.2006.874338

[3] Lo, Y.T. and Lee, S.W. (2013) Antenna Handbook: Theory, Applications, and Design. Springer Science & Business Media, Berlin.

[4] Lewin, L. (1951) Advanced Theory of Waveguides. Published for Wireless Engineer by Iliffe.

[5] Mazauric, V. (2011) Une approche variationnelle de lkélectromagnétisme.

[6] Tai, C.T. (1949) Application of a Variational Principle to Biconical Antennas. *Journal of Applied Physics*, **20**, 1076. http://dx.doi.org/10.1063/1.1698278

[7] Mailloux, R.J. (1969) Radiation and Near-Field Coupling between Two Collinear Open-Ended Waveguides. *IEEE Transactions on Antennas and Propagation*, **17**, 49-55. http://dx.doi.org/10.1109/TAP.1969.1139354

[8] Mailloux, R.J. (1969) First-Order Solutions for Mutual Coupling between Waveguides Which Propagate Two Orthogonal Modes. *IEEE Transactions on Antennas and Propagation*, **17**, 740-746. http://dx.doi.org/10.1109/TAP.1969.1139529

[9] Hockham, G.A. and Walker, G.H. (1973) Study of a Finite Phased Array Antenna. *3rd European Microwave Conference*, 1-4. http://dx.doi.org/10.1109/euma.1973.331796

[10] Bird, T.S. (1979) Mode Coupling in a Planar Circular Waveguide Array. *IEE Journal on Microwaves, Optics and Acoustics*, **3**, 172-180. http://dx.doi.org/10.1049/ij-moa.1979.0041

[11] Bird, T.S. (1990) Analysis of Mutual Coupling in Finite Arrays of Different-Sized Rectangular Waveguides. *IEEE Transactions on Antennas and Propagation*, **38**, 166-172. http://dx.doi.org/10.1109/8.45118

[12] Luzwick, J. and Harrington, R.F. (1982) Mutual Coupling Analysis in a Finite Planar Rectangular Waveguide Antenna Array. *Electromagnetics*, **2**, 25-42. http://dx.doi.org/10.1080/02726348208915155

[13] Arndt, F., Wolff, K.-H., Brunjes, L., *et al.* (1989) Generalized Moment Method Analysis of Planar Reactively Loaded Rectangular Waveguide Arrays. *IEEE Transactions on Antennas and Propagation*, **37**, 329-338.

[14] Macphie, R.H., Zaghloul, A., *et al.* (1980) Radiation from a Rectangular Waveguide with Infinite Flange—Exact Solution by the Correlation Matrix Method. *IEEE Transactions on Antennas and Propagation*, **28**, 497-503.

[15] Teodoridis, V., Sphicopoulos, T. and Gardiol, F.E. (1985) The Reflection from an Open-Ended Rectangular Waveguide Terminated by a Layered Dielectric Medium. *IEEE Transactions on Microwave Theory and Techniques*, **33**, 359-366.

[16] Baudrand, H., Tao, J.-W. and Atechian, J. (1988) Study of Radiating Properties of Open-Ended Rectangular Waveguides. *IEEE Transactions on Antennas and Propagation*, **36**, 1071-1077. http://dx.doi.org/10.1109/8.7219

[17] Li, J.-Y., Li, L.-W., Liang, C.-H., *et al.* (2001) Analysis of Finite-Slot Phased Arrays Fed by Rectangular Waveguides Using Method of Moments. *Microwave and Optical Technology Letters*, **29**, 132-136. http://dx.doi.org/10.1002/mop.1107

[18] Ghosh, B., Sinha, S.N. and Kartikeyan, M.V. (2010) Radiation from Rectangular Waveguide-Fed Fractal Apertures. *IEEE Transactions on Antennas and Propagation*, **58**, 2088-2093. http://dx.doi.org/10.1109/TAP.2010.2046835

[19] Serizawa, H. and Hongo, K. (2005) *Radiation from a Flanged Rectangular Waveguide. IEEE Transactions on Antennas and Propagation*, **53**, 3953-3962. http://dx.doi.org/10.1109/TAP.2005.859748

[20] Mongiardo, M. and Rozzi, T. (1993) Singular Integral Equation Analysis of Flange-Mounted Rectangular Waveguide Radiators. *IEEE Transactions on Antennas and Propagation*, **41**, 556-565. http://dx.doi.org/10.1109/8.222274

[21] Baudrand, H. and Bajon, D. (2002) Equivalent Circuit Representation for Integral Formulations of Electromagnetic Problems. *International Journal of Numerical Modelling: Electronic Networks, Devices and Fields*, **15**, 23-57. http://dx.doi.org/10.1002/jnm.430

[22] Aguili, T. (2000) Modélisation des composants SH F planaires par la méthode des circuits équivalents généralisés. Thesis Manuscript, National Engineering School of Tunis, Tunisia.

[23] Baudrand, H. and Aubert, H. (2003) L'Electromagnétisme par les Schémas Equivalents. Cepaduès Editions, Paris.

[24] Hajji, M., Mendil, S. and Aguili, T. (2014) A New Hybrid MoM-GEC Asymptotic Method for Electromagnetic Scattering Computation in Waveguides. *Progress in Electromagnetics Research B*, **61**, 197-210. http://dx.doi.org/10.2528/PIERB14101303

14

Negative Refractive Index Metamaterial Structure Using SRR by Incidenting the Light Horizontally

Ammar Armghan, Xinguang Hu, Shuai Yuan, Jinsong Xia*

Wuhan National Laboratory for Optoelectronics and School of Optical and Electronic Information, Huazhong University of Science and Technology, Wuhan, China

Email: ammar.armghan@gmail.com, *jsxia@hust.edu.cn

Abstract

Metamaterial structure based on split ring resonators (SRR) is proposed in order to produce a negative refractive index. For this structure we have used a new approach, instead of applying light perpendicularly incident. We apply horizontally incident input waves. A model of SRR is used to understand the behavior and its affects. We calculate the S-parameters using S-parameter analysis and the results for transmission, refractive index, permeability and permittivity of the structure is induced. The negative refractive index is found to be significantly dependent upon the width of the continuous wire as well as gap between resonators. Moreover, we study the effect of lattice constant on the electromagnetic response of the structure. It is expected that this work will provide useful information for design and fabrication of metamaterials with negative refractive index for in-plane applications.

Keywords

Metamaterials, Negative Refractive Index, Left Handed Material, Scattering Parameters

1. Introduction

Metamaterial is a composite structure made of metals and dielectrics. It implies that it is not a homogenous material, rather inhomogeneous artificial materials made of metal and dielectric materials. Metallic or dielectric structures in metamaterials are typically small; they are actually much smaller than the wavelength. Because of which, at a wavelength scale, we can assume an effectively homogenous material but with very different physi-

*Corresponding author.

cal properties in comparison to either metals or dielectrics. So, now we are creating artificial materials made of metallic and dielectric materials in a way that they have the desired property. The first quadrant has a positive ε and μ, this kind of materials is called double positive materials or natural dielectric materials. If an electromagnetic wave propagates in such a material, you will see a propagating wave with some loss or attenuation. In case of metals, the ε is negative and μ is positive, thus metals are also called epsilon negative materials. If electromagnetic wave propagates in a metal, it experiences huge attenuation that is why when discussing metals we discuss the skin depth effect of metal. We know that actually the waves will attenuate a lot into the metamaterials, so the actual skin depth effect for most of the electromagnetic waves is very small. In 1960s, Veselog theoretically proposed that materials having negative Mu and negative epsilon at the same time could produce a negative index of refraction and called them a left handed material [1]. In 1996, J. B. Pendry proposed that we could have a negative epsilon by arranging the metal wires in a simple cubic lattice. But the question still remained how we could get the negative Mu that was answered in 2000 by smith [2] [3]. Smith proposed a structure consisting of periodic array of split ring resonators and continuous wires that produced the negative permeability and negative permittivity [3].

In recent times, there has been massive development in the field of metamaterials. Negative refractive index (NRI) is the main focus point of current research. Within five years after Smith's proposal, the field of NRI has seen development at optical range by using different structures like paired nano-rods [4], nano-fishnet with circular voids [5], nano-fishnet with elliptical voids [6], and nano-fishnet with rectangular voids [7].

Metamaterials in THz regime are being used as absorbers [8], quarter waveplates [9], switches and modulators [10]-[15]. Metamaterial negative refractive index has numerous applications such as M-NRI for antennas [16], Superlens [17], and wireless power transfer [18] and for biomedical applications [19]-[26].

In this paper, we propose a novel structure of metamaterials, unit cell of complementary split ring resonators at 10THz frequency. This structure has been simulated by FDTD methods and the S-parameters. Then we derive the parameters *i.e.* refractive index, permittivity and permeability by using the S-parameters analysis technique discussed by smith [5]. The simulation results shows that by incidenting the light horizontally, negative Mu and negative epsilon are obtained for the proposed structure, which has the properties of Double Negative (DNG) metamaterials.

Most of the current metamaterials research is that light incident on the normal direction. But many researchers are trying to integrate devices with different function on single chip. Therefore, metamaterials to work with light in plane propagation direction is necessary for chip integration. However, few research of in plane light propagation is reported in case of metamaterials. In this paper, it studies a metamaterial structure for an in plane light propagation and studies its optical properties.

2. Design and Simulation

2.1. S-Parameter Extraction Technique

Scatter parameters (S-parameters), is usually use to describe the optical behavior of structures. S-parameters relate the incident light to reflected or transmitted light. The S-parameters are complex value numbers that represent reflection and transmission coefficients. In case of metamaterials, the S-parameters are directly proportional to the electric fields. So, we simply run the simulation and look at the S-parameters as transmitted filed divided by the incident field in direction and reflected field divided by the incident field in other direction.

Ideally, the fields would be recorded on the edge of the unit cell. In practice, the measurements are usually done at some distance from the metamaterial structure. Therefore, it is necessary to compensate the phase accumulated between metamaterial structure and measurement points. Different kinds of techniques have been developed in the past few years for extraction of S-parameters. In our simulation we are using the S-parameters extraction technique discussed by Smith *et al.* [3].

2.2. Simulation Technique

The structure was simulated FDTD method. Choosing the frequency unit in THz, S-parameter analysis tool was used to extract the S-parameters. Each calculation is based on 100 frequency samples.

The unit cell is shown in **Figure 1**. It consists of two materials: glass substrate and split ring resonator made of copper. This structure was simulated for 10 THz frequency range. The thickness of the substrate is 250 nm,

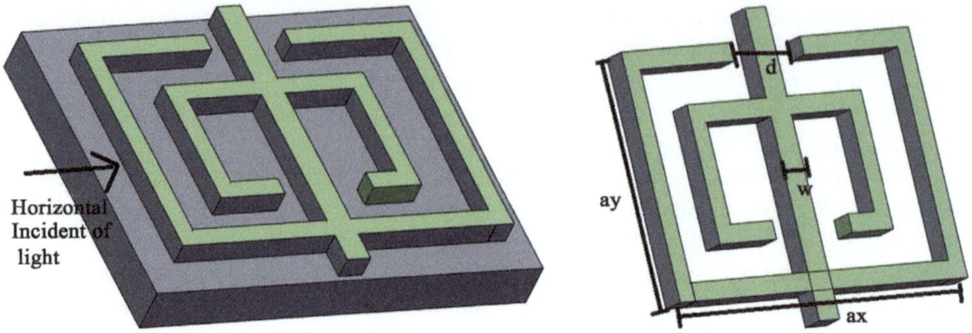

(a) (b)

Figure 1. (a) Shows the direction of the incident light and (b) shows the parameters.

width of metal wires is 140 nm, length of the metal wires of the outer ring is 2000 nm and the gap between the inner ring and the outer ring is 147 nm.

3. Results and Discussion

The calculated transmission spectra of the structure is shown in **Figure 2(a),** where the width of the wires are kept at 140 nm. As it can be seen in the figure, there exists two pass bands in the transmission spectra of the structure that are separated by a certain amount of frequency. The first transmission peak is at approximately 9.7 THz. This indicates that this pass band exhibits LH behavior, but the other one at around 13 THz is has RH behavior [27] [28]. The appearance of the LH behavior can also be confirmed by using the extracted effective parameters of the structure, as presented in **Figures 2(b)-(d)**. Obviously, both the effective permeability and permittivity are negative around the first peak, consequently the refractive index has a negative value ($n < 0$). However, the peak at around 13 THz exhibits the RH behavior because both permeability and permittivity are positive.

Moreover, there is a dependence of the structural parameters of the metal wire on the LH behavior. We found that the electromagnetic response of the structure is significantly affected by the width of the continuous wire. **Figure 3** shows the transmission spectra of the compound structure by varying the width of the continuous wire. For this purpose, the width of the wire pairs is kept at 140 nm while the width of the continuous wire is varied from 50 to 140 nm.

As can be seen, the transmission peak exhibiting the LH behavior is significantly improved with the decrease in width of the continuous wire. It should also be noted that the resonance is shifted with an increase in width of the continuous wire. To confirm this phenomenon, the effective permittivity, permeability and also refractive index of the structures are extracted from the scattering parameters, as shown in **Figures 3(b)-(d)**, respectively. As expected, the plasma frequency shifts towards a lower frequency by increase in width of the continuous wire, while the magnetic response remains roughly unchanged [24]. Thus, this explains the reason for the dependence of resonance peak on width of the continuous wire. As **Figure 3(d)** showing, the refractive index is decreasing with the decrease in width of the continuous wire.

Another interesting result is shown in **Figure 4**, where the LH behavior of the structure is also intensely affected by varying the gap *d* between the centers of the continuous wires. To study this phenomenon, the gap *d* was varied from 500 to 700 nm while the lattice constants ax and ay are 2200 nm and 1800 nm respectively. The effective permittivity, permeability and refractive index are extracted from simulation results. As it can be seen in the figure, the resonance is left-shifted by varying the gap d between the centers of the continuous wire. A reduction of negative refractive index appears when increasing the gap size.

We also studied the LH behavior dependency on the lattice constant. To study the effect of the lattice constant ay, ax is fixed at 2200 nm while ay is varied from 1500 to 2400 nm. The calculated effective parameters of the structure are shown in **Figure 5**. As ay is reduced, the resonant plasma frequency of the structure is reduced. If ay is reduced to 1500 nm the magnetic resonance frequency is still higher than the plasma frequency (at 10 THz). If ay is reduced any further than this point, the refractive index becomes positive eliminating the LH behavior of the structure. To study the effect of the lattice constant ax, ay was kept constant at 1800 nm while ax was varied from 2200 to 2600 nm. The calculated effective parameters are shown in **Figure 6**. As ax is increased, magnetic

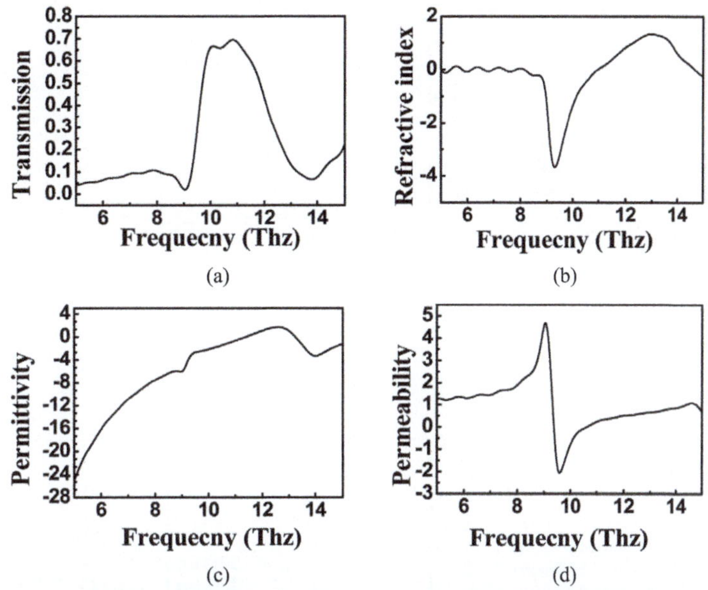

Figure 2. (a) Simulated transmission spectra of the structure, where the width of the wires are kept at 140 nm; (b)-(d) real part of the permittivity, permeability and refractive index extracted from the simulation data, respectively.

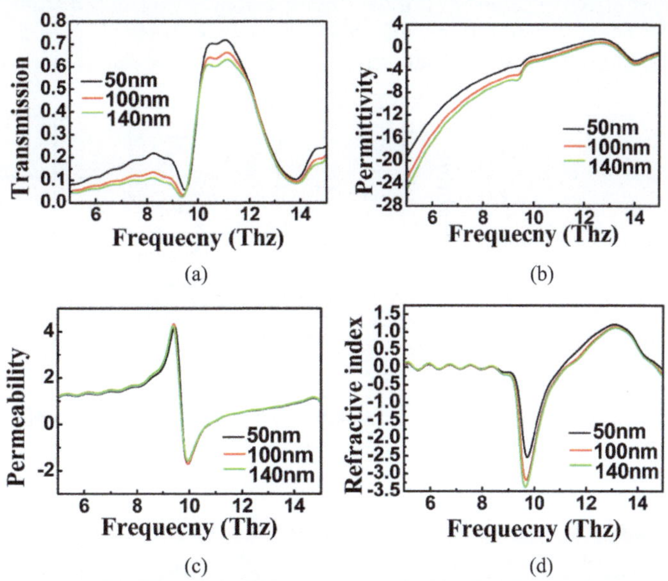

Figure 3. (a)Simulated transmission spectra of the structure with different widths of the continuous wires; (b)-(d) real part of the permittivity, permeability and refractive index, respectively.

resonance frequency of the structure is reduced while plasma frequency and the refractive index remain constant. Therefore, the refractive index is still negative.

The following **Table 1** shows a comparison of this structure with other structures reported in the literature, with respect to refractive indices.

In the above table we have compared our results with the other structures that are using perpendicular direction of the incident light, where as in our design we are using the horizontal direction of incidenting light and achieve the refractive index of −3.8. The results by using the horizontal direction of incidenting light are useful for designing novel devices for on chip integration by utilizing the optical properties of metamaterials.

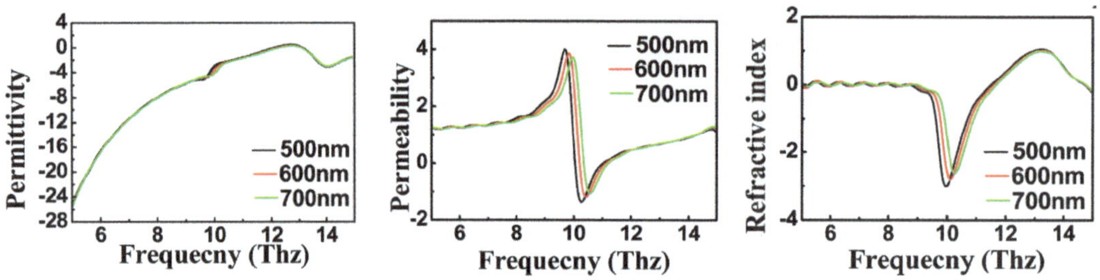

(a) (b) (c)

Figure 4. Extracted effective parameters of structure from the calculated scattering parameters: (a), (b) and (c) are the permittivity, permeability and refractive index, respectively. The lattice constant ax is 2200 nm, ay is 1800 nm.

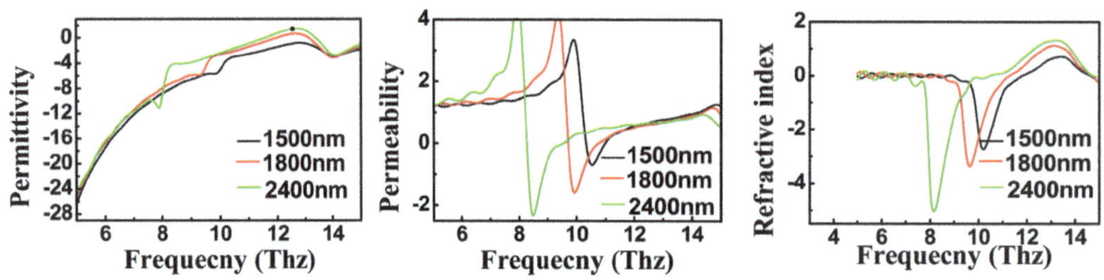

(a) (b) (c)

Figure 5. Extracted effective parameters of the structure from the calculated scattering parameters: (a), (b) and (c) are the permittivity, permeability and refractive index, respectively. The lattice constant ax is 2200 nm, and ay is varied from 1500 to 2400 nm.

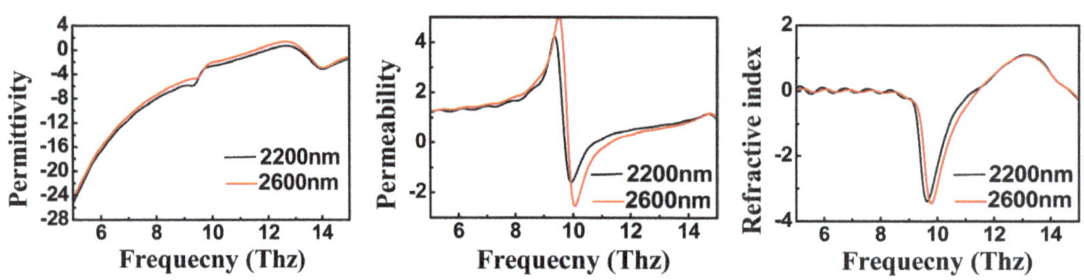

(a) (b) (c)

Figure 6. Extracted effective parameters of structure from the calculated scattering parameters: (a), (b) and (c) are the permittivity, permeability and refractive index, respectively. The lattice constant ay is 1800 nm, and ax is varied from 2200 to 2600 nm.

Table 1. Comparison between different structures.

References	Structures	Refractive index	Direction of pump
4	Paired nanorods	−0.3	Perpendicular
21	Nano-fishnet with circular rods	−2	Perpendicular
7	Nano-fishnet with rectangular rods	−1	Perpendicular
26	Metal-dielectric-metal	−1.05	Perpendicular
Our Design	Split ring resonators	−3.8	Horizontally

4. Conclusion

Simulation studies have been conducted on ring-type split ring resonators (SRR) to understand the properties of metamaterial structure with horizontally incident source. By incidenting the light along x-axis, the proposed structure shows a negative refractive index indicating the properties of double negative material. Moreover, it is observed that the width of the continuous wire as well as distance between the centers affects the left-handed behavior of the structure. Additionally, influence of lattice constant on the left-handed behavior of the structure is also studied. The results indicate strong dependence of the left-handed behavior on the lattice constants *ay*. The search results in our paper are useful for designing novel devices for on chip integration by utilizing the optical properties of metamaterials. In subsequent works, a cascaded version of the above detailed design can be placed on top of a silicon waveguide and can be simulated to extract parameters and thus behavioral pattern of light in such environment.

Acknowledgements

This work was supported by the National "863" Program of China (No. 2015AA016904), the Major State Basic Research Development Program of China (grant 2013CB632104 and 2013CB933303).

References

[1] Veselago, V.G. (1968) The Electrodynamics of Substances with Simultaneously Negative Values of ε and μ. *Soviet Physics Uspekhi*, **10**, 509-514. http://www.turpion.org/php/full/infoFT.phtml?journal_id=pu&paper_id=3699 http://dx.doi.org/10.1070/PU1968v010n04ABEH003699

[2] Pendry, J.B., Holden, A.J. Stewart, W.J. and Youngs, I. (1996) Extremely Low Frequency Plasmonics in Metallic Mesostructures. *Physical Review Letters*, **76**, 4773-4776. http://dx.doi.org/10.1103/PhysRevLett.76.4773

[3] Smith, D.R., Padilla, W.J., Vier, D.C., Nemat-Nasser, S.C. and Schultz, S. (2000) Composite Medium with Simultaneously Negative Permeability and Permittivity. *Physical Review Letters*, **84**, 4184-4187. http://dx.doi.org/10.1103/PhysRevLett.84.4184

[4] Shalaev, V.M., Cai, W., Chettiar, U., Yuan, H.-K., Sarychev, A.K., Drachev, V.P. and Kildishev, A.V. (2005) Negative Index of Refraction in Optical Metamaterials. *Optics Letters*, **30**, 3356-3358.

[5] Smith, D.R., Vier, D.C., Koschny, T. and Soukoulis, C.M. (2005) Electromagnetic Parameter Retrieval from Inhomogeneous Metamaterials. *Physical Review Letters*, **71**, 1-11.

[6] Zhang, S., Fan, W.J., Malloy, K.J., Brueck, S.R.J., Panoiu, N.C. and Osgood, R.M. (2006) Demonstration of Metal-Dielectric Negative-Index Metamaterials with Improved Performance at Optical Frequencies. *Journal of the Optical Society of America*, **23**, 434-438. http://dx.doi.org/10.1364/JOSAB.23.000434

[7] Dolling, G., Enkrich, C. and Wegener, M. (2006) Low-Loss Negative-Index Metamaterial at Telecommunication Wavelengths. *Journal of the Optical Society of America*, **31**, 1800-1802.

[8] Tao, H., Landy, N.I., Bingham, C.M., Zhang, X., Averitt, R.D. and Padilla, W.J. (2008) A Metamaterial Absorber for the Terahertz Regime: Design, Fabrication and Characterization. *Optics Express*, **16**, 7181-7188. http://dx.doi.org/10.1364/OE.16.007181

[9] Strikwerda, A.C., Fan, K., Tao, H., Pilon, D.V., Zhang, X. and Averitt, R.D. (2009) Comparison of Birefringent Electric Split-Ring Resonator and Meander Line Structures as Quarter-Wave Plates at Terahertz Frequencies. *Optics Express*, **17**, 136-149. http://dx.doi.org/10.1364/OE.17.000136

[10] Chen, H.T., Padilla, W.J., Zide, J.M.O., Gossard, A.C., Taylor, A.J. and Averitt, R.D. (2006) Active Terahertz Metamaterial Devices. *Nature*, **444**, 597-600. http://dx.doi.org/10.1038/nature05343

[11] Chen, H.T., Padilla, W.J., Cich, M.J., Azad, A.K., Averitt, R.D. and Taylor, A.J. (2009) A Metamaterial Solid-State Terahertz Phase Modulator. *Nature Photonics*, **3**, 148-151. http://dx.doi.org/10.1038/nphoton.2009.3

[12] Chen, H.T., Palit, S., Tyler, T., Bingham, C.M., Zide, J.M.O., O'hara, J.F., *et al.* (2008) Hybrid Metamaterials Enable Fast Electrical Modulation of Freely Propagating Terahertz Waves. *Applied Physics Letters*, **93**, Article ID: 091117. http://dx.doi.org/10.1063/1.2978071

[13] Chen, H.T., O'Hara, J.F., Azad, A.K. and Taylor, A.J. (2008) Experimental Demonstration of Frequency-Agile Terahertz Metamaterials. *Nature Photonics*, **2**, 295-298. http://dx.doi.org/10.1038/nphoton.2008.52

[14] Lapine, M., Powell, D., Gorkunov, M., Shadrivov, I., Marqués, R. and Kivshar, Y. (2009) Structural Tunability in Metamaterials. *Applied Physics Letters*, **95**, Article ID: 084105. http://dx.doi.org/10.1063/1.3211920

[15] Tao, H., Strikwerda, A.C., Fan, K., Padilla, W.J., Zhang, X. and Averitt, R.D. (2009) Reconfigurable Terahertz Metamaterials. *Physical Review Letters*, **103**, Article ID: 147401. http://dx.doi.org/10.1103/PhysRevLett.103.147401

[16] Chen, X., Huifeng, M., Yang, X., Cheng, Q., Jiang, W.X. and Cui, T.J. (2009) X-Band High Directivity Lens Antenna Realized by Gradient Index Metamaterials. *Proceedings of the Asia Pacific Microwave Conference*, Singapore, 7-10

December 2009, 793-797. http://dx.doi.org/10.1109/apmc.2009.5384269

[17] Lier, E. (2011) Metamaterial Lens Feed for Multiple Beam Antennas. United States Patent Application Publication, US 2011/0095953 A1. https://www.google.com/patents/US8576132

[18] Navarro-Cia, M., Beruete, M., Falcone, F., Sorolla, M. and Campillo, I. (2010) Antenna Directivity Enhancement Using a Metamaterial Parabolic Lens. *Proceedings of the Fourth European Conference on Antennas and Propagation* (*EuCAP*), Barcelona, 12-16 April 2010, 1-3. http://ieeexplore.ieee.org/xpl/login.jsp?tp=&arnumber=5505200&url=http%3A%2F%2Fieeexplore.ieee.org%2Fxpls%2Fabs_all.jsp%3Farnumber%3D5505200

[19] Zhang, X. and Liu, Z.W. (2008) Superlenses to Overcome the Diffraction Limit. *Nature Materials*, **7**, 435-441. http://dx.doi.org/10.1038/nmat2141

[20] Tao, Y. and Wang, G. (2012) Conformal Hyperthermia of Superficial Tumor with Left-Handed Metamaterial Lens Applicator. *IEEE Transactions on Biomedical Engineering*, **59**, 3525-3530. http://dx.doi.org/10.1109/TBME.2012.2218108

[21] Zhang, S., Fan, W.J., Panoiu, N.C., Malloy, K.J., Osgood, R.M. and Brueck, S.R.J. (2005) Experimental Demonstration of Near-Infrared Negative-Index Metamaterials. *Physical Review Letters*, **95**, Article ID: 137404. http://dx.doi.org/10.1103/physrevlett.95.137404

[22] Dolling, G., Wegener, M., Soukoulis, C.M. and Linden, S. (2007) Negative-Index Metamaterial at 780 nm Wave Length. *Journal of the Optical Society of America*, **32**, 53-55.

[23] Driscoll, T., Kim, H.T., Kim, B.J., Lee, Y.W., Jokerst, N.M., Palit, S., Smith, D.R., Di Ventra, M. and Basov, D.N. (2009) Memory Metamaterials. *Science*, **325**, 1518-1521. http://dx.doi.org/10.1126/science.1176580

[24] Vu, D.L., Pham, V.T., Do, T.V., Nguyen, T.T., Vu, T.T.T., Le, V.H. and Lee, Y.P. (2010) The Electromagnetic Response of Different Metamaterial Structures. *Advances in Natural Sciences*, **1**, 1-7.

[25] Xu, C. and Dong, J. (2010) Negative Refractive Index in Non-Resonance Spectrum Area. *COL*, **8**, 1067-1070.

[26] Zhong, M. (2014) Influence of Dielectric Layer on Negative Refractive Index and Transmission of Metal-Dielectric-Metal Sandwiched Metamaterials. *COL*, **12**, Article ID: 041601.

[27] Guven, K., Caliskan, M.D. and Ozbay, E. (2006) Experimental Observation of Left-Handed Transmission in a Bilayer Metamaterial under Normal-to-Plane Propagation. *Optics Express*, **14**, 8685-8693. http://dx.doi.org/10.1364/OE.14.008685

[28] Lam, V.D., Kim, J.B., Lee, S.J. and Lee, Y.P. (2008) Left-Handed Behaviour of Combined and Fishnet Structures. *Journal of Applied Physics*, **103**, Article ID: 033107. http://dx.doi.org/10.1063/1.2841726

15

A High Efficiency Doherty Power Amplifier for TV Band Applications

Mohamad Y. Abou-Shahine, Youssef Nasser, Karim Y. Kabalan

Electrical and Computer Engineering Department, American University of Beirut, Beirut, Lebanon
Email: mya26@aub.edu.lb, yn10@aub.edu.lb, kabalan@aub.edu.lb

Abstract

This paper presents a high efficiency Doherty power amplifier suitable for TV band applications. A class AB power amplifier is firstly implemented using a commercial GaN HEMT from Cree Incorporation, achieving a high power-added-efficiency of 77.78% and a 40.593 dBm output power with an associated gain of 21.65 dB. The Doherty amplifier has then been designed following the previous class AB scheme for the main amplifier and a class C scheme for the peak one. This amplifier attained a high power-added-efficiency of 81.94%, a 42.77 dBm output power, an associated gain of 21.32 dB, and an operating frequency bandwidth between 550 and 1000 MHz (58.06% fractional bandwidth) which made it suitable for TV band applications.

Keywords

Power Amplifier, Class AB, Doherty Power Amplifier, Efficiency, TV Band

1. Introduction

Recently, the design of RF front-ends in wireless communication for cognitive radio applications has become a hot research area. Highly efficient and linear power amplifier is a key component in these systems. However, the main problems of the design of power amplifier exist in their linearity region and their gain especially in multi-carrier systems (such as orthogonal frequency division multiplexing-OFDM) which are widely used in current wireless communication. Indeed, these systems present high peak-to-average power ratio (PAPR) caused by the large number of independent subcarriers with random phase and amplitude added together at the input of the amplifier which makes the latter impractical to work next to its saturation point [1]. Therefore, a power amplifier with high linearity and efficiency is great of importance. The simplest method is to back-off signals from the saturation region to the linear region at the cost of power efficiency. Another way may use predistortion methods or elimination and restoration techniques. However, these techniques which need additional components result in an increase in cost, size, and power dissipation. In order to solve these problems, a Doherty Power Amplifier (DPA) is the most promising candidate with high efficiency in power and bandwidth [2]-[8].

The Doherty amplifier operation can be achieved using two cells with a class-AB biased carrier amplifier cell and a class-C biased peak amplifier cell with respective input matching network and output matching network. The Doherty power Amplifier (DPA) operating principle is based on modulating the load of the active (carrier) device by using a second active peak device. The two power amplifiers are connected at the output through a quarter-wave transmission line to exploit the active load modulation concept performed by the peak amplifier on the main carrier one. This Doherty amplifier has high linearity and efficiency across the wideband signal.

In practice, Doherty amplifiers are the best candidates that can provide higher efficiency among all type of amplifiers. They are widely used for many applications where RF power amplifier efficiency is important. The efficiency of the Doherty amplifier can be improved if the efficiencies of the two amplifiers can be properly increased [4].

The Doherty technique is usually adopted to design power amplifiers for wireless systems and, in particular, in base stations, working in L-S-C Band for applications such as WiMAX, WLAN, cellular network, and others. Drain efficiencies up to 70% have been demonstrated for output powers between 5 W and 10 W [9]-[11], whereas drain efficiency up to 50% has been demonstrated for 250 W output power [12].

However, a well-known disadvantage of the DPA is its typical narrow bandwidth performance. Although significant improvements in bandwidth have been reported recently [13]-[17], the DPA is still fundamentally bandwidth limited by the quarter wave impedance inverter needed to obtain proper load modulation. This results in substantial frequency dependence at the back-off power region where the Power Amplifier (PA) is operating most of the time, and thereby narrow band performance for realistic signals.

In this paper, a class AB power amplifier and a high efficiency Doherty one with operating frequency at 800 MHz were developed and verified for the suitability for TV band applications. In comparison with previously designed power amplifiers, the proposed Doherty power amplifier attains higher efficiency and wider bandwidth. The proposed Doherty power amplifier reaches a power added efficiency of 81.94% with a 58.06% fractional bandwidth, so that it can be used as a high efficiency power amplifier for the TV band from 550 to 1000 MHz. The geometry, the design guidelines, and the results of the proposed amplifiers are presented. In addition, the characteristics of the designed amplifiers are investigated via Agilent Advanced Design System (ADS) software and verified using NI AWR Design Environment software.

2. Design, Simulation, and Results

2.1. Class AB Power Amplifier

The schematic of this proposed class AB amplifier is shown in **Figure 1**. The amplifier is simulated in both software, ADS and AWR. The active device employed in the AB amplifier is the CGH40010 from Cree Inc., a GaN HEMT with a typical output power of 10 W at the suggested drain bias voltage of 28 V. This transistor offers high efficiency, high gain, and wideband capabilities. The amplifier circuit was simulated with a constant drain supply voltage Vsup = 28 V and the gate bias voltage Vg = −2.1 V which represents a bias condition of a conventional class AB amplifier. The substrate parameters were set according to Rogers RO3203 substrate with dielectric constant $\varepsilon_r = 3.02$ and thickness h = 1.6 mm, on which the amplifier is designed.

To verify the accurate representation of the Cree CGH40010 model, the signal scattering parameters generated from ADS for the frequency 800 MHz were compared to the ones given in the datasheet. Close agreement, as presented in **Table 1**, is obtained.

After carrying out the stability test of this amplifier, it is found that the device is potentially unstable in the frequency range of interest. Indeed, the stability parameters, Δ, and the rollet factor k are calculated and found to be:

$$\Delta = 0.092 \angle -8.6^\circ$$

$$k = 0.163$$

This means that the device is potentially unstable. Stabilization methods were performed and series resistances were added to the input and the output of the transistor as shown in **Figure 1**. The input resistance is 3.5 Ohms and the output resistance is 32.5 Ohms. After adding these resistances, the stability parameters were recalculated and found to be:

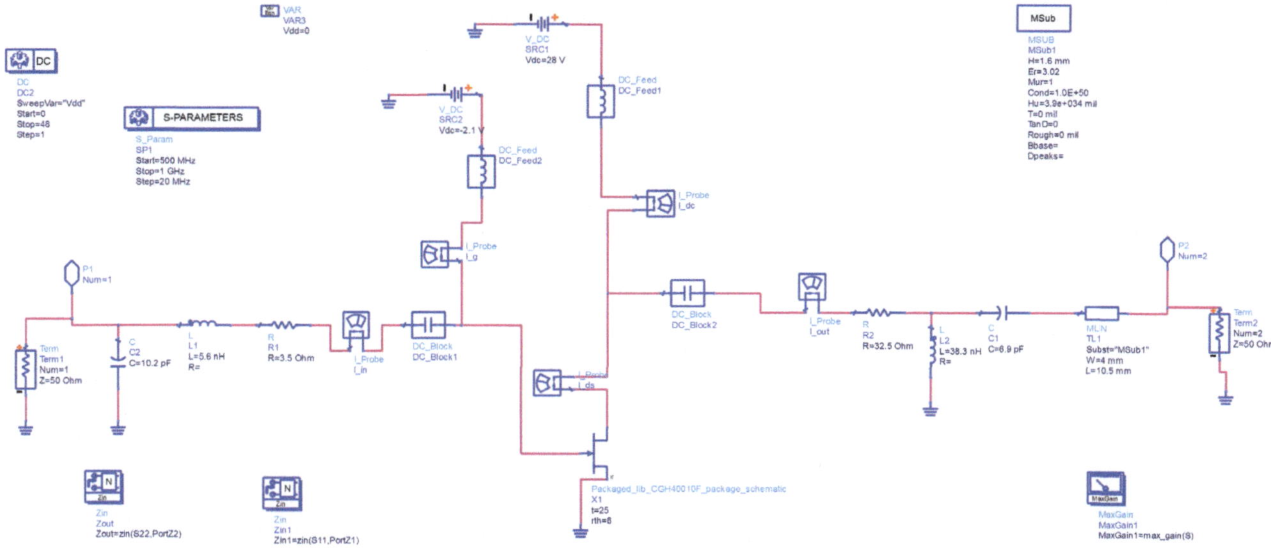

(a)

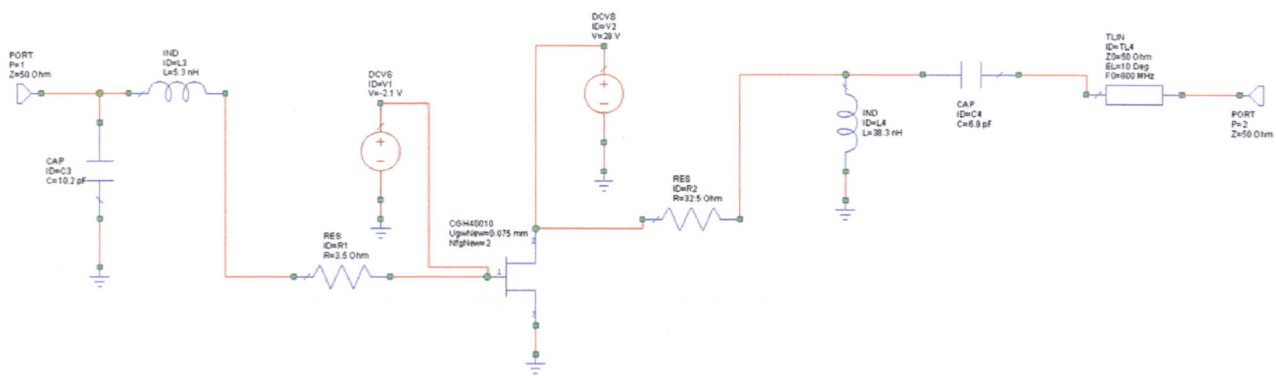

(b)

Figure 1. Schematic of the AB power amplifier with input and output stability resistors and matching circuits. (a) In ADS; (b) In AWR.

Table 1. S-parameters of the GaN transistor at 800 MHz.

	S11	S12	S21	S22
Simulation	0.911 $\angle$ −154°	0.023 $\angle$ 3.54°	11.933 $\angle$ 89.38°	0.298 $\angle$ −132.5°
Datasheet	0.894 $\angle$ −146.96°	0.029 $\angle$ 6.84°	11.58 $\angle$ 91.78°	0.317 $\angle$ −109.84°

$$\Delta = 0.162\ \angle -146.4^\circ$$

$$k = 1.57$$

After adding the stability resistances, the signal scattering parameters at 800 MHz had changed. The scattering parameters were simulated and found to be (at 800 MHz):

$$S11 = 0.772\ \angle -157.51^\circ$$

$$S12 = 0.016\ \angle -0.034^\circ$$

$$S21 = 8.057\ \angle\ 85.81^\circ$$

$$S22 = 0.17\ \angle -39.773^\circ$$

In order to maintain maximum power transfer, we need to match the impedance of the load to that of the source, so matching circuits are designed at the input and the output of the transistor. These circuits are accomplished by using passive networks and microstrip transmission lines as shown in **Figure 2**.

After completing the stability and the matching of the amplifier, the final signal scattering parameters at 800 MHz were simulated and found to be:

$$S11 = 0.056 \angle -119.8^\circ$$
$$S12 = 0.024 \angle -72.2^\circ$$
$$S21 = 12.073 \angle 13.6^\circ$$
$$S22 = 0.438 \angle -89.4^\circ$$

By feeding the amplifier with an optimum 800 MHz, 18.943 dBm input power, the output power of the amplifier circuit was found to be 40.593 dBm and the gain is 21.65 dB. The DC power is simulated and found to be:

$$Pdc = 28 \times 0.522 = 14.62 \text{ W}$$

Thus, the power added efficiency (PAE) is computed to be around 77.78%. All the results are summarized in **Table 2**.

The amplifier efficiency depends on the DC supply voltage and respectively on the input power. **Figure 3** shows the output power sweep curve as well as the corresponding PAE. As the DC voltage increases, the output power increases fairly linearly indicating that the amplifier has a constant gain. The efficiency also increases approximately linearly until it reaches the maximum at the optimum 28 volt DC supply voltage.

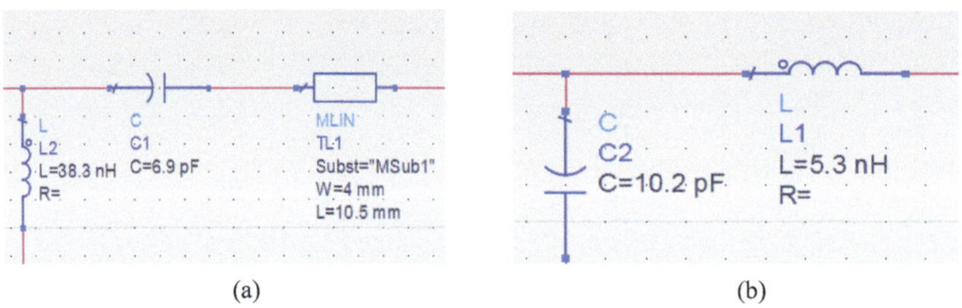

Figure 2. (a) Input matching network; (b) Output matching network.

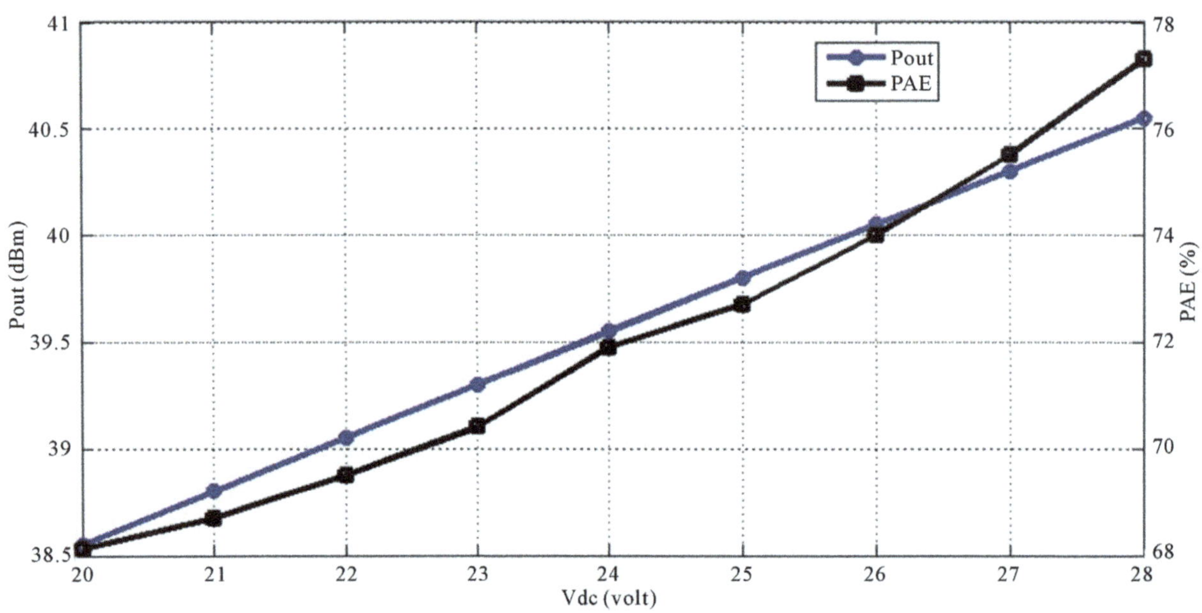

Figure 3. Output power and PAE versus DC voltage in the class AB amplifier.

Table 2. Final results of the class AB amplifier.

Input power (dBm)	DC power (W)	Output power (dBm)	Gain (dB)	PAE (%)	Drain efficiency (%)
18.943	14.62	40.593	21.65	77.78	78.43

The simulated reflection coefficient plots of the class AB amplifier are given in **Figure 4**. The amplifier is simulated via ADS and AWR software. It is clearly shown that this amplifier can operate from 650 MHz to 910 MHz, thus it is suitable for applications in the TV band. It is observed that the simulation results of the two software tools follow the same pattern with little difference.

The percentage bandwidth or the fractional bandwidth which is a measure of how wideband the amplifier is, is given by this equation:

$$\%BW = (f2 - f1)/((f1 + f2)/2)$$

The percentage bandwidth of this amplifier is 33.33%.

The power gain plot of the class AB amplifier is given in **Figure 5**. It is clearly shown that this amplifier has a good power gain above 19 dB for the operation frequency bandwidth.

2.2. Doherty Power Amplifier

As mentioned earlier, the Doherty amplifier procedure can be attained using two amplifiers with a class-AB biased carrier amplifier cell and a class-C biased peak amplifier cell with respective input matching network and output matching network. The two power amplifiers are connected at the output through a quarter-wave transmission line.

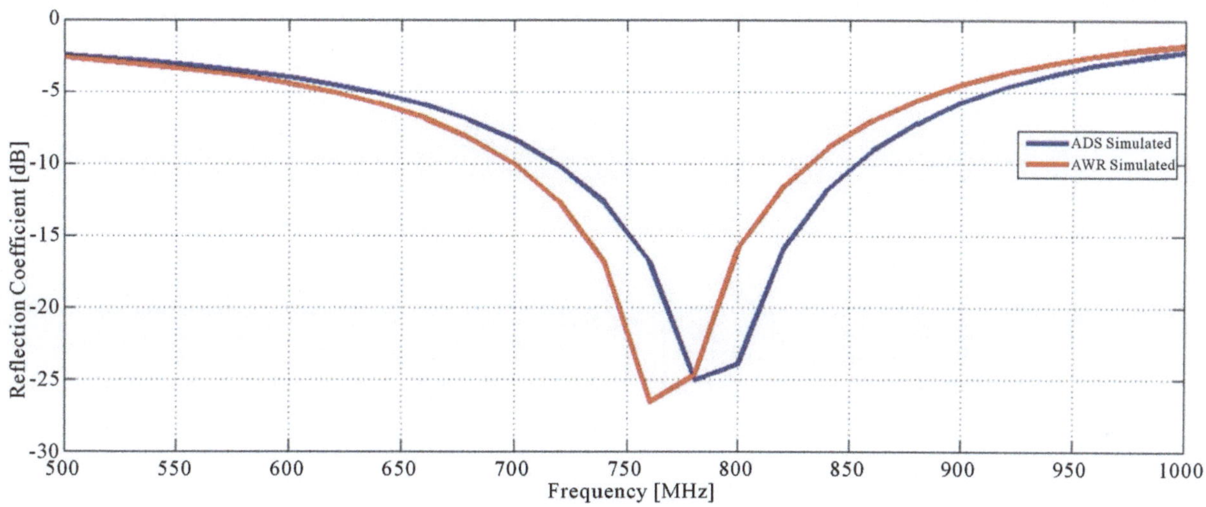

Figure 4. Simulated (ADS and AWR) reflection coefficient plot of the class AB amplifier.

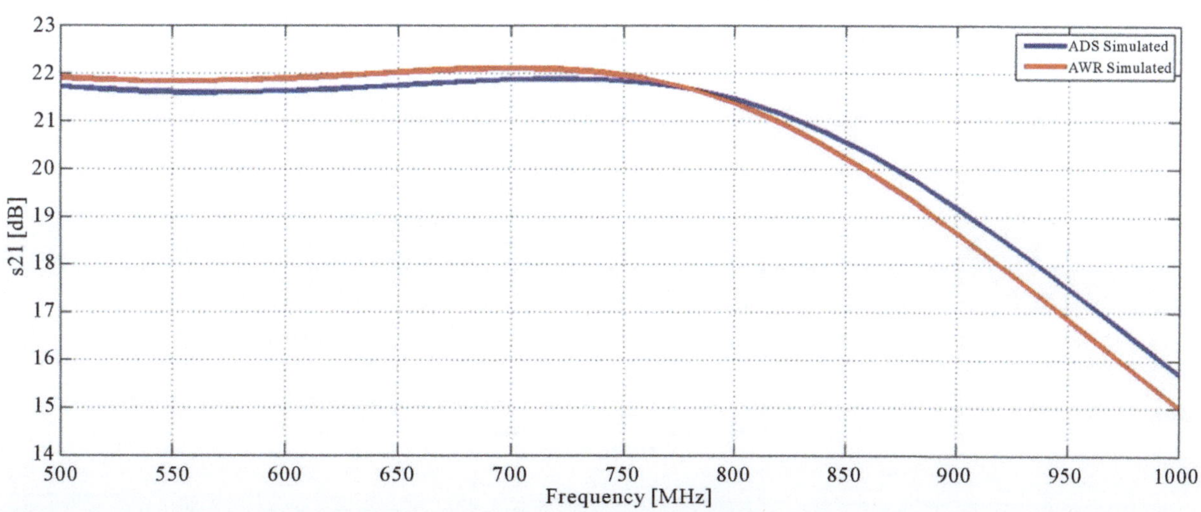

Figure 5. Simulated (ADS and AWR) power gain plot of the class AB amplifier.

Hereby is a detailed procedure used in designing a Doherty power amplifier followed with a flow chart, shown in **Figure 6**, which describes this procedure briefly.

Steps for Design of Doherty Amplifier:

1) The active devices (such as transistors) employed in the DPA are specified. They must operate on a wide range of frequencies and have an adequate output power.

2) These active devices are fully matched to ensure maximum power transfer, so input and output matching networks are implemented in the main and the peak amplifiers.

3) Stability of the main and the peak amplifiers are investigated to make sure the two amplifiers are unconditionally stable.

4) The two amplifiers are connected to each other following the well-known Doherty circuit with two transmission lines serving as impedance transformers in order to achieve the main load modulation.

5) An input power divider is inserted before the two amplifiers in order to split the power between the main and the peak amplifiers.

6) The DPA is then matched to attain maximum power transfer, so input and output matching circuits are designed for this purpose.

7) The stability of the DPA is examined and updated so that the DPA is unconditionally stable.

8) Finally, the circuit is fabricated on a primarily specified substrate.

The DPA configuration used in this paper is the well-known conventional AB-C scheme as shown in **Figure 7**, with a class AB amplifier used as a main stage and a class C amplifier used as a peak stage.

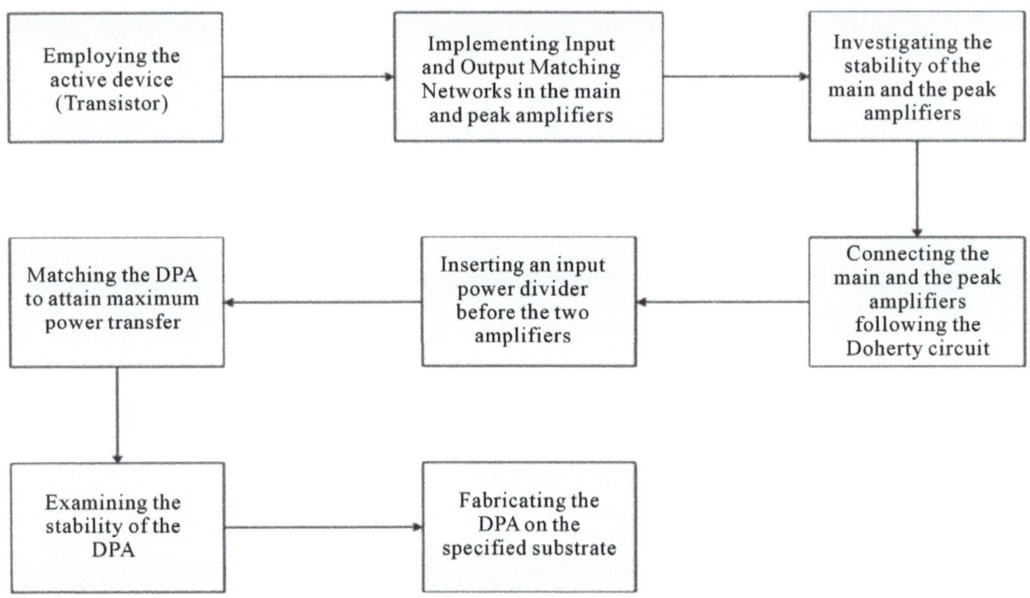

Figure 6. A flow chart of the steps of designing a Doherty power amplifier.

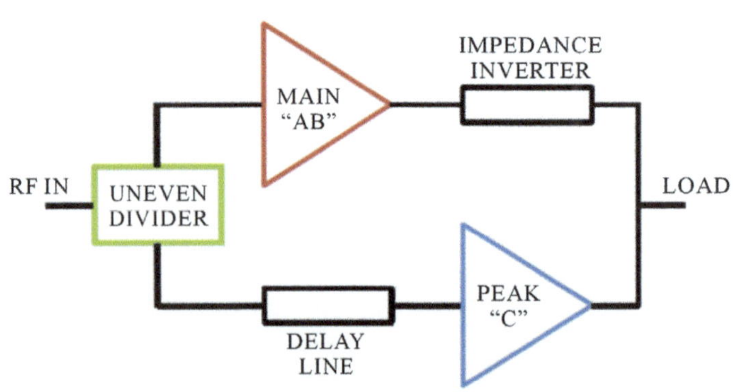

Figure 7. Block scheme of the designed AB-C DPA.

In our design, the previously designed class AB amplifier was used as the main amplifier. Hence, to complete the Doherty design, a class C amplifier is necessary as a peak amplifier. The schematic of this proposed class C amplifier is shown in **Figure 8**. The active device employed in the C amplifier is the ADL5602, an InGaP HBT with a single supply voltage of 5 V. This transistor offers high dynamic range and low noise figure in addition to that it is stable over frequency, temperature, and power supply. The amplifier circuit was simulated at 800 MHz using ADS with a supply voltage Vsup = 5 V and with input and output matching networks realized by passive elements as well as bias networks. The substrate parameters were set according to Rogers RO3203 substrate with dielectric constant $\varepsilon_r = 3.02$ and thickness h = 1.6 mm, in which the amplifier is designed on.

Figure 9 shows the simulated circuit schematic of the proposed AB-C DPA, where the main and the peak amplifiers are as mentioned before. The amplifier is simulated in both software, ADS and AWR. A broadband 90° hybrid coupler was used at the input in order to split the signals between the carrier and the peak amplifying paths. The electrical lengths of the delay and the impedance inverter microstrip lines were optimized to ensure maximum power transfer, maximum efficiency, and adequate output power.

After carrying out the stability test of the Doherty amplifier, it is found that the device is unconditionally stable in the frequency range of interest. The stability parameters, Δ, and the rollet factor k are calculated and found to be:

$$\Delta = 0.43 \angle -61.48^\circ$$

$$k = 1.4$$

In order to maintain maximum power transfer, matching circuits are designed at the input and the output of the transistor. These circuits are accomplished by using passive networks and microstrip transmission lines as shown in **Figure 10**.

Once the amplifier matching is complete, the final signal scattering parameters at 800 MHz were simulated and found to be:

$$S11 = 0.198 \angle -99.259^\circ$$

$$S12 = 0.035 \angle -55.086^\circ$$

$$S21 = 11.407 \angle 94.509^\circ$$

$$S22 = 0.175 \angle -90.038^\circ$$

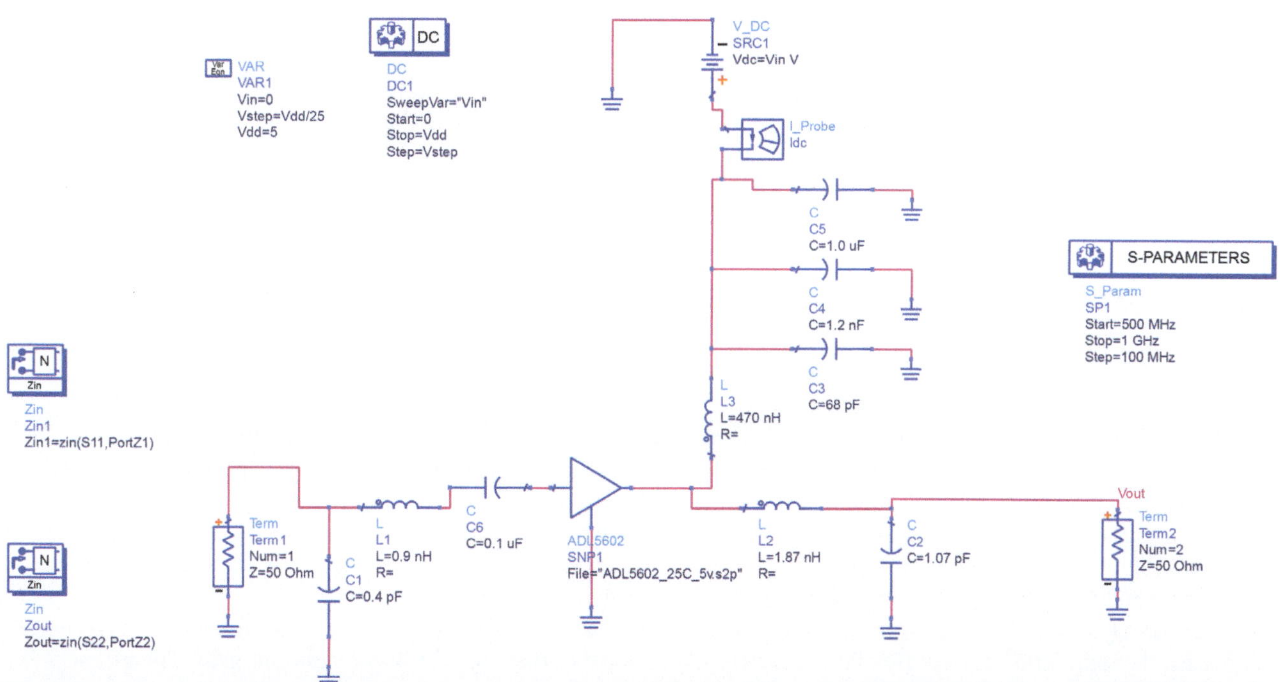

Figure 8. Schematic of the C power amplifier with bias and matching circuits.

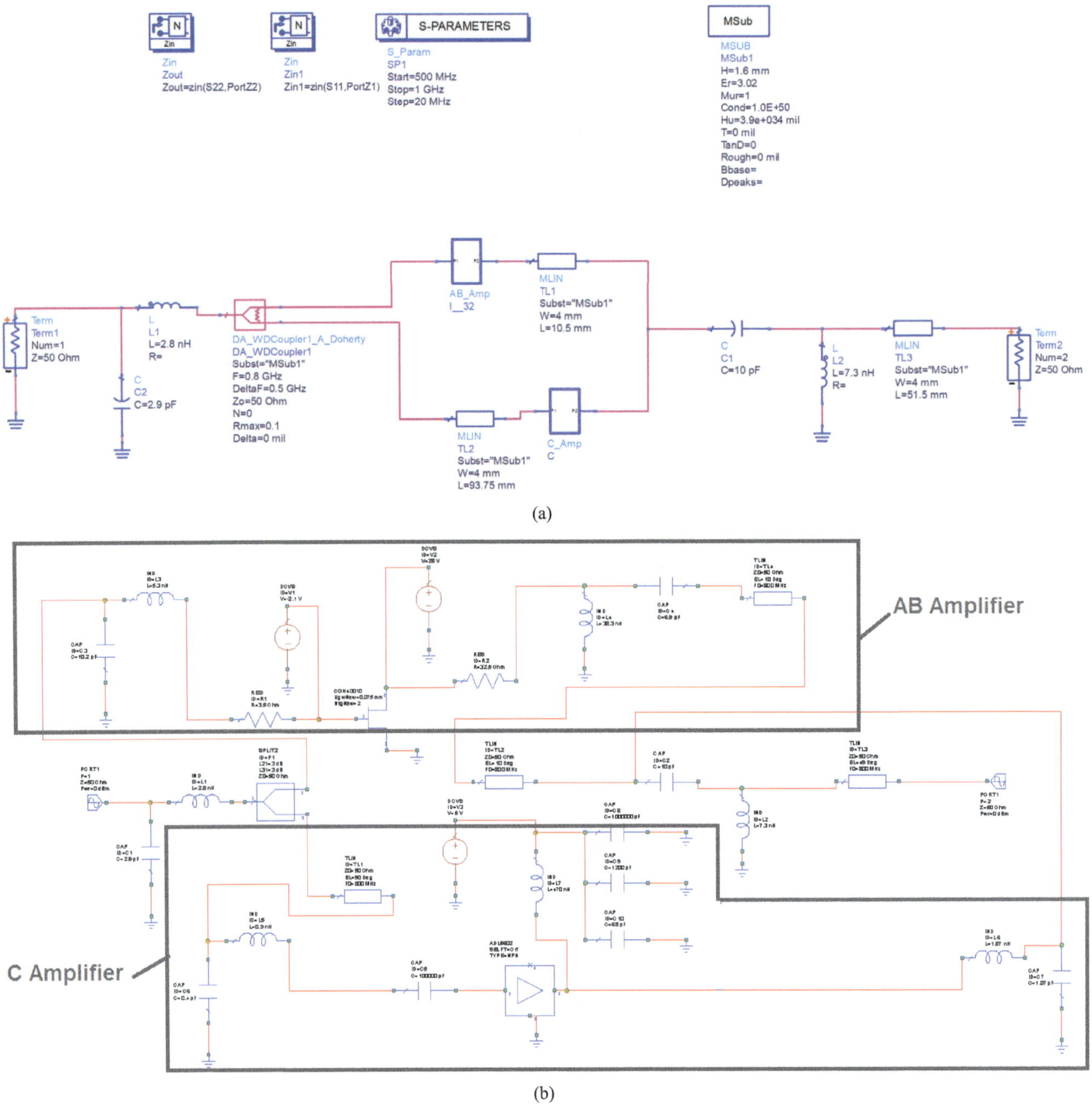

Figure 9. Schematic of the Doherty power amplifier with matching circuits. (a) In ADS; (b) In AWR.

By feeding the Doherty amplifier with an optimum 800 MHz, 21.45 dBm input power, the output power of the Doherty amplifier circuit was found to be 42.77 dBm and the gain is 21.32 dB. The DC power is simulated and found to be: Pdc = 22.92 W. Thus, the power added efficiency (PAE) is around 81.94%. All the results are summarized in **Table 3**. It is clearly shown that the efficiency of the DPA is higher than that of the class AB power amplifier. This is due to the Doherty principle implemented in this design and to the electrical lengths of the delay and impedance inverter transmission line previously mentioned.

Figure 11 shows the output power sweep curve as well as the corresponding PAE. As the DC voltage increases, the output power increases fairly linearly indicating that the amplifier has a constant gain. The efficiency also increases approximately linearly until it reaches the maximum at the optimum 28 volt DC supply voltage.

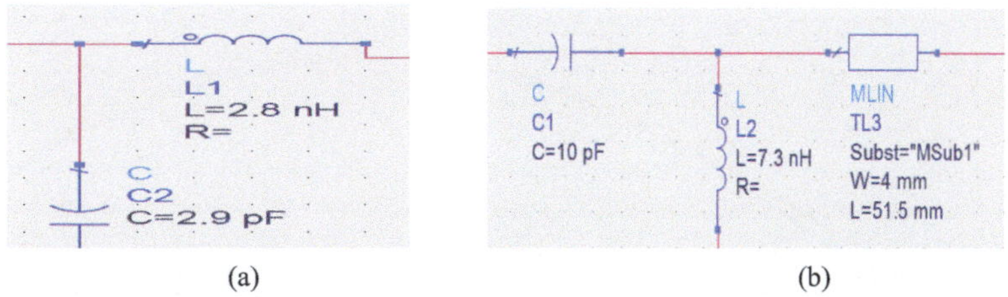

Figure 10. (a) Input matching network; (b) Output matching network.

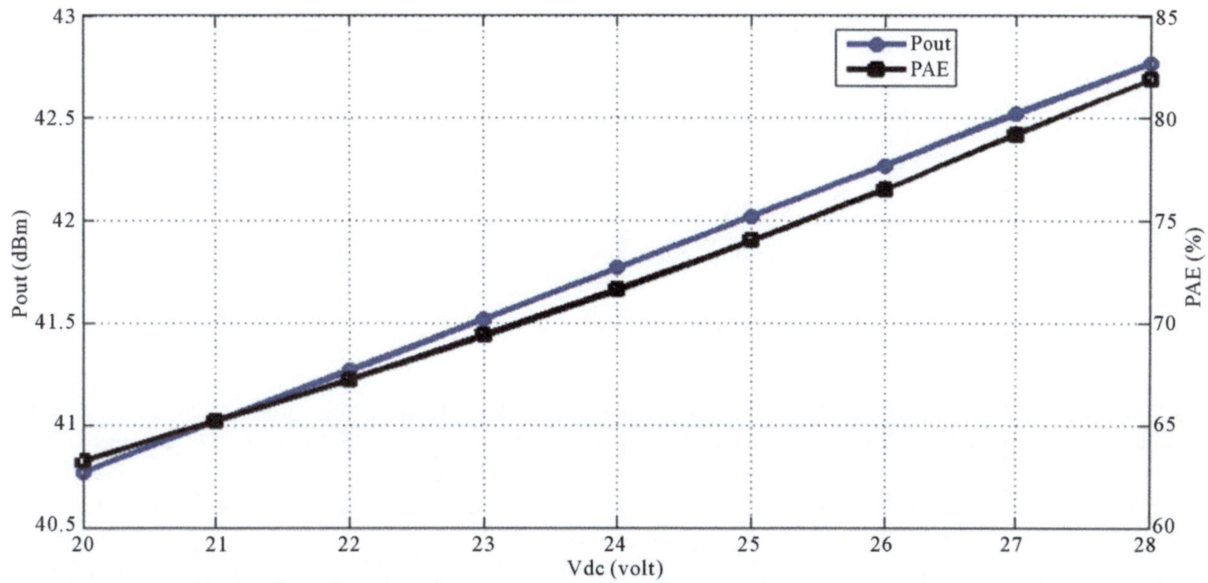

Figure 11. Output power and PAE versus DC voltage in the DPA.

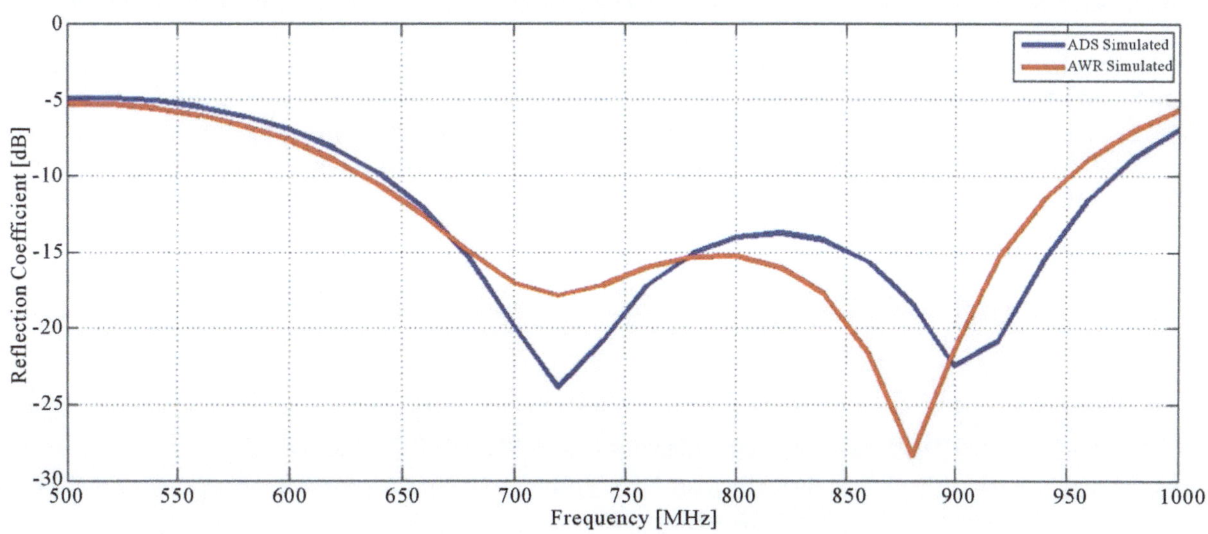

Figure 12. Simulated (ADS and AWR) reflection coefficient plot of the DPA.

Table 3. Final results of the Doherty power amplifier.

Input power (dBm)	DC power (W)	Output power (dBm)	Gain (dB)	PAE (%)	Drain efficiency (%)
21.45	22.92	42.77	21.32	81.94	82.55

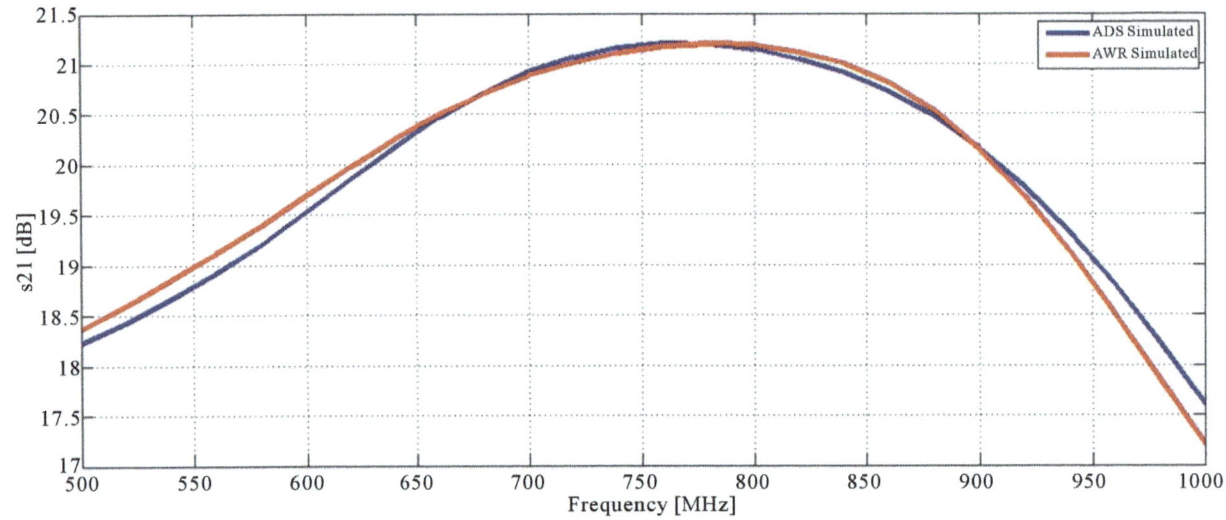

Figure 13. Simulated (ADS and AWR) power gain plot of the DPA.

The reflection coefficient of the DPA is given in **Figure 12**. It is clearly shown that this amplifier can operate from 550 MHz to 1000 MHz, thus it is suitable for all applications in the TV band.

Another advantage of the proposed Doherty amplifier rather than the efficiency increment is that the bandwidth has increased, and the proposed DPA can be suitable for more applications in the TV band. The percentage bandwidth of this DPA is 58.06% which is higher than that of the class AB power amplifier and higher than Doherty power amplifiers previously designed such as in [13], [15], and [17].

The power gain of the Doherty power amplifier is given in **Figure 13**. It is clearly shown that this amplifier has a good power gain above 20 dB for the operation frequency bandwidth.

3. Conclusion

A class AB power amplifier and a high efficiency Doherty one were implemented for the TV band applications. A class AB power amplifier was first designed using a commercial GaN HEMT from Cree Incorporation. The Doherty amplifier has been implemented, using the previous class AB amplifier as the main amplifier and a proposed class C amplifier as the peak one. This proposed DPA attained a high power-added-efficiency of 81.94%, a 42.77 dBm output power, an associated gain of 21.32 dB, and an operating frequency bandwidth between 550 and 1000 MHz which made it a convenient power amplifier that could be used in TV band applications.

Acknowledgements

The authors would like to thank the Lebanese National Council for Scientific Research for supporting this research work.

References

[1] Grebennikov (2005) RF Microwave Power Amplifier Design. McGraw-Hill, New York.

[2] Doherty, W.H. (1936) A New High Efficiency Power Amplifier for Modulated Waves. *Proceedings of IRE*, **24**, 1163-1182. http://dx.doi.org/10.1109/JRPROC.1936.228468

[3] Ahn, G., Kim, M., Park, H. and Jung, S. (2007) Design of a High-Efficiency and High-Power Inverted Doherty Amplifier. *IEEE Transactions on Microwave Theory and Techniques*, **55**, 1105-1111. http://dx.doi.org/10.1109/TMTT.2007.896807

[4] Bumman, K., Jangheon, K., Ildu, K. and Jeonghyeon, C. (2006) The Doherty Power Amplifier. *IEEE Microwave Magazine*, **7**, 42-50. http://dx.doi.org/10.1109/MW-M.2006.247914

[5] Choi, H.-J., Lim, J.-S., Jeong, Y.-C. and Kim, C.-D. (2006) Doherty Amplifier Using Load Modulation and Phase Compensation DGS Microstrip Line. *36th European Microwave Conference*, Manchester, 10-15 September 2006, 352-

355. http://dx.doi.org/10.1109/eumc.2006.281346

[6] Jung, J., Kim, U., Jeon, J., Kim, J., Kang, K. and Kwon, Y. (2005) A New "Series-Type" Doherty Amplifier for Miniaturization. *IEEE Radio Frequency Integrated Circuits* (*RFIC*) *Symposium*, 12-14 June 2005, 259-262.

[7] Raab, F.H. (1987) Efficiency of Doherty RF Power-Amplifier Systems. *IEEE Transactions on Broadcasting*, **BC-33**, 77-83. http://dx.doi.org/10.1109/TBC.1987.266625

[8] Kim, J., Cha, J., Kim, I. and Kim, B. (2005) Optimum Operation of Asymmetrical-Cells-Based Linear Doherty Power Amplifier—Uneven Power Drive and Power Matching. *IEEE Transactions on Microwave Theory and Techniques*, **53**, 1802-1809. http://dx.doi.org/10.1109/TMTT.2005.847073

[9] Kim, J., Moon, J., Woo, Y.Y., Hong, S., Kim, I., Kim, J. and Kim, B. (2008) Analysis of a Fully Matched Saturated Doherty Amplifier with Excellent Efficiency. *IEEE Transaction on Microwaves Theory and Techniques*, **56**, 328-338. http://dx.doi.org/10.1109/TMTT.2007.914361

[10] Lee, Y., Lee, M. and Jeong, Y. (2008) Unequal-Cells-Based GaN HEMT Doherty Amplifier with an Extended Efficiency Range. *IEEE Microwave and Wireless Components Letters*, **18**, 536-538. http://dx.doi.org/10.1109/LMWC.2008.2001015

[11] Markos, Z., Colantonio, P., Giannini, F., Giofrè, R., Imbimbo, M. and Kompa, G. (2007) A 6 W Uneven Doherty Power Amplifier in GaN Technology. *Proceedings of* 37*th European Microwave Conference*, Munich, October 2007, 1097-1100.

[12] Steinbeiser, C., Landon, T., Suckling, C., Nelson, J., Delaney, J., Hitt, J., Witkowski, L., Burgin, G., Hajji, R. and Krutko, O. (2008) 250 W HVHBT Doherty with 57% WCDMA Efficiency Linearized to −55 dBc for 2c11 6.5 dB PAR. *IEEE Journal of Solid-State Circuits*, **43**, 2218-2228. http://dx.doi.org/10.1109/JSSC.2008.2004456

[13] Kang, D., Kim, D., Cho, Y., Park, B., Kim, J. and Kim, B. (2011) Design of Bandwidth-Enhanced Doherty Power Amplifiers for Handset Applications. *IEEE Transactions on Microwave Theory and Techniques*, **59**, 3474-3483. http://dx.doi.org/10.1109/TMTT.2011.2171042

[14] Bathich, K., Markos, A.Z. and Boeck, G. (2011) Frequency Response Analysis and Bandwidth Extension of the Doherty Amplifier. *IEEE Transactions on Microwave Theory and Techniques*, **59**, 934-944. http://dx.doi.org/10.1109/TMTT.2010.2098040

[15] Bathich, K. and Boeck, G. (2012) Wideband Harmonically-Tuned GaN Doherty Power Amplifier. 2012 *IEEE MTT-S International Microwave Symposium Digest* (*MTT*), Montreal, 17-22 June 2012, 1-3. http://dx.doi.org/10.1109/MWSYM.2012.6259385

[16] Sun, G. and Jansen, R.H. (2012) Broadband Doherty Power Amplifier via Real Frequency Technique. *IEEE Transactions on Microwave Theory and Techniques*, **60**, 99-111. http://dx.doi.org/10.1109/TMTT.2011.2175237

[17] Qureshi, J.H., Li, N., Neo, W.C.E., van Rijs, F., Blednov, I. and de Vreede, L.C.N. (2010) A Wide-Band 20 W LMOS Doherty Power Amplifier. 2010 *IEEE MTT-S International Microwave Symposium Digest* (*MTT*), Anaheim, 23-28 May 2010, 1504-1507. http://dx.doi.org/10.1109/MWSYM.2010.5517561

16

A Method of Moment Approach in Solving Boundary Value Problems

Hilal M. El Misilmani[1], Karim Y. Kabalan[1], Mohamad Y. Abou-Shahine[1], Mohammed Al-Husseini[2]

[1]Electrical and Computer Engineering Department, American University of Beirut, Beirut, Lebanon
[2]Beirut Research and Innovation Center, Lebanese Center for Studies and Research, Beirut, Lebanon
Email: hilal.elmisilmani@ieee.org, kabalan@aub.edu.lb, mya26@aub.edu.lb, husseini@ieee.org

Abstract

Several available methods, known in literatures, are available for solving nth order differential equations and their complexities differ based on the accuracy of the solution. A successful method, known to researcher in the area of computational electromagnetic and called the Method of Moment (MoM) is found to have its way in this domain and can be used in solving boundary value problems where differential equations are resulting. A simplified version of this method is adopted in this paper to address this problem, and two differential equations examples are considered to clarify the approach and present the simplicity of the method. As illustrated in this paper, this approach can be introduced along with other methods, and can be considered as an attractive way to solve differential equations and other boundary value problems.

Keywords

Boundary Value Problem, Differential Equations, Method of Moment, Galerkin Method, Weight Coefficient

1. Introduction

The design and analysis of electromagnetic devices and structures before the computer invention were largely depending on experimental procedures. With the development of computers and programming languages, researches began using them to solve the challenging electromagnetic problems that could not be solved analytically. This led to a burst of development in a new field called computational electromagnetics (CEM), for which powerful numerical analysis techniques, including the Method of Moment (MoM), have been developed in this area in the last 50 years [1]. Harrington in [2] describes the MoM as a simple numerical technique used to convert integro-differential equations into a linear system that can be solved numerically using a computer. When the order of the equation is small, MoM can analytically solve the problem in a general and very clear manner.

Although MoM has been studied in a large number of publications, it has been mainly a part of graduate courses on computational electromagnetics, or aimed to help professionals apply MoM in their field problems. In fact, it is infrequent to see a work on MoM addressed to undergraduate students to aid their understanding of this method, and help them apply it to solve problems they face in their undergraduate courses.

A large number of publications address the use of MoM in solving various problems. Djordjevic and Sarkar in [3] showed that the inner product involved in MoM is usually an integral, which is evaluated numerically by summing the integrand at certain discrete points. Newman in [4] has presented three simple examples on the use of MoM in electromagnetics. These examples deal with the input impedance of a short dipole, a plane wave scattering from a short dipole, and two coupled short dipoles. The use of MoM in electromagnetic field has been introduced by Serteller *et al.* in [5]. This is done by presenting examples and software program, and also by giving the curriculum needed to quickly learn the basic concepts of numerical solutions.

As mentioned earlier, the idea in this paper is to introduce the approach in a simple and straightforward manner to make it understandable to students taking basic course in differential equation. First, the method is defined and illustrated, a conversion into respective integral equation is determined, and then the unknown function is expanded into a sum of weighted basis functions, where the weight coefficients are to be found. The Galerkin method, which selects testing functions equal to the basis functions, is adopted. The problem then becomes a system of linear equations, which is solved analytically or numerically to find the needed weight coefficients.

2. Familiarizing Students with the Method of Moments

As the Method of Moments is based on expanding the unknown solution of the differential equation into known expansion and testing functions that satisfy the boundary value constraints, it is advisable to first introduce this approach in the solution. Accordingly, the sough of solution is expanded into a sum of known function, each satisfying the boundary conditions of the problem, with unknown coefficients to be determined by the solution. This determination is done with the help of testing function chosen similarly as the expansion function for the simplicity of the solution. The resulting equation is then converted into a linear system of equations by enforcing the boundary conditions at a number of points. This resulting linear system is then solved analytically for the unknown coefficients. This approach is very simple and quite interesting when applied to differential equation of order less than 3, but it will get more complicated for equations of higher order.

Accordingly, it is advisable to start with some basic mathematical techniques for reducing functional equations to matrix equations. A deterministic problem is considered, which will be solved by reducing it to a suitable matrix equation, and hence the solution could be found by matrix inversion.

Simple examples using linear spaces and operators are used. At first, it is recommended to introduce MoM and define some terms related to first order non-homogeneous differential equation. The choice of this equation is important only for better understanding of the solution.

A general nth order linear differential equation, defined over a domain D, has the form:

$$a_n \frac{\mathrm{d}^n f(x)}{\mathrm{d}x^n} + a_{n-1} \frac{\mathrm{d}^{n-1} f(x)}{\mathrm{d}x^{n-1}} + \cdots + a_1 \frac{\mathrm{d}f(x)}{\mathrm{d}x} + a_0 f(x) = g(x) \quad (1)$$

In Equation (1), the coefficients $a_n, a_{n-1}, \cdots, a_1, a_0$ and $g(x)$ are known quantities, and $f(x)$ is the function whose solution is to be determined. Equation (1) can be written in the form of an operator equation:

$$L(f(x)) = g(x) \quad (2)$$

where L is the operator equation, operating on $f(x)$, and given by:

$$L = a_n \frac{\mathrm{d}^n \cdots}{\mathrm{d}x^n} + a_{n-1} \frac{\mathrm{d}^{n-1} \cdots}{\mathrm{d}x^{n-1}} + \cdots + a_1 \frac{\mathrm{d}\cdots}{\mathrm{d}x} + a_0 \quad (3)$$

The solution of Equation (1) is based on defining the inner product $\langle f, g \rangle$, a scalar quantity valid over the domain of definition of L, which is given by:

$$\langle f(x), g(x) \rangle = \int_D f(x) g(x) \mathrm{d}x \quad (4)$$

Similarly, we define:

$$\langle L(f(x)), g(x)\rangle = \int_D L(f(x))g(x)\,\mathrm{d}x \tag{5}$$

The first step in calculating the integral, using Method of Moments, is to expand f into a sum of weighted basis functions $f_1, f_2, f_3, \cdots$ in the domain of L, as:

$$f(x) = \sum_n \alpha_n f_n \tag{6}$$

Testing functions denoted $w_1, w_2, w_3, \cdots$ are defined in the range of L. These testing functions are used for all values of n. Using the inner product defined in (5), we obtain:

$$\sum_n \langle w_m, L(f_n)\rangle = \langle w_m, g(x)\rangle \quad \text{for } m = 1,2,3,\cdots \tag{7}$$

Expanding Equation (7) over the values of m and $n = 1,2,3,\cdots$, the following matrix equation is then obtained:

$$\begin{bmatrix} \langle w_1, L(f_1)\rangle & \langle w_1, L(f_2)\rangle & \cdots & \langle w_1, L(f_n)\rangle \\ \langle w_2, L(f_1)\rangle & \langle w_2, L(f_2)\rangle & \cdots & \langle w_2, L(f_n)\rangle \\ \cdots & \cdots & \cdots & \cdots \\ \langle w_m, L(f_1)\rangle & \langle w_m, L(f_2)\rangle & \cdots & \langle w_m, L(f_n)\rangle \end{bmatrix} \begin{bmatrix} \alpha_1 \\ \alpha_2 \\ \cdots \\ \alpha_n \end{bmatrix} = \begin{bmatrix} \langle w_1, g\rangle \\ \langle w_2, g\rangle \\ \cdots \\ \langle w_m, g\rangle \end{bmatrix} \tag{8}$$

In a simpler form:

$$[L_{mn}][\alpha_n] = [G_m] \tag{9}$$

and the solution for the unknown coefficients is then:

$$[\alpha_n] = [L_{mn}]^{-1}[G_m] \tag{10}$$

In our calculations, the test function w_m is chosen to be equal to the basis function f_n, which is known as Garlekin method. The determination of matrix $[L_{mn}]$ is straightforward, and its inverse is easy to obtain either analytically or numerically. Once this is done, the α_n coefficients are obtained, and the solution for f is found.

It is good to note here that choosing the appropriate basis/test function is necessary to get fast to the accurate solution.

2.1. Example 1

Considering the following second order differential equation defined by:

$$\frac{\mathrm{d}^2 f(x)}{\mathrm{d}x^2} = x^2 \tag{11}$$

defined over the domain $D = [0,1]$ with the following boundary conditions $f(0) = f(1) = 0$. Starting by choosing the basis function, let us choose:

$$f_n = x - x^{n+1} \tag{12}$$

It is clear from Equation (12) that the chosen basis function meets the boundary conditions and can be considered as a solution to the problem. Substituting Equation (12) into Equation (6), the left-hand side elements of Equation (7), which are the elements of the matrix $[L_{mn}]$, are found to be:

$$\begin{aligned} L_{mn} &= \langle w_m, L(f_n)\rangle = \int_0^1 \left(x - x^{m+1}\right)\left(\frac{-\mathrm{d}^2\left(x - x^{n+1}\right)}{\mathrm{d}x^2}\right)\mathrm{d}x \\ &= \int_0^1 \left(x - x^{m+1}\right)\left(n(n+1)x^{n-1}\right)\mathrm{d}x, \\ L_{mn} &= \frac{mn}{n+m+1}. \end{aligned} \tag{13}$$

In the same manner, we compute the elements of the matrix $[G_m]$, defined in Equation (7), which are found to be:

$$G_m = \langle w_m, g \rangle = \int \left(x - x^{m+1}\right) x^2 \mathrm{d}x = \frac{m}{4(m+4)} \tag{14}$$

Then, we start by choosing $N=1$, for which $n=m=1$ in L_{mn} and G_m, hence $L_{11} = 1/3$, $G_1 = 1/20$, $\alpha_1 = 3/20$, and $f(x)$ is given by:

$$f(x) = \frac{3}{20}\left(x - x^2\right) \tag{15}$$

It is clear from Equation (15), that the function $f(x)$ does not meet the original differential equation defined in (11). Accordingly, we need to increase the value of N.

Let $N=3$, and calculating the values of L_{mn}, G_m and α_n:

$$\begin{bmatrix} L_{11} & L_{12} & L_{13} \\ L_{21} & L_{22} & L_{23} \\ L_{31} & L_{32} & L_{33} \end{bmatrix} \begin{bmatrix} \alpha_1 \\ \alpha_2 \\ \alpha_3 \end{bmatrix} = \begin{bmatrix} g_1 \\ g_2 \\ g_3 \end{bmatrix}; \quad \begin{bmatrix} 1/3 & 1/2 & 3/5 \\ 1/2 & 4/5 & 1 \\ 4/5 & 1 & 9/7 \end{bmatrix} \begin{bmatrix} \alpha_1 \\ \alpha_2 \\ \alpha_3 \end{bmatrix} = \begin{bmatrix} 1/20 \\ 1/12 \\ 3/28 \end{bmatrix} \tag{16}$$

Hence, α_n could be calculated as:

$$\begin{bmatrix} \alpha_1 \\ \alpha_2 \\ \alpha_3 \end{bmatrix} = \begin{bmatrix} 0 \\ 0 \\ 1/12 \end{bmatrix} \tag{17}$$

Finally, calculating $f(x)$:

$$f(x) = \sum_n \alpha_n f_n = \alpha_1 f_1 + \alpha_2 f_2 + \alpha_3 f_3 = \frac{1}{12}\left(x - x^4\right) \tag{18}$$

The function $f(x)$ given in Equation (18) meets the boundary conditions defined in Equation (11), and accordingly it is the correct solution of the problem.

2.2. Example 2

Considering the following second order differential equation defined by:

$$\frac{\mathrm{d}^2 f(x)}{\mathrm{d}x^2} = 8x^2 + 2x \tag{19}$$

defined over the domain $D = [0,1]$ with the following boundary conditions $f(0) = f(1) = 0$. Starting by choosing the basis function, it was noted that the basis function used in example 1 could also be used here:

$$f_n = x - x^{n+1} \tag{20}$$

The chosen basis functions meet the boundary conditions and can be considered as a solution to the problem. $[L_{mn}]$ is calculated as in Example 1, since the same basis function is used, and it is given by:

$$L_{mn} = \frac{mn}{n+m+1} \tag{21}$$

In the same manner, we compute the elements of the matrix $[G_m]$, defined in Equation (7), which are found to be:

$$G_m = \langle w_m, g \rangle = \int \left(x - x^{m+1}\right)\left(8x^2 + 2x\right) \mathrm{d}x = \frac{m(8m+26)}{3(m+3)(m+4)} \tag{22}$$

Then, we start by choosing $N=1$ for which $n=m=1$. Accordingly, $L_{11} = -1/3$, $G_1 = 17/30$, $\alpha_1 = 21/30$, and $f(x)$ is given by:

$$f(x)=\frac{21}{30}\left(x-x^2\right) \tag{23}$$

It is clear from Equation (23), that the function $f(x)$ does not meet the original differential equation defined in Equation (19). Accordingly, we need to increase the value of N.

Let $N=3$, and calculating the values of L_{mn}, G_m, and α_n:

$$\begin{bmatrix} L_{11} & L_{12} & L_{13} \\ L_{21} & L_{22} & L_{23} \\ L_{31} & L_{32} & L_{33} \end{bmatrix}\begin{bmatrix} \alpha_1 \\ \alpha_2 \\ \alpha_3 \end{bmatrix}=\begin{bmatrix} g_1 \\ g_2 \\ g_3 \end{bmatrix}; \quad \begin{bmatrix} 1/3 & 1/2 & 3/5 \\ 1/2 & 4/5 & 1 \\ 4/5 & 1 & 9/7 \end{bmatrix}\begin{bmatrix} \alpha_1 \\ \alpha_2 \\ \alpha_3 \end{bmatrix}=\begin{bmatrix} 17/30 \\ 14/15 \\ 25/21 \end{bmatrix} \tag{24}$$

Hence, α_n could be calculated as:

$$\begin{bmatrix} \alpha_1 \\ \alpha_2 \\ \alpha_3 \end{bmatrix}=\begin{bmatrix} 0 \\ 1/3 \\ 2/3 \end{bmatrix} \tag{25}$$

Finally, calculating $f(x)$:

$$f(x)=\sum_n \alpha_n f_n=\alpha_1 f_1+\alpha_2 f_2+\alpha_3 f_3=\frac{1}{3}\left(x-x^3\right)+\frac{2}{3}\left(x-x^4\right)=\frac{2}{3}x^4+\frac{x^3}{3}-x \tag{26}$$

The function $f(x)$ given in Equation (26) meets the boundary conditions defined in Equation (19), and accordingly it is the correct solution of the problem.

3. Conclusion

As demonstrated, MoM approach could be easily used to solve mathematical problems and equations. It can be easily employed by undergraduate students. According to the type of the equation, the solution of the Moment Method will vary to accommodate for the change in the given problem.

References

[1] Gibson, W.C. (2008) The Method of Moments in Electromagnetics. Chapman & Hall/CRC, Taylor & Francis Group, UK.

[2] Harrington, R.F. (1968) Field Computation by Moment Methods. Krieger Publishing Co., Inc., Huntington.

[3] Djordjevic, A.R. and Sarkar, T.K. (1987) A Theorem on the Moment Methods. *IEEE Transactions on Antennas and Propagation*, **35**, 353-355. http://dx.doi.org/10.1109/TAP.1987.1144097

[4] Newman, E.H. (1998) Simple Examples of the Method of Moments in Electromagnetics. *IEEE Transactions on Education*, **31**, 193-200. http://dx.doi.org/10.1109/13.2311

[5] Serteller, N.F.O., Ak, A.G., Kocyigit, G. and Akinci, T.C. (2011) Experimental Study of Moment Method for Undergraduates in Electromagnetic. *Journal of Electronics and Electrical Engineering*, **3**, 115-118.

17

Collapse of Nonlinear Dust Sound Pulses in Cylindrical Dusty Plasma Waveguides

Volodymyr Grimalsky[1], Svetlana Koshevaya[1], Anatoliy Kotsarenko[2], Alvaro Zamudio-Lara[1]

[1]CIICAp, Autonomous University of State Morelos (UAEM), Cuernavaca, Mexico
[2]Faculty of Engineering, Autonomous University of Carmen (UNACAR), Ciudad Del Carmen, Mexico
Email: v_grim@yahoo.com, svetlana@uaem.mx, azamudio@uaem.mx, kotsarenko@gmail.com

Abstract

Nonlinear monopulses of dust sound waves in cylindrical dusty plasma waveguides bounded by a dielectric are investigated. The dusty plasma includes the positive ions as a light component and the negative dust as a heavy one. The dusty plasma with different masses of dust particles is considered. The set of hydrodynamic equations for the dust jointly with the Poisson one are used. The Boltzmann distribution is valid for the ions. The boundary conditions are applied at the smooth interface. When the moderate volume nonlinearity manifests, near the interface the variations of the dust concentration reach extremely high values, and the collapse of the dust sound waves occurs. In the cylindrical waveguides the collapse manifests at the values of the initial wave amplitudes essentially smaller than in the planar geometry. When the particles of lower masses are near the interface, the collapse realizes more rapidly.

Keywords

Dusty Plasma, Cylindrical Waveguide, Different Masses, Wave Collapse

1. Introduction

Last time the investigations of dusty plasmas are of great interest, not only due to various technological applications, but also because of atmosphere and ionosphere contaminations by a space trash and volcano activity [1]-[5]. An important property of dusty plasmas is the support of various waves and oscillations, both linear and nonlinear ones [3]-[8]. The dusty plasmas can be bounded and can be the waveguides for plasma waves. In the bounded plasma waves the oscillations of the surface charge take place [2] [6]. Even in the case when the volume nonlinearity is small, the variations of the dust concentration near interfaces can reach essential values [9]

[10] and cannot be considered as moderate. Earlier it demonstrated the sharp increase of the dust concentration in the thin layer near the boundary, which was called the collapse of the surface plasma waves [9]. The collapse of surface plasma wave was observed experimentally [11].

In the present paper the nonlinear monopulses of dust sound waves in the cylindrical dusty plasma waveguides bounded by a dielectric are investigated. The waves propagate along *OZ*-axis, see **Figure 1**.

The dusty plasma includes the positive ions as the light component and the negative dust as the heavy component. The set of hydrodynamic equations for the dust, namely, the continuity equation and the equation for the momentum jointly with the Poisson one for the electric field are used. The Boltzmann distribution is used for the ions. The electric and hydrodynamic boundary conditions are applied at the smooth interface dusty plasma-dielectric. The initial condition for the electric potential is monopulse-like, as well as for the perturbation of the dust concentration. The simulations have demonstrated that the wave collapse occurs near the interface at essentially smaller initial amplitudes, when compared with the planar case. The dynamics of the wave collapse depends on the distribution of the dust particles with different masses near the interface. When the particles of lower masses are near the interface, the collapse realizes more quickly.

2. Basic Equations

Consider the dusty cylindrical plasma waveguide, $0 < \rho < R$, bounded by a dielectric of the permittivity $\varepsilon > 1$ at $\rho > R$, see **Figure 1(a)**. The dusty plasma includes positively charged ions ($+e$) and negative charged dust particles ($-Q$). The temperature of ion fluid is T_i, one of the dust is T_d. It is assumed that $T_d \ll T_i$. Note that T_d is the temperature of the collective motion of the dust particles but not the temperature within each particle. The mass of the ion is m_i, the mass of the dust particle is $m_d \gg m_i$. Generally the dust particles can possess different masses in different regions: $m_d = m_d(r)$.

The set of hydrodynamic equations for the dust fluid concentration n_d and the velocity $\boldsymbol{v}_d$ is [2] [3] [12]-[14]:

$$\begin{aligned} &\frac{\partial n_d}{\partial t} + div(n_d \boldsymbol{v}_d) = 0, \\ &\frac{d\boldsymbol{v}_d}{dt} \equiv \frac{\partial \boldsymbol{v}_d}{\partial t} + (\boldsymbol{v}_d \cdot \nabla)\boldsymbol{v}_d = \frac{Q}{m_d}\nabla\varphi - \frac{T_d}{m_d n_d}\nabla n_d. \end{aligned} \tag{1}$$

The ion concentration n_i obeys the Boltzmann distribution:

$$n_i = n_{i0} \times \exp\left(-\frac{\varphi}{\varphi_{Ti}}\right), \quad \text{where} \quad \varphi_{Ti} = \frac{k_B T_i}{e}. \tag{2}$$

here n_{i0}, n_{d0} are unperturbed concentrations, in the neutral state there is $e \times n_{i0} = Q \times n_{d0}$.

Equation (1) are added by the Poisson equation for the electric potential φ (in absolute units):

$$\Delta\varphi + 4\pi e n_{i0} \times \exp\left(-\frac{\varphi}{\varphi_{Ti}}\right) = 4\pi Q n_d. \tag{3}$$

The basic equations have been reduced to the undimensional form. The following units have been used for prepare of undimensional values like φ_{Ti} for the potential, $l_n \equiv r_{Di} = \left(k_B T_i / 4\pi e^2 n_{i0}\right)$ for distances, $t_n = \left(m_{d0} / 4\pi Q^2 n_{d0}\right)^{1/2}$ for time. The parameter m_{d0} is some typical value of the dust particle mass. As the result the basic equations are:

$$\begin{aligned} &\frac{\partial n}{\partial t} + div(n\boldsymbol{v}) = 0, \quad \frac{\partial \boldsymbol{v}}{\partial t} + (\boldsymbol{v} \cdot \nabla)\boldsymbol{v} = \frac{1}{m}(\nabla\varphi - T \cdot \nabla \log n), \\ &\frac{\partial \vec{u}}{\partial t} + (\boldsymbol{v} \cdot \nabla)\boldsymbol{u} = \boldsymbol{v}; \quad m \equiv m(\rho - u_\rho), \quad \Delta\varphi + \exp(-\varphi) - n = 0. \end{aligned} \tag{4}$$

Here n, $\boldsymbol{v}$, $\boldsymbol{u}$ are undimensional dust concentration, velocity, and displacement; $T = (T_d/T_i) \times (e/Q) \ll 1$; the undimensional dust mass is $m \sim 1$. One can see that the dependence of mass on the coordinate results in an additional mechanism of nonlinearity. In calculations the value of $T \neq 0$ is used, because in the limiting case $T = 0$ it is necessary to consider the motion of the surface charge separately. Our results are independent on the value of T, when $T < 0.01$.

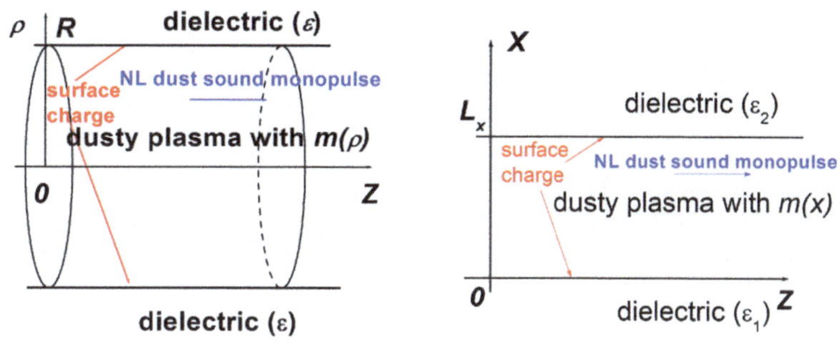

Figure 1. Geometry of the problem. Part (a) is the cylindrical dusty plasma waveguide, (b) is the planar one.

Within the dielectric at $\rho > R$ the Laplace equation is valid:

$$\Delta\varphi = 0. \tag{5}$$

The system possesses the axial symmetry. Therefore, the dust velocity $\boldsymbol{v}$ has only the two components v_ρ, v_z.

The volume equations are added by the boundary condition at $\rho = R$. Namely, the electric potential φ and the normal component of the electric induction D_ρ should be continuous at $\rho = R$:

$$\varphi(\rho = R-0) = \varphi(\rho = R+0), \quad \frac{\partial\varphi}{\partial\rho}(\rho = R-0) = \varepsilon\frac{\partial\varphi}{\partial\rho}(\rho = R+0). \tag{6}$$

Also, because it is considered the case $T \neq 0$, it should be $v_r(r = R\text{-}0) = 0$. Otherwise, in the limiting case $T \to 0$ the surface dust concentration should be considered separately. Note that the interface is assumed as smooth, so the tangential component of the dust velocity is $v_z(r = R-0) \neq 0$.

To solve the equations, the explicit 3-layer finite difference schemes have been applied for n, v_ρ, v_z, u_ρ. For the Poisson equation at $\rho < R$ the fast Fourier transform has been used:

$$\varphi(\rho,z) = \sum_l \Phi_l(\rho)\sin(k_l z); \; k_l = \frac{\pi l}{L_z}, \quad l = 1,2\cdots$$

$$\frac{1}{\rho}\frac{\mathrm{d}}{\mathrm{d}\rho}\left(\rho\frac{\mathrm{d}\Phi_l(\rho)}{\mathrm{d}\rho}\right) - \left(1+k_l^2\right)\Phi_l(\rho) = F_l; \tag{7a}$$

$$\text{where} \quad F(\rho,z) \equiv n - \left(\exp(-\varphi) - \varphi\right) = \sum_l F_l(\rho)\sin(k_l z);$$

Equation (5) for the potential φ within the dielectric at $\rho > R$ is:

$$\phi(\rho,z) = \sum_l \Psi_l(\rho)\sin(k_l z);$$

$$\frac{1}{\rho}\frac{\mathrm{d}}{\mathrm{d}\rho}\left(\rho\frac{\mathrm{d}\Psi_l(\rho)}{\mathrm{d}\rho}\right) - k_l^2\Psi_l(\rho) = 0. \tag{7b}$$

The proper solution of Equation (7b) that decreases at $\rho \to +\infty$ is

$$\Psi_l(\rho) = CK_0(k_l\rho) \tag{8}$$

Therefore, the boundary conditions (6) are rewritten as:

$$\frac{\mathrm{d}\Phi_l}{\mathrm{d}\rho} + \varepsilon k_l\Phi_l \cdot \frac{K_1(k_l R)}{K_0(k_l R)} = 0, \quad \rho = R-0. \tag{9}$$

In the center of the waveguide $\rho = 0$ the following condition is $\mathrm{d}F_l/\mathrm{d}r = 0$, because the radial component of the electric field is $E_\rho = 0$ there due to symmetry.

Equations (7a), (9) have been solved by the finite differences within the region $\rho < R$ with several iterations, which show a good convergence.

3. Results of Simulations

Earlier it was demonstrated that a manifestation of nonlinearity near the surfaces is more essential than in the volume [9] [10]. Moreover, it is possible to expect that an influence of the interface plasma—dielectric is more important in the cylindrical geometry than in the plane one, analogously to the wave collapse within framework of the nonlinear Schrödinger equation [15] [16].

Equations (4) (5) added by boundary conditions (6) have been simulated. It is possible to solve these equations under the proper initial conditions. Namely, the initial monopulses of almost rectangular shape and of moderate maximum values $|a_0| \ll 1$ have been considered:

$$\varphi(z,\rho,t=0) = -a_0 \exp\left(-\left(\frac{z-z_1}{z_0}\right)^6\right) < 0;$$

$$\tilde{n}(z,\rho,t=0) = \Delta\varphi + \exp(-\varphi) - 1 \approx a_0 \exp\left(-\left(\frac{z-z_1}{z_0}\right)^6\right) > 0; \tag{10}$$

$$v_z(z,\rho,t=0) = -\frac{\varphi}{(m(\rho))^{1/2}} > 0;$$

$$v_\rho(z,\rho,t=0) = u_\rho(z,\rho,t=0) = 0.$$

Here $\tilde{n} \equiv n-1$ is the perturbation of the dust concentration.

The following distribution of the masses of the dust particles has been used:

$$m(\rho) = m_1 + (m_2 - m_1)\cdot \exp\left(-\left(\frac{R-\rho}{\rho_q}\right)^4\right) \tag{11}$$

Namely, the values of the masses of the dust are m_1 and m_2 in the center of the waveguide and at the boundary, respectively; ρ_q is the scale of variation of the mass. The used parameters are: $z_0 = 10$, $R = 3$, $\rho_q = 1 - 0.1$, $\varepsilon = 4$. The results of simulations are tolerant to the change of the radius of the cylindrical waveguide $2 < R < 6$.

In **Figure 2** there are the dependencies of the maximum values of perturbations of the dust concentration $\tilde{n}$ on the time t in the cylindrical waveguide. The sharp peaks of the dust concentration near the surfaces are formed, which are localized both in longitudinal (z) and transverse (ρ) directions. The maximum values of $\tilde{n}$ are 40 - 50 times higher then in the volume. The nonlinearity ceases to be moderate near the surface. Moreover, under the levels of $|a_0| \geq 0.01$, see Equation (10), the calculations overflow. It is possible to interpret this phenomenon of the sharp growing and the narrowing of the dust concentration near the surface as a manifestation of the wave collapse [9] [11]. At the undimensional dust temperatures $T < 0.01$ the dynamics of surface dust sound waves does not depend on the value of this temperature.

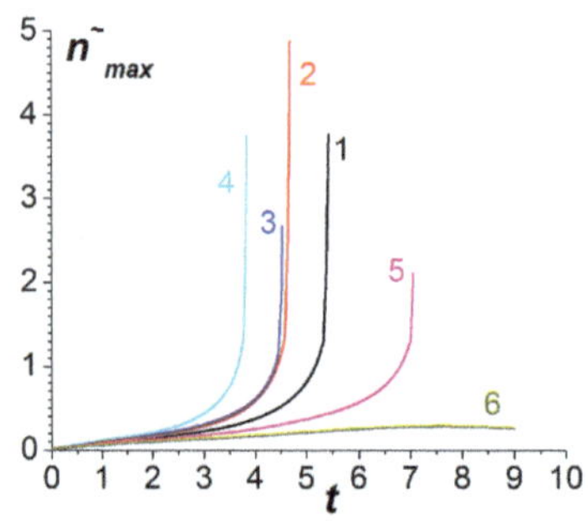

Figure 2. The dependencies of the maximum values of perturbations of dust concentration $\tilde{n}$ on the time t in the cylindrical waveguide. Curve 1 is for the initial amplitude $a_0 = -0.012$, constant masses $m_1 = m_2 = 1$; curve 2 is for $a_0 = -0.012$, $m_1 = 0.5$, $m_2 = 1$ (the lighter particles are within the volume), $\rho_q = 0.3$; curve 3 is for $a_0 = -0.012$, $m_1 = 1$, $m_2 = 0.5$ (the lighter particles are near the boundary), $\rho_q = 0.3$; curve 4 is for $a_0 = -0.012$, constant masses $m_1 = m_2 = 0.5$; curve 5 is for $a_0 = -0.01$, constant masses $m_1 = m_2 = 1$; curve 6 is for $a_0 = -0.008$, constant masses $m_1 = m_2 = 1$.

One can see that in the cylindrical geometry the wave collapse occurs at very low values of initial amplitudes of $\tilde{n}|a_0| \sim 0.01$. These values are one order lower than in the plane geometry [9] [10]. The analogous dependencies for the planar waveguide are presented in **Figure 3**. The thickness of the planar waveguide is $L_x = 6$, $\varepsilon_1 = \varepsilon_2 = 4$. The shape of the initial perturbation of the dust concentration $\tilde{n}$ is the same as in the cylindrical geometry, Equation (10). In the planar geometry the wave collapse occurs at higher initial amplitudes $|a_0| \sim 0.1$. Moreover, the tendencies to the collapse are different when the initial amplitudes are slightly above the threshold. Namely, the curves in **Figure 3**, planar geometry, are oscillatory, whereas in **Figure 2** the curves are similar to ones in the case of the explosive instability [17].

The wave dynamics depends on the values of the dust particle mass at the boundaries; see different curves in **Figure 2** and corresponding **Figure 4**, **Figure 5**. The wave collapse occurs at smaller times t when the lighter particles are near the boundary: $m_2 < m_1$. **Figure 4** and **Figure 5** are for the same masses of the particles near the center, $m_1 = 1$. The collapse occurs more quickly for the nonuniform distribution of masses $m_1 \neq m_2$ within the waveguide, compare **Figure 4** (greater masses of the dust particles near the boundary, $m_2 = 1$) and **Figure 5** (smaller masses near the boundary, $m_2 = 0.5$). The dynamics of the plasma wave below the collapse is presented in **Figure 6**. The perturbation of the dust concentration $\tilde{n}$ reaches its maximum value at some time moment and then decreases there.

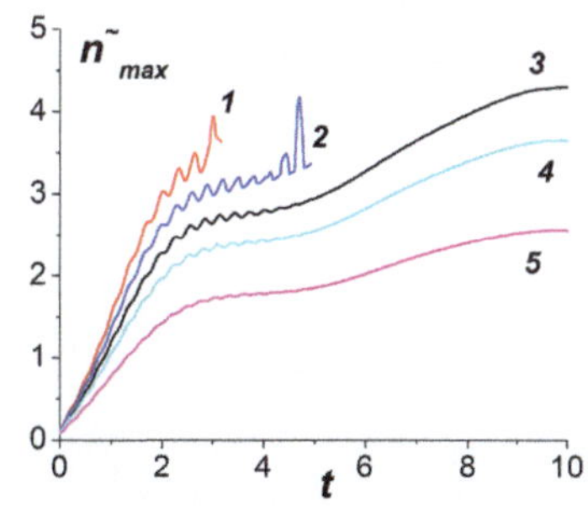

Figure 3. The dependencies of the maximum values of perturbations of dust concentration $\tilde{n}$ on the time t in the planar waveguide of the thickness $L_x = 6$. For all the cases the dust masses are constant $m_1 = m_2 = 1$. Curve 1 is for the initial amplitude $a_0 = -0.13$; curve 2 is for the initial amplitude $a_0 = -0.12$; curve 3 is for the initial amplitude $a_0 = -0.11$; curve 4 is for the initial amplitude $a_0 = -0.10$; curve 5 is for the initial amplitude $a_0 = -0.08$.

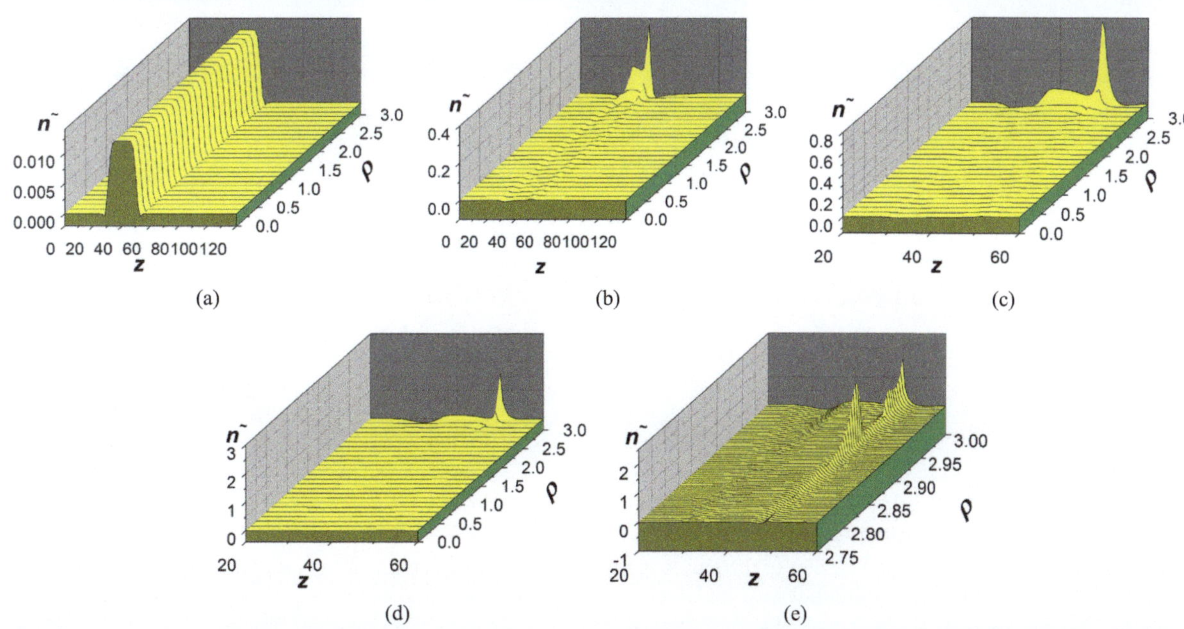

Figure 4. The spatiotemporal evolution of the perturbation of the dust concentration above the threshold for $a_0 = -0.012$, $m_1 = m_2 = 1$ (see curve 1 in **Figure 2**). Part (a) is at $t = 0$; part (b) is for $t = 4$; part (c) is for $t = 5$; parts (d), (e) are for $t = 5.42$, the general view and the details near the boundary.

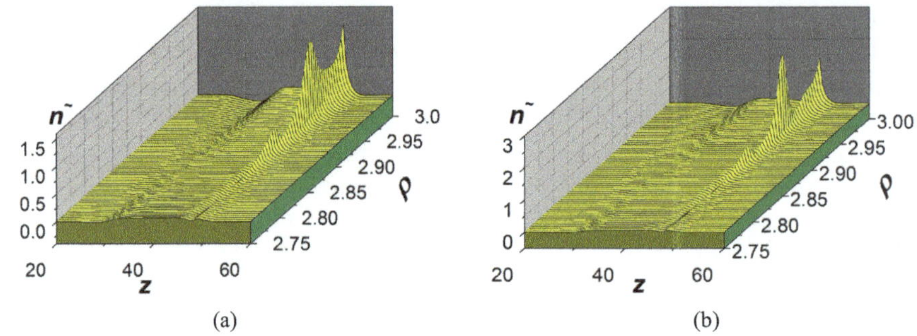

Figure 5. The spatiotemporal evolution of the perturbation of the dust concentration above the threshold for $a_0 = -0.012$, $m_1 = 1$, $m_2 = 0.5$ (see curve 3 in **Figure 2**). Part (a) is at $t = 4.5$; part (b) is for $t = 4.53$. The initial pulse is the same as in **Figure 4(a)**.

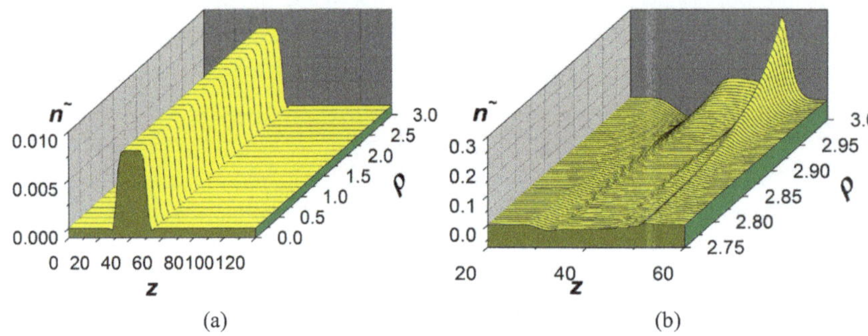

Figure 6. The spatiotemporal evolution of the perturbation of the dust concentration below the threshold (curve 6 in **Figure 2**). Part (a) is the initial distribution at $t = 0$; part (b) is at $t = 8$.

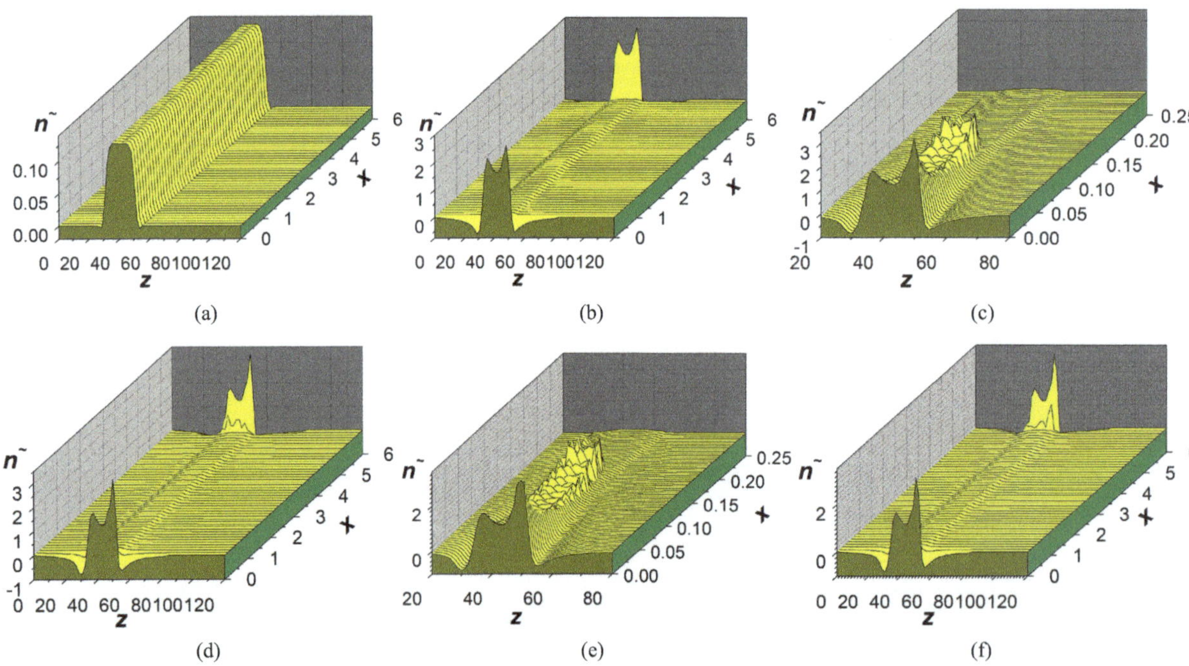

Figure 7. The spatiotemporal evolution of the pulses in the planar dusty waveguide. The initial data $a_0 = -0.12$ correspond to **Figure 3**, curve 1. Part (a) is the initial distribution at $t = 0$; part (b), is at $t = 2$; parts (c), (d) are at $t = 4$; parts (e), (f) are at $t = 4.5$.

The dynamics of wave collapse in the planar geometry is presented in **Figure 7**. There is some difference between the wave dynamics in the cylindrical and planar geometries. In the cylindrical geometry, **Figure 4**, there are two sharp peaks of the perturbation of the dust concentration in the transverse direction ρ. In the planar geometry, **Figure 7**, there are two peaks in the longitudinal direction z.

4. Conclusions

The evolution of initial rectangular pulses of dust sound waves in the cylindrical plasma waveguides with dielectrics has demonstrated the essential surface nonlinearity near the interface. Under low values of the input amplitudes the values of the dust concentrations near the surface dusty plasma-dielectric reach high values, and the formation of narrow peaks of the dust concentration occurs. The nonlinearity near the surface ceases to be moderate, the compression of these pulses both in longitudinal and transverse directions takes place, and the wave collapse can be observed. The dynamics of the wave collapse depends on the distribution of the masses of the dust particles within the waveguide.

The estimated temporal and spatial scales have demonstrated that an appearance of the investigated phenomena of the nonlinear wave collapse could be possible in the dust objects within the atmosphere and in laboratory plasmas.

Acknowledgements

The authors are grateful to SEP-CONACyT (Mexico) for a partial support of our work.

References

[1] Fortov, V.E., Ivlev, A.V., Khrapak, S.A., Khrapak, A.G. and Morfill, G.E. (2005) Complex (Dusty) Plasmas: Current Status, Open issues, Perspectives. *Physics Reports*, **421**, 1-103. http://dx.doi.org/10.1016/j.physrep.2005.08.007

[2] Shukla, P.K. and Mamun, A.A. (2002) Introduction to Dusty Plasma Physics. IoP Publisher, Bristol, 270. http://dx.doi.org/10.1887/075030653X

[3] Verheest, F. (2001) Waves in Dusty Space Plasmas. Kluwer, Dordrecht, 280.

[4] Shukla, P.K. and Mamun, A.A. (2003) Solitons, Shocks and Vortices in Dusty Plasmas. *New Journal of Physics*, **5**, 17.1-17.37.

[5] Popel, S.I., Kopnin, S.I., Kosarev, I.N. and Yu, M.Y. (2006) Solitons in Earth's Dusty Mesosphere. *Advances in Space Research*, **37**, 414-419. http://dx.doi.org/10.1016/j.asr.2005.12.003

[6] Ostrikov, K.N., Yu, M.Y. and Stenflo, L. (2000) Surface Waves in Strongly Irradiated Dusty Plasmas. *Physical Review E*, **61**, 782-787. http://dx.doi.org/10.1103/PhysRevE.61.782

[7] Kopnin, S.I. and Popel, S.I. (2005) Dust Acoustic Mode Manifestations in Earth's Dusty Ionosphere. *AIP Conference Proceedings*, **799**, 161-164. http://dx.doi.org/10.1063/1.2134590

[8] Popel, S.I., Golub', A.P., Losseva, T.V., Bingham, R. and Benkadda, S. (2001) Evolution of Perturbation in Charge-Varying Dusty Plasmas. *Physics of Plasmas*, **8**, 1497-1504. http://dx.doi.org/10.1063/1.1359743

[9] Grimalsky, V., Koshevaya, S., Kotsarenko, A. and Perez Enriquez, R. (2012) Collapse of Nonlinear Surface Dust Sound Waves in the Dusty Plasma-Dielectric System. *Physica Scripta*, **85**, 1-6. http://dx.doi.org/10.1088/0031-8949/85/03/035503

[10] Bustos De La Rosa, M.E., Grimalsky, V., Koshevaya, S. and Kotsarenko, A. (2013) Collapse of Nonlinear Dust Sound Pulses in Dusty Plasma Waveguides. *Progress in Electromagnetics Research Symposium Proceedings*, Stockholm, 12-15 August 2013, 1463-1467.

[11] Sarkar, S., Bose, M., Mukherjee, S. and Pramanik, J. (2013) Spatiotemporal Evolution of Dielectric Driven Cogenerated Dust Density Waves. *Physics of Plasmas*, **20**, Article ID: 064502. http://dx.doi.org/10.1063/1.4810804

[12] Infeld, E. and Rowlands, G. (2000) Nonlinear Waves, Solitons and Chaos. CUP, Cambridge, UK, 391. http://dx.doi.org/10.1017/CBO9781139171281

[13] Lifshitz, E.M. and Pitaevskii, L.P. (1981) Physical Kinetics. Pergamon, London, 452.

[14] Ostrovsky, L. and Potapov, A. (2002) Modulated Waves: Theory and Applications. The John Hopkins University Press, New York, 392.

[15] Berge, L. (1998) Wave Collapse in Physics: Principles and Applications to Light and Plasma Waves. *Physics Reports*, **303**, 259-370. http://dx.doi.org/10.1016/S0370-1573(97)00092-6

[16] Kivshar, Yu. and Agrawal, G. (2003) Optical Solitons. From Fibers to Photonic Crystals. Academic Press, New York, 540.

[17] Weiland, J. and Wilhelmsson, H. (1977) Coherent Non-Linear Interaction of Waves on Plasmas. Pergamon, Oxford, 353.

18

Bistability in Ephaptic Coupling as a Possible Memory Element Explored through a Colloid Interptation of Stereocilia Bistability

Michelangelo Rossetto

125 w 86 th st, New York, USA
Email: marossetto@libero.it

Abstract

When the processes of neurons are in close proximity they effectively couple to each other in a process termed Ephaptic coupling. This coupling occurs without the membranes touching when the space between them is some tens of nanometers. If the intra-membrane spacing has a mechanical bistability, which moves the membranes closer and further apart, the Ephaptic coupling will be turned on and off allowing one neuron to read, or not read the content of the other. This paper explores the possibilities of bistable ephaptic memory element by a study of an analogous system that operates at a much larger scale, the Stereocilia of the hair cell. Published measurement of hair cell Stereocilia force/displacement function shows both a negative slope region in the displacement function and tendency to express bistability. We show here how this negative slope region can arise through the exigency of colloid forces. An explanation of Colloid Theory, presented in a graphic form, shows how a colloid force function can be modified to match the measured hair cell cilia force function. The colloid force function is modified by a limiting function resulting from Stereocilia side links, structural details that tie together the hair cell Stereocilia clusters. Understanding how a limited, simple behavior such as Stereocilia bistability functions may point to a more general understanding of how bistability may underlie other areas of living organisms such as memory and computation.

Keywords

Ephaptic, Hair Cell, Stereocilia, Colloid, Bistability, Memory

1. Introduction

Ephaptic interactions in neural networks are now recognized as a significant modality that must be incorporated in the understanding of complex neural systems [1] [2]. This paper explores the possibilities that arise from bistability that is integrated into an ephaptic interaction. This bistable/ephaptic interaction, where the bistability moves the membranes involved closer and further apart, has the potential to serve as a memory element.

The importance of bistability in biology has been long recognized. Bistability plays a role in cell division, differentiation, cancer onset and apoptosis. These forms of bistability occur when a cell chooses one of two possible developmental paths depending on local conditions. They are slow acting. Time is measured in hours and days.

The colloid driven bistability we consider here is fast. It operates in microseconds. Colloid forces are electromagnetic forces resulting from the physical chemistry of cellular surfaces. A colloid force function results from the interaction of attractive and repulsive forces. The attractive forces are Van der Wall interactions where the constituent molecules are polarized. Repulsive forces result from the negatively charged electron clouds of nearby particles repelling each other. **Figure 5(a)** illustrates how the attractive/repulsive interaction results in the colloid function. The attractive and repulsive forces are inverse power functions with different exponents. Their interaction is seen to result in a function that has areas with both positive and negative slopes.

Measurement in university labs of colloid forces in the 1980s' uncovered the shape of the colloid function. It was, however, bedeviled by bistability during the measurements [3]. This bistability arose when the measuring probe presented a light load to the colloid system. In **Figure 5(b)** the jumps J and J' are expressions of the bistability that results from the negative slope region of the force curve. The part of the force curve that has a negative slope corresponds to the hump in the energy curve that separates two areas of minimum energy which are stable resting places. An exhaustive review of colloid measurement techniques and their difficulties was written by Craig in 1997 [4]. Israelachvilli wrote the definitive text on the subject of colloid interactions [5] [6].

Emulsions are colloid driven suspensions of, for example, fat droplets suspended in water or water droplets suspended in fat. They are stable when the interaction of the two substances rests in an energy minimum. Cosmetics and many food products are colloidal emulsions. Butter, for example, is a water in fat emulsion, while milk is a fat in water emulsion.

One of the earliest recognition of a colloidal system in a living biological function was the observation that clusters of vesicles found at nerve terminal synapse behave like an emulsion. A typical study of this phenomenon is found in a 1981 paper by Nir. S. from which the energy curve of **Figure 5(a)** was abstracted.

The vesicles clustered in nerve terminals adjust their spacing as an expression of their inter vesicle colloid interaction. Although the vesicles are spherical, when one considers the cross section formed by the common equatorial plane of two adjacent vesicles it is obvious that the cross section is the same as the cross section of two adjacent hair cell Stereocilia as shown in **Figure 3**. They both have similar dimensions, are bounded by similar membranes and have similar spacing [7]. These similarities are what lead to the consideration of hair cell Stereocilia to be working under the same influence of colloid forces as are the vesicles.

Two measurements, selected from Hair Cell literature, are used here in support of a theory of a colloid based bistable functionality in a living biological system. These two measurements are both quantitative. They are:

1) The measurement of a negative slope in the force/distance function of cilia which are accompanied by a mechanical bistability.

2) The measurement of the separation limit of adjacent cilia due to cilia side links which behave like a "door chain" that allows only limited separation.

We show how these measurements, taken together and integrated into a graphically presented colloid interaction, illustrate a plausible mechanism for the mechanical bistability observed in the Hair Cell Cilia and hypothesized for an ephaptic bistability.

1.1. Force/Distance Measurement

Martin measured the mechanical force versus distance function of hair cell cilia [8]. As shown in **Figure 1**, this measured curve shows that the hair cell cilia have an area of negative slope in their force versus distance function

These force measurements were made difficult by mechanical jumps, or bistability during the measurement. The area of negative slope can underlie the bistability observed in the hair cell cilia. Martin pointed out that this curve mimicked the characteristic curve of the tunnel diode, shown in **Figure 4**, and he suggested that knowledge accumulated in the study and application of the tunnel diode could be useful in interpreting hair cell activity. The tunnel diode can be used in electronic bistable circuits, electronic circuits that have two stable states.

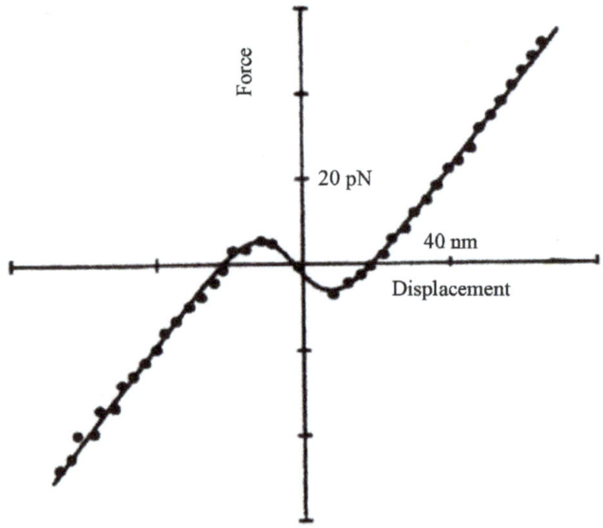

Figure 1. A simplified curve abstracted from the "force-displacement" function of the "cilia-kinocilium" bundle from *in vitro* hair cell measurements performed and presented by Martin [8]. This measurement uncovered a negative slope portion of the curve. It locates the active behavior of the hair cell, which includes bistability at the ciliated end of the cell. This observation led Martin to point out the functional similarity of the hair cell cilia and the operating curve of the Tunnel diode, an electronic semiconductor device. The negative slope part of the curve spans a distance of approximately 30 nM and a force range of about 20 pN.

1.2. Side Links Limit Cilia Separation

Nam studied the extra-cellular links in the vestibular hair bundle. **Figure 2** and **Figure 3** show the cilia side links schematically.

These links constrain, or limit, to nanoscale dimension the spacing between cilia. After careful measurements Nam stated that the side links experience zero tension when they are slack. When they reach the limit of their extendibility they rigidly prevent any further increase in the spacing between cilia, just as a door chain does [9]. Nam summarized this behavior graphically and I have redrawn it, with a scale based on his measurements and the dimensions in Martins measurements, in **Figure 5(c)**.

1.3. Bistability in the Tunnel Diode

When the tunnel diode [10] is biased with the constant current (I bias) it has the two stable voltage states, X and Y, which express its bistability as shown in **Figure 4**. Martins measured "force-displacement" curve of Hair Cell Cillia, [8] which contained a segment with a negative slope, can be expected to function in its mechanical domain with a bistability in a manner analogous to the electrical behavior of the tunnel diode.

2. Model Theory

By showing how a colloid force function can be modified by the limiting function resulting from cilia side links we demonstrate how a new function is formed. This function has the same form as Martins measured cilia function and has the same form as the tunnel diode curve.

The theory of hair cell activity presented here is based on the area of physical chemistry that deals with colloid interactions. Interactions between colloid particles are an electromagnetic phenomenon. The combined van der Waal attractive dipole forces and the electrostatic repulsion experienced by the orbiting electrons of adjacent particles as they approached each other underlies colloid interactions. This model considers the cilia of hair cells to be colloid particles whose electromagnetic surface interactions underlie the bistable behavior observed in the course of Hair Cell Cilia measurements. It describes the requirements for a device that marshals colloid forces in

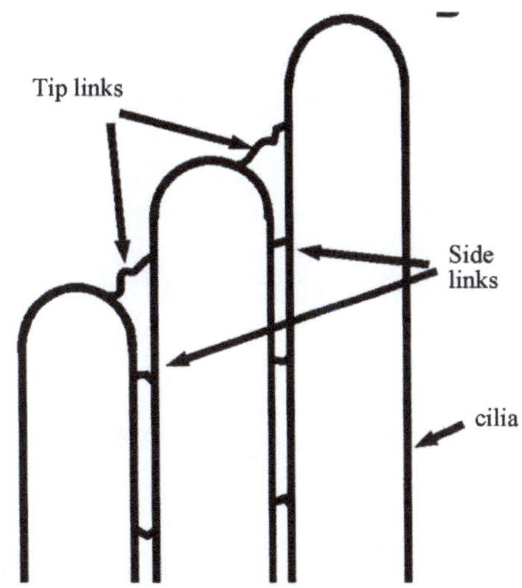

Figure 2. A schematic side view of adjacent cilia showing the typical placement of the side links that keep the cilia from moving too far apart.

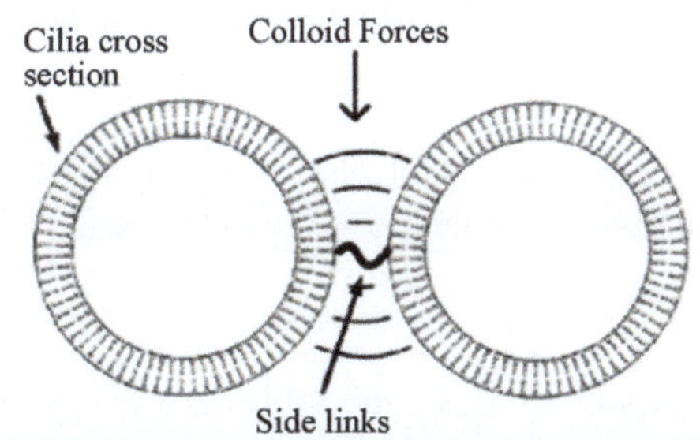

Figure 3. A schematic cross sectional view of two adjacent cilia showing how the side links limit the spacing between cilia and how the colloid forces activate the inter cilia interactions. This figure expands on **Figure 2** of [7] by the inclusion of the side links.

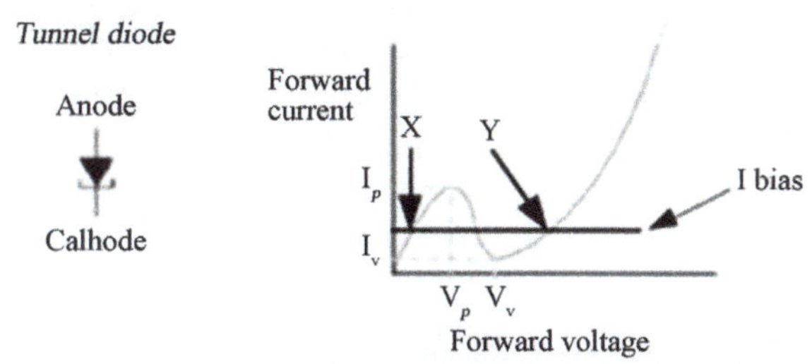

Figure 4. The functional curve of the tunnel diode, an electronic device which can be used as an amplifier, oscillator, or bistable memory element. Martin pointed out its similarity to the measured hair cell force curve. When the tunnel diode is biased with a constant current, "I bias" in this figure, the diode has two stable states. These two stable states are labeled X and Y in the figure.

the realization of an active mechanical device which exhibits a mechanical bistability that can exist in a living biological organism. Background and a historical perspective of the application of colloid theory to hair cell cilia can be found in a previous paper [7].

2.1. Colloid Energy

Figure 5(a) shows both Vw, the van der Waal attractive force, and Ve, the electrostatic repulsive force. They are both inverse power functions with different exponents in the energy domain. The van der Waal forces are the attraction resulting from dipole moments. The electrostatic repulsive forces result from the orbital electrons of adjacent particles getting close enough to feel each other. The interaction of the attractive van der Waal and repulsive electrostatic forces is equivalent to a subtraction of two similar large numbers. This is a process that emphasizes small differences. In this case the subtraction of the differing ripples produced by the two inverse power functions with different exponents, results in the complex shape of the resulting combined energy function Vt of **Figure 5(a)**.

A change of the distance between two colloid particles will subject them to energy changes of the type shown in **Figure 5(a)**, which is adapted from a study of vesicles [6]. This figure presents in simple graphic form the DLVO energy relations that determine the basic behavior of a typical lyophobic colloid system. The abscissa (x) is separation, with zero distance, or contact, at the left of the figure. Vt is the resultant energy of interaction which combines Van der Waals attraction Vw and the electrostatic repulsion Ve.

2.2. Colloid Forces

Figure 5(b) presents the slope of the energy curve Vt, dV/dx, which is the force experienced by closely opposed colloid particles. Colloid particles will congregate in an energy minima if it is deep enough, that is deeper than the thermal energy of the particles [6]. Note that at the energy minimum the force is zero. This is a stable resting place on the force curve.

The part of the Force curve important to a colloid theory of the hair cell, is the part of the curve of **Figure 5(b)** that has a negative slope. The part of the force curve that has a negative slope is like the negative portion of the measured cilia force function shown in **Figure 1**. Force curves similar to the curve of **Figure 5(b)** have often been drawn from careful measurements of colloid systems. Craig presents a review of the field of colloid force measurements, the difficulties and the instruments used to make them [4].

2.3. Limiting Function

The links between cilia limit the separation that is possible between these elements. This limiting of the inter cilia spacing described by Nam [9], is represented by the curve of **Figure 5(c)**. When the limit function is combined with the force function, the limited force function will be as shown in **Figure 5(d)**. The placement of the limiting function breakpoint is consistent with the quantitative measurements listed in the results section.

2.4. The Limited Force Function

Figure 5(d) shows how the limit function modifies the colloid force function.

When the cilia are close to each other the links between adjacent colloid particles are slack. The forces they experience are the Colloid forces. When the links are stretched to their limit, the Limit Function quickly dominates. The result is that the Limited Force Function, shown in **Figure 5(d)**, now resembles both the form of the tunnel diode curve of **Figure 4** and the form of the measured force-distance curve of the cilia of the hair cell of **Figure 1**. The Points labeled A and B in **Figure 5(d)** are the two stable states that underlie the bistable functionality. They represent two stable points in the Limited Force curve.

3. Results

The limiting of the cilia spacing, or more generally, the limiting of the spacing between colloid elements, is the key to the concept of colloid mediation of the bistability as seen in hair cell cilia.

After careful measurements Nam stated that the side links experience zero tension when they are slack. When they reach the limit of their extendibility they rigidly prevent any further increase in the spacing between cilia, just as a door chain does. Nam [9] found the length of the links when fully extended, and therefor the break point for this limiting action is 78 nM in the system he studied. The break point of 78 nM fits quite well with the measured active range of 30 Nm measured by Martin in the Hair Cell.

4. Discussion

Laboratory measurement of colloid force functions reported by Craig [4] and Horn [3] often showed bistable

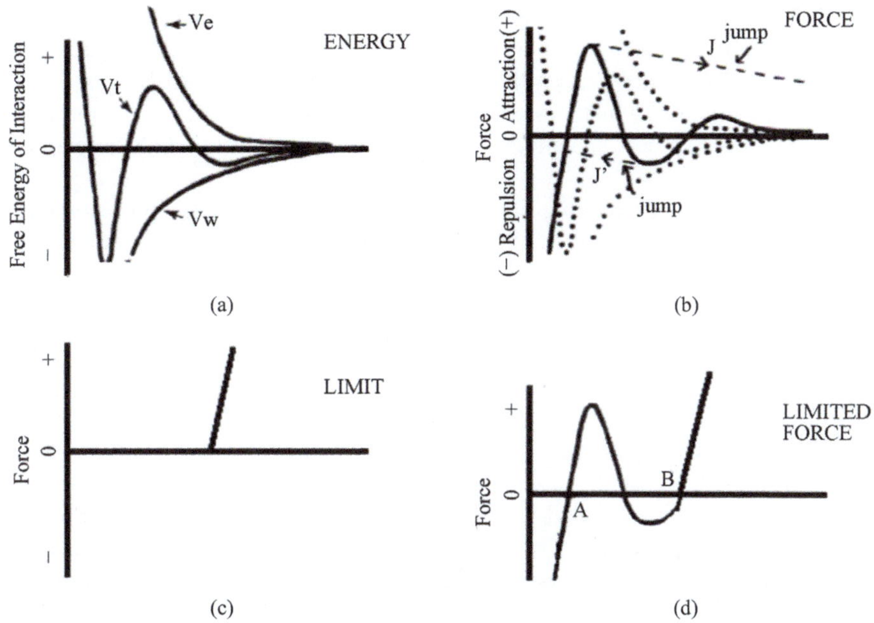

Figure 5. (a) this figure (first presented as Figure 3A in ref [6] is a summery of basic colloid theory showing the free energy of interaction Vt that results from the combined Van der Walls attraction, Vw, and the electrostatic repulsion, Ve, experienced by adjacent colloid particles. The Van der Wall force, Vw, is the consequence of the attraction of molecular dipoles in adjacent particles. The electrostatic repulsion, Ve, arises when the orbital electrons of adjacent particles get close enough to feel each other. The abscissa, X, represents the distance between the colloid particles. Contact, zero distance, is at the left of the figure. (b) The slope of the energy curve Vt, dV/dx, which is the force experienced by adjacent colloid particles as a function of separation (x) was drawn from Figure 3B of ref [7]. It represents a graphic summery of the colloid force relations that are explored in this paper. Note that at the energy minimum of (a), the force curve, shows a stable resting place where the colloid particles experience neither repulsion nor attraction. (c) The side link behavior expressed graphically. Showing how the links are like the ropes that tie a boat to a dock. When the rope is slack there is no force in the rope, when the boat moves away from the dock, the rope reaches its limit and becomes tense. The swiftly rising curve limits how far apart the cilia are to each other. (d) The combined interaction of the colloid force function and the limit function showing how the resulting combined function has the same form as the measured cilia force function of **Figure 1** and the tunnel diode curve of **Figure 4**.

behavior when the colloid system was lightly loaded by the measuring apparatus. This bistability is shown by the jumps J and J' in the Force function of **Figure 5(b)**.

Martin's measurements of the hair cell cilia force function shown in **Figure 1** were irritatingly accompanied by bistable jumps during the measurements. Bistable behavior is seen in lightly loaded Hair Cell cilia [8]. Notice that in Martin's curve there are no data points in the section with a negative slope. This is the unstable zone which engenders jumps from one stable zone to the other.

Cillia functional similarity to the tunnel diode operating curve includes bistability. Martin reported his observations of a "tunnel diode like" curve and bistability but did not have an attempt to define how the organism achieves this result. This is the question this paper addresses. Is the mechanism behind bistability in the hair cell natures use of colloid forces? By bringing together hair cell measurements reported in the biophysical literature and colloid theory this paper presents arguments to support a colloid view of hair cell cilia functionality.

While the theory of colloid mediated bistability is developed here to explain the observed behavior of the Hair Cell Cilia, the process described is not limited to the Hair Cell Cilia. Colloid forces will develop whenever two cellular membranes have the appropriate close spacing. If the membranes of adjacent neurons or their processes are close enough to experience colloid forces and are tied together with a limiting link, similar to the side links seen between Hair Cell Cilia, a spatial bistability will appear between these membranes. This spatial bistability would be well placed to modulate Ephaptic coupling between neurons [11].

The bistability observed in Stereocilia operates in a dimensional space where microns are the measure of Stereocilia dimensions. The area available for a colloid interaction when two dendrites pass closely is limited. This area will be orders of magnitude smaller than the Hair Cell Stereocilia area involved in a colloid interaction. Yet this small patch is just as capable of bistability as is the Hair Cell if the spacing is appropriate and there is a limiting fiber. There are hints of fibers spanning the space between adjacent dendritic membranes in some early SEM images [12] [13].

If this small area common to two passing dendrites is functioning as a Ephaptic coupling and is at the same time bistable we have a unitary memory element. The two bistable states allow either an ephaptic coupling or no coupling.

If nature has chosen to fill it's heed for a resettable memory element by exploiting the bistability inherent in a colloid system, a two state ephaptic coupling mechanism may be a contender.

5. Conclusions

If the Stereocilia of the hair cell are indeed behaving as a colloid system, as the data quoted here support, it is a model for a process that may be present in areas of an organism that need memory and computation.

This paper presents a theory of Hair Cell functionality that is based on published biophysical measurements and Colloid Theory. It builds on a summary of both hair cell structure and basic Colloid theory that can be found in an earlier paper [7], which showed how colloid theory could be applied to understanding some early hair cell measurements.

The importance of this paper is twofold.

First it demonstrates that published experimental measurements, considered individually, may not lead to an explanation of the process being measured. A perspective that allows for the simultaneous view of the results of disparate work, either in the same field or in different fields, may engender novel insights. By considering Biophysical measurements in conjunction with colloid theory new insights unfold.

Secondly by recognizing that a phenomenon that is expressed on two grossly disparate scales can be studied in the more accessible larger scale. Bistability that is measurable in Hair Cell cilia can give insight toward the understanding of the less accessible behavior of a possible Ephaptic bistability. An Ephaptic colloid bistability memory element, if it exists, would allow not only remembering but also forgetting. A colloid driven bistability could be both set and reset by the attraction and repulsion of nearby electric fields. Colloid forces are electromagnetic forces.

6. A Testable Prediction

If the vibratory motion of the cilia bundle is operating under the influence of the electro magnetic forces of a colloid system it should be susceptible to external magnetic fields. This makes it a candidate for the magnetic sensitivity which underlies the compass used by migrating animals. A test for this sensitivity could be made by observing the resting discharge of, for instance, a lateral line hair cell stitch as the surrounding magnetic field is rotated.

References

[1] Vroman, R., Klassen, L. and Kamermans, M. (2013) Ephaptic Communication in the Vertebrate Retina. *Frontiers in Human Neuroscience*, **7**, 612.

[2] Anastassiou, C.A. and Koch, C. (2014) Ephaptic Coupling Endogenous Electric Field Activity: Why Bother? *Current Opinion in Neurobiology*, **31**, 95-103.

[3] Horn, R.G. and Israelachvilli, J.N. (1981) Direct Measurement of Structural Forces between Two Surfaces in a Nonpolar Liquid. *The Journal of Chemical Physics*, **75**, 1400-1411. http://dx.doi.org/10.1063/1.442146

[4] Craig, V.S. (1997) A Historical Review of Surface Force Measurements Techniques. *Colloids and Surfaces A*, **129-130**, 75-93. http://dx.doi.org/10.1016/S0927-7757(97)00029-0

[5] Israelachvilli, J. (1985) Intermolecular and Surface Forces. Academic Press, London.

[6] Nir, S.J., Bentz, J. and Duzgunes, N. (1981) Two Modes of Reversible Vesicle Aggregation: Particle Size and the DLVO Theory. *Journal of Colloid and Interface Science*, **84**, 266-269. http://dx.doi.org/10.1016/0021-9797(81)90284-8

[7] Rossetto, M. (2003) Colloid View of Hair Cell Cilia: A Selective Review. *Colloid and Surfaces B*, **29**, 257-263. http://dx.doi.org/10.1016/S0927-7765(03)00004-3

[8] Martin, P., Mehta, A.D. and Hudspeth, A.J. (2000) Negative Hair-Bundle Stiffness Betrays a Mechanism for Mechanical Amplification by the Hair Cell. *Proceedings of the National Academy of Sciences of the United States of America*, **97**, 12026-12031. http://dx.doi.org/10.1073/pnas.210389497

[9] Nam, J., Cotton, J., Peterson, E. and Grant, W. (2006) Mechanical Properties and Consequences of Stereocilia and Extracellular Links in Vestibular Hair Bundles. *Biophysical Journal*, **90**, 2786-2785. http://dx.doi.org/10.1529/biophysj.105.066027

[10] Roy, D.K. (1997) Tunneling and Negative Resistance Phenomena in Semiconductors. Pergamon Press, Elmsford.

[11] Anastassiou, C.A., Perin, R., Markram, H. and Kich, C. (2011) Ephaptic Coupling of Cortical Neurons. *Nature Neuroscience*, **14**, 217-223. http://dx.doi.org/10.1038/nn.2727

[12] Gulley, R.L. and Rees, T.S. (1981) Cytoskeletal Organization at the Postsynaptic Complex. *The Journal of Cell Biology*, **91**, 298-302.

[13] Landis, D.M. and Reese, T.S. (1983) Cytoplasmic Organization in Cerebellar Dendritic Spines. *The Journal of Cell Biology*, **97**, 1169-1178.

19

Antenna Array Pattern Synthesis via Coordinate Descent Method

Yuanhao Wang[1], Xiaoxi He[1], Jiangning Wang[1], Sergey Berezin[2], Wolfgang Mathis[1]

[1]Institut für Theoretische Elektrotechnik, Gottfried Wilhelm Leibniz Universität Hannover, Hannover, Germany
[2]Applied Mathematics Department, Saint-Petersburg State Polytechnical University, St. Petersburg, Russia
Email: ywang@tet.uni-hannover.de, servberezin@yandex.ru

Abstract

This paper presents an array pattern synthesis algorithm for arbitrary arrays based on coordinate descent method (CDM). With this algorithm, the complex element weights are found to minimize a weighted L_2 norm of the difference between desired and achieved pattern. Compared with traditional optimization techniques, CDM is easy to implement and efficient to reach the optimum solutions. Main advantage is the flexibility. CDM is suitable for linear and planar array with arbitrary array elements on arbitrary positions. With this method, we can configure arbitrary beam pattern, which gives it the ability to solve variety of beam forming problem, e.g. focused beam, shaped beam, nulls at arbitrary direction and with arbitrary beam width. CDM is applicable for phase-only and amplitude-only arrays as well, and furthermore, it is a suitable method to treat the problem of array with element failures.

Keywords

Antenna, Phased Array, Optimization, Pattern Synthesis, Coordinate Descent Method

1. Introduction

Comparing conventional antenna, antenna array provides advantages in flexibility and versatility. By changing the complex weights (including amplitude and phase) on each array element, the radiation pattern can be controlled and reconfigured. The antenna array pattern synthesis problem consists of finding weights that satisfy a set of specifications on the beam pattern [1]. Several most famous approaches can solve certain array pattern synthesis problems analytically, such as Schelkunoff method [2], the Dolph-Chebyshev synthesis [3], and the Taylor method [4]. Over the last several decades, many numerical techniques have been proposed. There are a few methods often used e.g. Convex Optimization, Genetic Algorithms (GA) [5], Differential Evolution Algorithm (DE) [6], Particle Swarm Optimization (PSO) [7] and Taguchi's Method [8] [9].

As an advantage of convex optimization, once an array pattern synthesis problem is convex, the global optimum must exist [1]. The most difficult part is reforming of a nonconvex array pattern synthesis problem into a convex problem. So far many array pattern synthesis problems have already been solved with convex optimization. Focused beam pattern with fix nulls and upper bound constraints over all side lobes can be solved with convex optimization [1]. In [10], a shaped beam array pattern synthesis problem was solved with sequential convex optimizations. If the amplitude-response error is linearly approximatable, it can be solved with convex optimization as well [11]. Some array pattern synthesis problems e.g. radiation pattern with lower bound constraints or synthesis with phase-only array cannot be solved with convex optimization method [1].

While the usage of convex optimization is restricted to the class of problems that can be transformed to be convex, the metaheuristic algorithms, e.g. GA, DE and PSO, can provide much more flexibility. They are not restricted in terms of types of array pattern synthesis problem. Furthermore, these evolutionary algorithms are able to deal with a huge parameter set without getting trapped in the local minima [12]; thus such a stochastic process is better suitable for large size arrays. In [13] GA was used to optimize the large thinned arrays in order to obtain low SLL (Side Lobe Level). The GA has no need to calculate derivatives and search from many points instead of a single point [5]. Further articles [14]-[17] present GA used for array pattern synthesis with different fitness functions and different requirements of the pattern. Differential Evolution Algorithm (DE) belongs to evolutionary algorithm as well. Comparing with the GA, the DE has no need for coding of the parameters and used different cross-over strategies. DE uses very few control parameters, and is easy to implement with any sort of program language. Many modified versions of the standard DE have been published in [12] [18]-[20]. PSO is similar in some ways to GA and DE, but requires less calculation and generally fewer lines of code. An overview about PSO in electromagnetics has been offered in [21]. In [22] the results of GA and PSO solving array pattern synthesis problems are compared. Many variants of the PSO specialized for difference array pattern synthesis problems have been discussed in [23]-[25].

In this article, an array pattern synthesis algorithm for arbitrary arrays based on coordinate descent method (CDM) is presented. CDM has been first mentioned in [26] for solving smooth unconstrained minimization problems. As an efficient optimization algorism, CDM is already used for many areas e.g. the autofocus system of SAR-Radar [27] [28], tomography [29] and digital image processing [30]. CDM minimizes a function with respect to a single parameter while holding the remaining parameters constant. Comparing with other direct search methods, the descent direction of CDM is predefined once the coordinate system is given. This property is meaningful for array pattern synthesis problems, because every complex amplitude of array element can be seen as a predefined search direction (or one dimension) of a N-dimension space, which describes the domain of a array pattern function.

This paper is an illustration of the utility of CDM for array pattern synthesis problems. In Section 2, a briefly analytical description of CDM is presented, along with the formulation of the array pattern synthesis problems and corresponded fitness function. It is shown in Section 2.4, how the applications of adaptive weights function accelerate the optimization process and help achieving better results. Example of the applications of the CDM in different array pattern synthesis problems with comparisons to known algorithms is shown in Section 3. Then there is the conclusion in Section 4.

2. Problem Formulation and Resolution

2.1. Coordinate Descent Method

Coordinate descent methods were among the first optimization schemes suggested for solving smooth unconstrained minimization problems. The main advantage of these methods is the simplicity of each iteration, both in generating the search direction and in performing the update of variables [31]. A coordinate descent method can be described as follow:

We have $\boldsymbol{a} = (a_1, a_2, \cdots, a_N) \in \mathbb{C}^N$ and $f : \mathbb{C}^N \to \mathbb{R}$.

The optimization problem for $f(\boldsymbol{a})$ is

$$\hat{\boldsymbol{a}} = \arg\min_{\boldsymbol{a} \in \mathbb{C}} f(\boldsymbol{a}). \tag{1}$$

As the optimization in (1) has no closed-form solution, we resort to an iterative method based on coordinate descent. In coordinate descent optimization, each parameter is optimized in turn, while holding all other parameters fixed. Because the parameters are interrelated, optimization over the entire set of parameters must be performed a number iterations. We start with arbitrary $\boldsymbol{a}^{(0)} = \left(a_1^{(0)}, \cdots, a_N^{(0)}\right) \in \mathrm{dom}f$ as initial value. Let $\hat{a}_n^{(i)}$ denote the optimum value of nth coordinate (the nth element) of the ith iteration; then, the coordinate descent estimate at the next iteration $i+1$ is defined as

$$\forall i \in [1,\infty) \subset \mathbb{N} : \hat{a}_n^{(i+1)} = \arg\min_{\boldsymbol{a} \in \mathbb{C}} f\left(\hat{a}_1^{(i+1)}, \cdots, \hat{a}_{n-1}^{(i+1)}, a_n, \hat{a}_{n+1}^{(i)}, \cdots, \hat{a}_N^{(i)}\right). \tag{2}$$

Now compute the $f^{(i+1)} = f\left(\hat{a}_1^{(i+1)}, \cdots, \hat{a}_N^{(i+1)}\right)$ and compare it with $f^{(i)} = f\left(\hat{a}_1^{(i)}, \cdots, \hat{a}_N^{(i)}\right)$. While $\left\| \dfrac{f^{(i+1)} - f^{(i)}}{f^{(i+1)}} \right\| > \epsilon$, which is a predefined value as quality criterion, repeat the procedure until $\left\| \dfrac{f^{(i+1)} - f^{(i)}}{f^{(i+1)}} \right\| \leq \epsilon$.

2.2. The Linear Array Pattern

For the sake of clarity, our array pattern synthesis problem is described for a linear array. Consider an antenna array composed of N isotropic elements placed at arbitrary and known locations $x_1, \cdots, x_N \in R$. The power pattern in far field is given by

$$G(\boldsymbol{a}, \theta) = \left\| \sum_{n=1}^{N} a_n \mathrm{e}^{\mathrm{j}kx_n \cos(\theta)} \right\|^2, \tag{3}$$

where $a_n \in \mathbb{C}$ is the complex amplitude at the nth element with $a_n = w_n \mathrm{e}^{j\phi_n}$, where w_n is the real amplitude and ϕ_n is the phase at nth element. $k = \dfrac{2\pi}{\lambda}$ is the wave number for a wavelength λ and θ denotes the polar angle.

2.3. The Fitness Function

The goal of array pattern synthesis problem is to find optimal parameters including array weights so that the designed beam pattern satisfies a set of specifications [1]. In this paper we reach the desired beam pattern via minimization of a fitness function f, whose similar form has been introduced in [32].

For a power pattern we define $f : \mathbb{C}^N \to [0,\infty) \subset \mathbb{R}$

$$f\left(\boldsymbol{a}^{(i)}\right) := \int_0^{\pi} \left(G\left(\boldsymbol{a}^{(i)}, \theta\right) - S(\theta)\right)^2 \mathrm{d}\theta. \tag{4}$$

$G\left(\boldsymbol{a}^{(i)}, \theta\right)$ converges to $S(\theta)$ in the senses of $f \cong 0$ by i passes the entire given interval. Now we rewrite a_n into its complex form $a_n = w_n \mathrm{e}^{\mathrm{j}\phi_n}$. According to (3) the fitness function (4) can be reformed to

$$f(\boldsymbol{w}, \boldsymbol{\phi}) = \int_0^{\pi} \left(\left\| \sum_{n=1}^{N} w_n \mathrm{e}^{\mathrm{j}(C_n(\theta) + \phi_n)} \right\|^2 - S(\theta) \right)^2 \mathrm{d}\theta \tag{5}$$

$$\text{with } C_n(\theta) = kx_n \cos(\theta), \boldsymbol{w} \in \mathbb{R}^N, \boldsymbol{\phi} \in \mathbb{R}^N.$$

For the n-th element the fitness function is given by

$$f(w_n, \phi_n) = \int_0^{\pi} \left(\left\| w_n \mathrm{e}^{\mathrm{j}(C_n(\theta) + \phi_n)} + R(\theta) \right\|^2 - S(\theta) \right)^2 \mathrm{d}\theta = \int_0^{\pi} (\xi)^2 \, \mathrm{d}\theta \tag{6}$$

with

$$\xi = w_n^2 + w_n R(\theta) \mathrm{e}^{-\mathrm{j}(C_n(\theta) + \phi_n)} + w_n R(\theta)^* \mathrm{e}^{\mathrm{j}(C_n(\theta) + \phi_n)} + R(\theta) R(\theta)^* - S(\theta) \tag{7}$$

where $R(\theta)$ denotes sum of the remaining terms in $\sum_{n=1}^{N} w_n e^{j(C_n(\theta)+\phi_n)}$, which are not depending on w_n and ϕ_n and $()^*$ denotes complex conjugate.

Due to following important characters of $G(w_n,\phi_n,\theta)$

$$G(\theta)\big|_{w_n,\phi_n} = G(\theta)\big|_{-w_n,\phi_n\pm\pi} \tag{8}$$

$$G(\theta)\big|_{w_n,\phi_n} = G(\theta)\big|_{w_n,\phi_n\pm 2\pi} \tag{9}$$

it is easy to prove

$$f(w_n,\phi_n) = f(-w_n,\phi_n \pm \pi) \tag{10}$$

$$f(w_n,\phi_n) = f(w_n,\phi_n \pm 2\pi) \tag{11}$$

as shown in **Figure 1**. That means we can use a surjection

$$p: A \to B \tag{12}$$

with

$$A =: \{(w_n,\phi_n) | w_n,\phi_n \in \mathbb{R}\} \tag{13}$$

$$B =: \{(w_n,\phi_n) | w_n \in [0,+\infty), \phi_n \in [0,2\pi)\}, \tag{14}$$

which do not change $G(\theta)$ and $f(w_n,\phi_n)$, to cover the entire $f(w_n,\phi_n)$. Thus the new domain of $f(w_n,\phi_n)$ is $\mathbf{dom} f := \{w_n \in [0,+\infty), \phi_n \in [0,2\pi)\}$. Because $f(w_n)$ and $f(\phi_n)$ are independent of each other and *f* is a quadratic function, there is only one $f_{\min}(w_n,\phi_n)$ in the new $\mathbf{dom} f$ and $\hat{a} = \arg\min f(w_n,\phi_n)$ can be solved numerically.

2.4. The Weight of Fitness Function

A crucial advantage of CDM is that a weight function $P(\theta)$ can be used, or more precisely, multiplied to the fitness function (4), in order to save the needed time and steps of the optimization process or achieve better results:

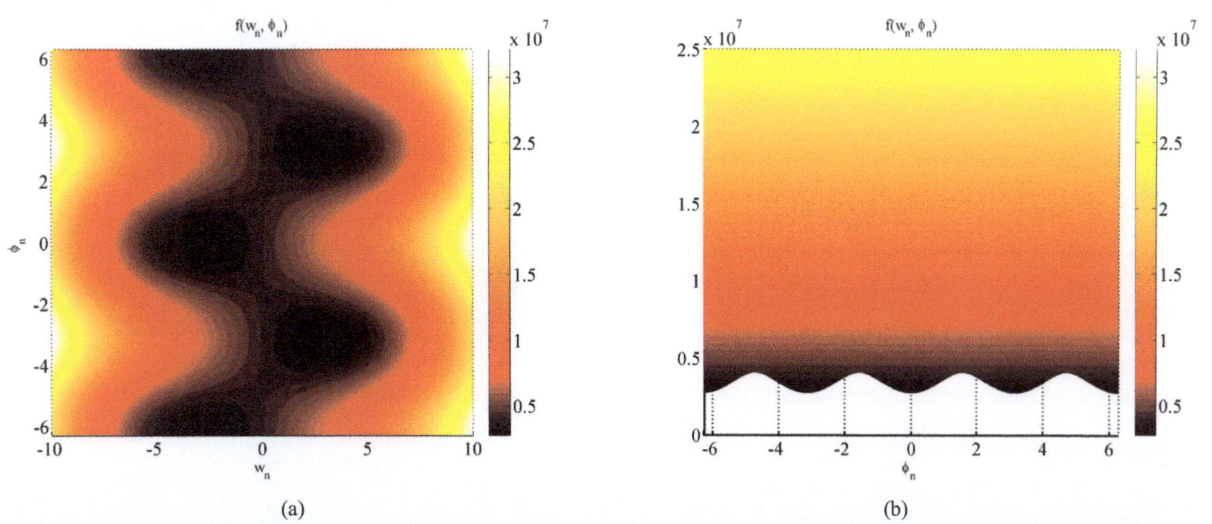

Figure 1. $f(w_n,\phi_n)$ by arbitrary iteration for a linear array (a) w_n,ϕ_n-view (horizontal); (b) $\phi_n, f(w_n,\phi_n)$-view (vertical).

$$f(\boldsymbol{a}) := \int_0^{\pi} \left(G(\boldsymbol{a},\theta) - S(\theta)\right)^2 \cdot P(\theta)\,\mathrm{d}\theta. \tag{15}$$

Because of the mathematical property of CDM, an arbitrary function $P(\theta) > 0$ can be multiplied to the fitness function, functioning as a weight function, without the need of changing any other part of the algorithm. Hence, an algorithm can be implemented to modify the fitness function during the optimization process by means of varying the weight function $P(\theta)$. With the application of different types of weight function $P(\theta)$ (or the algorithm to calculate the weight function $P(\theta)$), this array pattern synthesis algorithm can be exclusively used for different optimization problems, in order to acquire maximal speed and best results.

It should be emphasized that the effect of the usage of the weight function will be different according to the type of optimization problems. For problems which can obtain a perfect matching, e.g. Dolph-Chebyshev pattern (**Figure 4(a)**), the application of the weight function will only alter the speed of the optimization process, without changing the result, as the fitness function will always converge to zero. But when dealing with optimization problems which does not have a perfect matching, e.g. the problem of array with element failures (**Figure 8**), using the weight function can both change the speed of the process and the resulted beam pattern. This character of weight function can be quite useful when a improved result is required, but it also means that the weight function should be determined carefully when not expecting a huge change in the result.

A simple example is given here: when synthesizing the Dolph-Chebyshev pattern with CDM (shown in **Figure 4(a)**), the convergence is quite slow after the SLL started to descend. The reason can be considered as that at the side lobe area, the $G(\boldsymbol{a},\theta)$ and $S(\theta)$ are both considerable small and thus their difference gives less feedback to the fitness function. Here weight functions like $P(\theta) = \frac{1}{G(\boldsymbol{a},\theta)}$ or $P(\theta) = \frac{|G(\boldsymbol{a},\theta) - S(\theta)|}{G(\boldsymbol{a},\theta)}$ can be employed, in order to enlarge the weights of the SLL in the fitness function. Thus the convergence is observably accelerated, which can be seen in **Figure 2(a)**, that the SLL descend much faster when using the weight function. It is shown in **Figure 2(b)**, how the SLL descend when matching a shaped beam using the same weight functions. In this example the optimization problem is from [33], which is a perfect matching optimization problem. The beam pattern of this example is shown in **Figure 6(a)**.

Another example for the optimization problems without perfect matching is given here with the desired function being a rectangular function (**Figure 7(a)**). The ripple of the shaped main beam should not exceed $0.1\ \text{dB}$, while the SLL less than $-28\ \text{dB}$. The algorithm to calculate the adaptive weight function is described with a flow diagram in **Figure 3**.

3. Numerical Results

Examples of the array pattern synthesis problem using CDM are presented and compared with PSO, GA and DE. All simulations are carried out on Intel® Xeon® CPU E5-2687W v2 @ 3.40 GHz with 256 GB RAM computer. All simulations use the fitness function (4) with different weights. In the following we use

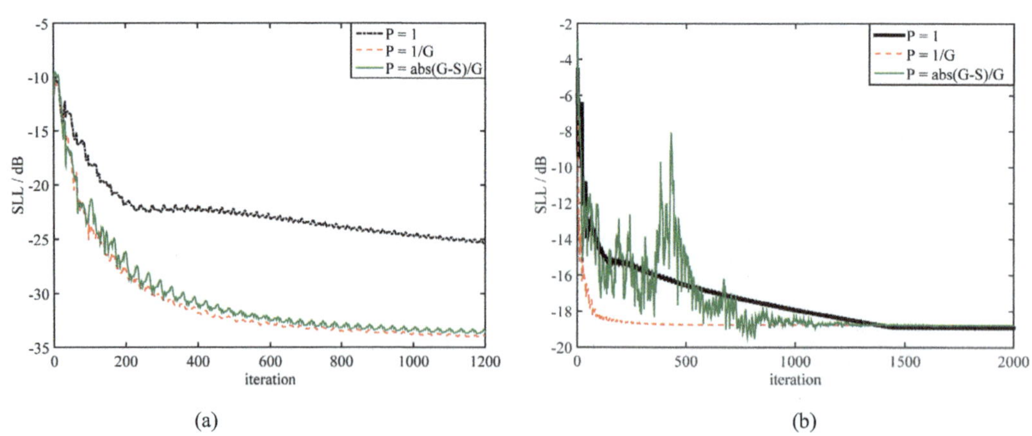

Figure 2. Investigations of relationships between SLL and weights. (a) Dolph-Chebyshev beam pattern; (b) Shaped beam pattern.

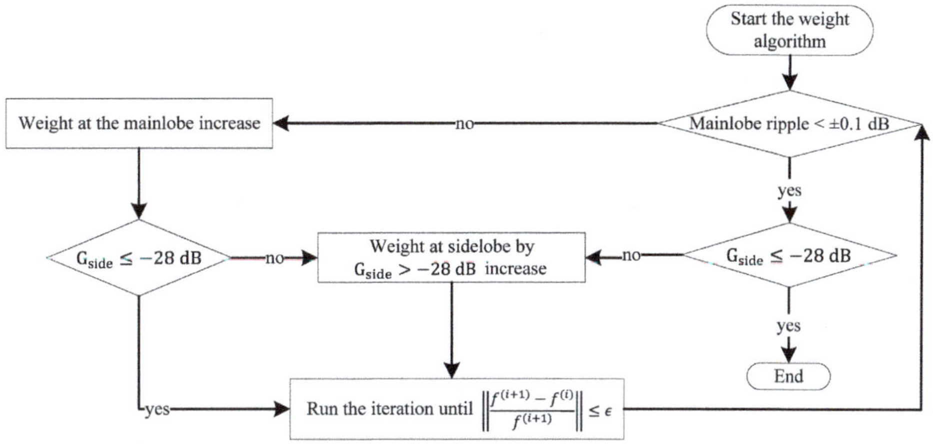

Figure 3. Flow process chart for a non-perfect matching (shaped beam pattern in **Figure 7(a)**).

$F(w,\phi) = \frac{1}{1+f(w,\phi)}$ as quality criterion, because $F(w,\phi) \in [0,1]$ is more suited for representation in diagram as $f(w,\phi) \in [0,\infty)$.

3.1. Dolph-Chebyshev Synthesis

In this example the desired pattern is a Dolph-Chebyshev pattern with 32 elements and SLL by -35 dB. Equal distance d between two elements is half wavelength. In **Figure 4(a)** original Dolph-Chebyshev pattern and three other patterns from CDM, PSO and DE are shown. The desired function $S(\theta)$ is an original Dolph-Chebyshev, thus it provides a perfect matching. The main lobs from all methods do not much differ from the original one. The pattern from CDM shows the best match of the SLL (perfect matching). This can be explained with the amplitude and phase distribution which is shown in **Figure 5**. **Figure 4(b)** represents the convergence of the fitness function over the time. $F(w,\phi)$ by using CDM converges to 0.9999 in 5.8 seconds, while by using DE 1500 seconds are needed to achieve the same results. The optimization by using PSO, which converges to 0.9999 in 76 seconds, is neither get best results nor a short calculation time.

3.2. Shaped Beam Synthesis

The synthesis of shaped beams is a classical problem since it has many applications ranging from radar and remote sensing to communication systems [34]. Two examples are shown, one is a perfect matching, the other one is a non-perfect matching. The first example is shown in **Figure 6(a)**. This pattern has been represented in [33]. All three algorithms can synthesize a very similar beam pattern as the desired pattern. However, only the CDM succeed a perfect matching. The time of optimization by using CDM is 4.5 seconds for a $F(w,\phi) \geq 0.9$, while the PSO 35 seconds and DE 1200 seconds needed for the same result.

The second example (**Figure 7(a)**) is a non-perfect matching, because the desired function $S(\theta)$ is a rectangular function, which means that it is not possible to achieve a perfect matching with finite elements. For this synthesis the weighting algorithm in **Figure 3** has been used to achieve the requirements which is described in subsection 3. The progress of convergence $F(w,\phi)$ is shown in **Figure 7(b)**. It converges at first at high speed and then become stable after 50 seconds. Since the optimization is non-perfect matching, the $F(w,\phi)$ is not possible to reach one.

3.3. The Array with Element Failures

Beam pattern reconfiguration of an antenna array with faulty elements has practical importance for antennas which cannot be repaired immediately. The possibility of reconfiguration of the antenna array with faulty elements has been considered by several authors over the years [35]-[39]. In this example we take the results from [36] as a reference. In this case the elements with Nr. 2, Nr. 5 and Nr. 6 of 32 elements Dolph-Chebyshev array

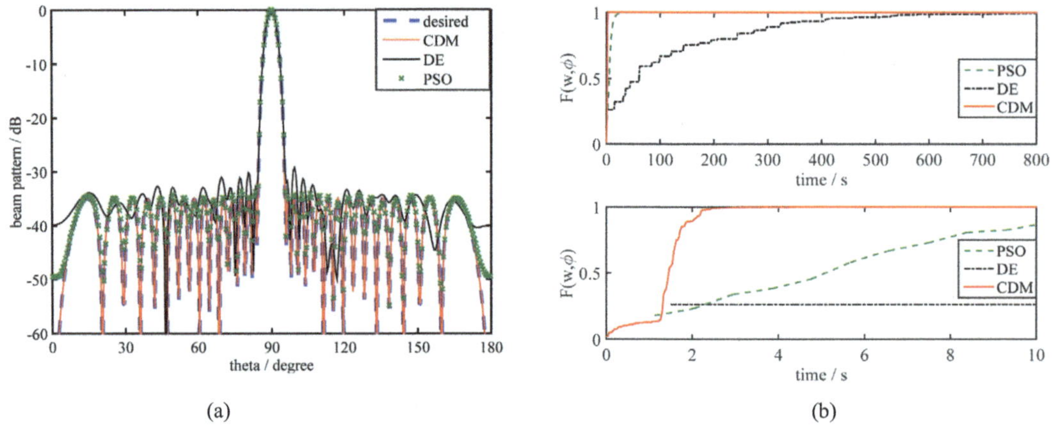

Figure 4. Synthesize a Dolph-Chebyshev beam pattern with three different optimization algorithms.

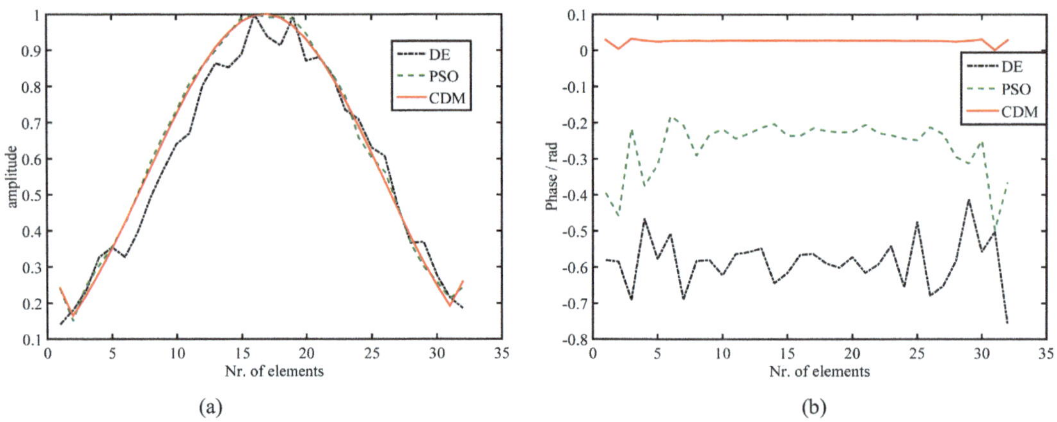

Figure 5. Comparison the results of optimizations of the Dolph-Chebyshev beam pattern (perfect matching).

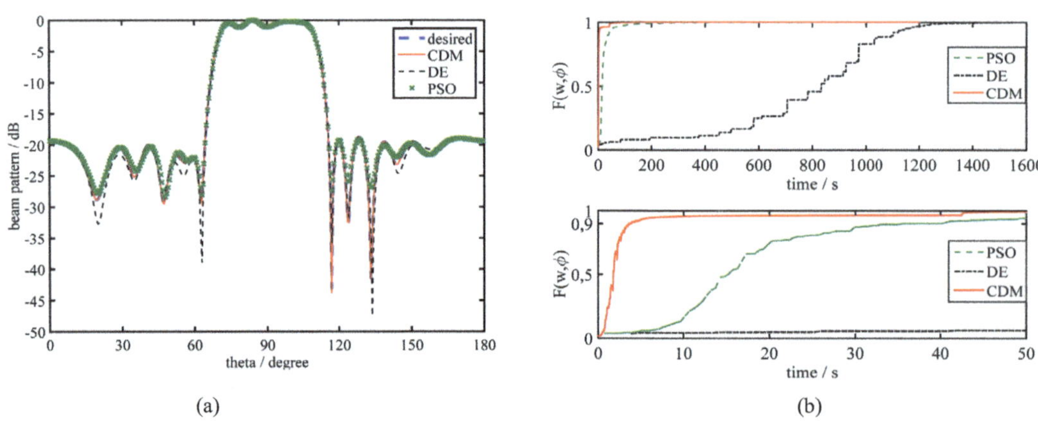

Figure 6. Synthesize a shaped beam pattern with three different optimization algorithms (perfect matching).

are failed, while the faulty elements do not radiate at all ("on-off" fault [40]). The result is shown in **Figure 8**. The half-power beam width of the results from CDM is 4.79°, which is 5.36° with GA. The SLL from CDM is -35.3 dB, which satisfies the original Dolph-Chebyshev condition as well.

3.4. The Planar Array

In this example the minimization of the SLL of a 8×8 array with uniform element distance half wavelength is presented. In this case the half power beam width should keep 30°, while the SLL as low as possible. The result is shown in **Figure 9(b)**. The SLL has sunk from -12.5 dB to -21 dB.

4. Conclusion

An iterative method, which is based on the coordinate descent method to synthesize most of the current beam

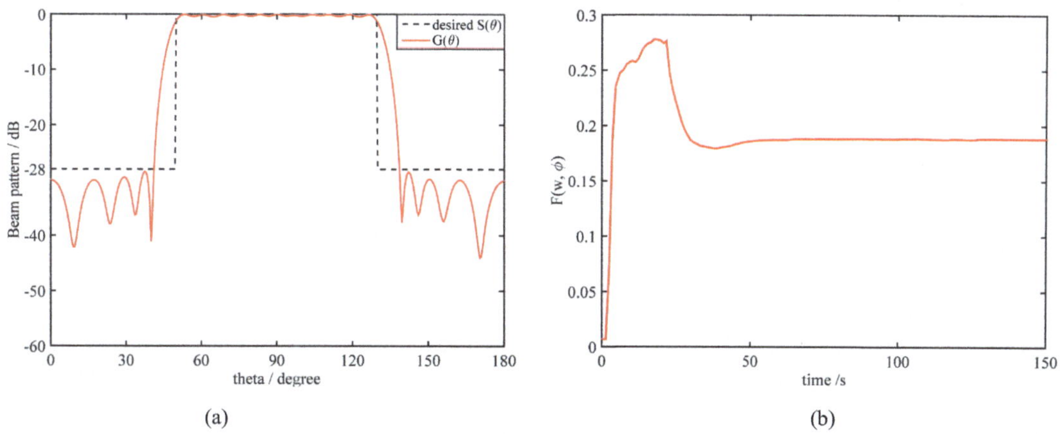

Figure 7. Synthesize a shaped beam pattern with CDM (non-perfect matching).

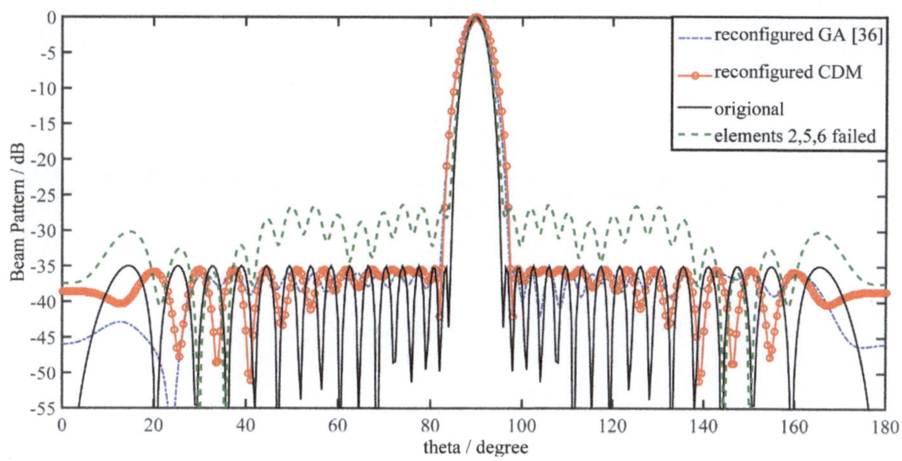

Figure 8. Dolph-Chebyshev array with 3rd, 5th und 6th elements failed.

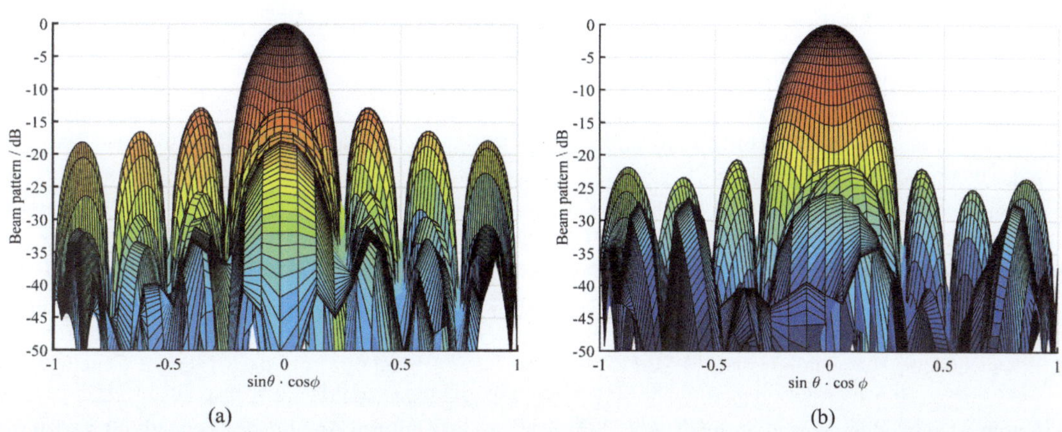

Figure 9. Beam pattern synthesis for planar array with CDM.

patterns from linear to planar array, is presented above. CDM allows changing the fitness function with a multiplication of a weight function $P(\theta)$. The function $P(\theta)$ can be changed any time during the optimization procedure depending on the requirement of the desired beam pattern. Therefore, we believe that it is worth investing more research in this area, in order to find more efficient and effective weight functions, and specialize the method for more array pattern synthesis problems. Due to the L_2 norm, the fitness function converges by every iteration, so the algorithm is very efficient to reach the optimum solution. Compared with other methods like PSO, GA and DE, CDM is accurater and faster, also easier to implement.

Acknowledgements

We thank the Editor and the referee for their comments.

References

[1] Lebret, H. and Boyd, S. (1997) Antenna Array Pattern Synthesis via Convex Optimization. *IEEE Transactions on Signal Processing*, **45**, 526-532. http://dx.doi.org/10.1109/78.558465

[2] Schelkunoff, S.A. (1943) A Mathematical Theory of Linear Arrays. *Bell System Technical Journal*, **22**, 80-107. http://dx.doi.org/10.1002/j.1538-7305.1943.tb01306.x

[3] Dolph, C.L. (1946) A Current Distribution for Broadside Arrays Which Optimizes the Relationship between Beam Width and Side-Lobe Level. *Proceedings of the IRE*, **34**, 335-348. http://dx.doi.org/10.1109/JRPROC.1946.225956

[4] Taylor, T.T. (1955) Design of Line-Source Antennas for Narrow Beamwidth and Low Side Lobes. *Transactions of the IRE Professional Group on Antennas and Propagation*, **3**, 16-28. http://dx.doi.org/10.1109/TPGAP.1955.5720407

[5] Goldberg, D.E. (1989) Genetic Algorithms in Search, Optimization, and Machine Learning. Addison-Wesley Professional, Boston.

[6] Storn, R. and Price, K. (1997) Differential Evolution—A Simple and Efficient Heuristic for Global Optimization over Continuous Spaces. *Journal of Global Optimization*, **11**, 341-359. http://dx.doi.org/10.1023/A:1008202821328

[7] Kennedy, J. and Eberhart, R. (1995) Particle Swarm Optimization. *IEEE International Conference on Neural Networks*, **4**, 1942-1948. http://dx.doi.org/10.1109/icnn.1995.488968

[8] Taguchi, G., Chowdhury, S. and Wu, Y. (2004) Taguchi's Quality Engineering Handbook. Wiley-Interscience, Hoboken.

[9] Weng, W.-C., Yang, F. and Elsherbeni, A.Z. (2007) Linear Antenna Array Synthesis Using Taguchi's Method: A Novel Optimization Technique in Electromagnetics. *IEEE Transactions on Antennas and Propagation*, **55**, 723-730. http://dx.doi.org/10.1109/TAP.2007.891548

[10] Fuchs, B., Skrivervik, A. and Mosig, J.R. (2013) Shaped Beam Synthesis of Arrays via Sequential Convex Optimizations. *IEEE Antennas and Wireless Propagation Letters*, **12**, 1049-1052. http://dx.doi.org/10.1109/LAWP.2013.2280043

[11] Nongpiur, R.C. and Shpak, D.J. (2014) Synthesis of Linear and Planar Arrays with Minimum Element Selection. *IEEE Transactions on Signal Processing*, **62**, 5398-5410. http://dx.doi.org/10.1109/TSP.2014.2350966

[12] Subhashini, K.R. and Praveen Kumar, A.T. (2014) Comparative Analysis of Linear and Nonlinear Pattern Synthesis of Hemispherical Antenna Array Using Adaptive Evolutionary Techniques? *International Journal of Antennas and Propagation*, **2014**, Article ID: 987140. http://dx.doi.org/10.1155/2014/987140

[13] Haupt, R.L. (1994) Thinned Arrays Using Genetic Algorithms. *IEEE Transactions on Antennas and Propagation*, **42**, 993-999. http://dx.doi.org/10.1109/8.299602

[14] Haupt, R.L. and Johnson, J.M. (1999) Dynamic Phase-Only Array Beam Control Using a Genetic Algorithm. *Proceedings of the* 1*st NASA/DoD Workshop on Evolvable Hardware*, Pasadena, 19-21 July 1999, 217-224. http://dx.doi.org/10.1109/EH.1999.785456

[15] Marcano, D. and Duran, F. (2000) Synthesis of Antenna Arrays Using Genetic Algorithms. *IEEE Antennas and Propagation Magazine*, **42**, 12-20. http://dx.doi.org/10.1109/74.848944

[16] Ares-Pena, F.J., Rodriguez-Gonzalez, J.A., Villanueva-Lopez, E. and Rengarajan, S.R. (1999) Genetic Algorithms in the Design and Optimization of Antenna Array Patterns. *IEEE Transactions on Antennas and Propagation*, **47**, 506-510. http://dx.doi.org/10.1109/8.768786

[17] Elkamchouchi, H.M. and Hassan, M.M. (2014) Array Pattern Synthesis Approach Using a Genetic Algorithm. *IET Microwaves, Antennas & Propagation*, **8**, 1236-1240. http://dx.doi.org/10.1049/iet-map.2013.0718

[18] Kurup, D.G., Himdi, M. and Rydberg, A. (2003) Synthesis of Uniform Amplitude Unequally Spaced Antenna Arrays Using the Differential Evolution Algorithm. *IEEE Transactions on Antennas and Propagation*, **51**, 2210-2217.

http://dx.doi.org/10.1109/TAP.2003.816361

[19] Goudos, S.K., Siakavara, K., Samaras, T., Vafiadis, E.E. and Sahalos, J.N. (2011) Self-Adaptive Differential Evolution Applied to Real-Valued Antenna and Microwave Design Problems. *IEEE Transactions on Antennas and Propagation*, **59**, 1286-1298. http://dx.doi.org/10.1109/TAP.2011.2109678

[20] Guney, K. and Basbug, S. (2013) Null Synthesis of Time-Modulated Circular Antenna Arrays Using an Improved Differential Evolution Algorithm. *IEEE Antennas and Wireless Propagation Letters*, **12**, 817-820. http://dx.doi.org/10.1109/LAWP.2013.2271273

[21] Robinson, J. and Rahmat-Samii, Y. (2004) Particle Swarm Optimization in Electromagnetics. *IEEE Transactions on Antennas and Propagation*, **52**, 397-407. http://dx.doi.org/10.1109/TAP.2004.823969

[22] Boeringer, D.W. and Werner, D.H. (2004) Particle Swarm Optimization versus Genetic Algorithms for Phased Array Synthesis. *IEEE Transactions on Antennas and Propagation*, **52**, 771-779. http://dx.doi.org/10.1109/TAP.2004.825102

[23] Khodier, M.M. and Christodoulou, C.G. (2005) Linear Array Geometry Synthesis with Minimum Sidelobe Level and Null Control Using Particle Swarm Optimization. *IEEE Transactions on Antennas and Propagation*, **53**, 2674-2679. http://dx.doi.org/10.1109/TAP.2005.851762

[24] Donelli, M., Martini, A. and Massa, A. (2009) A Hybrid Approach Based on PSO and Hadamard Difference Sets for the Synthesis of Square Thinned Arrays. *IEEE Transactions on Antennas and Propagation*, **57**, 2491-2495. http://dx.doi.org/10.1109/TAP.2009.2024570

[25] Roy, S., Martinez, S.Z., Coello Coello, C.A. and Sengupta, S. (2012) A Multi-Objective Evolutionary Approach for Linear Antenna Array Design and Synthesis. *Proceedings of the* 2012 *IEEE Congress on Evolutionary Computation* (*CEC*), Brisbane, 10-15 June 2012, 1-8. http://dx.doi.org/10.1109/CEC.2012.6252989

[26] Zangwill, W. (1969) Nonlinear Programming: A Unified Approach. Prentice-Hall, Englewood Cliffs.

[27] Kragh, T.J. (2006) Monotonic Iterative Algorithm for Minimum-Entropy Autofocus? *Proceedings of the Adaptive Sensor Array Processing Workshop*, Lexington, MA, 6-7 June 2006.

[28] Ash, J.N. (2012) An Autofocus Method for Backprojection Imagery in Synthetic Aperture Radar. *IEEE Geoscience and Remote Sensing Letters*, **9**, 104-108. http://dx.doi.org/10.1109/LGRS.2011.2161456

[29] Bouman, C.A. and Sauer, K. (1996) A Unified Approach to Statistical Tomography Using Coordinate Descent Optimization. *IEEE Transactions on Image Processing*, **5**, 480-492. http://dx.doi.org/10.1109/83.491321

[30] Shen, H.F., Zhang, L.P., Huang, B. and Li, P.X. (2007) A MAP Approach for Joint Motion Estimation, Segmentation, and Super Resolution. *IEEE Transactions on Image Processing*, **16**, 479-490. http://dx.doi.org/10.1109/TIP.2006.888334

[31] Nesterov, Y. (2012) Efficiency of Coordinate Descent Methods on Huge-Scale Optimization Problems. *SIAM Journal on Optimization*, **22**, 341-362. http://dx.doi.org/10.1137/100802001

[32] Perini, J. and Idselis, M. (1971) Note on Antenna Pattern Synthesis Using Numerical Iterative Methods. *IEEE Transactions on Antennas and Propagation*, **19**, 284-286. http://dx.doi.org/10.1109/TAP.1971.1139919

[33] Bucci, O.M., Mazzarella, G. and Panariello, G. (1989) Reconfigurable Arrays by Phase-Only Control. *Proceedings of the IEEE Antennas and Propagation Society International Symposium*, San Jose, 26-30 June 1989, 142-145. http://dx.doi.org/10.1109/APS.1989.134633

[34] Elliott, R.S. (1981) Antenna Theory and Design. Prentice-Hall, Englewood Cliffs.

[35] Peters, T.J. (1991) A Conjugate Gradient-Based Algorithm to Minimize the Sidelobe Level of Planar Arrays with Element Failures. *IEEE Transactions on Antennas and Propagation*, **39**, 1497-1504. http://dx.doi.org/10.1109/8.97381

[36] Yeo, B.-K. and Lu, Y. (1999) Array Failure Correction with a Genetic Algorithm. *IEEE Transactions on Antennas and Propagation*, **47**, 823-828. http://dx.doi.org/10.1109/8.774136

[37] Zainud-Deen, S.H., Ibrahem, M.S., Sharshar, H.A. and Ibrahem, S.M.M. (2004) Array Failure Correction with Orthogonal Method. *Proceedings of the Twenty-First National Radio Science Conference*, Cairo, 16-18 March 2004, B7-1-9.

[38] Mitilineos, S.A. and Capsalis, C.N. (2005) On Array Failure Mitigation Using Genetic Algorithms and a Priori Joint Optimization. *IEEE Antennas and Propagation Magazine*, **47**, 227-232. http://dx.doi.org/10.1109/MAP.2005.1599213

[39] Keizer, W.P.M.N. (2007) Element Failure Correction for a Large Monopulse Phased Array Antenna with Active Amplitude Weighting. *IEEE Transactions on Antennas and Propagation*, **55**, 2211-2218. http://dx.doi.org/10.1109/TAP.2007.902008

[40] Bucci, O.M., Capozzoli, A. and D'Elia, G. (2000) Diagnosis of Array Faults from Far-Field Amplitude-Only Data. *IEEE Transactions on Antennas and Propagation*, **48**, 647-652. http://dx.doi.org/10.1109/8.855482

20

Design of Integrated Triple Band Notched for Ultra-Wide Band Microstrip Antenna

Yahya S. H. Khraisat, Ahmad Bataineh, Malak Angor

Electrical and Electronics Department, Al-Balqa' Applied University/Al-Huson University College, Irbid, Jordan
Email: Yahya_khr@bau.edu.jo, a_bataineh75@bau.edu.jo, dr.angormalak@bau.edu.jo

Abstract

Ultra-wideband (UWB) is one of the recent topics that received a great concern from academia and industry. However, UWB found many difficulties to be standardized due to the overlay working that made UWB an important potential interference source to many licensed and unlicensed spectrum throughout the band 3.1 to 10.6 GHz. This paper demonstrates the design of integrated triple band notched for UWB Microstrip antenna. We simulated UWB short range systems which require low power and these are built using inexpensive digital components. We proposed a compact triple band notched CPW (Co-planar Waveguide) fed Micro strip Antenna (MSA) for UWB. This band-notched antenna has rejection characteristics at 3.2 GHz (for Wi-MAX band 3.16 to 3.32 GHz), at 5.5 GHz (for WLAN 2 band—5.3 to 5.72 GHz) and at 7.9 GHz (for ITU band 7.72 GHz to 8.13 GHz). The simulation was done using IE3D simulator.

Keywords

Ultra Wide Band Antenna, Microstrip Antenna, Partial Ground Plane, Band-Notched

1. Introduction

In 2002, the Federal Communications Commission (FCC) allocated the ultra-wideband (UWB) frequency range from 3.1 - 10.6 GHz for unlicensed UWB applications. The limitation of equivalent isotropically radiated power (EIRP) in band emission does not exceed −41.3 dBm/MHz for extremely wide impedance bandwidth [1]. Printed planar monopole antennas have been designed to operate in UWB systems [2]. There are some wireless communication applications which have already occupied frequencies in the UWB band such as the wireless local area network (WLAN) a.k.a. IEEE 802.11 a/g and HIPERLAN/2 WLAN which operate at 5.15 - 5.35 GHz and 5.725 - 5.825 GHz, respectively and IEEE 802.16 (3.3 - 3.8 GHz) [3]. In addition, the worldwide interoperability for microwave access (WiMAX) has also operated covering the frequency from 5.25 - 5.85 GHz in some countries [4].

One way to suppress these interfering signals is to use a spatial filter such as a frequency selective surface above the antenna. However, this approach requires too much space and is expensive [5]. To disband this problem, the UWB antenna with build-in frequency notch structure is often chosen. A lot of functions have been used for frequency rejection in UWB antenna designs. The most popular technique to provide this character is embedded a slot on the patch or ground plane [6]. However, in order to achieve monopole antenna designing, a large ground plane is required, which could not be suitable for compact wireless devices. Some techniques are used to reject undesired frequencies, for examples, placing the parasitic strips on the opposite side of the radiating element [7], inserting two strips in the printed slot antenna to create two rejection bands [8], and embedding a C-shaped slit into the fed element with a parasitic strip in slot antenna [9]. However, the antennas have more complicated structures to implement than others. In [10], the band-notched slot antennas were studied, in which two types of narrow slits on the exciting stubs were used for two antennas, and two parasitic strips were placed in the rectangular slot for another one [11]. Although, the antenna design is capable of undesired frequency rejection, the parasitic strips lead to more complex structure.

2. Antenna Design

First a Primitive antenna is taken. This antenna consists of a beveled rectangular radiating patch and a CPW (coplanar waveguide) type feed structure **Figure 1**. The essence of this design strategy is that three notching elements are embedded onto the primitive patch antenna to produce band-stop filtering function at those above mentioned frequencies. Notch elements are meticulously selected and embedded onto the antenna [12].

The band notches are introduced in order to stop the function of the antenna in that particular frequency range. By this the interference between the UWB system and the narrow band system is reduced to a great extent. Introduction of the band notches helps us to avoid the use of the band stop filters and hence reducing the cost and complexity of the antenna. Now-a-days demand is for miniaturized technology and MSA helps us in achieving that.

2.1. Primitive Antenna

Antenna Description

Measurements of the primitive antenna are shown in **Table 1**.

Primitive antenna consists of a rectangular patch (trimmed) and trimmed ground plane to enhance the antenna's broadband performance and also this arrangement increases the flow of surface current through the feedline and concentrates the surface current around the bottom of the radiating patch. First a rectangular patch is designed using simulation program IE3D and then it is trimmed by removing triangular shaped parts using the extrude feature. Triangular shaped parts are also removed from the rectangular shaped ground plane in order to smooth the surface current flow. Radiating patch is made using PEC (Perfect Electrical Conductor) material.

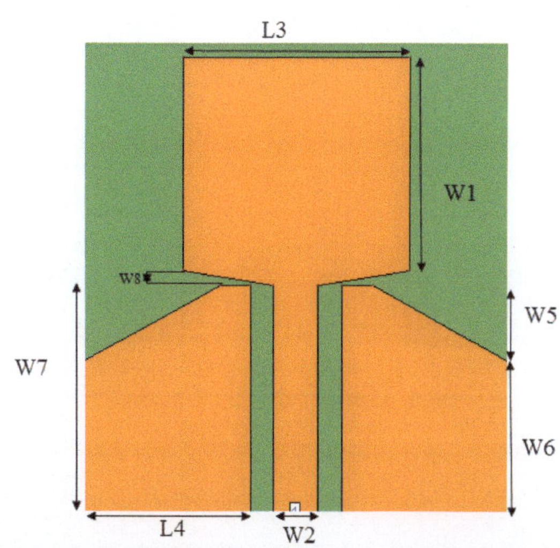

Figure 1. Geometry of CPW fed planar antenna.

- Dielectric material used here has relative permittivity = 4.4.
- Thickness of substrate = 1.6 mm.
- Length of substrate = 28 mm.
- Width of substrate = 31 mm.
- Tangent Loss = 0.003 [8].

2.2. Antenna with C-Slot (for Band-Notch at 3.2 GHz)

Figure 2 shows the antenna design with C—slot to produce band—notched at 3.2 GHz.

Antenna Description:

Measurements of the C-shaped slot at 3.5GHz is shown in **Table 2**.

FCC (Federal Communication Commission) approved and authorized the 3.1 GHz to 10.6 GHz band as the UWB. But Wi-MAX operates in the range of 3.16 GHz to 3.32 GHz and interferes with UWB devices. So instead of using a band-stop filter at the receiver antenna we have etched a C-shaped slot on the radiating patch in order to facilitate band-rejection facility around 3.2 GHz, so that the interference is minimized. The S—parameters of this design is shown in **Figure 3**.

The 3-D radiation pattern is shown in **Figure 4**.

2.3. Antenna with C-Slot (for Band-Notch at 5.5 GHz)

Figure 5 shows the geometry of the antenna with C—slot to produce band—notch at 5.5 GHz.

Table 1. Measurements of the primitive antenna (without any slots).

• Parameters	• Dimensions
• W1	• 14 mm
• W2	• 3 mm
• W5	• 5 mm
• W6	• 10 mm
• W7	• 15 mm
• W8	• 1 mm
• L3	• 15 mm
• L4	• 11 mm

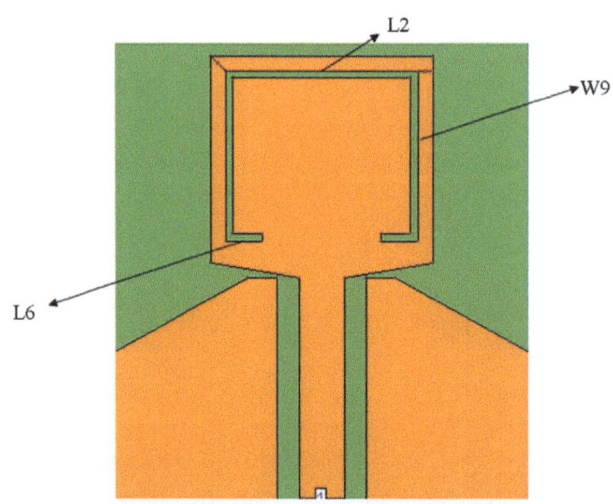

Figure 2. Geometry of antenna with C-slot to produce band-notch at 3.2 GHz.

Table 2. Measurements of the C-shaped slot (band notch at 3.5 GHz).

• Parameters	• Dimensions
• L2	• 13 mm
• L6	• 2.5 mm
• W4	• 11.5 mm
• W9	• 11.5 mm
• Thickness of slot	• 0.5 mm

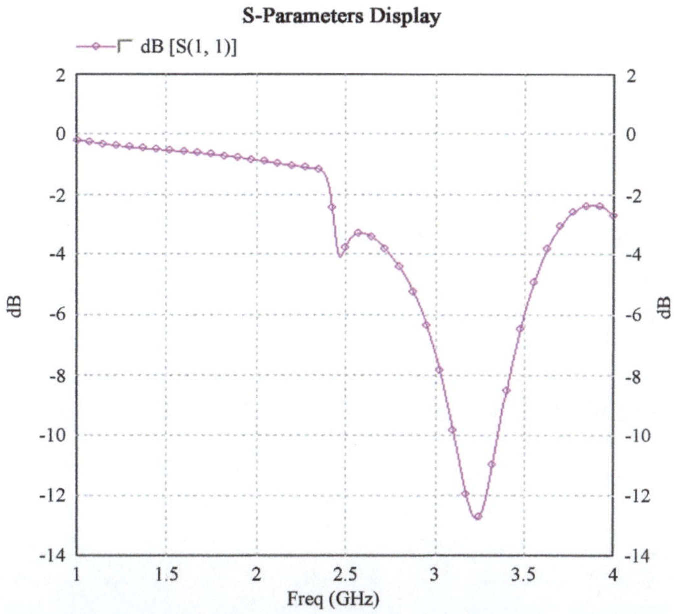

Figure 3. Band-notch at 3.2 GHz return loss.

Figure 4. 3D radiation pattern for 3.2GHz with max gain 3.28.

Antenna Description:

Measurements of the C-shaped slot at 5.5 GHz is shown below in **Table 3**.

WLAN 2 operates from 5.3 GHz to 5.72 GHz, interferes with the UWB systems. In order to minimize the interference we have etched a C-shaped slot on the radiating patch to create a notch around 5.5 GHz.

The S-parameters of this design is shown in **Figure 6**.

The 3-D radiation pattern is shown in **Figure 7**.

2.4. Antenna with Two CSRRs

Figure 8 shows the antenna design with a pair of CSRRs at 7.9 GHZ.

Antenna Description:

Measurements of the two CSRRs at 7.9 GHz is shown in **Table 4**.

ITU band operates in the range of 7.72 GHz to 8.13 GHz and possess a threat to UWB systems. By etching 2 CSRRs on the ground plane we have created the required band notch in the ITU band.

The S-parameters of this design is shown in **Figure 9**.

The 3-D radiation pattern is sown in **Figure 10**.

2.5. Antenna with Triple Band-Notch Features

Antenna Description:

To achieve band notch controllability we have taken following steps:

1) Assigning a distinct rejection band for each notching element.

2) Controlling the shape of each element and their optimized positions in order to minimize coupling.

3) Combination of each notch element on the primitive UWB antenna is aimed for providing triple band rejection function at Wi-MAX (3.16 - 3.32 GHz), WLAN 2 (5.3 - 5.72 GHz) and ITU 8 GHz (7.72 - 8.13 GHz).

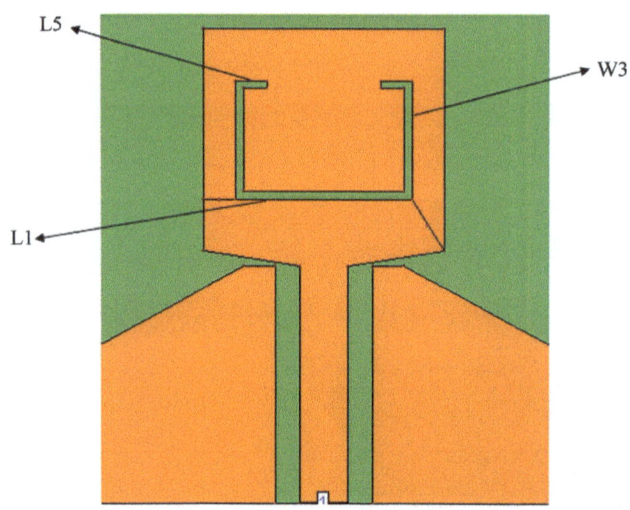

Figure 5. Geometry of antenna with C-slot to produce band-notch at 5.5 GHz.

Table 3. Measurements of the C-shaped slot (band notch at 5.5 GHz).

Parameters	Dimensions
L1	11 mm
L5	2 mm
W3	7.5 mm
Thickness of the slot	0.5 mm

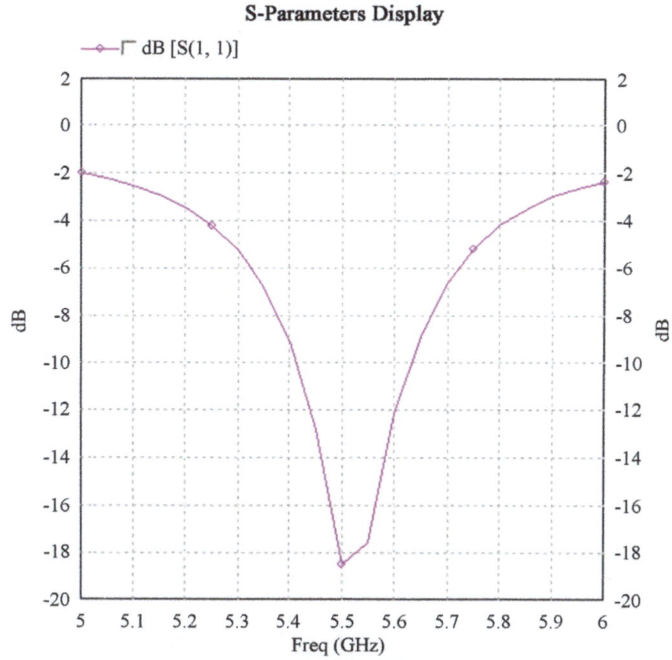

Figure 6. Band-notch at 5.5 GHz return loss.

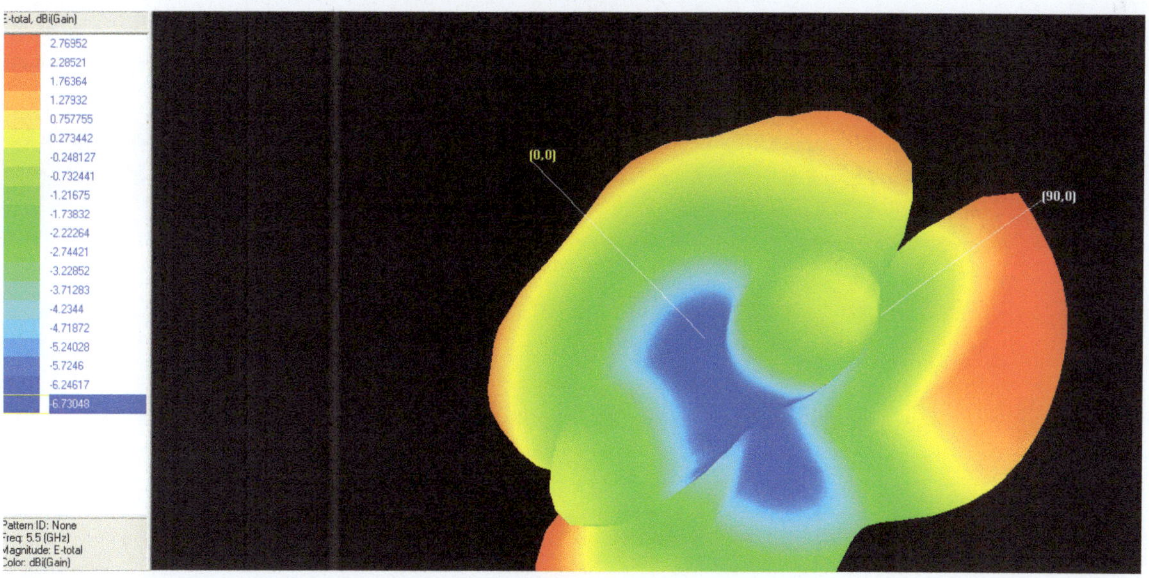

Figure 7. 3D radiation pattern for 5.5 GHz with max gain 2.76 dB.

Design Details:

The measurements of the antenna as well as the measurements of the two C-shaped slots and pair of CSRRs is shown below in **Table 5**.

The band notch operations are achieved by etching 2 C-shaped slot in the rectangular metal radiating patch and by etching a pair of CSRRs in the ground plane (as shown in the **Figure 11**). It is found that by adjusting the total length of the C-shaped slot to be approximately half wavelength of the desired notched frequency, a destructive interference can take place, causing the antenna to be non responsive at that particular frequency. It is very easy to tune the notch centre frequency with the change in the total length of the C-shaped slot. More experiments were carried out on the length of the C-shaped slot using the simulation program IE3D.

Figure 12 shows the return loss for triple band notched antenna.

Table 4. Measurements of the 2 CSRRs (Complementary Split Ring Resonator) providing us with a band notch at 7.9 GHz.

• Parameters	• Dimensions
• L8	• 4.5 mm
• L7	• 1.5 mm
• L9	• 2.5 mm
• L10	• 0.5 mm
• W10	• 4.5 mm
• W11	• 2 mm
• Thickness of slot	• 0.5 mm

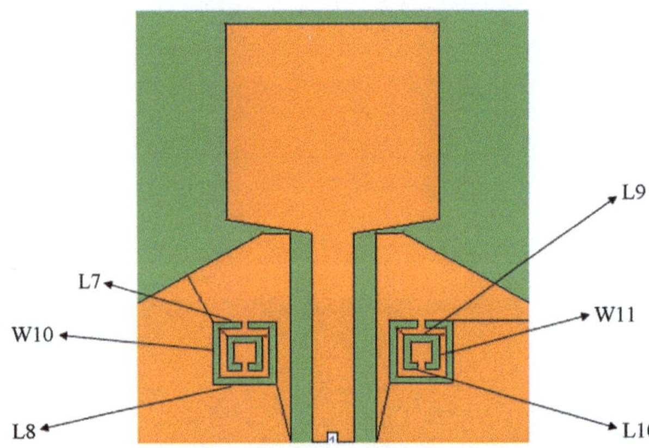

Figure 8. Geometry of antenna with a pair of CSRRs (notch at 7.9 GHz).

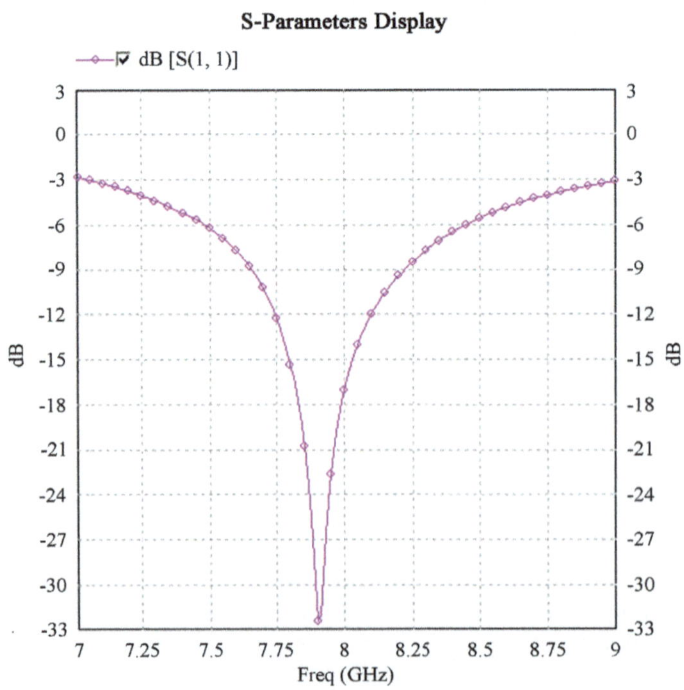

Figure 9. Pair of CSRRs (notch at 7.9 GHz) return loss.

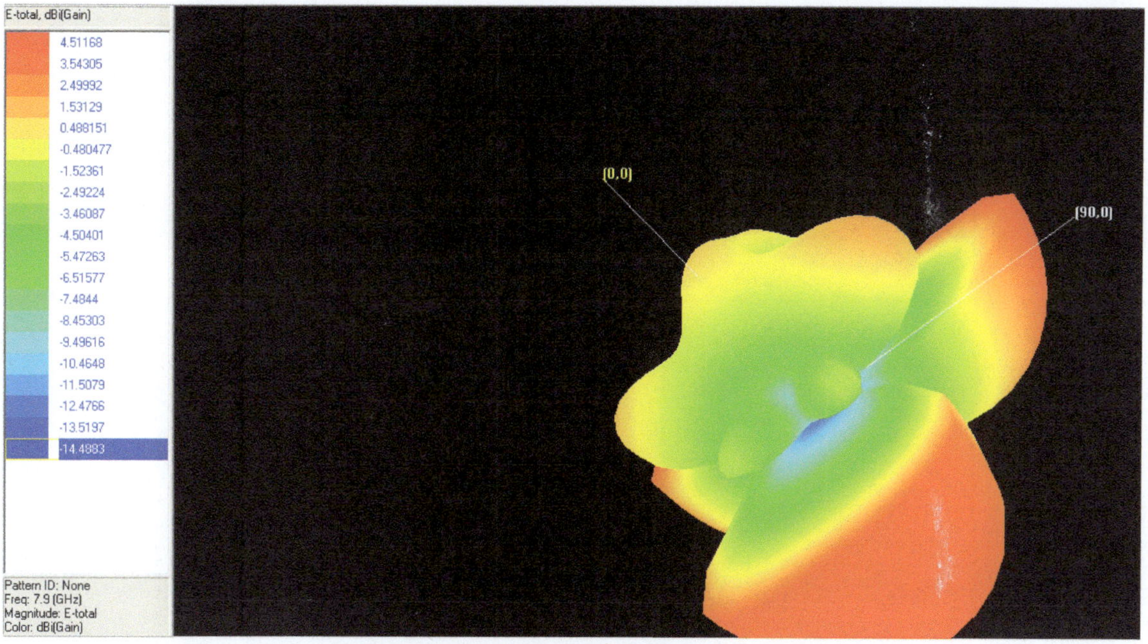

Figure 10. 3D radiation pattern for 7.9 GHz with max gain 4.51 dB.

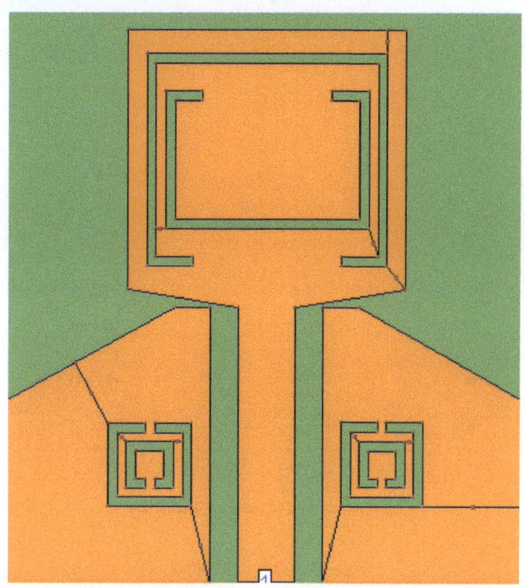

Figure 11. Geometry of antenna with triple band notch.

This Return Loss graph belongs to the triple band notch antenna, with the band notches at 3.2 GHz, 5.5 GHz and 7.9 GHz. The band notches at the frequency 3.2 GHz and 7.9 GHz is small because of the cross coupling caused due to the integration of all the 3 slots at a time in one antenna.

The 3-D radiation pattern for 3.2GHz is shown in **Figure 13**.

The 3-D radiation pattern for 5.5 GHz is shown in **Figure 14**.

The 3-D radiation pattern for 7.9GHz is shown below in **Figure 15**.

Summary of triple band notched for UWB MSA is shown below in **Table 6**.

3. Conclusions

This paper proposed design of simple rectangular microstrip antenna for UWB applications. The antenna is capable of achieving multi notched-band. The following points summarized our design:

Table 5. the above table gives us the measurements of the antenna as well as the measurements of the 2 C-shaped slots and a pair of CSRRs.

• Parameters	• Dimensions
• W1	• 14 mm
• W2	• 3 mm
• W3	• 7.5 mm
• W4	• 11.5 mm
• W5	• 5 mm
• W6	• 10 mm
• W7	• 15 mm
• W8	• 1 mm
• W9	• 11.5 mm
• W10	• 4.5 mm
• W11	• 2 mm
• L1	• 11 mm
• L2	• 13 mm
• L3	• 15 mm
• L4	• 11 mm
• L5	• 2 mm
• L6	• 2.5 mm
• L7	• 1.5 mm
• L8	• 4.5 mm
• L9	• 2.5 mm
• L10	• 0.5 mm
• Thickness of the slot	• 0.5 mm

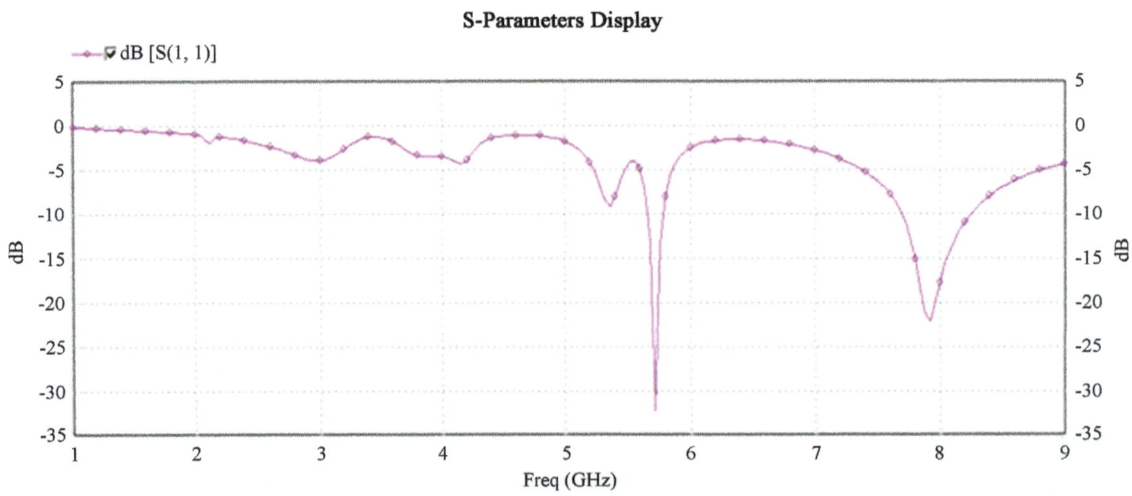

Figure 12. Return loss for triple band notched antenna.

Table 6. Summary for triple band notched for UWB MSA.

• Excitation	• Maximum gain	• Return loss
• 3.2 GHz	• −1.32 dB	• −3 dB
• 5.5 GHz	• 1.10 dB	• −33 dB
• 7.9 GHz	• 5.11 dB	• −23 dB

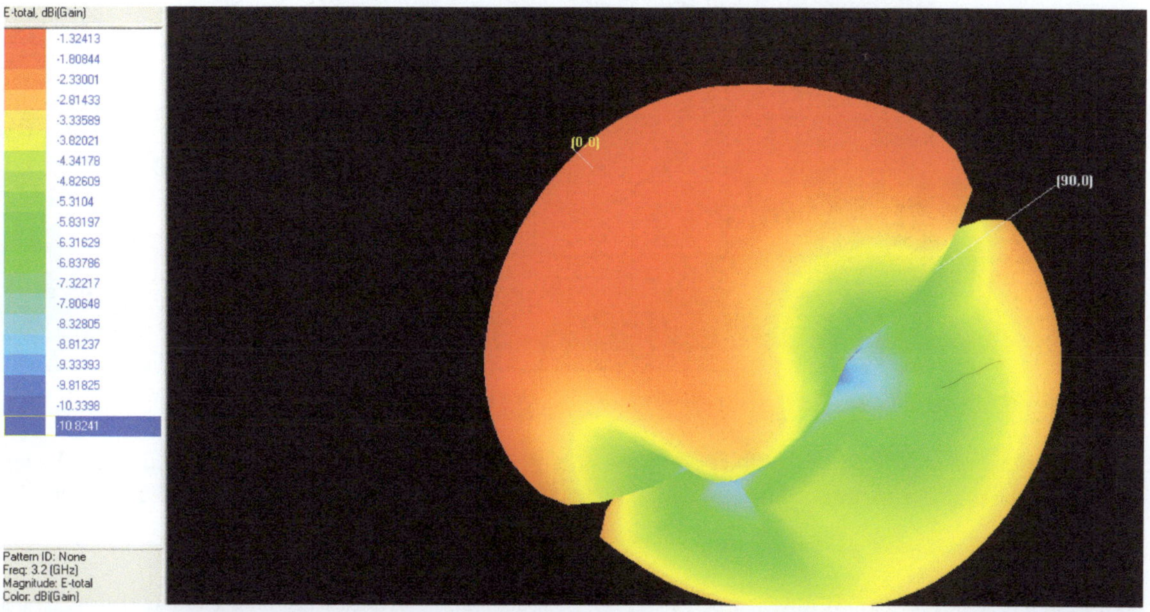

Figure 13. Return loss for triple band notched antenna.

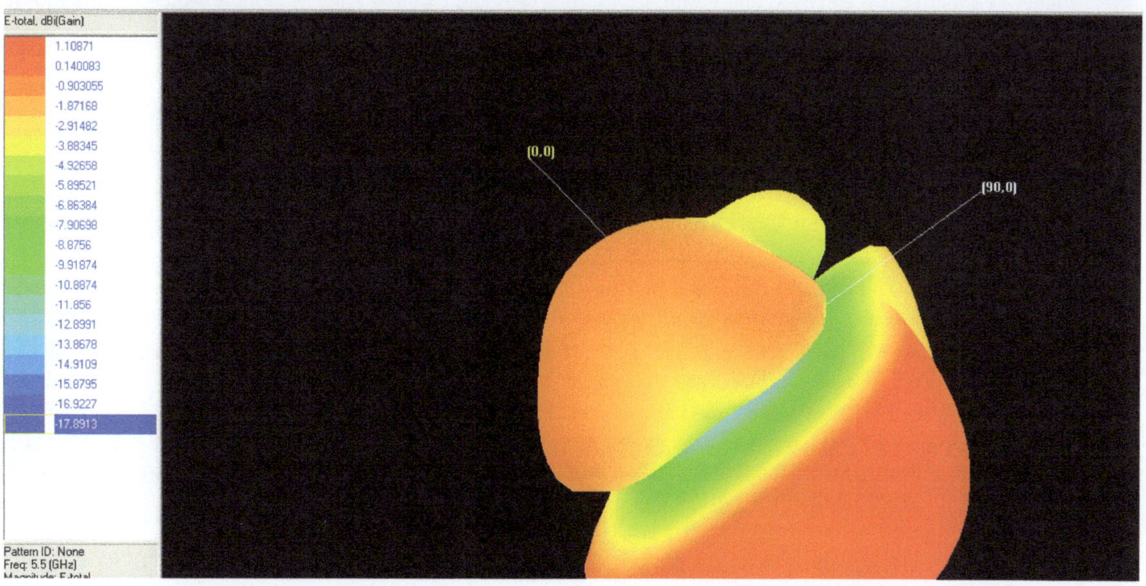

Figure 14. Return loss for triple band notched antenna.

1) To minimize the potential interferences between the UWB system and the narrow band systems, a compact CPW-fed planar UWB antenna with triple band rejection features was designed. First on the basic antenna individual notches are designed and their band-notch properties are studied. Then all the three notches are embedded onto the primitive antenna. While integrating all the notch elements utmost care has been taken to minimize the

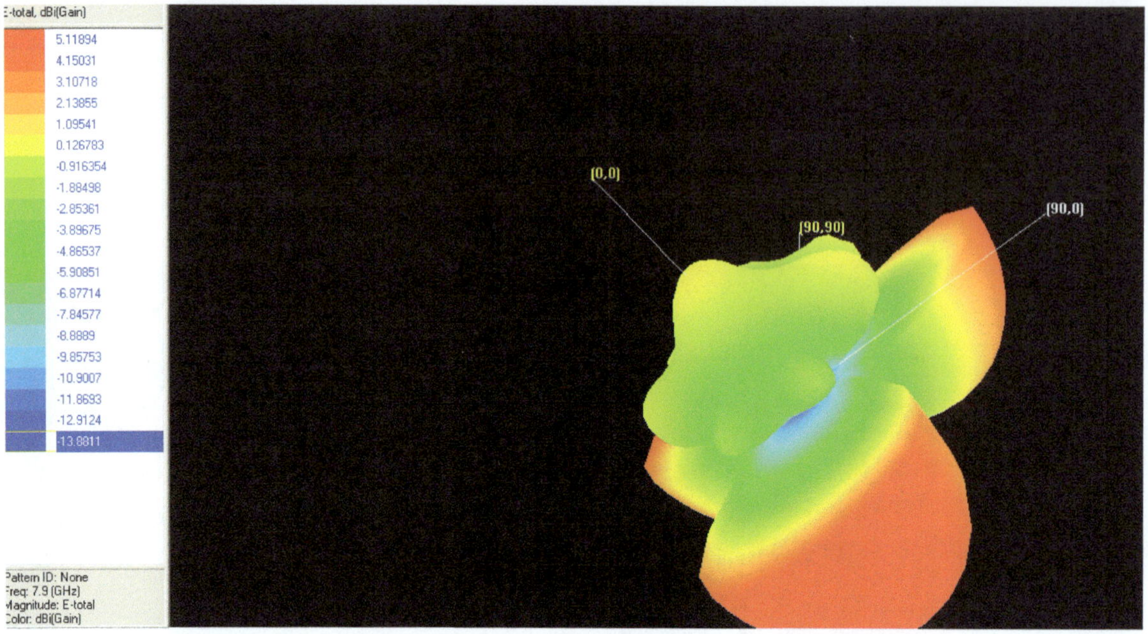

Figure 15. 3D radiation pattern for 7.9 GHz with max gain 5.11 dB.

cross-coupling among them; so that their operation doesn't get hampered by the presence of other notch elements.

2) Antenna operates in the specified 3.1 - 10.6 GHz range and the notches are at 3.2 GHz, 5.5 GHz and 7.9 GHz.

References

[1] Commission, F.C. (2002) Revision of Part 15 of the Commission's Rules Regarding Ultra-Wideband Transmission Systems, First Report and Order. ET Docket 98-153, FCC 02-48, 100-118.

[2] Liang, J., Chiau, C.C., Chen, X. and Parini, C.G. (2004) Printed Circular Disc Monopole Antenna for Ultra-Wideband Applications. *Electronics Letters*, **40**, 1246-1247. http://dx.doi.org/10.1049/el:20045966

[3] Latré, B., Braem, B., Moerman, I., Blondia, C. and Demeester, P. (2011) A Survey on Wireless Body Area Networks. *Wireless Networks*, **17**, 1-18. http://dx.doi.org/10.1007/s11276-010-0252-4

[4] Choi, S.H., Park, J.K., Kim, S.K. and Park, J.Y. (2004) A New Ultra-Wideband Antenna for UWB Applications. *Microwave and Optical Technology Letters*, **40**, 399-401. http://dx.doi.org/10.1002/mop.11392

[5] Mardani, H., Ghobadi, C. and Nourinia, J. (2010) A Simple Compact Monopole Antenna with Variable Single- and Double-Filtering Function for UWB Applications. *IEEE Antennas and Wireless Propagation Letters*, **9**, 1076-1079. http://dx.doi.org/10.1109/LAWP.2010.2091391

[6] Li, W.-M., Ni, T., Zhang, S.-M., Huang, J. and Jiao, Y.-C. (2011) UWB Printed Slot Antenna with Dual Band-Notched Characteristic. *Progress in Electromagnetics Research Letters*, **25**, 143-151. http://dx.doi.org/10.2528/PIERL11061702

[7] Dong, Y.D., Hong, W., Kuai, Z.Q. and Chen, J.X. (2009) Analysis of Planar Ultrawideband Antennas with On-Ground Slot Band-Notched Structures. *IEEE Transactions on Antennas and Propagation*, **57**, 1886-1893. http://dx.doi.org/10.1109/TAP.2009.2021910

[8] Zhang, M., Yin, Y.-Z., Ma, J., Wang, Y., Xiao, W.-C. and Liu, X.-J. (2010) A Racket-Shaped Slot UWB Antenna-coupled with Parasitic Strips for Band-Notched Application. *Progress In Electromagnetics Research Letters*, **16**, 35-44. http://dx.doi.org/10.2528/PIERL10060910

[9] Chen, W.-S. and Ku, K.-Y. (2008) Band-Rejected Design of the Printed Open Slot Antenna for WLAN/WiMAX Operation. *IEEE Transactions on Antennas and Propagation*, **56**, 1163-1169. http://dx.doi.org/10.1109/TAP.2008.919192

[10] Gao, G.-P., Mei, Z.-L. and Li, B.-N. (2010) Novel Circular Slot UWB Antenna with Dual Band-Notched Characteristic. *Progress in Electromagnetics Research C*, **15**, 49-63. http://dx.doi.org/10.2528/PIERC10032703

[11] Lin, Y.-C. and Hung, K.-J. (2006) Compact Ultrawideband Rectangular Aperture Antenna and Band-Notched Designs. *IEEE Transactions on Antennas and Propagation*, **54**, 3075-3081. http://dx.doi.org/10.1109/TAP.2006.883982

[12] Pandavand, D. and Sethi, S. (2013) Design of Compact Plannar Ultra-Wide Band Antenna with Band-Notched Characteristic. Thesis.

21

Comparison of Excitation of Acoustic-Electromagnetic Wave in Piezoelecric Crystal and Crystal with Potential of Deformation

S. Koshevaya, V. Grimalsky, Y. Kotsarenko, J. Escobedo-Alatorre

Autonomous University of Morelos State, CIICAp, Cuernavaca, Mexico
Email: svetlana@uaem.mx

Abstract

In this article, the comparison of excitation in high frequencies of acoustic-electromagnetic wave in piezoelecric crystal and crystal with potential of deformation GaAs is investigating. Possible mechanisms of coupling different hybrid waves are the piezoeffect and the deformation potential. As a model it is analyzing a film of crystal places between two symmetrical substrates with the other materials without an acoustic contact. This film includes 2D electron gas with a high negative differential conductivity and uniform initial distribution of electrons. The hybrid acoustic-electromagnetic wave and hybrid space charge wave interact. Amplification of space charge wave takes place due to negative differential conductivity in GaAs. This amplification of space charge waves is causing the amplification of acoustic-electromagnetic wave. It is to show that the symmetric modes, emerging as transverse ones, interact more effectively with the space charge waves. Another important result is the following: at the frequencies $f \approx 10$ GHz, the excitation efficiency of acoustic-electromagnetic wave with transverse displacement due to piezoeffect is more effective, but at higher frequencies, the deformation potential is dominating.

Keywords

Piezoelectric Crystal, Deformation Potential, Acoustic-Electromagnetic and Space Charge Waves

1. Introduction

The amplification of the acoustic-electromagnetic waves is very important problem [1]. A possible solution of this problem is the resonant coupling of acoustic waves with the microwave electric field of space charge waves (SCW) in materials possessing negative differential conductivity (NDC) like GaAs [2] [3]. Namely, under propagation in the bias electric field higher than the critical value for observing negative differential conductivity (NDC), the SCW is subject to amplification, and its microwave electric field can achieve high values. Due to piezoeffect or deformation potential, this microwave electric field excites hypersonic acoustic-electromagnetic waves (AEW). In [2] [3], it is demonstrated that this excitation has a resonant character with respect to the frequency and the thickness of the GaAs film. The critical value of bias electric field in GaAs is $E_c = 3.5$ kV/cm that it is good for analysis of amplification in different mechanisms like piezoeffect or deformation potential so as other materials have very big critical field like crystal InP and GaN.

The present paper is focused on theoretical research of excitation of acoustic-electromagnetic waves (AEW) in GaAs films of a sub-micron thickness placed between non-piezoelectric substrates. The spatial increments of SCW due to NDC have been calculated for two mechanisms like piezoeffect and deformation potential. An amplification of SCW is possible for frequencies from 10 GHz for case of piezoeffect. For deformation potential the frequencies range increase from 10 GHz to 30 GHz, so the mechanism of deformation potential is more useful in high frequencies.

The thin GaAs film plays the role in a transducer of acoustic electromagnetic waves. The amplification has a resonant character, and the intensities of energy may reach the values of 10 W/cm^2. The realization of the good amplification has showed that usefulness of hybrid waves in high frequencies range more $f \approx 10$ GHz is very important for application in communication, medicine and control systems. The analysis of basic mechanisms of amplification is given in this article. The resonance excitation and amplification of acoustic-electromagnetic wave with space mode in structure of GaAs is analyzing for two cases of piezoeffect and deformation potential, but the amplification of high frequencies are realizing only in case of deformation potential.

2. Model of Amplification Due to Piezoeffect

Let us to consider a thin film of n-GaAs located between two mediums, the substrates are from material without piezoeffect, and the transversal wave passed along z axis. The component of displacement U is in direction of the y axis. The film consists of a 2D gas with a high negative differential mobility. The value L is the length of the element (along axis OZ), h is the thickness of film (see **Figure 1**). This film has a substrates with depth d. The hybrid wave is decreasing in the substrate, then, all its energy is in the very thin film with 2D electron beam. The effective size of the waveguide is describing by the parameter $2H = 2(h + d/2)$, d is the depth of substrate.

Let us to analyze the simplest transversal in displacements u mode using elastic theory equation with the piezoeffect [1]. It is necessary to use the variable electric field $E = -\frac{\partial \varphi}{\partial z}$ (in direction of axes Z) determined by potential φ,

$$\frac{1}{s^2}\frac{\partial^2 u}{\partial t^2} = \Delta u + \Gamma \Delta \frac{\partial u}{\partial t} - \frac{2\beta}{\rho s^2}\frac{\partial^2 \varphi}{\partial x \partial z} \qquad 1(a)$$

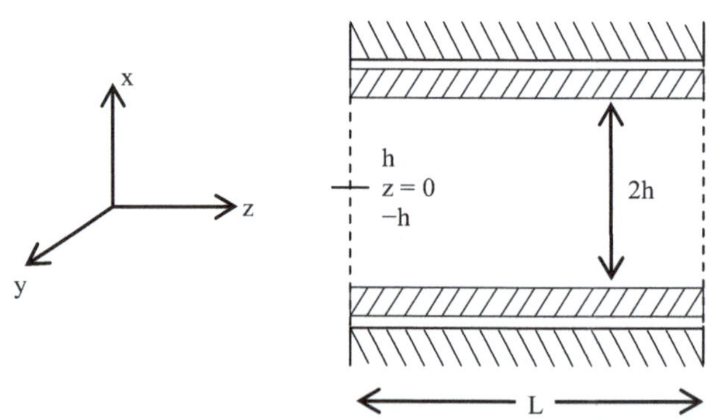

Figure 1. Geometry of waveguide for computing simulation of amplification acoustic-electromagnetic wave where 2 h is the transversal size of guide in x direction and L is the length of the guide. In direction y the model of guide has size going to infinitive value.

$$-\Delta\varphi + \frac{\beta}{\varepsilon_0 \varepsilon}\frac{\partial^2 \varphi}{\partial x \partial y} = 0 \qquad 1(b)$$

$$\left[\frac{\partial u}{\partial x} + -\frac{\beta}{\rho s^2}\cdot\frac{\partial \varphi}{\partial z}\bigg| x = +, -h/2\right] = 0\,, \qquad 1(c)$$

where $s \approx 3 \times 10^5$ cm/s is the sound velocity, β is the module of coefficient of piezoeffect, ρ is the density parameter, φ is the potential determined the electric field $\mathrm{grad}\varphi = \vec{E}$ [1]-[5].

The boundary conditions are in (Equation 1(c)). In the case of free very thin film it is $\frac{h}{d/2+h} \prec\prec 1$. In this case it is necessary to consider the viscosity coefficient Γ for the simulation. The wave process shows that the amplification realizes if the synchronism between the acoustic-electromagnetic wave with transversal displacement mode and the space charge wave [5],

$$k_0 \cong \sqrt{\frac{\omega_0^2}{s^2} - \frac{(2m+1)^2 \pi^2}{4H^2}} = \frac{\omega_0}{v_0} \qquad (2)$$

where ω_0 is the frequency of synchronism, s, v_0 is sound and space charges waves velocities, respectively; index $m(1,2,3,\cdots) \succ\succ 1$), H must have also an optimal value.

Resonance condition $\cos gH \cong 0$ must hold true,
where

$$g = \frac{\omega_{cr}}{s} = \frac{(2m+1)\pi}{2H}.$$

Parameters g and ω_{cr} are given by

$$g = \frac{\omega_{cr}}{s} = \frac{(2m+1)\pi}{2H}$$

and

$$\omega_{cr}^2 = \frac{\pi^2 s^2}{4H^2}(2m+1)^2.$$

These conditions for the amplification are better for the symmetric modes if the displacement

$$u_n \propto B\cdot\sin gx\cdot e^{i(\omega t - kz)}, \phi_n \propto \frac{ik\pi}{g}\cdot B\cdot\cos gx\cdot e^{i(\omega t - kz)}.$$

Boundary conditions are taken in case of the absent of the acoustic contact of the film with the environment (Equation 1(c)).

In the simulation it is to use Equation (1) and the resonance condition (2). It is necessary to take into account the 2D electron gas of n-GaAs thin film and the equations for slow changing amplitudes of acoustic-electromagnetic hybrid wave with potential φ. The computer modeling shows the effective amplification of waves in the presence of the negative differential mobility (conductivity). The following parameters have been chosen $n_0 \approx 10^{15}$ cm^{-3}, the electron concentration in the film, $v_0 \approx 2 \times 10^7$ cm/s, the electron velocity, $L = 0.1$ cm, the length of the film $H = 1$ μm. The intensity of wave can reach value of 10 W/cm^2 at microwave frequencies $\omega = 5 \times 10^{10}$ - 2×10^{11} s^{-1}. In the **Figures 2-4**, the amplification of the acoustic-electromagnetic wave in microwave range is to present.

Traditional construction for realization of acoustic-electromagnetic amplification is in **Figure 3**. This construction is for high frequencies $f \approx 10^{11}$ Hz ranges too.

Additionally, in simulations it is considering the electric variable field distribution in input antenna like Gauss impulse.

$$E^{\sim}(z,y,t) = E_{10}\exp\left(-((z-z_1)/z_0)^2 - ((y-y_1)/y_0)^2\right)\times\exp\left(-((t-t_1)/t_0)^2\right)$$

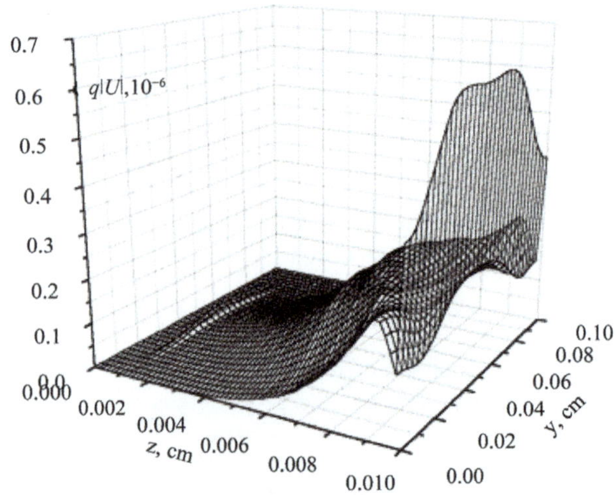

Figure 2. Distribution of the amplitude of acoustic deformation $q|U|$ along the film in direction z for time t = 4 ns. The resonant excitation is given at frequency $\omega = 10^{11}$ s^{-1} of the hybrid acoustic-electromagnetic mode, and the size of the waveguide is H = 1 μm.

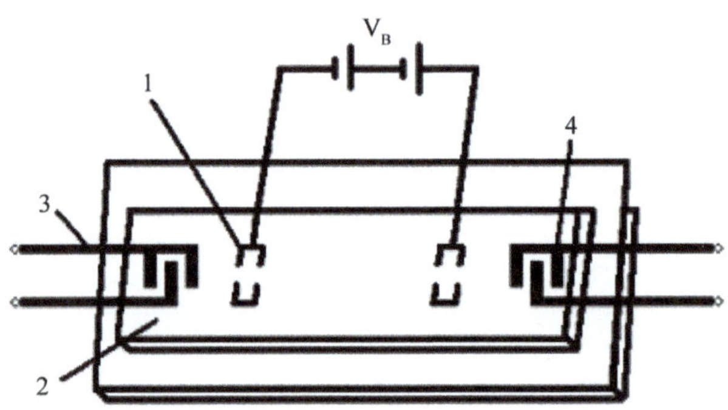

Figure 3. Traditional structure for exiting of acoustic-electromagnetic waves. It is necessary to prepare the cathode and anode 1, and the input 3 and the output 4 antennas; the film of n-GaAs is in i-GaAs substrate.

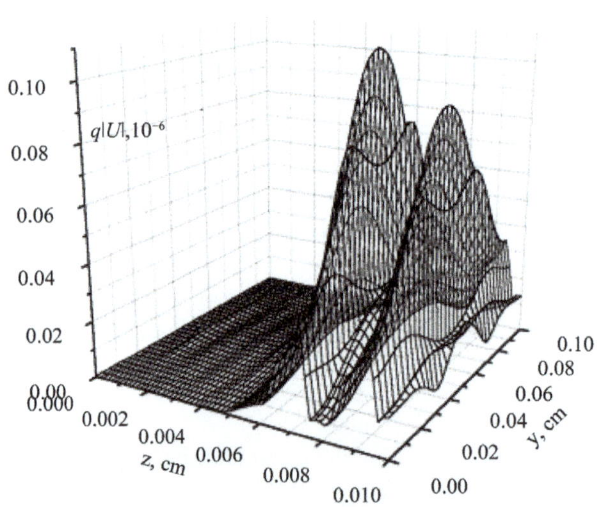

Figure 4. Distribution of the displacements deformation amplitude along the film in direction z at time t = 4 ns with a size of wave guide H = 0.1 μm.

Electric variable field distribution in input antenna has parameters: $z_1 = 0.001$ cm, $z_0 = 0.0001$ cm; $y_1 = 0.05$ cm, $y_0 = 0.02$ cm; $t_1 = 2.5$ ns, $t_0 = 1.25$ ns; $E_{10} = 200$ V/cm and $E_B = 4.56$ kV/cm, the drift field, $n_0 = 10^{11}$ cm^{-2}, the 2D concentration of electron gas, and q, the transverse wave number (index m equal 5) of the acoustic-electromagnetic mode.

Excitation of the acoustic-electromagnetic wave due to deformation potential is in the Lamb's acoustic mode [1], which possess longitudinal component of mechanic displacement, has the electric field too and interacts with space charge waves due to the deformation potential. The equation for the elastic displacement components of the Lamb's wave [1] $u = (u_1, 0, u_3)$ takes a form.

$$\rho\frac{\partial^2 u_1}{\partial t^2} = c_{11}\frac{\partial^2 u_1}{\partial x^2} + c_{44}\frac{\partial^2 u_1}{\partial z^2} + (c_{12}+c_{44})\frac{\partial^2 u_3}{\partial x \partial z} + G\Delta\frac{\partial u_1}{\partial t} + \frac{1}{3}\Xi_u^L\left(\frac{\partial n^L}{\partial y} + \frac{\partial n^L}{\partial z}\right),$$
$$\rho\frac{\partial^2 u_3}{\partial t^2} = c_{11}\frac{\partial^2 u_3}{\partial z^2} + c_{44}\frac{\partial^2 u_3}{\partial x^2} + G\Delta\frac{\partial u_3}{\partial t} + (c_{12}+c_{44})\frac{\partial^2 u_1}{\partial x \partial z} + \Xi_d^\Gamma\frac{\partial n^\Gamma}{\partial z} + \left(\Xi_d^L + \frac{1}{3}\Xi_u^L\right)\frac{\partial n^L}{\partial z} + \frac{1}{3}\Xi_u^L\frac{\partial n^L}{\partial y} \quad (3)$$

Here $\Xi_{d,u}^{\Gamma,L}$ are the components of tensor of deformation potential; c_{ij} are the elastic constants [1] [5] [6], the term $G\Delta\partial u/\partial t$ is due to dissipation, as in the Equation (3). Note that the partial concentrations n^L, n^Γ for the valleys Γ, L are present in the Equations (3).

As in the case of piezoactive transverse modes, the effective excitation of acoustic-electromagnetic waves with displacement like in Equations (3) takes place near cut-off frequency. It is shown that the symmetric modes emerging as transverse ones interact more effectively with the space charge waves. A distribution of the displacements near cut-off is

$$u_1 \approx \sin\left(\frac{\omega}{s_l}x\right)\exp\left(i(\omega t - kz)\right);\ u_3 \approx 0;\ \frac{\omega}{s_l} \approx \frac{\pi}{2h}(2n+1),\ n = 0,1,2,\cdots, \quad (4)$$

here $s_l \approx 5.5 \times 10^5$ cm/s is the longitudinal sound velocity in GaAs [7].

The computer modeling in the base of Equations (3) and (4) is the same like for case the piezoeffect. This modeling has demonstrated that the distributions in the plane of the film of excited acoustic-electromagnetic waves are very similar to those obtained for piezoeffect. But it is shown ¡ that the efficiency of excitation due to deformation potential becomes dominating at higher frequencies $f > 30$ GHz.

3. Conclusion

The resonant interaction of space charge waves with acoustic-electromagnetic waves in thin GaAs films can be used for effective in microwave wave range. At the frequencies $f < 30$ GHz efficiency of excitation of acoustic-electromagnetic due to piezoeffect is more effective, but at higher frequencies, deformation potential is dominating. An intensity of acoustic-electromagnetic waves can reach 10 W/cm^2 at microwave frequencies range $f =$ 10 GHz ÷ 30 GHz. To increase a frequency of excited wave to millimeter range, it is possible to choose the resonant excitation of acoustic-electromagnetic waves at the second harmonic of the input signal.

Acknowledgements

Authors thank SEP-CONACyT for the support of our work.

References

[1] Dieulesaint, E. and Royer, D. (2000) Elastic Waves in Solids. Springer, New York.

[2] Grimalsky, V., Gutierrez-D., E., Garcia-B., A. and Koshevaya, S. (2004) Resonant Excitation of Microwave Acoustic Modes in n-GaAs Film. *Proceedings of International. Conference on Microelectronics ICM*-2004, Tunis, 6-8 December 2004, 72-75. http://dx.doi.org/10.1109/icm.2004.1434209

[3] Grimalsky, V., Gutierrez-D., E., Garcia-B., A. and Koshevaya, S. (2006) Resonant Excitation of Microwave Acoustic Modes in n-GaAs. *Microelectronics Journal*, **37**, 395-403. http://dx.doi.org/10.1016/j.mejo.2005.06.003

[4] Mitin, V.V., Kochelap, V.A. and Stroscio, A. (1999) Quantum Heterostructures: Microelectronics and Optoelectronics. Cambridge University Press, Cambridge.

[5] Shur, M. (1989) GaAs Devices and Circuits. Plenum Press, New York.

[6] Carnez, B. (1980) Modeling of Submicrometer Gate FET Including Effect of Nonstationary Dynamics. *Journal of Applied Physics*, **51**, 784-790.

[7] Koshevaya, S.V., Grimalsky, V.V., Garcia-B., A. and Díaz-A., M.F. (2006) Amplification of Hypersonic by GaAs Crystals. *Ukrainian Journal of Physics*, **51**, 594-598.

22

Synoptic Analysis of Epidemiologic Evidence of Glioma Risk from Mobile Phones

Norbert Leitgeb

Institute of Health Care Engineering, Graz University of Technology, Graz, Austria
Email: norbert.leitgeb@tugraz.at

Abstract

Concern about brain cancer risks from mobile phone use, in particular of glioma, is mainly based on epidemiologic studies reporting on an increased risk estimate (odds ratio, OR) in particular of long-term and/or heavy users. Concern is enhanced by contradictory results and alarming conclusions from some epidemiologic studies. In a new synoptic approach all reported data from epidemiological studies published since 2001 were analyzed. This approach provided new insight with regard to the suspected link between mobile phone use and glioma. Two quite different data pools could be identified with numerous studies from one single research group opposing all other national and international studies. It could be shown that in spite of puzzling differences of their data pattern, in dependence of the number of exposed cases both data pools exhibit the similar trend of ORs towards reduced glioma risk with increasing statistical power, finally even converging towards a reduced risk, although from either side of the zero-risk line. While in the pooled data a potential long-term risk could be masked by reassuring short-term data the synoptic analysis in dependence of dose surrogates such as cumulated use time, cumulated call time or cumulated number of calls the seemingly increased long-term risk of glioma could be identified as an effect from low statistical power. In spite of worrying differences among some epidemiological studies, overall, the synoptic analysis of the entire body of data supports reassuring rather than alarming conclusions on glioma risks from mobile phone use even in long-term and/or heavy users.

Keywords

Health Risk, Dose Response, Carcinogenicity, Electromagnetic Field, Telecommunication

1. Introduction

Based on epidemiologic studies mainly at mobile phone frequencies around 1 - 2 GHz the International Agency for Research on Cancer (IARC) of the World Health Organization (WHO) has concluded there was "limited evidence" that radio frequency (RF) electromagnetic fields (EMF) are carcinogenic to humans and classified them as possibly carcinogenic to humans class 2B [1]. This challenges existing exposure limits for general public's health protection [2] [3] and further enhances existing public concern about brain cancer risk from mobile phones.

In the RF range the situation regarding epidemiologic studies is characterized by two opposing positions. Several studies from the Swedish group of Hardell *et al.* (SHG) report on consistent elevated risks of brain cancer [4]-[9] which opposes the numerous other studies, including the multinational Interphone study (IPS) with 13 countries participating [10], and all other national studies (ONS) including Sweden, which report on a lack of carcinogenic evidence up to 10 years mobile phone use but conclude there is some uncertainty with regard to specific cancers such as glioma at prolonged exposure [11]-[25]. This raised further concern and increased the need for clarification.

In this study it is shown that conclusions on suspected long-term risks are possible by a synoptic view on the existing body of evidence, because fortunately, it has become common practice that papers on epidemiologic studies include tables with extensive numbers of detailed results in terms of risk estimates (odds ratios, OR) which have been calculated for a large variety of different scenarios such as kind of tumor, tumor location, laterality, radiation source (e.g. mobile phone, wireless, DECT), region (e.g. urban or rural), magnitude of exposure and dose metric (e.g. subscription time, use time, latency time, regular use, cumulated use time cumulated call time, cumulated number of calls). However, interestingly, so far authors as well as reviewers including risk assessment bodies draw their conclusions based on only one or few selected (usually worst-case) OR values per study [1] [24] [25]. So far, little use has been made of the entire pool of available data.

Therefore, in a new approach of synoptic analysis of all ORs published so far, a puzzling fact could be identified. In contrast to the ELF range, where the ORs reported with childhood leukemia fitted well into one common data pool [10], a quite different scenario was found in the RF range [11]: The synoptic analysis identified a striking difference between reported results of mobile phone brain cancer studies of one particular research group, namely the SHG group which exhibits an implausible pattern compared to all other data, namely of IPS and ONS [13]-[25] which were consistent within themselves. This indicated major differences in SHG's studies with regard to methodology and/or data analysis. With the synoptic analysis it could be already shown that ORs critically depend on the underlying number of exposed cases [10] [11]. Moreover, surprisingly, in spite of the puzzling contradictions and the different data patterns, with increased number of exposed cases the data pools on brain cancer of both groups, namely of SHG as well as of all other studies (IPS and ONS) headed towards the same OR value—although from different sides of the zero-risk line. Even more, the ORs converged towards a reduced risk (OR = 0.8) and, hence, were reassuring rather than alarming [11]. However, since this synoptic analysis did not yet differentiate between various types of brain cancer, it could be argued that existent risks of particular cancer types could have been masked by lacking risks of others. Therefore, this paper presents the results of the analysis on a specific brain cancer type, namely on glioma, with particular emphasis on the suspected long-term risk.

2. Method

In the new synoptic approach all reported ORs of available epidemiologic studies on mobile phone use and risk of glioma published from 2001 until 2014 were pooled and analyzed. Studies were identified by a literature search in the data bases Pubmed, FEMU and IEEE Xplore. The studies were included without further selection such as with regard to size of a study or exposure metric. None of the available studies was rejected. Therefore, the entire body of available data was analyzed without any further restrictions. In case ORs were given both unadjusted and adjusted for cofactors, the latter were used.

The synoptic analysis should allow getting an overview on the existing body of evidence, identifying potential discrepancies and outliers, finding dependencies on parameters, if existent, and testing the null hypothesis. Consequently, the pool of data was to investigate in particular with regard to:

1) the null hypothesis that there is no causal relationship between mobile phone use and glioma;

2) a potential dependence of ORs on the number of exposed cases N_{exp} used for OR calculation as surrogate for the statistical power;

3) a potential dependence of ORs on the accumulated exposure (dose) in particular with regard to a potential long-term glioma risk. Since studies report on different estimated dose metrics such as cumulated use time

(CUT), cumulated call time (CCT) and cumulated number of calls (CNC), analyses were made on each of them.

Usually, epidemiological studies use quite different cut-off points to classify the accumulated exposure (dose). For synoptic analysis this allowed generating various data points to quantitatively analyze a potential dose dependence rather than comparing "heavy users" (which are differently defined among the studies) with others. However, the dose response analysis does not account for the different statistical power of the various ORs. Therefore, in addition to time trend analyses, the data were also separated into different data pools, related to low and high dose, respectively, to account for the different statistical power and check whether ORs in dependence on N_{exp} converged, and if so, towards which endpoints.

3. Results

Overall, 16 epidemiologic studies on mobile phone use and glioma with 523 reported OR values could be analyzed. (Please note that due to overlap of similar OR values, not all data points can be identified in the presented figures.) **Figure 1** presents an overview on the pooled data of all glioma studies of SHG as well as of ONS and IPS. It can be seen that ORs and their variance clearly depend on the number of exposed cases N_{exp} which was already encountered in the pooled brain cancer data [11]. Like the overall brain cancer data also glioma data of ONS and IPS exhibit the well-known funnel-shaped pattern with a wide variance extending to either side of the zero risk line OR = 1 at low N_{exp}, comprising three orders of magnitudes from OR = 0.3 up to OR = 30. It can be seen that there are some outliers (related to results from [13]). With increasing N_{exp} the variance decreases and ORs become lower and lower, and finally converge towards OR = 0.8 which indicates a reduced risk.

In contrast to this and similar to brain cancer results SHG data considerably differ from all other data with regard to their variance and the data pattern. The overwhelming majority of ORs is larger than 1, hence ORs are at only one side of the zero risk line (OR = 1) which is not plausible from a statistical point of view. Consequently, the majority is indicating an increased risk (**Figure 1**). However, it is interesting to note, that in spite of the puzzling pattern, with increasing N_{exp} also SHG data converge to the same endpoint as ONS and IPS namely towards reduced risk. However, with regard to the obvious methodological difference and bias, SHG data had to be excluded from the following analysis of a potential dose-response.

Since **Figure 1** includes all glioma data irrespective the duration or intensity of mobile phone use, it does not yet allow concluding on any dose response. Therefore, this was analyzed in a second step based on different dose surrogates. **Figure 2** shows the time trend of pooled ORs of IPS and ONS in dependence on the *cumulated use years* (CUY) as derived from questionnaires in terms of years of use or latency time until diagnosis. The quadratic regression curve across the time trend data remains mostly below the zero-risk line (OR = 1) and only beyond 10 years approaches zero risk (OR = 1). The quadratic regression coefficient amounts to 0.028. Therefore, in spite of the reassuring results up to 10 years it could be argued that the final slight increase towards OR = 1 could indicate a potential long-term effect. However, the time trend suffers from a major disadvantage, namely, that it gives the same weight to all ORs irrespective their statistical power (N_{exp}). To account for this, the entire OR data pool was separated into two dose-related sub-pools, one with CUY lower than 5 years and another with CUY larger or equal 10 years (**Figure 3**). Although the number of data points is lower, it can be clearly seen that both sub-pools exhibit a quite similar pattern. In both cases with increasing N_{exp} the risk estimates decrease. None of the sub-pools tends towards an increased risk. On contrary, they converge towards reassuring endpoints (OR = 0.8 and OR = 1, respectively).

The same was done with another surrogate of cumulated exposure, namely the (estimated) *cumulated call time* (CCT). For its calculation, three estimates are necessary, namely user's estimates on their use time, their average number of calls and their average call time. Similar to CUY, the time trend exhibits no increased risk, however, again ORs increase towards the largest CCT value even to ORs of clearly increased risk (**Figure 4**). The quadratic regression coefficient amounts to 0.112. Again, interpretation needs some caution due to the fact that the same weight is given to all OR values irrespective their statistical power. Therefore, the data pool was again separated into two sub-pools with CCTs lower and higher than 500 h, respectively. Now, similar to CUY in dependence of N_{exp} the patterns of both sub-pools clarifies the seeming long-term effect: Both patterns show the well-known decrease of ORs with increasing statistical power (N_{exp}). Again, none of the sub-pools tends towards an increased risk (**Figure 5**). Now, the seemingly increased risk at highest CCT as indicated by the time trend can be explained. It is caused by the high variance and some offset of ORs at low N_{exp} rather by a long-term effect.

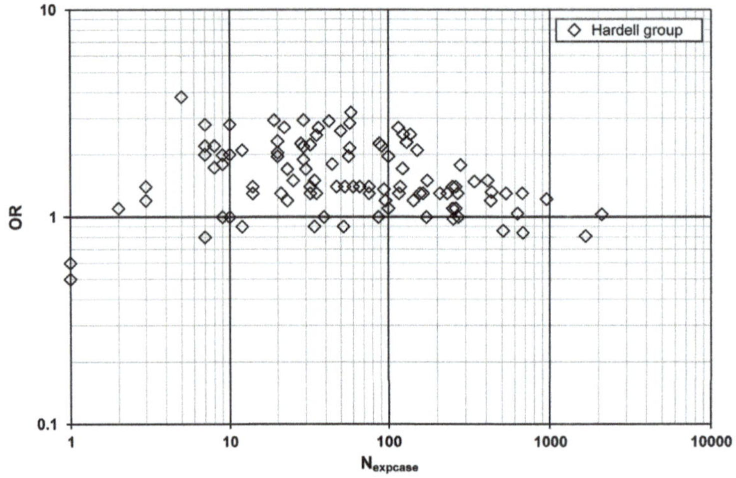

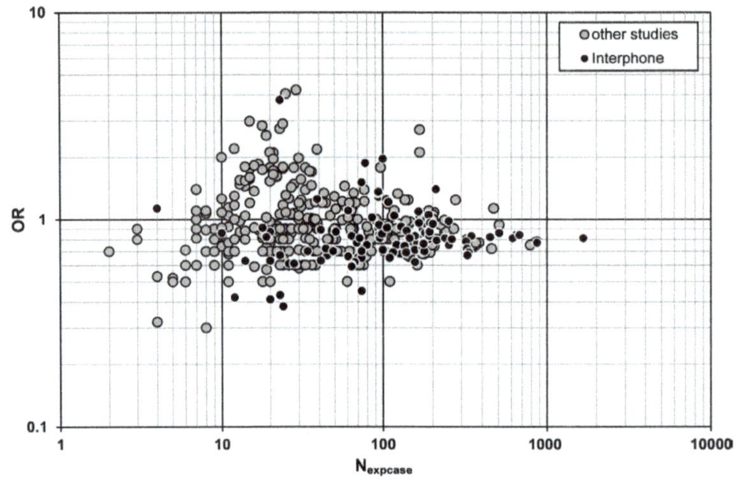

Figure 1. Reported odds ratios (OR) of epidemiological case-control studies on glioma and mobile phone use over the associated number of exposed cases N_{exp}; data of the Swedish Hardell group (above) and results of all other studies (below); diamonds: Swedish Hardell studies, grey points: all other national studies; black points: interphone study.

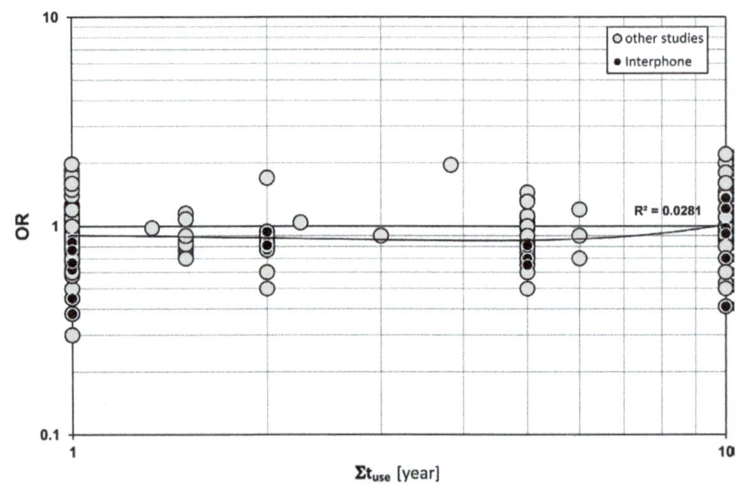

Figure 2. Reported odds ratios (OR) of epidemiological case-control studies on glioma and mobile phone use (without SHG data) over the cumulated use time t_{use}, grey points: all other national studies; black points: interphone study.

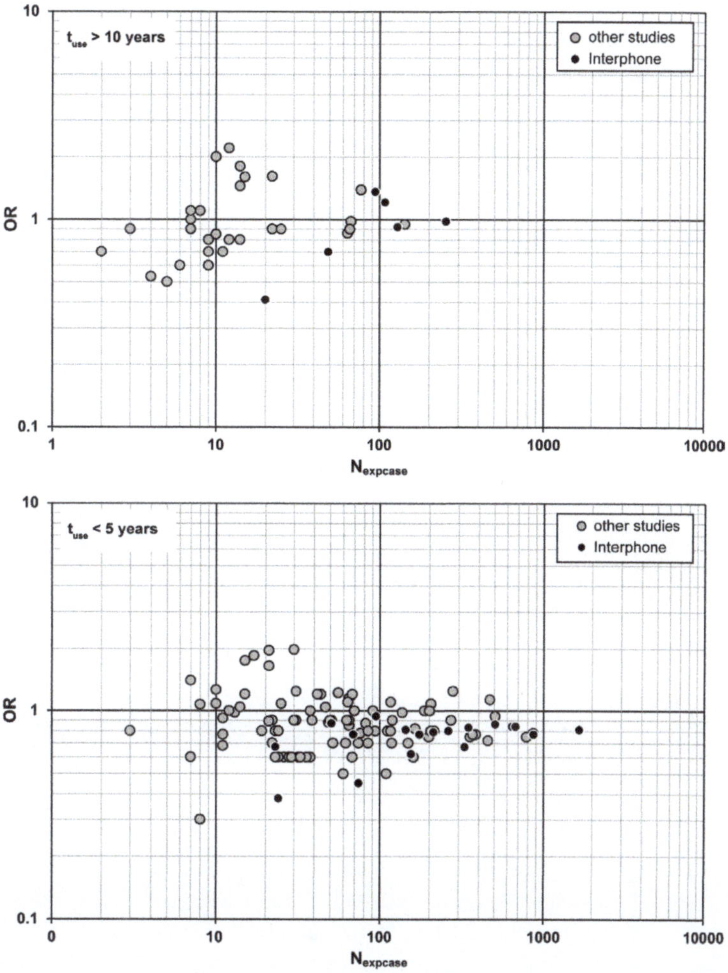

Figure 3. Reported odds ratios (OR) of epidemiological case-control studies on glioma and mobile phone use (without SHG data) over the associated number of exposed cases N_{exp}; use time larger than 10 years (above), use time larger lower than 5 years (below); grey points: all other national studies; black points: interphone study.

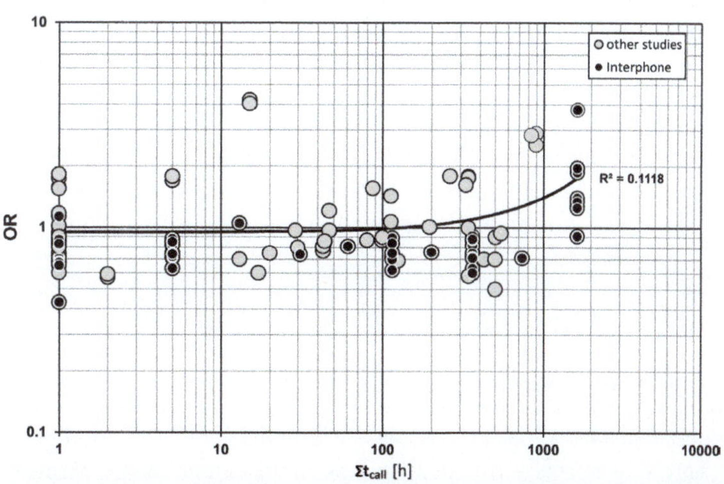

Figure 4. Reported odds ratios (OR) of epidemiological case-control studies on glioma and mobile phone use (without SHG data) over the cumulated call time t_{call}; grey points: all other national studies; black points: interphone study.

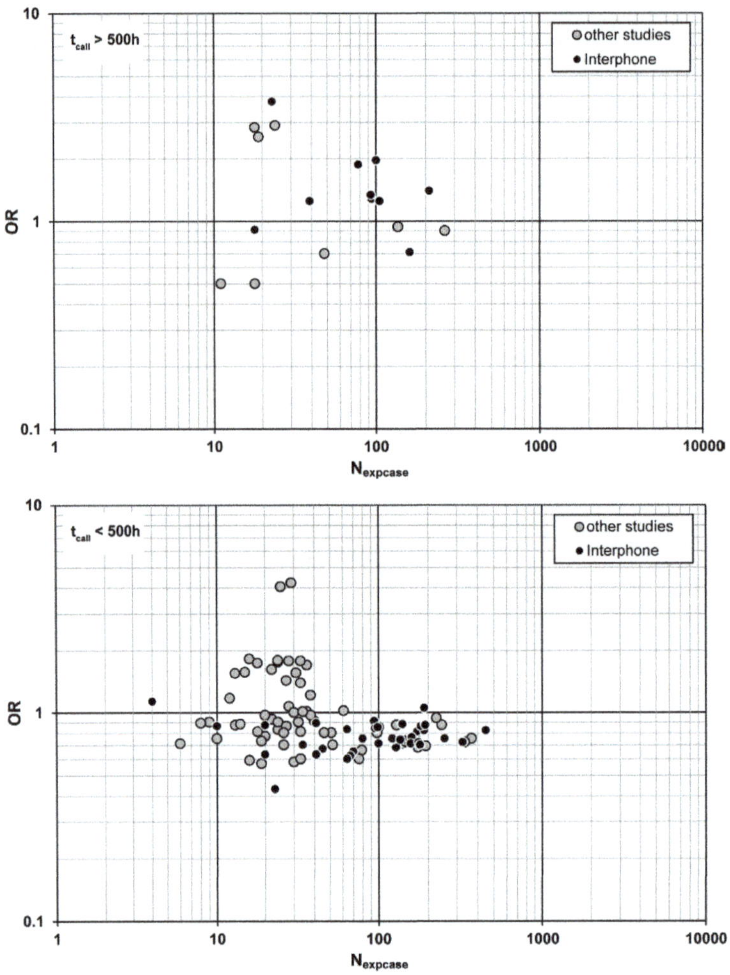

Figure 5. Reported odds ratios (OR) of epidemiological case-control studies on glioma and mobile phone use (without SHG data) over the associated number of exposed cases N_{expD} with cumulated call time larger than 500 h (above) and lower than 500 h (below); grey points: all other national studies; black points: interphone study.

The analysis of the third dose surrogate, the *cumulated number of calls* (CNC) as derived from questionnaires also confirmed the previous results (**Figure 6**). The time trend regression line stays in the reduced risk region at OR = 0.83 until about 8000 calls and increases to OR = 1 towards the highest number of calls. The check of the impact of low N_{exp} was made with two sub-pools of high and low number of calls, namely with $N_{call} > 10,000$ and $N_{call} < 5000$, respectively. In agreement with the results of the other dose surrogates the patterns exhibited the well-known funnel-shaped data distribution with high OR variance at low N_{exp} and again with a trend towards reduced risk both for results with high and low number of calls. Again, this does not indicate an increased risk of heavy and/or long-term users (**Figure 7**).

4. Discussion

It is interesting to note that reported risk estimates receive different public and scientific attention, depending on whether they indicate increased or decreased risk. This is of particular importance with mobile phone use and glioma, where the consistent OR values below 1 are readily interpreted as an effect of bias rather than an EMF effect just because protective effects were postulated as "implausible" while harmful effects seemingly are not. This is in spite of the fact that there are no established mechanisms for either assumption. Such a "plausibility bias" explains why for most dose estimates consistent results with ORs lower 1 indicating reduced risks gain little attention while the single OR above 1 for estimated "heavy" users dominates risk perception—in spite of the

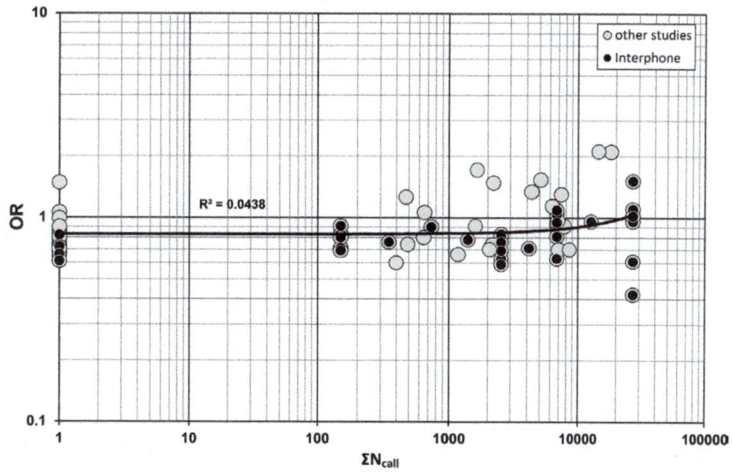

Figure 6. Reported odds ratios (OR) of epidemiological case-control studies on glioma and mobile phone use (without SHG data) over the cumulated number of calls N_{call}; grey points: all other national studies; black points: interphone study.

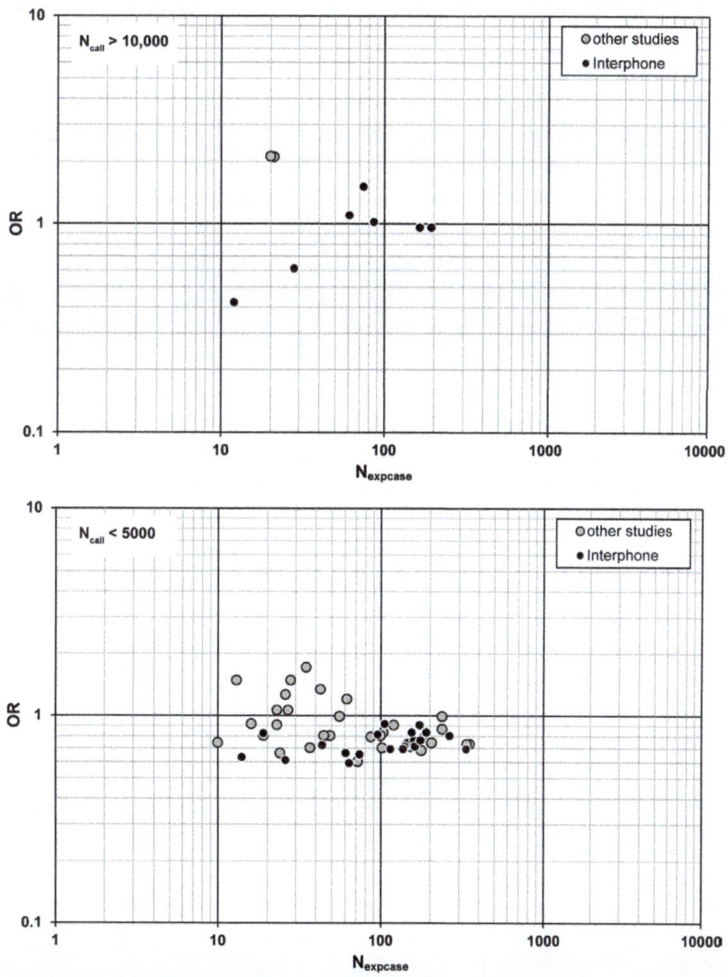

Figure 7. Reported odds ratios (OR) of epidemiological case-control studies on glioma and mobile phone use (without SHG data) over the associated number of exposed cases N_{exp} with cumulated numbers of calls larger than 10,000 (above) and lower than 5000 (below); grey points: all other national studies; black points: interphone study.

fact that it is most prone to uncertainties such as due to unreliable exposure assessment in the presence of technological changes over the years, unknown background exposure, inadequate and even implausible dose estimates, different recall bias of cases (or their proxis) and controls, in addition, there are biased selection of cases, biased participation rates, insufficient assessment of individual cancer-related confounders and different criteria used in studies for classifying exposure groups.

It is well known but needs to be repeated that risk estimates of epidemiological studies indicate just an association rather than a proof of causality. Whatever the OR value may be, whether there is a causal relationship needs to be proven in a second step. Consequently, ORs indicating a reduced glioma risk are not implausible by themselves. Not believing does not mean not existing. Therefore, analysis should be open for any outcome and likewise for ORs higher than and lower than 1. Risk estimates below 1 do not justify à priory negligence but may be explained also by other causes than methodological bias such as lifestyle, increased brain activity and, hence, cerebral blood flow during a call, yet unidentified co-factors—or others including potential subtle beneficial thermal EMF effects.

In spite of many ORs lower 1 there is concern about a potential long-term risk. This concern is mainly driven by controversial results of the SHG versus all other national and international studies and further supported by the (relative) OR increase at suspected long-term and/or heavy users irrespective the fact that even in this case ORs may stay at or below 1, and irrespective the fact that long-term results suffer from the highest uncertainty and the highest probability of bias. So far, by opposing studies with different outcome (and conclusions) the question whether or not a long-term health risk exists could not be convincingly answered.

The new approach to synoptically look at all available data was triggered by the common habit of epidemiologists, to publish tables over tables with ORs (some papers report over 100 individual ORs) without discussing these results in detail. Frequently, this detailed information (which is welcome as such) remains ignored, even if it contains statistical significant results. Usually only one or few ORs are picked out of the published bulk of data (and/or the pool of unpublished results) and are presented as "the" result without any convincing rational for the particular selection. Even in reviews epidemiologic studies are characterized by only one (or few) "representative" OR. At the end, without convincing reasoning it is concluded whether there is a risk or not. In addition, this (common) kind of reviewing suffers from another major flaw: In concentrating on selected ORs from different studies, each of the study-representing ORs is given the same weight—in spite of different number of exposed cases and hence, different statistical power. It is no surprise that this necessarily leads to conclusions that may be challenged.

The new synoptic approach overcomes these deficiencies and accounts for the different reliability of calculated ORs. It allows bringing the various published ORs into perspective. It also makes it possible to clarify the seeming increase of OR with dose and identifying it as an effect of the low number of associated exposed cases. It is no surprise that heavy users are less numerous than average users and, consequently, with low N_{exp} the uncertainty of associated ORs is increased. It is not self-evident and important to note, that with increasing N_{exp} ORs tend towards a final endpoint, in particular indicating no or even reduced risk. It is interesting that data exhibit the same trend also at every dose surrogate irrespective their different reliability and different meaning for example, high number of calls can be accumulated by short-term intense use as well as long-term occasional use. The reliability of dose surrogates differs. On the one hand, CUY (cumulated use years) might be assessed with more reliability because it may be based on hard facts such as subscriber's contracts (although this does not yet allow differentiating the intensity of use). On the other hand, CNC (cumulated number of calls) requires a guess about the average number of calls over many years while CCT (cumulated call time) needs a further guess about the average duration per call. The limited reliability of dose data is demonstrated by the existence of obvious implausible guesses (such as 12 h call time per day) as demonstrated in the Interphone study.

The synoptic approach and the use of every reported OR irrespective the quality of a study could be challenged by arguments such as:

- *Pooling studies with quite different methodology is not justified.* This argument ignores that the synoptic approach must not be confused with meta-analyses. While these have their merit on their own, the joint analysis of studies with the same design still stays with their methodological weaknesses and biases. In contrast, in the presence of a body of data a synoptic analysis is just what scientist do in a first step, namely, to get a general overview over the available evidence. In a second step, it allows checking consistency and whether the data patterns provide additional insight—or not. After decades of epidemiological studies and a meanwhile huge amount of available data, a "more of the same" strategy would hardly contribute to new insights.

The existent body of information allows already concluding that new studies with the same design adding some new data points to the pool of existent data will hardly change the overall image.

- *Pooling of ORs from studies with different methodology does not make sense.* In fact, the contrary is true. It allows concluding on the consistency of the data and identifying inconclusive results, if there are any. The benefit of the synoptic view could already be demonstrated very clearly [10] [11]. If all epidemiological studies were reliable, in spite of existing variance in the results, overall, the results should fit in a consistent pattern. It was surprising how clearly it was possible to identify the puzzeling SHG cluster at first glance, which is just neither plausible by itself, nor plausible compared to all the other studies (**Figure 1**). Such a situation was not found in the ELF range after 36 years of epidemiological research [10].
- *Pooling ORs of different exposure scenarios and dose parameters is no valid approach.* This argument ignores, that of course sub-pooling can be done (and was done) also with data related to different scenarios, in addition to the overall pooled data (see **Figures 2-7**). In addition, the overall synoptic approach is very useful for testing the null hypothesis: If there is no causal relationship, different exposure scenarios or exposure metrics would not count and consequently, the related data would fit into a common and consistent pattern converging towards a final value. If not, the associated data patterns would differ in some way or another and convergence, if seen, may be towards a risk estimate unequal 1. In fact, with regard to glioma, the pattern(s) confirm the null hypothesis: Even when data were analyzed separately and in dependence of dose, the resulting patters still agreed fairly well.
- *Pooling of studies of different quality degrades the conclusions.* In fact, weighing the evidence by the quality of studies and excluding inadequate studies from analysis is a common and valuable approach in risk assessment. However, it suffers from the fact that it makes it necessary to justify the dismissal and acceptance of studies which may be challenged by stakeholders. In addition, selection may be prone to bias. This is demonstrated by the fact that different assessment groups such as ICNIRP and WHO versus the Bioinitiative group give quite different weight to the same studies. Therefore, exclusion of studies feeds the public controversy between "believers" and "non-believers" of mobile phone health risks. In contrast to this, the synoptic approach does not need selection and, hence does not need justifying any selection criterion. Even if single studies may not be accessible, this would not degrade the outcome in the presence of a sufficient data pool.
- *Not excluding dubious studies might impair the conclusion.* This argument does not hold, since few additional data points will hardly change the overall pattern. In addition, the quality criteria of research funds are such that the design of (expensive) large studies is reviewed very critically while small studies may not have been assessed at all but usually involve small numbers of cases and, hence report on ORs in an N_{exp} range with high variance and do not critically influence the synoptic analysis. Furthermore, outliers can easily be identified by their implausibility compared to the body of evidence. It must be noted that implausible ORs are not restricted to poor studies only. Even reputed studies may include some implausible ORs such as presented in **Figure 1** with two OR outliers at $N_{exp} = 140$.
- *Multiple papers of the same study generate disproportionate numbers of ORs and hence may degrade the outcome.* Fortunately, this argument does not hold. In fact, papers of the same basic data report on similar ORs. Consequently, such ORs overlap, hence, do not change the pattern. (Overlap is the reason why not all analyzed data points can be identified in the presented figures.) Results, that do not fit in the overall cluster can be identified and checked for consistency and plausibility (such as the SHG data, **Figure 1**).

The detailed analysis of the glioma data showed, that irrespective their inherent differences none of the various dose parameters did indicate an increased risk if the statistical power of ORs (indicated by N_{exp}) was taken into account. Furthermore, it allowed identifying an obvious bias in the SHG data and an their implausible distribution. However, it could be demonstrated that at the end with increasing N_{exp} even SHG data finally converge towards the same finding, namely lack of increased glioma risk.

Hopefully the clarification by synoptic analysis can contribute to a less emotional debate and the acceptance of the fact that overall, with regard to mobile phone and glioma risk the available evidence is reassuring rather than alarming.

5. Conclusions

This synoptic analysis of all available epidemiological data on mobile phone use and glioma risk allows the following conclusions:

Similar to the overall brain cancer results [26]-[28] glioma data allow identifying two quite different data pools, one reported from one single group only with alarming results however, with an implausible pattern and an obvious offset which indicates a methodological difference rather than a biological effect. Interestingly, with increasing statistical power even this data pool converges towards lack of increased risk.

The analysis of different dose-related parameters and the pattern of all pooled data do not support the assumption of a glioma risk of long-term and/or heavy mobile phone users.

With increasing statistical power the risk estimates converge towards a reduced health risk. However, in lack of a convincing EMF interaction mechanism the indication of a reduced health risk estimate does not justify concluding on a causal relationship with EMF and a beneficial effect either.

Apart from differences in detail, a look on all available data including SHG, IPS and ONS, consistently does not indicate an increased mobile phone glioma risk, neither short-term nor long-term. The available evidence is reassuring rather than alarming.

Conflict of Interest

The author declares no conflict of interest.

References

[1] IARC (2011) IARC Monographs on the Evaluation of Carcinogenic Risks to Humans. Volume 102. Non-Ionizing Radiation, Part 2: Radio Frequency (RF) Electromagnetic Fields. Lyon, 2011.

[2] ICNIRP (1998) Guidelines for Limiting Exposure to Time-Varying Electric, Magnetic and Electromagnetic Fields (up to 300 GHz). *Health Physics*, **74**, 494-522.

[3] IEEE (2005) IEEE Standard for Safety Levels with Respect to Human Exposure to Radio Frequency Electromagnetic Fields (up to 300GHz).

[4] Hardell, L. and Carlberg, M. (2013) Using the Hill Viewpoints from 1965 for Evaluating Strengths of Evidence of the Risk for Brain Tumors Associated with the Use of Mobile and Cordless Phones. *Reviews on Environmental Health*, **28**, 97-106. http://dx.doi.org/10.1515/reveh-2013-0006

[5] Hardell, L., Carlberg, M., Söderqvist, F. and Hansson Mild, K. (2013) Case-Control Study of the Association between Malignant Brain Tumors between 2007 and 2009 and Mobile and Cordless Phone Use. *International Journal of Oncology*, **43**, 1833-1845.

[6] Carlberg, M., Hansson Mild, K. and Hardell, L. (2013) Use of Mobile Phones and Cordless Phones Is Associated with Increased Risk for Glioma and Acoustic Neuroma. *Pathophysiology*, **20**, 85-110. http://dx.doi.org/10.1016/j.pathophys.2012.11.001

[7] Hardell, L., Carlberg, M. and Hansson Mild, K. (2011) Pooled Analysis of Case-Control Studies on Malignant Brain Tumors and the Use of Mobile and Cordless Phones and Deceased Subjects. *International Journal of Oncology*, **38**, 1465-1474. http://dx.doi.org/10.3892/ijo.2011.947

[8] Hardell, L., Carlberg, M. and Hansson Mild, K. (2006) Pooled Analysis of Two Case-Control Studies on Use of Cellular and Cordless Telephones and the Risk for Malignant Brain Tumors Diagnosed in 1997-2003. *International Archives of Occupational and Environmental Health*, **79**, 630-639.

[9] Hardell, L., Carlberg, M. and Hansson Mild, K. (2005) Use of Cellular Telephones and Brain Tumor Risk in Urban and Rural Areas. *Occupational and Environmental Medicine*, **62**, 390-394. http://dx.doi.org/10.1136/oem.2004.017434

[10] Leitgeb, N. (2014) Childhood Leukemia Not Linked with ELF Magnetic Fields. *Journal of Electric and Magnetic Analysis and Applications*, **6**, 174-183. http://dx.doi.org/10.4236/jemaa.2014.67017

[11] Leitgeb, N. (2014) Synoptic Analysis of Epidemiologic Evidence of Brain Cancer Risks from Mobile Communication. *JEMAA*, **6**, 413-424.

[12] Interphone Study Group (2010) Brain Tumor Risk in Relation to Mobile Telephone Use: Results of the INTERPHONE International Case-Control Study. *International Journal of Epidemiology*, **39**, 675-694. http://dx.doi.org/10.1093/ije/dyq079

[13] Coureau, G., Bouvier, G., Lebailly, P., Fabbro-Peray, P., Gruber, A., Leffondre, K., Guillamo J.-S., Loiseau, H., Mathoulin-Péklissier, S., Salamon, R. and Baldi, I. (2014) Mobile Phone Use and Brain Tumors in the CERENAT Case-Control Study. *Occupational and Environmental Medicine*, **71**, 514-522. http://dx.doi.org/10.1136/oemed-2013-101754

[14] Swerdlow, A.J., Feychting, M., Green, A.C., Kheifets, L. and Savitz, D. (2011) Mobile Phones, Brain Tumors and the

Interphone Study. Where Are We Now? *Environmental Health Perspectives*, **119**, 1534-1538. http://dx.doi.org/10.1289/ehp.1103693

[15] Takebayashi, T., Varsier, N., Kikuchi, Y., Wake, K., Taki, M., Watanabe, S., Akiba, S. and Yamaguchi, N. (2008) Mobile Phone Use, Exposure to Radiofrequency Electromagnetic Field, and Brain Tumor. A Case-Control Study. *British Journal of Cancer*, **98**, 652-659. http://dx.doi.org/10.1038/sj.bjc.6604214

[16] Hours, M., Bernardt, M., Montestrucq, L., Arslan, M., Bergerer, A., Deltour, I. and Cardis, E. (2007) Cell Phones and Risk of Brain and Acoustic Nerve Tumors: The French INTERPHONE Case-Control Study. *Revue d'Epidémiologie et de Santé Publique*, **55**, 321-332.

[17] Schlehofer, B., Schlaefer, K., Blettner, M., Berg, G., Böhler, E., Hetinger, I., Kunna-Grass, K., Wahrendorf, J. and Schüz, J. (2007) Environmental Risk Factors for Sporadic Acoustic Neuroma (Interphone Study Group Germany). *European Journal of Cancer*, **43**, 1741-1747. http://dx.doi.org/10.1016/j.ejca.2007.05.008

[18] Hepworth, S.J., Schoemaker, M.J., Muir, K.R., Swerdlow, A.J., van Tongeren, M.J. and McKinney, P.A. (2006) Mobile Phone Use and Risk of Glioma in Adults: Case-Control Study. *BMJ*, **332**, 883-887. http://dx.doi.org/10.1136/bmj.38720.687975.55

[19] Klaeboe, L., Blaasaas, K.G. and Tynes, T. (2006) Use of Mobile Phones in Norway and Risk of Intracranial Tumors. *European Journal of Cancer Prevention*, **16**, 158-164.

[20] Lahkola, A., Auvinen, A., Raitanen, J., Schoemaker, M.J., Christensen, H.C., Feychting, M., Johansen, C., Klaeboe, L., Lönn, S., Swerdlow, A.J., Tynes, T. and Salminen, T. (2006) Mobile Phones Use and Risk of Glioma in 5 North European Countries. *International Journal of Cancer*, **120**, 1769-1775.

[21] Lönn, S., Ahlbom, A., Christensen, H.C., Johansen, C., Schüz, J., Edström, S., Henrikksson, G., Lundgren, J., Wennerberbg, J. and Feychting, M. (2006) Mobile Phone Use and Risk of Parotid Gland Tumor. *American Journal of Epidemiology*, **164**, 637-643. http://dx.doi.org/10.1093/aje/kwj242

[22] Schüz, J., Böhler, E., Berg, G., Schlehofer, B., Hettinger, I., Schlaefer, K., Wahrendorf, J., Kunna-Grass, K. and Blettner, M. (2006) Cellular Phones, Cordless Phones and the Risk of Glioma and Meningioma (Interphone Study Group Germany). *American Journal of Epidemiology*, **163**, 512-520.

[23] Lönn, S., Ahlbom, A., Hall, P. and Feychting, M. (2005) Long-Term Mobile Phone Use and Brain Tumor Risk. *American Journal of Epidemiology*, **161**, 526-535. http://dx.doi.org/10.1093/aje/kwi091

[24] Auvinen, A., Hietanen, M., Luukkonen, R. and Koskela, R.S. (2002) Brain Tumors and Salivary Gland Cancers among Cellular Telephone Users. *Epidemiology*, **13**, 356-359. http://dx.doi.org/10.1097/00001648-200205000-00018

[25] Inskip, P.D., Tarone, R., Hatch, E., Wilcosky, T.C., Sjapiro, W.R., Selker, R.G., Fine, H.A., Black, P.M., Loeffler, J.S. and Kinet, M.S. (2001) Cellular Telephone Use and Brain Tumors. *New England Journal of Medicine*, **34**, 79-86. http://dx.doi.org/10.1056/NEJM200101113440201

[26] Ahlbom, A., Feychting, M., Green, A., Kheifets, L., Savitz, D.A. and Swerdlow, A.J. (2009) Epidemiologic Evidence on Mobile Phone and Tumor Risk: A Review. *Epidemiology*, **20**, 639-652. http://dx.doi.org/10.1097/EDE.0b013e3181b0927d

[27] Kundi, M. (2009) The Controversy about a Possible Relationship between Mobile Phone Use and Cancer. E*nvironmental Health Perspectives*, **117**, 306-324. http://dx.doi.org/10.1289/ehp.11902

[28] Lagorio, S. and Röösli, M. (2014) Mobile Phone Use and Risk of Intracranial Tumors: A Consistency Analysis. *Bioelectromagnetics*, **35**, 79-90. http://dx.doi.org/10.1002/bem.21829

23

Measurable Dielectric Permittivity Range for TE and TM Modes in a Shielded Dielectric Resonator

Eduardo Javier Paez[1], Roberto Cessare Callarotti[2], Yuande Sanchez[3]

[1]Centre for Quantum Technologies, National University of Singapore, Singapore City, Singapore
[2]Instituto Venezolano de Investigaciones Científicas (IVIC), Caracas, Venezuela and Universidad del Turabo, Gurabo, Puerto Rico
[3]Fundación Instituto de Ingeniería, Caracas, Venezuela
Email: ejpaezb@nus.edu.sg, robercallarotti@gmail.com, sanchez@fii.gob.ve

Abstract

Dielectric resonator methods constitute one of the most useful techniques for the measurement of electromagnetic material properties in the microwave frequency range. Several geometric configurations are used for this purpose and, in the present paper, we consider the case of a dielectric rod enclosed in a cylindrical metallic enclosure. To carry out dielectric measurements in this system it is necessary to know the highest permittivity constant value for which the resonance condition still can be attained into the cavity. Using an approach based on magnetic and electric Hertzian potentials we have derived the set of TE and TM modes for the relevant geometry described and, then we have calculated the valid dielectric permittivity constant range of measurements for low-loss materials in a cylindrical cavity using a simple resonance frequency condition. Finally, we present a simple application of this method in order to determine the dielectric permittivity constant of heavy oil with 11 API.

Keywords

Dielectric Resonator, Hertzian Potential, Cylindrical Cavity, Electromagnetic Characterization, Dielectric Permittivity Constant Measurement

1. Introduction

When dielectric measurements are performed in a shielded dielectric resonator, the shift of the resonant frequency caused by the presence of the sample (with a given dimension and dielectric permittivity constant), must be taken into account. The electromagnetic field must satisfy the boundary conditions required at the metal shielding (usually assumed as a perfect electric conductor) and at the material sample, thus obtaining a standing wave inside the cavity. The resonant frequency of this configuration decreases with respect to the resonant frequency of the empty cavity because the electric field increases due to the presence of a low loss dielectric material, so that forcing the wavelength to be smaller in order to achieve the border condition at the metal enclosure. For any specific combination of the diameter and the dielectric permittivity constant (see **Figure 1**) of the central material there is a cut-off frequency which below that it is possible to have a resonant condition. Above this frequency the cavity will exhibit a high impedance at the excitation port so that any electromagnetic mode practically disappears. Thereby, it is necessary to know the value of the highest dielectric permittivity constant which could be measured in the cavity for a given radius of the central rod.

In this paper we employ the formalism of the Hertzian potentials to find the electromagnetic solution of this configuration. Then we determine the limit of the real dielectric permittivity constant by numerically solving the nonlinear electromagnetic equations, thus obtaining the wave number in both materials inside the cavity (dielectric rod and air) and independently calculating the frequency inside those volumes.

Hertzian potentials have been used to solve different electromagnetic problems: in the study of the properties of aperture array systems [1], non-linear waveguides [2], Green's functions for multilayered media [3] and electromagnetic wave interaction with nanodevices [4]. They have also been applied for the determination of TE and TM modes of circular cylindrical cavities using magnetic-type and electric-type Hertzian potentials respectively [5]. **Figure 1** shows the case of a cylindrical dielectric resonator enclosed by a metal shield, where b is the radius of the outer cylinder, a is the inner resonator radius and d is the height (length of the structure). The configuration can be regarded as a cylindrical waveguide enclosing a central sample of radius a, and terminated by perfectly conducting planes. The general solution for the axial $\boldsymbol{E}$ field in TM modes was discussed in [6], proposing a general solution for the axial $\boldsymbol{E}$ field for TM modes.

In [7] the solution of this system was obtained by solving directly Maxwell's equations and substituting the corresponding zero axial component of magnetic or electric field for TE or TM mode respectively. A similar approach was presented in reference [8]. The aforementioned methods have the disadvantage that the cut-off wave number appears in the denominator of the total power flow equation implying a contradiction since it means that the power will vanish when the radius is very small. This contradiction is solved by including a factor in the numerator which depends on the cut off wave number, as proposed by [9] for a similar system.

The electromagnetic field in the material sample (region 1) and in air (region 2) are determined by using the relevant Hertzian potentials in their general mathematical form. The proper boundary conditions are then applied to the derived equations which can then be solved numerically. In practice, the resonant frequency and the system quality factor are measured then to determine the electromagnetic properties of materials [6] [7] [10]-[12].

In this paper we first present a brief introduction to Hertz's potentials, and then we apply this theory to the classical case of a cylindrical waveguide [5]. We then discuss the waveguide which includes the central cylindrical sample. Finally we discuss the usable range of this experimental arrangement in the determination of dielectric and magnetic properties.

2. Hertzian Potentials

In a charge free region with constant permittivity, the divergence of $\boldsymbol{E}$ is zero, thus implying that this field can

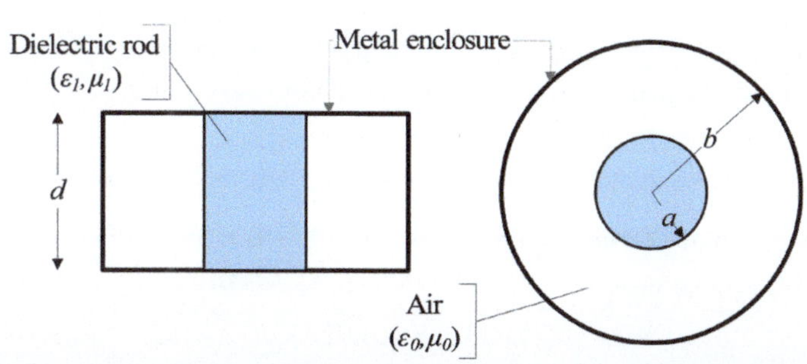

Figure 1. Shielded dielectric resonator.

be expressed as the curl of an auxiliary vector potential. This is the base for Hertz potential formulation [13]. For a homogeneous, isotropic, and charge-free region, $\boldsymbol{E}$ can be determined from the curl of an auxiliary vector potential Π_m defined as magnetic-type Hertzian potential. Thus,

$$\boldsymbol{E} = -j\omega\mu\nabla\times\Pi_m \tag{1}$$

Then from Maxwell's equations,

$$\begin{aligned} \nabla\times\boldsymbol{H} &= j\omega\varepsilon\boldsymbol{E} \\ \nabla\times\boldsymbol{H} &= k^2\nabla\times\Pi_m \\ \boldsymbol{H} &= k^2\Pi_m + \nabla\phi_m \end{aligned} \tag{2}$$

where ϕ_m is an auxiliary scalar potential, and $k^2 = \omega^2\mu\varepsilon$.

Now, we can use $\nabla\times\boldsymbol{E} = -j\omega\mu\boldsymbol{H}$, and substituting (1), we obtain:

$$\boldsymbol{H} = \nabla\times\nabla\times\Pi_m \tag{3}$$

Using the following identity $\nabla\times\nabla\times = \nabla\nabla\cdot - \nabla^2$ and combining (2) and (3), we obtain:

$$\nabla\nabla\cdot\Pi_m - \nabla^2\Pi_m = k^2\Pi_m + \nabla\phi_m \tag{4}$$

Since ϕ_m is an auxiliary scalar potential we can select $\phi = \nabla\circ\Pi_m$, obtaining the following Helmholtz equation,

$$\nabla^2\Pi_m + k^2\Pi_m = 0 \tag{5}$$

A second Helmholtz equation can be obtained from the fact that the divergence of $\boldsymbol{H}$ is always zero. This fact can be expressed in terms of the curl of another vector potential Π_e called electric-type Hertzian potential. As in the previous case, the same analysis applies for the TM modes where the electric-type Hertzian potential should be used [13]. The magnetic field can be expressed as,

$$\boldsymbol{H} = j\omega\varepsilon\nabla\times\Pi_e \tag{6}$$

Using Maxwell's equations and combining with (6), we obtain,

$$\boldsymbol{E} = \nabla\times\nabla\times\Pi_e = k^2\Pi_e + \nabla\phi_e, \tag{7}$$

where Π_e satisfies the Helmholtz wave equation,

$$\nabla^2\Pi_e + k^2\Pi_e = 0 \tag{8}$$

3. TE and TM Modes for the Cylindrical Cavity Resonator

3.1. Solution for TE Modes

In the case of TE modes, if we observe the magnetic-type Hertzian potential, the only way to generate a zero component in longitudinal axis of $\boldsymbol{E}$ field is that the Hertzian potential has zero components in radial and azimuthal direction. It means that magnetic-type Hertzian potential can be expressed like:

$$\Pi_m = \psi_m(\rho,\phi)\sin(\beta z)\hat{\boldsymbol{a}}_z \tag{9}$$

In the case of resonators the perfect conductor boundary conditions at $z=0$ and $z=d$, the solution in terms of a stationary $\sin(\beta z)$ dependence can satisfy such boundary conditions when β takes the discrete values given by

$$\beta = p\pi/d \tag{10}$$

Due to this fact, when the cavity geometry coincides with the coordinate system, we can use the separation variable method in order to achieve the boundary conditions [14]. Then,

$$\psi_m(\rho,\phi) = R(\rho)P(\phi) \tag{11}$$

Substituting (11) in (5), we obtain the Bessel differential equation

$$\rho^2\left(\frac{1}{R}\frac{\partial^2 R}{\partial\rho^2}\right)+\rho\left(\frac{1}{R}\frac{\partial R}{\partial\rho}\right)+\frac{1}{P}\left(\frac{\partial^2 P}{\partial\varphi^2}\right)+\rho^2 k_c^{\,2}=0 \tag{12}$$

A possible P dependence with ϕ is spatially harmonic: $e^{-jk\phi}$. This implies that k_ϕ is an integer: n. Therefore, Equation (12) becomes,

$$\rho^2\left(\frac{\partial^2 R}{\partial\rho^2}\right)+\rho\left(\frac{\partial R}{\partial\rho}\right)+R\left(\rho^2 k_c^{\,2}-n^2\right)=0 \tag{13}$$

The differential equation shown above is known as the Bessel Differential Equation, whose general solution is,

$$R(\rho)=BJ_n(k_c\rho)+CY_n(k_c\rho), \tag{14}$$

where, J_n is the Bessel function of first kind and Y_n is the Bessel function of second kind (or Newman function). The real constants B and C arise as the most general solution for R. The solution for the magnetic-type Hertzian potential in region the second region is,

$$\boldsymbol{\Pi}_m(\rho,\varphi,z)=\left[BJ_n(k_c\rho)+CY_n(k_c\rho)\right]\cos(n\phi)\sin(\beta z)\hat{\boldsymbol{a}}_z \tag{15}$$

The Helmholtz wave Equation (5) is valid for both regions of the shielded resonator, with the requirement that in the enclosed section $(0\le\rho\le a)$ the solution for the magnetic-type Hertzian potential only the Bessel function AJ_n is allowed (where A is another real constant), because Y_n in the center of the cavity goes to infinity. This enclosed section is characterized by a permittivity constant ε_1 and a permeability μ_1. For the second region $(a\le\rho\le b)$ the permittivity constant is ε_0, the permeability is μ_0, and the E and H fields are given by Equations (2) and (3).

It is important to note that $\beta_1=\beta_2=p\pi/d$. Therefore, the cut-off wave numbers are given by,

$$k_{c1}^2=\omega_1^2\mu_1\varepsilon_1-\beta^2, \tag{16}$$

$$k_{c2}^2=\omega_2^2\mu_0\varepsilon_0-\beta^2. \tag{17}$$

The boundary conditions of the tangential electric field at $\rho=b$ and $\rho=a$ and the continuity of the tangential magnetic field at $\rho=b$ provide the following equations

$$B\left[\frac{n}{b}J_n(k_{c2}b)-k_{c2}J_{n+1}(k_{c2}b)\right]+C\left[\frac{n}{b}Y_n(k_{c2}b)-k_{c2}Y_{n+1}(k_{c2}b)\right]=0. \tag{18}$$

$$A\mu_1\left[\frac{n}{a}J_n(k_{c1}a)-k_{c1}J_{n+1}(k_{c1}a)\right]=B\mu_0\left[\frac{n}{a}J_n(k_{c2}a)-k_{c2}J_{n+1}(k_{c2}a)\right]+C\mu_0\left[\frac{n}{a}Y_n(k_{c2}a)-k_{c2}Y_{n+1}(k_{c2}a)\right] \tag{19}$$

$$Ak_{c1}^2J_n(k_{c1}a)=k_{c2}^2\left[BJ_n(k_{c2}a)+CY_n(k_{c2}a)\right] \tag{20}$$

Solving for B and C in (18), substituting it in (19) and (20), and then dividing side by side these two expressions, we can obtain the following relation,

$$\frac{k_{c2}^2\mu_1\left[\frac{n}{a}J_n(k_{c1}a)-k_{c1}J_{n+1}(k_{c1}a)\right]}{k_{c1}^2\mu_2 J_n(k_{c1}a)}=\frac{\alpha\left[\frac{n}{a}J_n(k_{c2}a)-k_{c2}J_{n+1}(k_{c2}a)\right]+\left[\frac{n}{a}Y_n(k_{c2}a)-k_{c2}Y_{n+1}(k_{c2}a)\right]}{\left[\alpha J_n(k_{c2}a)+Y_n(k_{c2}a)\right]} \tag{21}$$

where

$$\alpha=-\frac{\left[\frac{n}{b}Y_n(k_{c2}b)-k_{c2}Y_{n+1}(k_{c2}b)\right]}{\left[\frac{n}{b}J_n(k_{c2}b)-k_{c2}J_{n+1}(k_{c2}b)\right]}. \tag{22}$$

By making $\omega_1=\omega_2=\omega$, from Equations (16) and (17) we obtain the following relation:

$$\frac{k_{c1}^2+\beta^2}{k_{c2}^2+\beta^2}=\frac{\mu_1\varepsilon_1}{\mu_0\varepsilon_0}. \tag{23}$$

Equations (21) and (23) give the relation between the electromagnetic parameters of the unknown medium and the resonant frequency. The cut off wave number can be determined if the dielectric permittivity constant is known or viceversa. This system is solved numerically using the Matlab function *fsolve*. The solution given in Equation (21) reduces to the simple form given in [15] and [6] when $n=0$. The possible sources of uncertainty present in this configuration are the exact dimensions of the cavity, losses in the wall's finite conductivity and resolution of the measurement equipment [16]-[20]. Equations (21) and (23) reduce to the limit case of a simple cavity resonator filled with a single material when $a=b$ or when the electromagnetic parameters of both mediums are equal. Because the frequency variable was canceled in Equations (21) and (23), some solutions could be added or eliminated. The solution obtained should be checked by using $\omega_1=\omega_2$.

3.2. Solution for TM Modes

The treatment is similar for TM modes but using electric-type Hertzian potential as was presented on page 3. The general solution for this potential is in principle the same as in (15). ***E*** and ***H*** fields can be obtain using (6) and (7) and after applying the electromagnetic edge conditions, it gives,

$$\frac{k_{c2}^2\varepsilon_1\left[\frac{n}{a}J_n(k_{c1}a)-k_{c1}J_{n+1}(k_{c1}a)\right]}{k_{c1}^2\varepsilon_2 J_n(k_{c1}a)}=\frac{\alpha\left[\frac{n}{a}J_n(k_{c2}a)-k_{c2}J_{n+1}(k_{c2}a)\right]+\left[\frac{n}{a}Y_n(k_{c2}a)-k_{c2}Y_{n+1}(k_{c2}a)\right]}{\left[\alpha J_n(k_{c2}a)+Y_n(k_{c2}a)\right]} \tag{24}$$

where,

$$\alpha=-\frac{Y_n(k_{c2}b)}{J_n(k_{c2}b)} \tag{25}$$

It is also worth pointing out that for this given circular symmetry of the set of ***E*** and ***H*** fields, which is easily appreciable in the corresponding Hertzian potentials (15), we can expect pure TE and TM as stated in [21].

4. Convergence Range for Dielectric Measurements

Previously, we have mentioned that the frequency in both regions (air and dielectric) must be equal in order to represent a real resonant condition. However since we set $\omega_1=\omega_2$ in Equations (16) and (17), this condition should be verified after obtaining the parameters from the Equations (21) and (23). The previous argument is the final condition to probe that the numerical solution satisfies the resonance condition inside the cavity, otherwise, it represents a non-real solution and it is not possible to support a resonant mode inside the cavity.

We now exemplify our statement with an example. Consider a resonant frequency calculation for TE and TM modes in a circular cylindrical cavity resonator with a central dielectric rod. In the **Table 1** it is considered the case of two empty cavities designed for the TE_{011} and TM_{011} modes respectively with a resonant frequency of 5 GHz, being twice the radius a of the length *d*. See the **Figure 1**.

For a given dielectric permittivity constant/radius of the central cylinder, the resonant frequency is calculated as solution of the corresponding TE or TM equations set. Thus, for a selected working mode order n and a fixed radius of the dielectric rod, the resonant frequency decreases when we increase the dielectric permittivity constant. Then we reach a value of permittivity where the wave numbers k_{c1} and k_{c2} correspond to different frequency values according to Equations (21) and (23). When this happens we face a non-physical solution. This implies that the existent mode inside the cavity could correspond to higher order modes. This is the usable range of a particular resonant mode in a specific cavity. In the limit, the resonant frequency begins to be complex which really implies that it is an attenuate mode. The higher order modes for all design are given in the **Table 2**.

In the following we have determined the usable range for the Cavity 1 and Cavity 2. This is shown in **Figures 2-4** for the set of modes 011, 021 and 031 respectively, where we have determined the dependence of the resonance frequency with respect to the dielectric permittivity constante of the central sample under measurement for a fixed radius. In the case of TE modes, Cavity 1, the cutoff relative dielectric permittivity constant are $\varepsilon_{r1}=1.6$, $\varepsilon_{r1}=4.6$ and $\varepsilon_{r1}=10.1$ respectively. The respective values for TM modes, Cavity 2, are

Table 1. Cavity design for TE modes.

Cavity	Frequency (MHz)	Radius (cm)	Length (cm)
Cavity 1	5000	7.023	3.512
Cavity 2	5000	6.420	3.210

Table 2. Higher order modes of the TE/TM_{011} for cavity 1: 5.000 GHz and cavity 2: 5.000 GHz.

Mode	Cavity 1 Frequency (MHz)	Cavity 2 Frequency (MHz)
TE/TM_{011}	5000.0	5000.0
TE/TM_{021}	6398.7	6215.7
TE/TM_{031}	8124.0	7947.9

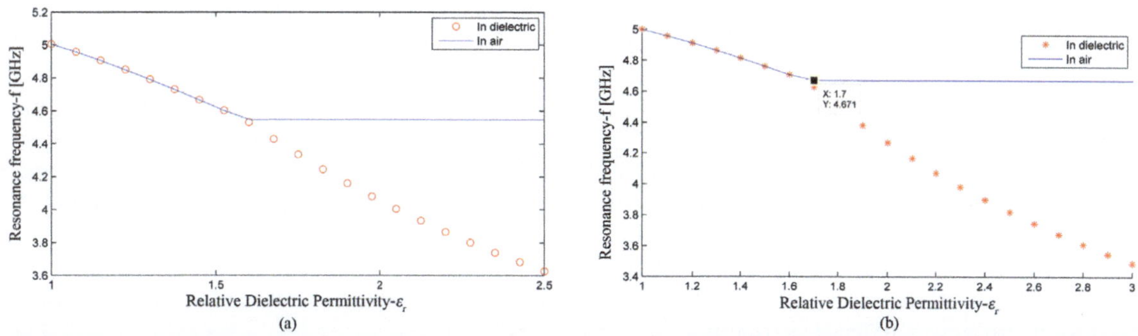

Figure 2. Usable range for dielectric measurement for (a) Cavity 1 TE_{011}; (b) Cavity 2 TM_{011}. In both cases it can be appreciated the breaking point of the resonant condition for this particular mode. TM_{011} mode has a higher measurable range for dielectric permittivity constant than TE_{011} mode.

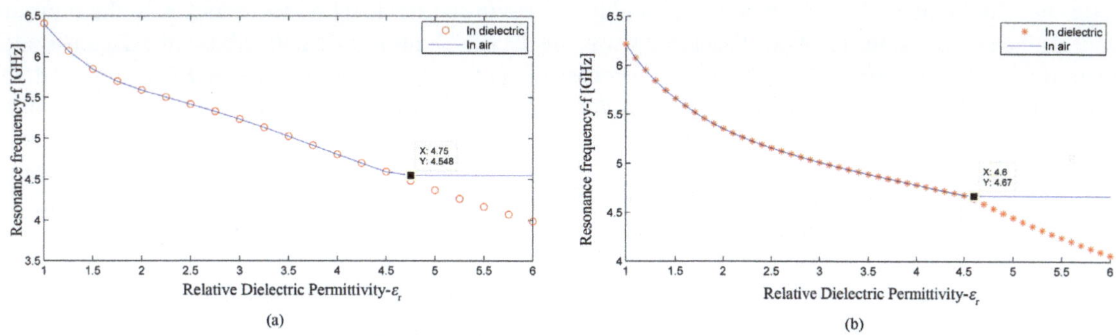

Figure 3. Usable range for dielectric measurement for: (a) Cavity 1 TE_{021}; (b) Cavity 2 TM_{021}. It can be noticed that the measurable range is higher than for the previous mode. In this case both ranges are of similar length.

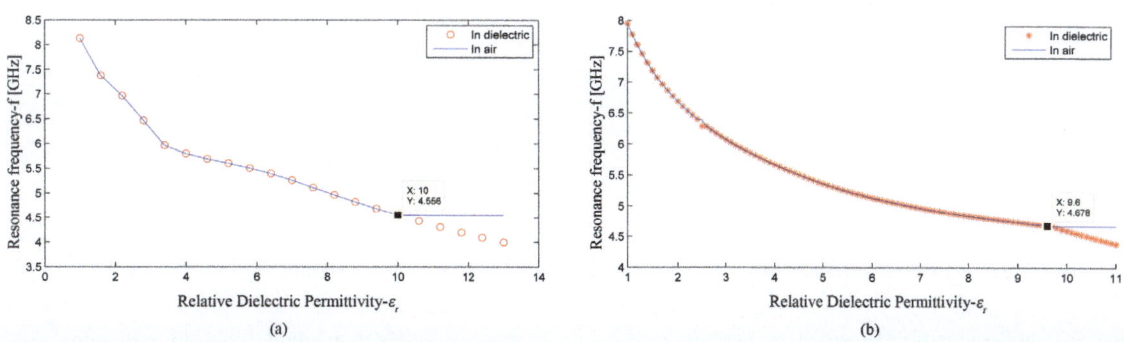

Figure 4. Usable range for dielectric measurement for: (a) Cavity 1 TE_{031}; (b) Cavity 2 TM_{031}. The usable range for the TE mode is higher than the respective for the TM mode.

$\varepsilon_{r1} = 1.7$, $\varepsilon_{r1} = 4.6$ and $\varepsilon_{r1} = 9.6$, which are in general lower than the previous case of TE modes.

Notice in general how the dielectric range increases for higher resonant modes, however compromising the measurement of the tangent loss due to lower quality factors because cavity losses.

For instance, observe that it is not possible to use the TE_{011} mode in the cavity 1 to characterize materials with permittivity values higher than 1.6 using a dielectric rod of 3 cm. However this situation can be solved if the dielectric rod diameter is reduced and, as consequence, the permittivity constant range increases. **Figure 3** shows the new usable range using a dielectric rod of radius 2 cm for both Cavity 1 and Cavity 2.

This method is valid for each configuration cavity-resonant mode-dielectric rod, however due to the non-linearity of Equations (21) and (23), or equivalently (24) and (25) for TM modes, it results difficult to predict the percentage of increment in the valid usable range for dielectric measurements due to a reductions in the dielectric rod's diameter.

Here it is easily observable how the usable range is increased, allowing to carry out measurements of higher permittivity values. This method is in general valid for any mode and resonant frequency.

Some authors do not use TM modes because any air gap between the top surface of the central cylindrical sample and the top surface of the external cavity can distort deeply the resonant frequency [15].

One important aspect worth mentioning is that there are some flattened sections, portion with horizontal trends, of the curves shown in the **Figures 2-4**. Those portions represent regions of less sensitivity of the permittivity regarding the frequency measurement. This region is not a recommendable region to measure properties due to higher expected uncertainty associated to the measurement according results given in [19] and [20]. However the method shown in this article to determine the usable range of a given cavity can be used to avoid those regions of higher associated uncertainty, simply changing the diameter of the dielectric rod and hence the position in the curve of the expected permittivity under measurement. This result may be of the interest for metrology in dielectric permittivity.

5. Experimental Realization

In this section we explored the previous results by using a double-cylindrical dielectric resonator, diameter is 140 ± 0.1 mm and height 36 ± 0.1 mm resonating at 5 GHz in the mode TE_{011} to characterize the permittivity constant of heavy oil with 11 API. The cavity was made of Aluminium C330 whose maximum dielectric conductivity is 21×10^6 S/m. To determine the effective permittivity constant we used the method of the critical points [22] with the help of a Vector Network Analyzer Anritsu, model 37269D, 40 MHz - 40 GHz. Posteriorly, the theory of mixing of concentric cylinders was used to determine the real permittivity constant of heavy oil. As the holder for petroleum sample we selected quartz due to its low dielectric losses and a very well known real permittivity constant of 3.850 [25]. The selection of its diameter was done taking into consideration the analysis presented in this article. The resulted resonant frequency was 4.881 GHz due to imperfections in constructions of the cavity.

Based on references [23]-[25] we expected a value to be lower than 4. It can be noticed in the **Figure 5** that an adequate radius for this purpose is 3 cm. It is not recommended a large radius due to the simple fact that the

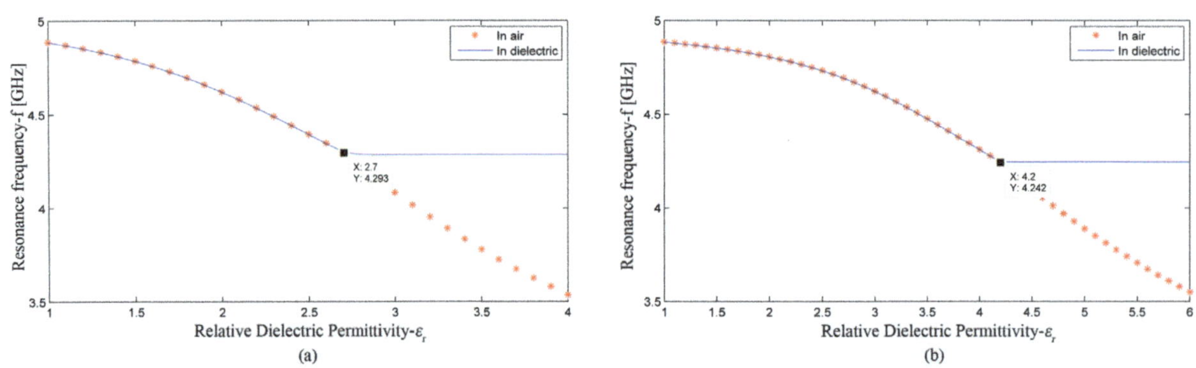

Figure 5. Usable range for dielectric measurement for a resonant cavity at 5 GHz in the mode TE_{011} for two radius: (a) 1 cm; (b) 3 cm. It is obvious that the recommended one is 3 cm which provides enough range for the characterization. Both radius where selected in this analysis according material availability.

resonant frequency of the cavity is lowered implying an electromagnetic characterization at another frequency than the required.

The real effective dielectric determined using the method of the critical points is 2.688 and the corresponding for heavy oil is 2.356 in total agreement with the tend of the previous results known for lower frequencies.

6. Conclusion

In this paper we presented a general methodology which may be useful to determine the usable dielectric range for material characterization. Due to the presence of the central cylinder, a displacement of the resonant frequency occurs proportional to the permittivity constant of the material until the resonant pattern of the mode under study can no longer exist. It means that for a given cavity and fixed rod's radius to be located at the center of the cavity there exists a maximum value of measurable permittivity constant. No resonant conditions can be satisfied for higher permittivity values. The best alternative in order to increase the range of permittivity constant that can be measured reducing the radius of the central dielectric rod or choosing a higher resonant mode with the disadvantage that cavity loses can affect the results by changing the resonant frequency and the quality factor. We have shown that the behavior of the resonant frequency versus the dielectric constant should be taken into account when selecting the best appropriate region with high sensitivity in order to ensure the repeatability of measurement and to reduce the uncertainty associated to the measurement. Finally we have shown with a simple example the electromagnetic characterization of heavy oil with 11 API selecting a proper radious for a petroleum holder made of quartz of purity 99%.

Acknowledgements

This work was supported by FONACIT under the project No. 2011001317. We wish to thank to Roque Rodríguez for his contribution in this work.

References

[1] Love, D.C. and Rothwell, E.J. (2006) A Mode-Matching Approach to Determine the Shielding Properties of a Doubly Periodic Array of Rectangular Apertures in a Thick Conducting Screen. *IEEE Transactions on Electromagnetic Compatibility*, **48**, 121-133.

[2] Massaro, A., *et al.* (2009) Design and Modeling of GaAs/AlGaAs Nonlinear Waveguides by Hertzian Potential Formulation. *European Microwave Conference*, Rome, 29 September 2009-1 October 2009, 719-722.

[3] Yla-Oijala, P., Taskinen, M. and Sarvas, J. (2001) Multilayered Media Green's Functions for Mpie with General Electric and Magnetic Sources by the Hertz Potential Approach. *Progress in Electromagnetics Research*, **33**, 141-165. http://dx.doi.org/10.2528/PIER00120802

[4] Tartarini, D. and Massaro, A. (2011) GPU Approach for Hertzian Potential Formulation Tool Oriented For Electromagnetic Nanodevices. *Progress in Electromagnetics Research M*, **17**, 135-150.

[5] Collin, R.E. (1960) Field Theory of Guided Waves. 2nd Edition, IEEE Antennas and Propagation Society, USA.

[6] Gershon, D., Calame, J.P., Carmel, Y. and Antonsen, T.M. (2000) Adjustable Resonant Cavity for Measuring the Complex Permittivity of Dielectric Materials. *Review of Scientific Instruments*, **71**, 3207-3209. http://dx.doi.org/10.1063/1.1304865

[7] Chen, L.J. and Lue, J.T. (1998) The Transition from the d- to s-State Due to Thermal Fluctuation for High-Tc Superconductors as an Evidence from the Microwave Penetration-Depth Measurement. *IEEE Transactions on Microwave Theory and Techniques*, **46**, 1251-1256.

[8] Pozar, D.M. (2005) Microwave Engineering. 3rd Edition, John Wiley & Sons, Inc., Hoboken.

[9] Kukharchik, P.D., Serdyuk, V.M. and Titovitsky, J.A. (2008) Diffraction of Hybrid Modes in a Cylindrical Cavity Resonator by a Transverse Circular Slot with a Plane Anisotropic Dielectric Layer. *Progress in Electromagnetics Research*, **3**, 73-94. http://dx.doi.org/10.2528/PIERB07112502

[10] Shen, Z.Y., *et al.* (1992) High-Tc-Superconductor-Sapphire Microwave Resonator with Extremely High Q-Values up to 90 K. *IEEE Transactions on Microwave Theory and Techniques*, **40**, 2424-2431.

[11] Liu, J.H., Chen, C.L., Lue, H.T. and Lue, J.T. (2003) A New Method Developed in Measuring the Dielectric Constants of Metallic Nanoparticles by a Microwave Double-Cavity Dielectric Resonator. *IEEE Microwave and Wireless Components Letters*, **13**, 181-183. http://dx.doi.org/10.1109/LMWC.2003.811668

[12] Yan-Shian, Y., Juh-Tzeng, L. and Zhi-Ren, Z. (2005) Measurement of the Dielectric Constants of Metallic Nanoparti-

cles Embedded in a Paraffin Rod at Microwave Frequencies. *IEEE Transactions on Microwave Theory and Techniques*, **53**, 1756-1760.

[13] Stratton, J.A. (1941) Electromagnetic Theory. McGraw-Hill Boook Company, USA.

[14] Geyi, W. (2008) Time-Domain Theory of Metal Cavity Resonator. *Progress in Electromagnetics Research*, **78**, 219-253. http://dx.doi.org/10.2528/PIER07090605

[15] Ong, C.K., Varadan, V.V. and Varadan, V.K. (2004) Microwave Electronics Measurement and Materials Characterization. John Wiley & Sons Ltd., Chichester.

[16] Dester, G.D., Rothwell, E.J., Havrilla, M.J. and Hyde, M.W. (2010) Error Analysis of a Two-Layer Method for the Electromagnetic Characterization of Conductor-Backed Absorbing Material Using an Open-Ended Waveguide Probe. *Progress in Electromagnetics Research B*, **26**, 1-21. http://dx.doi.org/10.2528/PIERB10080506

[17] Krupka, J., Gregory, A.P., Rochard, O.C., Clarke, R.N., Riddle, B., *et al.* (2001) Uncertainty of Complex Permittivity Measurements by Split-Post Dielectric Resonator Technique. *Journal of the European Ceramic Society*, **21**, 2373-2676. http://dx.doi.org/10.1016/S0955-2219(01)00343-0

[18] Riddle, B., Baker-Jarvis, J. and Krupka, J. (2003) Complex Permittivity Measurements of Common Plastics over Variable Temperatures. *IEEE Transactions on Microwave Theory and Techniques*, **51**, 727-733.

[19] Paez, E., Azpurua, M.A., Tremola, C. and Callarotti, R.C. (2012) Uncertainty Estimation in Complex Permittivity Measurements by Shielded Dielectric Resonator Technique Using the Monte Carlo Method. *Progress in Electromagnetics Research B*, **41**, 101-119. http://dx.doi.org/10.2528/PIERB12041306

[20] Paez, E., Azpurua, M.A., Tremola, C. and Callarotti, R.C. (2012) Uncertainty Minimization in Permittivity Measurements in Shielded Dielectric Resonators. *Progress in Electromagnetics Research M*, **26**, 127-141. http://dx.doi.org/10.2528/PIERM12082811

[21] Harrington, R.F. (1961) Time-Harmonic Electromagnetic Fields. John Wiley and Sons, Hoboken, 220.

[22] Sun, E.Y. and Chao, S.H. (1995) Unloaded Q Measurement—The Critical-Points Method. *IEEE Transactions on Microwave Theory and Techniques*, **43**, 1983-1986. http://dx.doi.org/10.1109/22.402290

[23] Callarotti, R.C. and Paez, E.J. (2014) Microwave Dielectric Properties of Heavy Oil and Heating of Reservoirs. *SPE Energy Resources Conference*, Port of Spain, 9-11 June 2014, 127-141. http://dx.doi.org/10.2118/169937-MS

[24] Callarotti, R.C. and Paez, E.J. (2013) Efficient Eigenvalue Numerical Solutions for Time Dependent Linear Systems. *Computational Methods and Experimental Measurements*, **16**, 43-54.

[25] Paez, E.J., Callarotti, R.C., Azprua, B.M. and Rodrguez, R.J. (2013) Medida de la permitividad de lquidos de bajas prdidas en un resonador dielctrico blindado en alta frecuencia (4881 GHz). In 10 *Congreso internacional de metrologa elctrica*, Buenos Aires, September 2013, Vol. 10.

24

A Low Profile Four-Way Directional Antenna with Dual Polarization

Rohani Bakar[1], Shumpei Fuse[1], Hiroyuki Arai[1], Yuki Inoue[2], Keizo Cho[3]

[1]Department of Electrical and Computer Engineering, Yokohama National University, Yokohama, Japan
[2]Research Laboratories, NTT DOCOMO INC, Yokosuka, Japan
[3]Department of Electrical, Electronics and Computer Engineering, Chiba Institute of Technology, Chiba, Japan
Email: harshaz002@yahoo.com, fuse.shumpei@sharp.co.jp, arai@ynu.ac.jp

Abstract

This paper presents a low profile dual polarized directional antenna composed of loop and dipole arrays mounted on a ground plane with each loops and dipoles being fed independently. Each loop antenna is paired with a reflector while each dipole antenna is paired with a director and a reflector. The proposed antenna is intended for an indoor base station (BS) with resonance frequency of 2.4 GHz and capable of producing four orthogonal directional pattern with downward elevation angle equals to 30˚ and half power bandwidth (HPBW) less than 80˚ in both vertical and horizontal polarization. The reflection characteristics of the loop and dipole arrays are less than −10 dB and the mutual coupling between the vertical and horizontal polarization elements is nearly less than −20 dB. In later progress, the dipole antenna was substituted with printed dipole antenna to achieve a better performance. Both the calculated and measured results demonstrated that the desired radiation patterns were achieved, and the measured results agreed well with the calculated ones. Consequently, a low profile antenna with a thickness of 0.16 λ (20 mm) having the expected radiation pattern is successfully designed.

Keywords

Directional Antenna, Dual Polarization, Dipole Antenna, Loop Antenna, Reflector

1. Introduction

The communication systems today is facing an ever-increasing demand on providing a wireless system that is the capable of enabling higher voice and data rates while at the same times supporting higher capacity users. Especially for indoor communication systems, various researches regarding channel capacity enhancement were done to oversee this situation [1]-[4]. High channel capacity can be developed using MIMO technology, and a handful of studies on maximizing channel capacity have been reported [5]-[7]. In multi antennas environment, optimizing the radiation pattern and polarization are very effective especially in enhancing the channel capacity. For example, Dietrich *et al.* [8] empirically demonstrated that a low correlation coefficient was obtained using pattern and polarization diversity. An optimized radiation pattern can be delivered by adjusting the electrical length of parasitic elements of Yagi-Uda thus enhancing channel capacity and improving antenna performance as reported by Honma *et al.* [9]. In addition, Kyritsi *et al.* [10] showed that the channel capacity of a hybrid polarization system outperformed a single polarization system due to the independence of the propagation paths of the two polarizations. Qin *et al.* [11] also reported that the capacity of the system could be improved significantly by using antennas radiating different polarizations compared with antennas with the same polarization.

One of an important design consideration aside from high performance characteristics is miniaturization. A low profile structure is desirable because of easy installation and low costing. However, a notable problem regarding miniaturization of antenna is mutual coupling. As the arrangement of the antenna elements especially the spacing between the elements will be narrow, the mutual coupling between the elements will increase instead. One of the methods to address this issue is by using a matching network [12]. However, this method will cease to unnecessary when we can design an antenna that is small enough in size without reducing the channel capacity by effectively utilizing the polarization [13]-[17]. At the moment however, only few studies have examined miniaturization with consideration of radiation patterns, particularly the design of low profile antenna with directive patterns.

In this paper, we present a novel low profile directional antenna intended for an indoor base station (BS). Although few researches have been reported regarding this topic [13]-[16], this antenna is designed with an emphasis on radiation pattern instead. And unlike the MIMO antenna that is previously designed considering radiation pattern in [17], our priority is to produce a more directive pattern in both vertical and horizontal plane. Previously, we have evaluated a BS MIMO antenna mounted on the ceiling for downlink transmission and demonstrated that directive radiation patterns with a tilted angle in downward direction are effective for enhancing signal-to-noise ratio (SNR) and channel capacity compared with a sleeve antenna configuration [18]. The MIMO antenna used in [18], which 30 mm in height was constructed using cavity-backed slot antenna as vertical element and dipole antenna as horizontal element was proposed in [19].

Although dual polarization was achieved in [19], partitions needed to be used inside the cavity in 4 elements adaptation to suppress mutual coupling. In this paper however, we utilize a dipole as the horizontal element and loop antenna as the vertical element with a much lower height of 20 mm. By using a loop, we eliminated the need for cavity and partition and the design became much simpler and with low mutual coupling too. The antenna configuration also was fed independently, which eradicated the need for complex feeding circuit such as phase shifter for example. The design also provides a space in the middle of the structure where we can utilize it according to desirable applications. Although we have already reported this antenna characteristic briefly in [20], we did not elaborated in details on the process of the structure itself. Therefore, in this paper we explain the process and present the summary of the structure, operation principle and antenna characteristic performance of the proposed antenna.

2. Antenna Design

Channel capacity can be enhanced efficiently by utilizing the radiation pattern and the polarization of designed antenna [9]. In this paper we propose an antenna configuration consists of arrays of loops and dipoles, with operating frequency of 2.4 GHz that gives the lambda, λ equals to 125 mm. The proposed antenna has dual polarization elements, vertical and horizontal with downward elevation angle at 30° and half power bandwidth (HPBW) less than 80°. Each of the arrays will be discuss in details sequently.

2.1. Vertical Polarization Element

In the initial stage of the research, we designed a single loop with a reflector as shown in **Figure 1(a)**. A loop was mounted on a square ground plane with dimension 220 × 220 mm, 55 mm from the edge of one side of the ground plane to make room for other arrays. The height and the length of the loop antenna are 15 mm and 34 mm respectively. The loop structure is actually similar to two monopoles with similar radiation pattern as shown in **Figure 1(b)**, which shows the radiation pattern when the loop was fed without a reflector. But unlike mono-

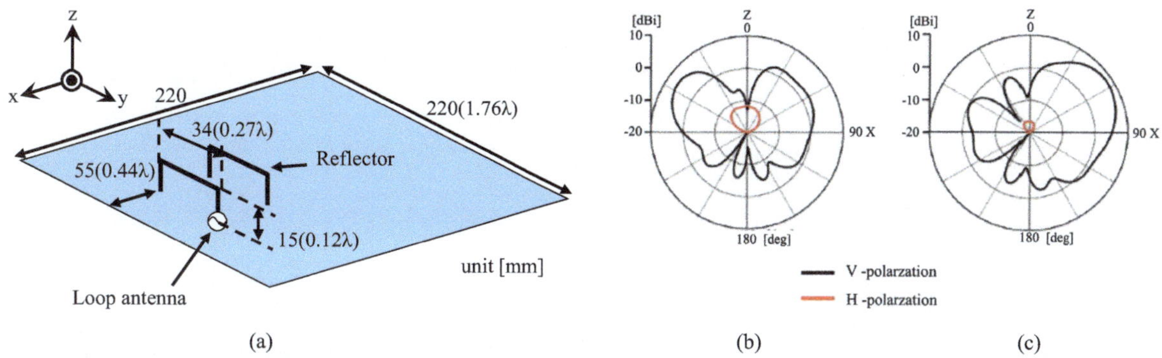

Figure 1. (a) Layout of a single loop antenna with a reflector mounted on a ground plane. Radiation pattern of a single loop antenna in the xz-plane (b) without and (c) with a reflector. V- and H-polarization represent vertical and horizontal polarization respectively.

pole, simply by adding a reflector with the same height to the loop antenna we managed to suppress and shape the radiation pattern into beam-like pattern and tilt it into an angle. From **Figure 1(c)** we confirmed that not only the pattern has been tilted to 56°, the HPBW also has been reduced to less than 80° with the gain of 8.9 dBi at resonance frequency of 2.4 GHz. The black line in both **Figure 1(b)** and **Figure 1(c)** also show that the loop array gives the vertical polarization element to the structure.

Another three pairs of loop and reflector arrays were added and arranged symmetrically to each other as shown in **Figure 2(a)** produce another 3 additional patterns, realizing 4 beams in vertical polarization in 4 different directions. Each loop was denoted as 1 to 4 indicating their respective ports where they were fed independently but simultaneously. Comparing the result between **Figure 1(c)** and **Figure 2(c)**, we found similar patterns are obtained where there is only slight difference in performance. Similar results are also obtained for other loop arrays (port 2, 3 and 4) with low mutual coupling at 2.4 GHz between each port as shown in **Figure 2(b)**. From this graph we confirmed that the resonance frequency has increases and mutual coupling became stronger when the identical configurations were arranged in four directions. The reflection characteristic of $S_{11}$is less than −10 dB at 2.4 GHz, while the mutual coupling S_{21} and S_{31} between adjacent and facing elements are less than −30 dB and approximately −20 dB, respectively. The reason for the observed low mutual coupling in **Figure 2(b)** is due to the suppression of backward radiation in the 270° direction by the reflector.

These output performances are able to be maintained simply by adjusting the length of the loop antenna and the distance between this feeding element and the edge of the ground plane to 60 mm, without altering the height of the element. As a result, an antenna of 15 mm in height representing vertical polarization element is achieved. By just adding a reflector, a directional radiation pattern is achieved while at the same time, a lower mutual coupling compared to without reflector is delivered as confirmed in **Figure 2(b)**.

2.2. Combination of Vertical and Horizontal Polarization Elements

Successfully confirming the characteristics of vertical polarization element with desired results, the next following step was to add the horizontal element into the structure to produce dual polarization elements as targeted. This was done initially by extracting just one element of the loop arrays (port 1) from the four arrays shown in **Figure 2(a)** and added a dipole antenna, denoted as port 5 to serve as horizontal element in the structure as illustrated in **Figure 3(a)**. Combination of dipole antenna and another antenna such as slot or even loop is not exactly a new approach in designing a dual polarization antenna especially when dipole antenna easily deliver good performance with simple configuration as horizontal element. However in this design, the loop and dipole antenna is fed independently to produce dual polarization elements which can be achieve by using just a simple feeding circuit. A distance of D_r was set between the dipole and the loop while adjusting the length of the dipole to match the input impedance at resonance frequency. The reflection characteristic S_{55} of the dipole as a function of D_h, which represents the height of the dipole from the ground plane are shown in **Figure 3(c)**. As D_h becomes smaller, the resonance frequency shifts to a higher frequency making it more difficult to match the impedance properly. When D_h is greater than 15 mm, which a distance higher than the vertical polarization elements given by the loop array, the performance becomes better where S_{55} read below-10 dB. From these results we con-

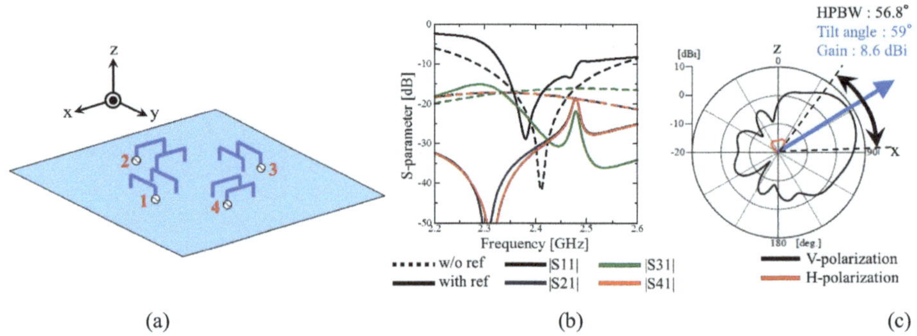

Figure 2. (a) Layout of four loop arrays mounted on a ground plane; (b) Reflection characteristics for port 1 and mutual coupling between ports; (c) Radiation pattern for port 1 in xz plane.

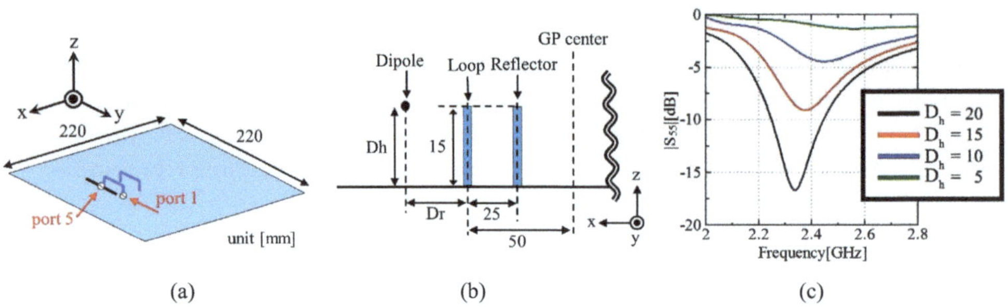

Figure 3. (a) Layout of a dual-polarized antenna constructed using a loop array (port 1) and a single dipole antenna (port 5) mounted on a ground plane; (b) Two-dimensional view in the yz plane; (c) Reflection characteristics of dipole antenna for each D_h values raging from 5 to 20 mm.

cluded that to obtain a good reflection characteristic, we need to set the dipole at a height of between 18 and 20 mm.

Figure 4 shows the reflection characteristics for the loop and dipole array evaluated at several D_r values ranging from 5 mm to 35 mm, which represents the distance between the loop and dipole antenna at the D_h fixed at 18 mm. This parameter analysis is important to confirm the influence of both vertical and horizontal elements against each other. From **Figure 4(a)** we confirmed that although the loop antenna has a component parallel to the dipole, the reflection characteristics show that the element resonant at 2.4 GHz regarding the value of D_r and the impedance is matched well below −10 dB. From these results we concluded that the added dipole element only has a very small effect on the loop element. This is one of the important benefits of using dipole as horizontal polarization element aside from simple and easy configuration. With respect to the dipole, the resonance frequency increases and shifted to higher value as the D_r gets smaller. Although the dipole antenna has been influenced a little bit by the loop array, from this parameter analysis we were able to conclude that a D_r greater than 25 mm is preferable where the reflection characteristic read at below −10 dB.

Upon completing the analysis, we proceeded to add other 3 dipoles antennas to the structure in **Figure 2(a)** to achieve 4 radiation patterns in 4 directions for horizontal element similar to vertical element process before as illustrated in **Figure 5**. Vertical polarization element is constructed using four loop antenna and reflectors, while four dipole antennas with added directors were used for horizontal polarization element. A dipole without a director gave a half circle untilted pattern that would not satisfy our target. Similar to reflector, a director was added into the structure to change the radiation pattern to the desired angle. A few parameters of the elements were optimized to produces the best performance of this structure. Both the height of the loop and dipole arrays was maintained at 15 mm and 18 mm respectively. The length of the loop antenna and reflector is 35 mm and 40 mm while for dipole antenna and director is 55 mm and 53 mm respectively. **Figure 6(a)** and **Figure 6(b)** show the radiation pattern for the loop array and dipole array calculated at port 1 and port 5 respectively. Although both the patterns tilted at the same angle, the HPBW of the dipole is larger and the gain is slightly lower than the loop. From **Figure 6(c)** also we confirmed that the reflection characteristic for dipole, S_{55} is higher than −10 dB. From these results we concluded that the HPBW and reflection characteristic of the dipole need to be improved.

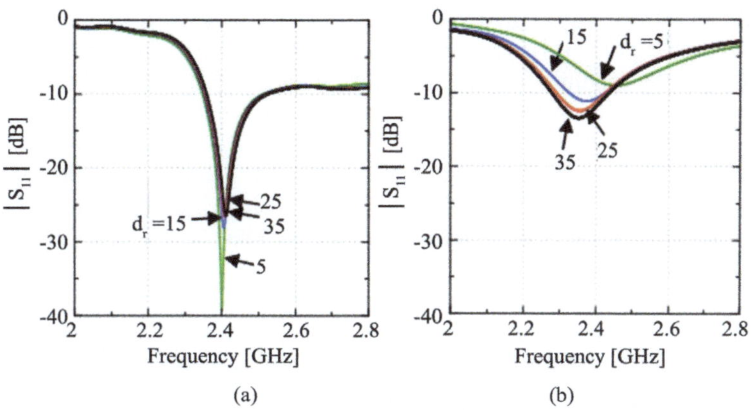

Figure 4. Reflection characteristic of (a) the loop and (b) dipole arrays for several D_r values.

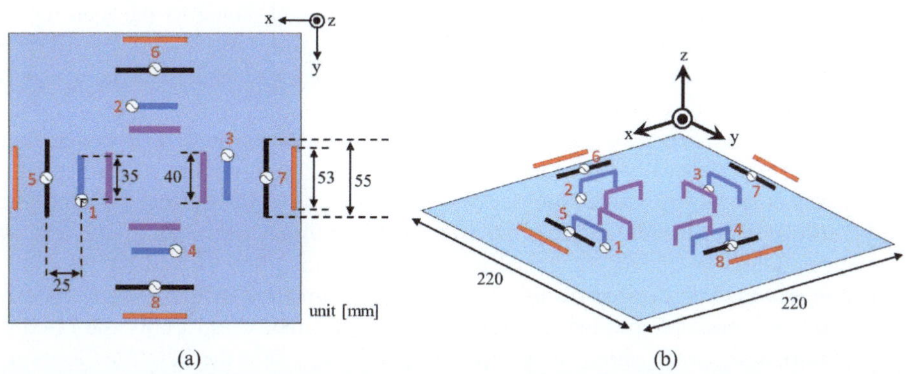

Figure 5. Layout of antenna structure consists of four loops and dipoles arrays in (a) two-dimensional view and (b) three-dimensional view.

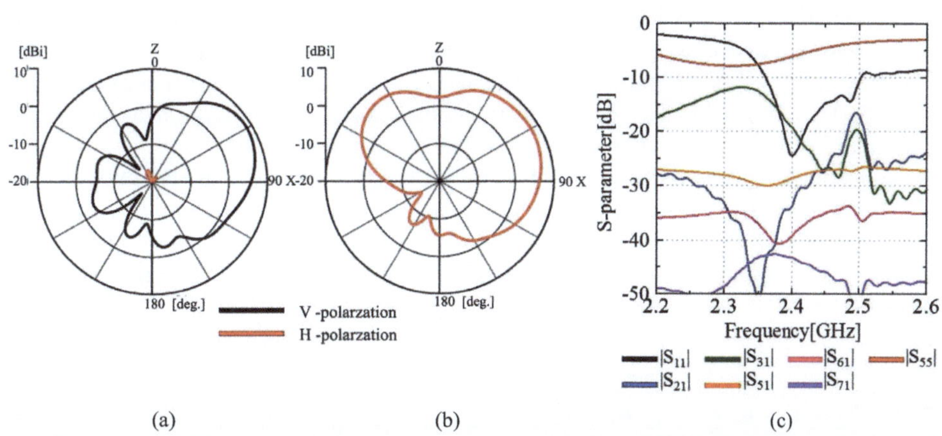

Figure 6. Radiation patterns of (a) loop and (b) dipole antennas. (c) Reflection characteristics for the antenna configuration.

The results also show that dipole array exhibits a back lobe problem as opposed to radiation pattern for loop array.

When only the feeding element of the dipole was mounted, it radiation pattern forms a half circle shape pointing to z-direction due to the reflection from the ground plane. To obtain obliquely upward beam tilted, we first consider mounting director, which location was optimized by analyzing a few parameters. We confirmed that a director operated well when the distance between a director and a feeding element was 30 mm. For the distance between a feeding element of the dipole and that of loop, more than 25 mm was found to be desirable, as indicated from the results in **Figure 4(b)**. However, to improve the undesirable performance of dipole array, a reflector is added to each dipole array. In consideration of the above results, the director was positioned at the edge of the ground, and the feeding element were positioned 30 mm inside the edge of ground plane. Subsequently, reflector was also mounted to improve the HPBW that is larger than the desired value when we used structure in **Figure 5**.

Figure 7(a) shows the reflector was positioned 30 mm from the dipole antenna, exactly on the top of the loop antenna. All the parameters for the loop arrays were maintained while the height of the dipole arrays was adjusted to 20 mm to match the impedance properly at 2.4 GHz. The radiation patterns show in **Figure 7(b)** shows an increasing in cross polarization for the loop although other parameters are maintained. On the other hands, the tilted angle for dipole becomes worse, reading at 79° even though the HPBW has improved slightly to 78.8° and the gain maxed at 9.1 dBi as show in **Figure 7(c)**. We also confirmed that there is a back lobe problem from the figure. Again, parameters with low elevation angle combining with slightly poor S_{55} and big back lobe have to be improved to produce a good antenna performance.

After simulating the structure, we were able to decide the working position for the reflector. When the reflector of the dipole antenna was positioned outside the loop antenna, nearer to the edge of the ground plane than the loop antenna, impedance matching is difficult to achieve and unfortunately the tilted angle is larger than the desired value. However, when the reflector was moved and positioned closer to the feeding element of the loop antenna, horizontal polarization is generated in dipole array. Therefore, reflectors of the dipole antennas need to be located inside the feeding elements of the loop antennas, positioned nearer to the center of the ground than the feeding elements of the loop antennas.

Figure 8 shows the optimized configuration of the proposed antenna with 4 arrays of loops and dipoles arranged in 4 directions symmetrically towards each other with port assignments as numbered. The vertical polarization element was constructed with a loop antenna and by adding a reflector, a directional radiation pattern with tilted angle of 60° is produced. On the other hand, the horizontal polarization element was constructed using a dipole antenna, a director and a reflector. Similar to loop array, director and reflector are used to shape and direct the radiation pattern into desired results. Each of the parameters of these elements especially the length and the position, was optimized to produce a good antenna performance especially the tilted radiation pattern in both vertical and horizontal polarization. In **Figure 8(b)**, we show that the height of loop arrays was set at 15 mm while the height of dipole arrays was fixed at 20 mm. Furthermore, we illustrate the adjustment of the reflector for the dipole towards the center of the ground plane to optimize the antenna performance.

The added reflector gave benefit to dipole where the reflection characteristic is improved and the back lobe problem is eliminated as shown in **Figure 9(b)**. The distance between the reflector and the loop is just 0.04 λ (5 mm) which is very small as shown in **Figure 8(b)**. Interestingly, the added reflector influenced the performance of the dipole array only, without affecting the performance of the loop array at all considering the small distance between the two. This was achieved because in radiation pattern for the loop array, the null pointed towards z-direction is small enough and not affected by the existence of the reflector.

Figure 9 shows the performance of the optimized antenna structure calculated using port 1 for loop array and port 5 for dipole array. **Figure 9(a)** and **Figure 9(b)** show the radiation patterns in the xz plane for both loop and dipole arrays respectively with the radiation patterns tilted at similar angle. For both the loop and dipole arrays, cross polarizations are sufficiently low, and directional radiation patterns with suppressed backward radiation

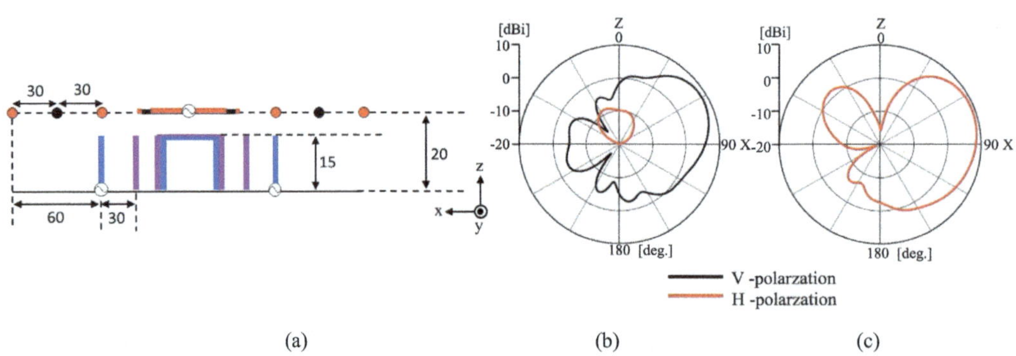

Figure 7. (a) Two-dimensional view of xz plane for added reflector in dipole array. Radiation patterns for (b) loop (port 1) and (c) dipole (port 5) in xz plane.

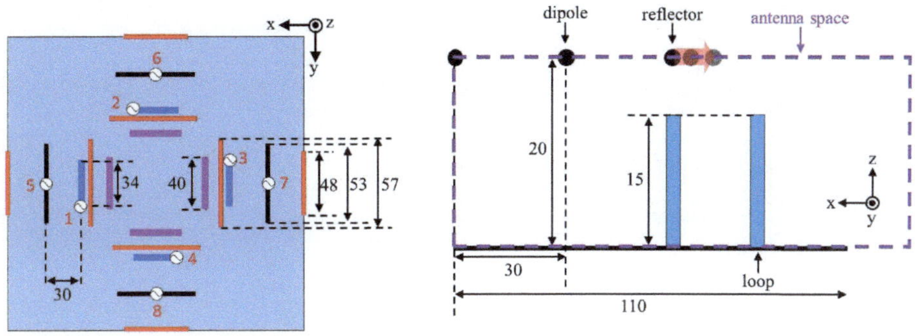

(a) (b)

Figure 8. (a) Two-dimensional view of optimized structure of proposed antenna; (b) Reflector position after the adjustment.

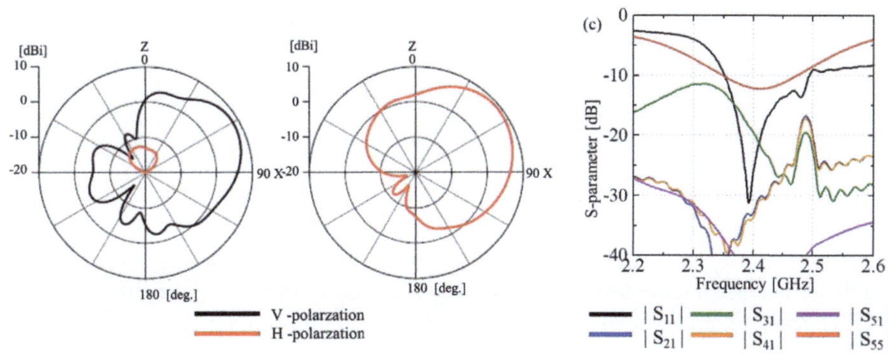

(a) (b) (c)

Figure 9. Radiation patterns of (a) loop array and (b) dipole array. (c) Reflection characteristics of the optimized antenna configuration.

are obtained. The HPBWs for loop and dipole arrays are 57.1° and 70.9° respectively, while the tilted angles of the beam and gains are 61° and 52°, and 8.4 dBi and 9.1 dBi respectively. The radiation patterns for the other ports also display the same characteristics due to the symmetrical property of the structure.

On the other hand, the reflection characteristics result as illustrated in **Figure 9(c)** shows that the reflection characteristics for the loop array (S_{11}) and dipole array (S_{55}) are less than −10 dB at 2.4 GHz. Furthermore, the mutual coupling between the vertical and horizontal polarization elements (S_{51}) is less than −30 dB at resonance frequency, which is sufficiently low due to the orthogonal polarization. For the mutual coupling between vertical polarization elements, although S_{31} is slightly higher than −20 dB, S_{21} and S_{41} is less than −30 dB which are sufficiently low as the backward radiation are suppressed by the reflectors of the loop arrays. **Table 1** presents the comparison between the loop antenna and dipole antenna to summarize the performance of the proposed optimized configuration.

3. Design Improvement and Measurement Results

In **Figure 8** the feeding line structure of the dipole antenna was not considered for simplicity especially when we want to make a prototype out of the design. However, it is necessary to examine a more detailed simulation model to accurately evaluate the antenna performance characteristics. Therefore, in the final stage of the design we replaced the dipole antenna structure which is a straight wire with a printed dipole antenna as illustrated in **Figure 10(a)**. A dipole antenna was printed on both side of a glass epoxy with dielectric substrate of $\varepsilon_r = 4.8$ and a thickness of 1.6 mm. On the ground side of the substrate, the conductor increases in width form 4 to 12 mm forming a tapered shape, functioning as balun as shown in **Figure 10(b)**. The height of the printed dipole antenna was maintained at 20 mm as for dipole antenna.

Figure 10(c) shows a fabricated model of the proposed antenna configuration with printed dipole replacing the dipole as horizontal polarization element. The location was as before as shown in **Figure 8(a)** and the

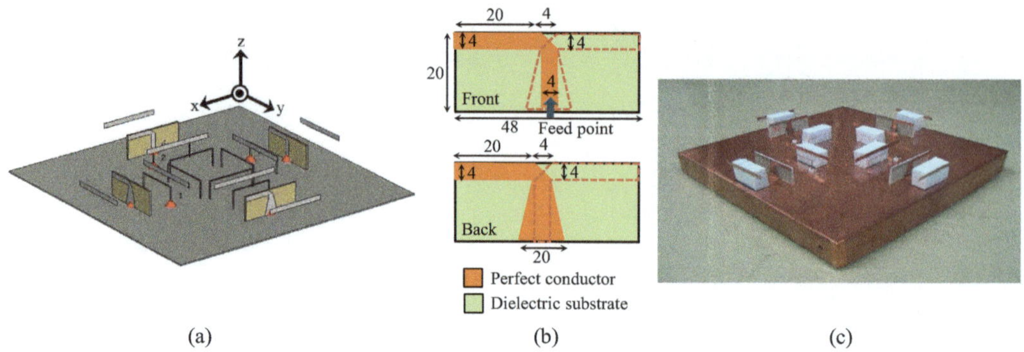

Figure 10. (a) Layout of the final proposed antenna configuration with dipole antenna being replaced with the printed dipole antenna; (b) The front and back view of the printed dipole antenna; (c) Photo of the fabricated antenna design.

Table 1. Performance comparison between loop arrays (port 1) and dipole arrays (port 5).

	HPBW [°]	Tilted angle [°]	Gain [dBi]
Loop	57.1	61.0	8.4
Dipole	70.9	52.0	9.1

dimension of each parameter is shown in **Figure 10(b)**. The feeding elements and reflectors for the loop arrays are 1 mm in thickness, while the directors and reflectors for dipole antennas were 4 mm instead. All of the elements were made of perfect conductors, with the height of the loop and dipole arrays were 15 and 20 mm respectively. In the fabricated antenna shown in **Figure 10(c)**, a copper plate was used for the ground plane, loop antennas, directors, and reflectors. All the feeding for both loops and dipoles antenna were conducted by feed pins from beneath the ground. The directors and reflectors of dipole antenna were easily fixed using the styrene foam to make them stay in the designated position and height.

After fabricating the proposed antenna, we measured it for both reflection characteristic and radiation pattern to compare it with calculated results. **Figure 11(a)** and **Figure 11(b)** show the comparison of reflection characteristics between the calculated (a) and measured (b) results for the proposed antenna. From these graphs we confirmed that at resonance frequency of 2.4 GHz, both the loop array (S_{11}) and dipole array (S_{55}) readings are below −10 dB in both calculated and measured data. In term of mutual coupling between the vertical and horizontal elements (S_{51}), as expected the measured value is increased considering the feeding structures of the dipole antennas. However, the reading of approximately −20 dB at 2.4 GHz is sufficiently low and acceptable. Moreover, although the mutual coupling between the facing vertical polarization elements (S_{31}) is a little bit higher in the calculated data, we confirmed that all of the isolation characteristics including mutual coupling between the adjacent vertical elements are nearly less than −20 dB in the measured data.

Finally, we compared the calculated and measured radiation patterns in the xz plane for both loop and dipole arrays as shown in **Figure 11(c)** and **Figure 11(d)**. The above patterns are for calculated results while the below patterns are for measured results. From these results we found that the cross polarizations are sufficiently low and the directional radiation patterns with suppressed backward radiations are obtained. According to the calculation, the tilted angles of the loop and dipole arrays are 61° and 52° respectively and HPBWs are 55.5° and 70.9° respectively. Although the calculated radiation patterns are normalized with the maximum value, the gains for the loop and dipole elements are 8.0 dBi and 8.2 dBi respectively. Notably, the measured radiation patterns are similar to the calculated radiation patterns. The tilted angles for the loop and dipole arrays are 63° and 50° respectively and he HPBW are 61° and 78° respectively.

As indicated by these results, the desired radiation patterns are achieve based on the design guidelines of 60° to 80° HPBW and a 60° tilted angle, and a low profile directional antenna with a thickness of 0.16 λ (20 mm) has been achieved.

4. Conclusions

We have proposed a low profile dual polarized directional antenna, consisting of loop and dipole arrays for an

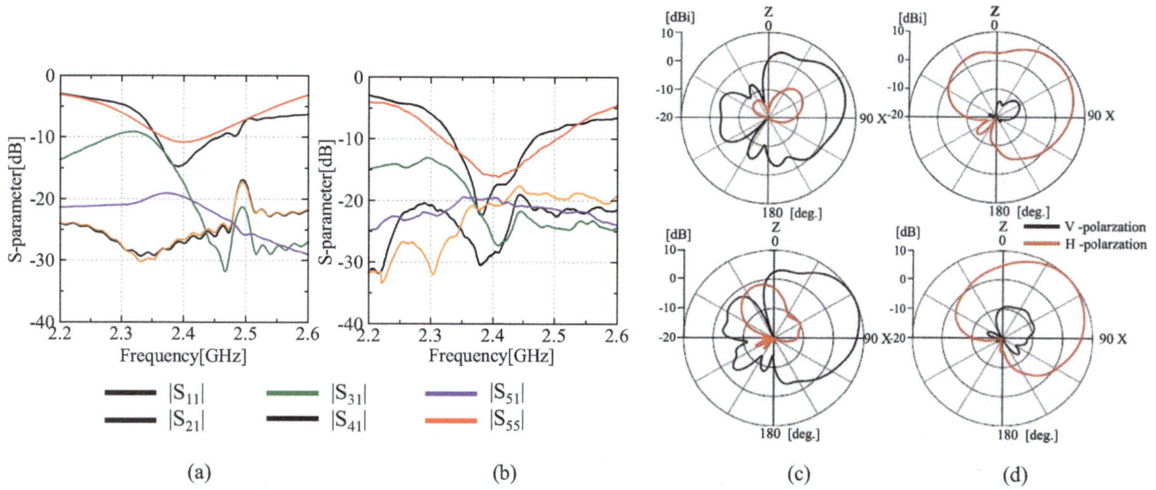

Figure 11. Reflection characteristics of the (a) calculated and (b) measured proposed antenna. Radiation patterns for (a) loop array (port 1) and (b) dipole array (port 5) with both calculated (above two) and measured (below two) results.

indoor base station. The proposed antenna has four directional radiation patterns in four orthogonal directions in the horizontal plane. A narrow HPBW of approximately 60° to 80° and a tilted angle from the z-direction close to 60° were set as the target parameters of each radiation pattern. Both the calculated and measured results demonstrated that the desired radiation patterns were achieved, and the measured radiation pattern agreed well with the calculated ones. The reflection characteristics of the loop and dipole antennas are less than −10 dB and mutual coupling between vertical polarization elements, and between vertical and horizontal polarization elements is nearly less than −20 dB in the measured data. Consequently, we successfully developed a low-profile directional antenna with a thickness of only 0.16 λ (20 mm).

In this paper, the proposed antenna design is capable of transmitting up to eight beams, and up to four beams with vertical or horizontal polarization depending on the state of the mobile station. In addition, the proposed antenna can be used as a pattern diversity antenna, and vertical and horizontal polarizations can be radiated in the identical directions. Thus, the potential applications of the proposed antenna cover a wide range of possibilities.

References

[1] Cho, K. and Inoue, Y. (2010) Influence of Cross-Polarization Characteristics on Indoor MIMO Performance Using a Dual-Polarized Base Station Antenna. 2010 *IEEE International Symposium on Antennas and Propagation and CNC-USNC/URSI Radio Science Meeting—Leading the Wave*, AP-S/URSI 2010.

[2] Sugimura, D., Arai, M., Sakaguchi, K., Araki, K. and Sotoyama, T. (2011) A Study on Beam Tilt Angle of Base Station Antennas for Base Station Cooperation Systems. *IEEE International Symposium on Personal, Indoor and Mobile Radio Communications*, *PIMRC*, Toronto, 11-14 September 2011, 2374-2378. http://dx.doi.org/10.1109/pimrc.2011.6139945

[3] Wang, P., Wang, H., Ping, L. and Lin, X. (2011) On the Capacity of MIMO Cellular Systems with Base Station Cooperation. *IEEE Transactions on Wireless Communications*, **10**, 3720-3731. http://dx.doi.org/10.1109/TWC.2011.083111.101154

[4] Arai, H., Abe, K., Takemura, N. and Mitsui, T. (2013) Dual-Polarized Switched Beam Antenna with Variable Phase Shifter. 2013 *International Workshop on Antenna Technology* (*iWAT*), **2**, 19-22. http://dx.doi.org/10.1109/IWAT.2013.6518289

[5] Sayeed, A.M. and Raghavan, V. (2007) Maximizing MIMO Capacity in Sparse Multipath with Reconfigurable Antenna Arrays. *IEEE Journal of Selected Topics in Signal Processing*, **1**, 156-166. http://dx.doi.org/10.1109/JSTSP.2007.897057

[6] Quist, B.T. and Jensen, M.A. (2009) Optimal Antenna Radiation Characteristics for Diversity and MIMO Systems. *IEEE Transactions on Antennas and Propagation*, **57**, 3474-3481. http://dx.doi.org/10.1109/TAP.2009.2026922

[7] Matthaiou, M., Sayeed, A.M. and Nossek, J.A. (2010) Maximizing LoS MIMO Capacity Using Reconfigurable Antenna Arrays. 2010 *International ITG Workshop on Smart Antennas* (*WSA*), Bremen, 23-24 February 2010, 14-19.

http://dx.doi.org/10.1109/WSA.2010.5456446

[8] Dietrich, C.B., Dietze, K., Nealy, J.R. and Stutzman, W.L. (2001) Spatial, Polarization, and Pattern Diversity for Wireless Handheld Terminals. *IEEE Transactions on Antennas and Propagation*, **49**, 1271-1281. http://dx.doi.org/10.1109/8.947018

[9] Honma, N., Nishimori, K., Takatori, Y., Ohta, A. and Kubota, S. (2007) Proposal of Compact MIMO Terminal Antenna Employing Yagi-Uda Array with Common Director Elements. *IEEE International Symposium on Antennas and Propagation*, Honolulu, 9-15 June 2007, 2405-2408.

[10] Kyritsi, P., Cox, D.C., Valenzuela, R.A. and Wolniansky, P.W. (2002) Effect of Antenna Polarization on the Capacity of a Multiple Element System in an Indoor Environment. *IEEE Journal on Selected Areas in Communications*, **20**, 1227-1239. http://dx.doi.org/10.1109/JSAC.2002.801225

[11] Qin, P.Y., Guo, Y.J. and Liang, C.H. (2010) Effect of Antenna Polarization Diversity on MIMO System Capacity. *IEEE Antennas and Wireless Propagation Letters*, **9**, 1092-1095. http://dx.doi.org/10.1109/LAWP.2010.2093116

[12] Wallace, J.W. and Jensen, M.A. (2004) Mutual Coupling in MIMO Wireless Systems: A Rigorous Network Theory Analysis. *IEEE Transactions on Wireless Communications*, **3**, 1317-1325.

[13] Andrews, M.R., Mitra, P.P. and de Carvalho, R. (2001) Tripling the Capacity of Wireless Communications Using Electromagnetic Polarization. *Nature*, **409**, 316-318. http://dx.doi.org/10.1038/35053015

[14] Getu, B.N. and Andersen, J.B. (2005) The MIMO Cube—A Compact MIMO Antenna. *IEEE Transactions on Wireless Communications*, **4**, 1136-1141. http://dx.doi.org/10.1109/TWC.2005.846997

[15] Das, N.K., Inoue, T., Taniguchi, T. and Karasawa, Y. (2004) An Experiment on MIMO System Having Three Orthogonal Polarization Diversity Branches in Multipath-Rich Environment. *Proceedings of the* 2004 *IEEE* 60*th Vehicular Technology Conference*, Los Angeles, 26-29 September 2004, 1528-1532.

[16] Chiu, C.Y., Yan, J.B. and Murch, R.D. (2007) Compact Three-Port Orthogonally Polarized MIMO Antennas. *IEEE Antennas and Wireless Propagation Letters*, **6**, 619-622. http://dx.doi.org/10.1109/LAWP.2007.913272

[17] Sarrazin, J., Mahé, Y., Avrillon, S. and Toutain, S. (2008) Investigation on Cavity/Slot Antennas for Diversity and MIMO Systems: The Example of a Three-Port Antenna. *IEEE Antennas and Wireless Propagation Letters*, **7**, 414-417. http://dx.doi.org/10.1109/LAWP.2008.2000830

[18] Uchida, D., Arai, H., Inoue, Y. and Cho, K. (2010) A Low-Profile Dual-Polarized Directional Antenna for Enhancing Channel Capacity in Indoor MIMO Systems. *IEICE Transactions on Communications*, **E93-B**, 2570-2577. http://dx.doi.org/10.1587/transcom.E93.B.2570

[19] Suyama, A., Uchida, D., Arai, H., Inoue, Y. and Cho, K. (2009) A Dual Polarization 8-Port MIMO Antenna with Dipole and Slot Antennas. IEICE Technical Report, Macao University, Macao, 113-116.

[20] Uchida, D., Fuse, S., Arai, H., Inoue, Y. and Cho, K. (2010) A Low-Profile Dual-Polarized Directional Antenna for an Indoor MIMO Transmission. *Proceedings of the* 2010 *IEEE Antennas and Propagation Society International Symposium* (*APSURSI*), Toronto, 11-17 July 2010, 1-4.

25

An Entropy Approach to the Natures of the Electric Charge and Magnetic Flux

Salama Abdelhady[1*], Mohamed S. Abdelhady[2]

[1]Faculty of Energy Engineering, Aswan University, Aswan, Egypt
[2]Department of Mechanical Engineering, Beni-Suef University, Beni-Suef, Egypt
Email: [*]Salama_Abdelhady@hotmal.com

Abstract

As a result of investigating analytically and experimentally the interchangeability between thermal, electrical and magnetic energies, it is proved in the presented study that the electric and magnetic energies have an equivalent grade and nature as the thermal energy. So, the natural flow of electric charges and magnetic flux is associated also by increase of entropy of universe, similar to the heat, and has the same nature of waves as the electromagnetic heat radiation. Casting the Maxwell's wave equation into an energy frame of reference that replaces the time in wave equations by entropy, it is possible to represent the electric charges and magnetic flux as electromagnetic waves of electric and magnetic potentials. Such results lead to modify exclusively the definition of the physical nature of the electric charge and magnetic flux which is unavailable in literature.

Keywords

Heat, Electric Current, Magnetic Flux, Entropy, Electromagnetic Waves, Maxwell's Wave Equation

1. Introduction

Generally, flows in thermodynamic systems are driven by forces; *i.e.* flows and forces occur in conjugate pairs. That is, heat flow is driven by differences in temperature between the system and its surroundings, volume flow by differences in pressure, electric charge flow by differences in electrical potential, magnetic flux is a flow driven by magnetic potential, and mass flow by differences in concentration [1].

The second law of thermodynamics provides a rule that describes the direction of change in a system in the absence of external forces. It depends on natural evidence that heat flows from warm objects to cold objects, that

[*]Corresponding author.

objects fall downward in a gravity field, that electric current flows from high potential to low potential, the magnetic flux emerges from magnets of high magnetic potential to iron rods of zero magnetic potential and solutes diffuse from regions of high concentration into regions of low concentration. So, the second law is a rule which captures these facts in a remarkably concise way [2]. Essential to the second law of thermodynamics is the idea of a *reversible* flow. A flow is reversible when it is driven by an infinitesimal force, *i.e.* a force which is so close to zero that a small change in the magnitude of the force at the appropriate place can reverse the direction of the flow [3]. In real systems, flows are almost always irreversible, for example flow of heat into an ice cube from a glass of warm water which occurs between a substantial temperature difference, or the flow of electric current through a lamp which occurs across a substantial electrical potential difference or the flow of magnetic flux from a magnet of high magnetic potential to piece of iron. Such processes are irreversible and lead to increase the entropy of the universe [4].

In the following analysis, the reversible change of heat into electric current will be studied as implemented through the definition of thermoelectric effects [5]. Then, it will be studied the interchangeability of heat into magnetic flux and the interchangeability of electric current into magnetic flux and vice versa through mathematical and experimental analysis. Such similarities will be introduced to modify the second law of thermodynamics to include the electric and magnetic energies as sources of increase of entropy similar to heat. Accordingly, it will be tried modifications of the Maxwell's wave equations to define the nature of the electric charge and magnetic flux as electromagnetic waves, similar to the heat flux, but of different potentials.

2. Interchangeability of Heat and Electric Current

Thermoelectricity is defined as the direct conversion of heat into electric energy, or vice versa [5]. The thermoelectric effect is also defined in literature as a phenomenon by which a temperature difference is directly converted to electric potential and vice versa [6]. Such definitions indicate in general a reversible conversion of electricity into heat and vice versa. It means also the difference between the flow of heat and of the flow of electric current is the potential of each flow. Replacing the thermal potential by an electric potential converts the flow of heat into a flow of electric current or vice-versa. Accordingly, both flows should have the same nature as proved by discoveries of Joule, Seebeck, Peltier, and Thomson. According to Joule's law, a conductor carrying a current generates heat at a rate proportional to the product of the resistance (R) of the conductor and the square of the flowing current (I) [7]. The German physicist Thomas J. Seebeck discovered in the 1820s that if a closed loop is formed by joining the ends of two strips of dissimilar metals and the two junctions of the metals are at different temperatures, an electromotive force, or voltage, arises that is proportional to the temperature difference between the junctions [8]. Such circuit of this type is called a thermocouple; a number of thermocouples connected in series are called a thermopile [9]. In 1834, the French physicist Jean C. A. Peltier discovered an effect inverse to the Seebeck effect: If a current passes through a thermocouple, the temperature of one junction increases and the temperature of the other decreases, so that heat is transferred from one junction to the other [10]. The rate of heat transfer is proportional to the current and the direction of transfer is reversed if the current is reversed. The Scottish scientist William Thomson (later Lord Kelvin) discovered in 1854 that if a temperature difference exists between any two points of a current-carrying conductor, heat is either evolved or absorbed depending upon the material [11]. If heat is absorbed by such a circuit, then heat may be evolved if the direction of the current or of the temperature gradient is reversed [12]. It had been shown that the Seebeck effect is a result of the combined Peltier and Thomson effects. Many devices based on thermoelectric effects are used to measure temperature, generate heat from electric current or generate electricity from flowing heat [13].

The equivalence of electric energy and thermal energy as concluded from the definition of the thermoelectric effect is also proved by the applications of the Seebeck effect in thermoelectric generators and of Peltier effect in thermoelectric refrigerators [14]. However, the similarity of the natures of heat and electric current is indicated also by the analogy of the laws that characterize their flows through conductors where the thermal potential is analogous to the electric potential and the flow of electric charges is analogous to the flow of heat.

The heat transfer by conduction is expressed as follows [15]:

$$\frac{\dot{Q}}{A}=\frac{k}{\Delta x}\left(T_1-T_2\right)=\frac{1}{k/\Delta x}\left(T_1-T_2\right)=\frac{1}{R_t}\left(T_1-T_2\right) \tag{1}$$

Similarly, the charge transferred is expressed by Ohm's law as [16]:

$$I = \frac{\dot{Q}_e}{A} = \frac{1}{R}(V_1 - V_2) \tag{2}$$

3. Interchangeability of Heat and Magnetic Flux

A simple experiment had been elaborated to prove interchangeability of heat and magnetic (or mechanical) energy, and the equivalence of the grades of the thermal energy, or heat, and the magnetic energy. **Figure 1** shows a permanent magnet immersed in an insulted water basin and was allowed to attract polished iron balls along an inclined smooth glass plane. The temperature of water was initially recorded. As the magnet performed a mechanical work, a decrease in the water temperature was measured by a digital thermometer **Figure 2**. By applying the first law of thermodynamics on the magnet-water closed-system [17], we get the following equation:

$$C_F \frac{\Delta t}{\Delta \tau} = n m_b g s \sin \varphi \tag{3}$$

In this process, the heat is equally converted into magnetic work through the introduced magnet during a continuous process. So, the difference between the heat and the magnetic energy is replacing the thermal potential by magnetic potential which attracts the iron-balls. Similar results are identified through the work in the field of

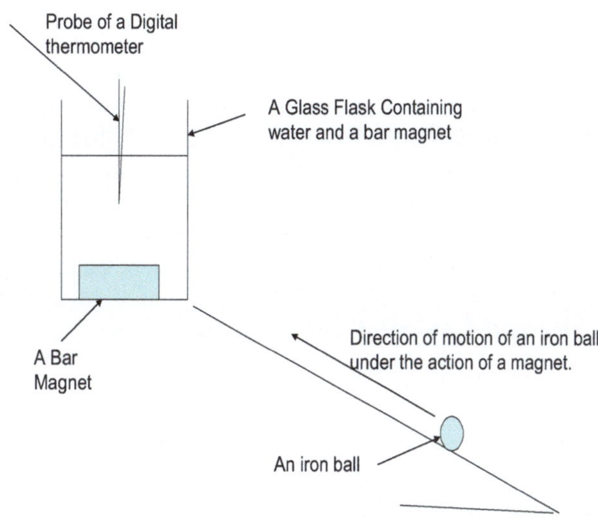

Figure 1. Measurement of magnet's work during attraction of iron balls.

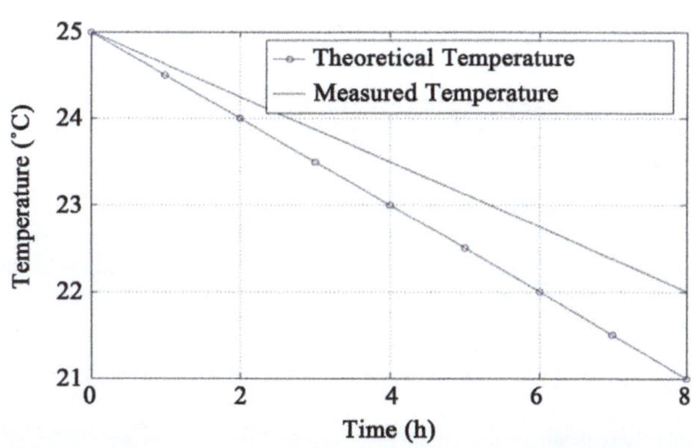

Figure 2. Measurement and calculation results of the system temperature. (Experimental data: thermal capacity of the flask = 1.26 kJ/ deg, mass of each iron ball = 0.3 kg, number of ttracted balls per minute = 42 balls, s = 20 and φ= 30 deg.).

the magnetocaloric effect which is defined as the heating or cooling of a magnetic material due to the application of a magnetic field [18]. The magnetic potential in such applications is replaced by thermal potential. This effect has been called adiabatic demagnetisation for years which is considered as a phenomenon and represents a practical application of the magnetocaloric effect in magnetic materials [19]. So, the thermal energy can be converted into mechanical energy and vice versa. Such results prove that the heat has an equivalent grade as to the magnetic energy and both should be similar.

4. Interchangeability of Electric Energy and Magnetic Energy

Reviewing Faraday's experiment of magnetic induction, it is possible to prove the similarity of the natures of the flow of electric current and the flow of magnetic flux. **Figure 3** shows Faraday's apparatus that demonstrates interactions between magnetic and electric fluxes [20]. When the shown switch is closed, a magnetic field is emitted from the coil on the top part of the iron ring, called as a primary coil, and is transmitted to the coil on the bottom part of the ring which is called as a secondary coil. The galvanometer is used to detect any current induced in the secondary coil. It was found that each time the switch is closed; the galvanometer detects a current in one direction in the secondary coil. Each time the switch is opened, the galvanometer detects also a current in the opposite direction. Interestingly, if the switch remains closed or open for any length of time, there is no current through the galvanometer.

Available literature defines the electric current as a flow of electrons and defines the magnetic flux as the quantity of magnetic field that penetrates an area at right angles [21]. According to this definition of electric current; the flow of its electrons in the primary coil should induce, by analogy between the electric current and magnetic flux, a flow of some other particles in the iron ring. Similarly, if the magnetic flux that passes through the secondary coil is defined as the quantity of magnetic field that penetrates an area at right angles, then it should not regenerate a flow of particles in the secondary coil as both have different natures. So, the demonstrated Faraday's experiment shows that the conversion of electric energy into magnetic energy in the primary coil is done by replacing the electric potential of the electric energy by magnetic potential and the conversion of the magnetic energy into electric energy in the secondary coil is done by replacing the magnetic potential by electric potential. It means that both fluxes should have similar natures so that both can be interchanged alternatively.

5. Entropy and the Second Law of Thermodynamics

The previously proved interchangeabilities of heat, electric current and magnetic flux prove the equivalence of their grades and the similarity of their natures. However, the second law defines a new state function, the *entropy* that may describe the irreversibilities of heat flow when dealing with the entropy of the universe rather than the entropy of a system as a state function. For reversible processes the heat flow is expressed as the product of temperature, as the force that derives the heat flow, times the increase of the entropy of the system associated by the heat flow [22].

$$\delta Q_{th} = TdS \tag{4}$$

Equation (4) is used to represent the heat added to a system T-s property diagram seen in **Figure 4**.

However, the total changes in entropy (system plus surroundings) for reversible process where the same heat is transferred through infinitesimally small difference [23]

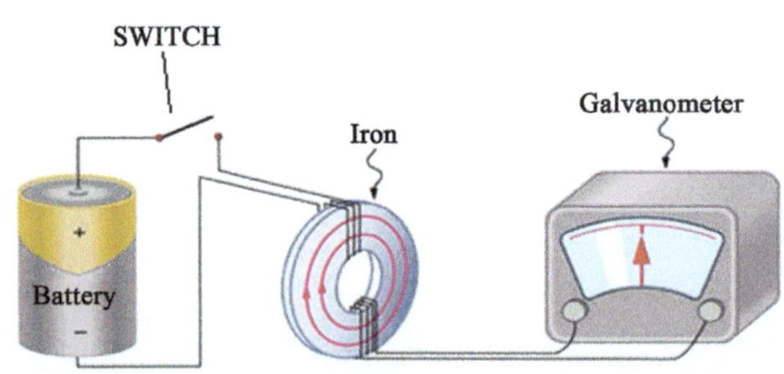

Figure 3. Faraday's apparatus for demonstrating the interactions between electric and magnetic energies [11].

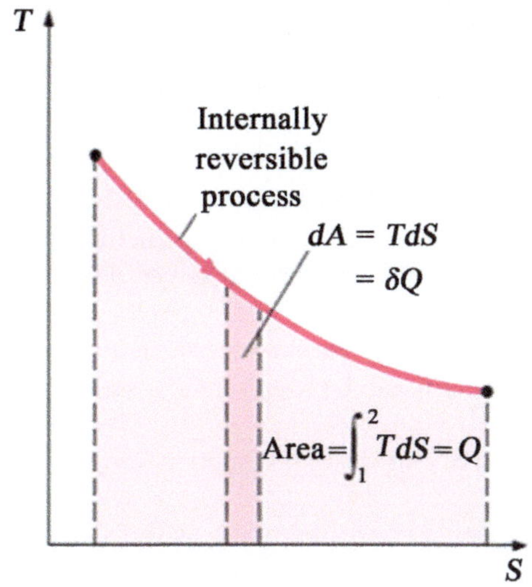

Figure 4. Heat representation in a T-s property diagram.

$$T_{\text{system}} \cong T_{\text{surroundings}},$$

$$\Delta S_{\text{total}} = \Delta S_{\text{system}} + \Delta S_{\text{surroundings}} = \left(\frac{Q}{T}\right)_{\text{system}} - \left(\frac{Q}{T}\right)_{\text{surroundings}} = 0 \tag{5}$$

So, the total change in entropy for a reversible process is zero as the entropy is a state function. However, for real processes where the temperature of the system should be higher than the temperature of surroundings [24]

$$T_{\text{system}} > T_{\text{surroundings}} \tag{6}$$

So, we get

$$\Delta S_{\text{tot}} = \Delta S_{\text{surroundings}} + \Delta S_{\text{system}} = \frac{Q}{T_{\text{surroundings}}} - \frac{Q}{T_{\text{system}}} > 0$$

So, in any physical process, the entropy S for an isolated system never decreases; that is, we have

$$\Delta S \geq 0 \quad \text{always} \tag{7}$$

Reviewing the electric energy, we find also that the current will flow from high potential to low potential.

According to the similarity of heat, electric energy and magnetic energy, such principle of increase of entropy should include also the increase of entropy of the universe due to transfer of all these energies. So, we should the following relation between the electrical potential of the system " E_{system} "and of the surroundings " $E_{\text{surroundings}}$ ", similar to the relation in Eq. (6) between the temperatures of the system and surroundings:

$$E_{\text{system}} > E_{\text{surroundings}} \tag{8}$$

Accordingly; there is an increase of entropy of the universe associated by the flow of electric current and, by similarity, of magnetic flux that leads also to apply Equation (7) [24].

So, the second law should consider the energies similar to the heat as the flow of electric charges and magnetic flux as sources of similar irreversibilities. In other words, the flow of electric charge and magnetic flux should be associated also by an increase of entropy of a system and expressed by a similar equation as (4). Hence, the reversible transfer of electric current should be expressed as the product of the electric potential times the increase of entropy associated by the flow of electric charge and the magnetic flux as the product of the magnetic potential times the increase of entropy associated by the flow of magnetic flux according to the following equations:

$$\delta Q_{\text{elec}} = E dS \tag{9}$$

$$\delta B_{\text{magnetic}} = H dS \tag{10}$$

Equations (9) and (10) were the foundations that were previously introduced to find a fundamental energy equation that embraces the flow of heat, electric current and magnetic flux which was applied in many citations in the following form [25]-[28]:

$$TdS + EdS + BdS = dU + pdV \tag{11}$$

According to Equations (4), (9), (10), and (11) entropy can be defined as a measure of energizing, or disordering, the molecules of a system by thermal, electric or magnetic energies. The postulate that the flow of electric current is associated by flow of entropy was early discovered by Einstein and his colleagues [29]. However, as the heat is represented in the T-s diagram by the area down the process line as seen in a T-s property diagram as shown in **Figure 4** [4], the electric energy can be represented also by an area in an E-s property diagram as seen in **Figure 5** and the magnetic energy can be represented also by the area in an H-s property diagram as seen in **Figure 6**.

However, relying on the definition of entropy of a system in terms of its statistical concept according to Ludwig Boltzmann as follows [30]:

$$S = k_B \cdot \ln \Omega \tag{12}$$

k_B : Boltzmann's constant = 1.38×10^{-23}, Ω is the number of arranging the microstates of a system at specified total internal energy. According to Equation (12), Boltzmann introduced a statistical concept of entropy. He showed also that the arrow of time was statistical in origin and he sought to derive the Second Law of Ther-

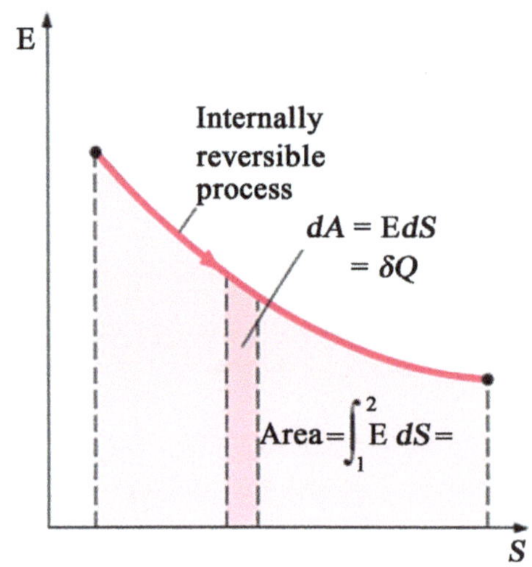

Figure 5. Electric Energy representation in an E-s property diagram.

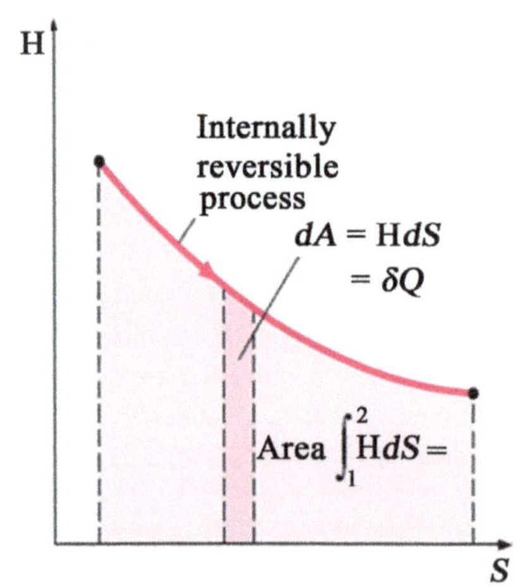

Figure 6. Magnetic energy representation in a H-s property diagram.

modynamics from the statistical behavior of a large number of molecules obeying the simple laws of mechanics, namely the conservation of momentum and energy. Such analysis was given the name statistical mechanics [31]. Considering such point of view, the task of explaining the arrow of time only through entropy generation due to reversible and irreversible processes looks like the task of finding some simple relations which obliquely "cause" entropy to increase with high probability. According to the statistical definition of entropy as introduced by Boltzmann, the reversible flow of thermal, electric and magnetic energies should lead also to the increase entropy [32].

6. The Maxwell's Equations in an Energy Frame of Reference

The representation of energy, or heat, as flow of electromagnetic waves is defined mathematically by Maxwell's equations. However, it is expressed in literature in terms of flow of electric energy wave in an E-t plane, and magnetic energy wave in an H-t plane as follows [33]:

$$\left(\nabla^2 - \frac{1}{c^2}\frac{\partial^2}{t^2}\right)E = 0 \tag{13}$$

$$\left(\nabla^2 - \frac{1}{c^2}\frac{\partial^2}{t^2}\right)H = 0 \tag{14}$$

where c is the speed of light in the medium, in vacuum $c = c_0 = 299{,}792{,}458$ meters per second. Replacing the time in the previous Maxwell's equations by entropy, Maxwell's equations may be represented in an energy frame of reference composed of the coordinates E-H-s as seen in **Figure 7**. The areas scanned by the electric and magnetic waves in such frame express actually the flow of electric and magnetic energies as components of the flowing electromagnetic wave as seen in **Figure 5** and **Figure 6**. So, the modified form of Maxwell's relation casted into an energy frame of reference that replaces the time by an energy property, the entropy, will be as follows [34]:

$$\left(\nabla^2 - \frac{1}{c^2}\frac{\partial^2}{s^2}\right)E = 0 \tag{15}$$

$$\left(\nabla^2 - \frac{1}{c^2}\frac{\partial^2}{s^2}\right)H = 0 \tag{16}$$

Replacement of time by entropy sustains Eddington argument of the arrow of time as indication of the direction of progressive increase of random elements in nature and his lengthy conclusion that such randomness is a property of entropy alone which is also a measure of randomness in time [35]. Such fuzzy approach realizes a dream of Tesla (1856-1943) who searched the transformation of the wave equations as described by James Clerk Maxwell (1831-1879) into an energy frame of reference that expresses the flow of energy associated by the flow of electromagnetic waves [36].

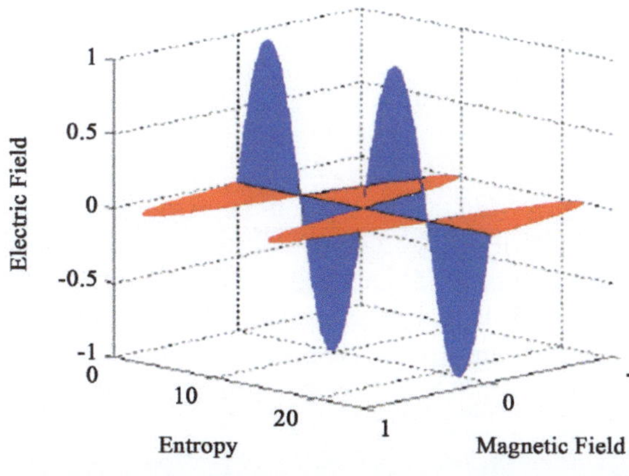

Figure 7. Flow of electromagnetic waves in an energy frame of reference that shows the electric energy flow in an electric field—entropy plane and the magnetic energy flow into a magnetic field—entropy plane [34].

According to the indicated representation of the electromagnetic waves in the introduced coordinate system, the bounded or shaded areas by the wave lines in the E-s and H-s planes of **Figure 7** represent the imparted flow of energy. Denoting the imparted energy per an electromagnetic wave by the symbol "Ћ" Joule/wave, it is possible to mathematically estimate such energy or shaded areas as follows [37]:

$$\text{Ћ} = \int E dS + \int H dS \tag{17}$$

Accordingly, it is possible to calculate the energy flow per unit time as follows:

$$\dot{W} = \text{Ћ}\nu \quad \text{Watt} \tag{18}$$

"ν" is the frequency of the electromagnetic wave, or the number of waves per unit time. This means that the energy quantization is stemmed mainly from its nature as a wave. Such understanding is a direct conclusion of representing the flow of electromagnetic waves in the introduced energy frame of reference.

7. The Nature of Electric Charge and Magnetic Flux

The previously proved interchangeabilities between thermal energy and electric energy, thermal energy and magnetic energy and between electric energy and magnetic energy prove that the heat flux, electric current and magnetic flux should have the same nature. So, if the primary form of heat flux is a flow of thermal radiation as electromagnetic waves of thermal potential, we may postulate the electric current is a flow of electromagnetic waves of electric potential and the magnetic flux as electromagnetic waves of magnetic potential.

The representation of heat, as flow of electromagnetic waves, is defined mathematically by Maxwell's equations and is represented graphically in **Figure 7**. According to the discussed similarity of flow of thermal radiation as defined by the modified Maxwell's wave equation Equations (15) and (16) and the electric current; the electric charge can be defined as a special solution of the Maxwell's equations as flow of electromagnetic waves whose electric component oscillates initially around a specific positive or negative potential "$+/-\Delta E$," while its magnetic component oscillates around zero potential. Such solution is presented mathematically as follows [38]:

$$E(r,s) = \varphi(r,s) + / - \Delta E \tag{19}$$

$$H(r,s) = \psi(r,s) \tag{20}$$

Equations (19) and (20) represent a solution of Maxwell's electromagnetic wave equations of specific initial conditions that define the flow of electric charges as a flow of ionized electromagnetic waves which have a non-zero electric potential. Such solution can be represented graphically as seen in **Figure 8**. In case of **Figure 8**, the net energy imparted per wave will be energy of net negative potential- ΔE , in other words a negatively charged energy, which can be calculated as follows:

$$Q_{el} = \int_0^{2\pi} E dS + \int_0^{2\pi} H dS \tag{19}$$

Such definition of electric charge or current as electromagnetic waves of electric potential found plausible explanation of the discovered Tesla's "Radiant Energy" or Tesla's "Dark Waves" as normal transfer of electric current, of wave-nature, through air by the high electric potential of Tesla's tower [39]. Such postulated nature of electric charge succeeds also in solving the duality confusion in the field of electrodynamics [40].

According to the discussed similarity of the natures of the flow of thermal radiation as defined by the modified Maxwell's wave equation Equations (15) and (16) and the magnetic flux; the magnetic flux can be defined as a special solution of such equations by defining magnetic flux as flow of electromagnetic waves whose magnetic component oscillates initially around a specific positive or negative potential "$+/-\Delta H$ ", while its electric component oscillates around zero potential as seen in **Figure 9**. Such solution is presented mathematically as follows:

$$E(r,s) = \varphi(r,s) \tag{17}$$

$$H(r,s) = \psi(r,s) + / - \Delta H$$

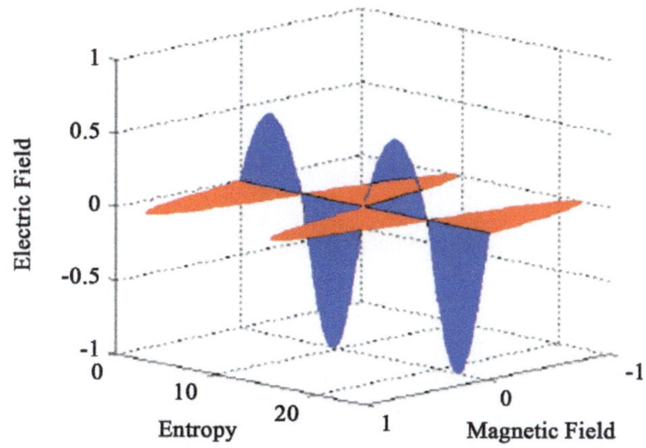

Figure 8. Flow of electric charges as E. M. waves of negative electric potential.

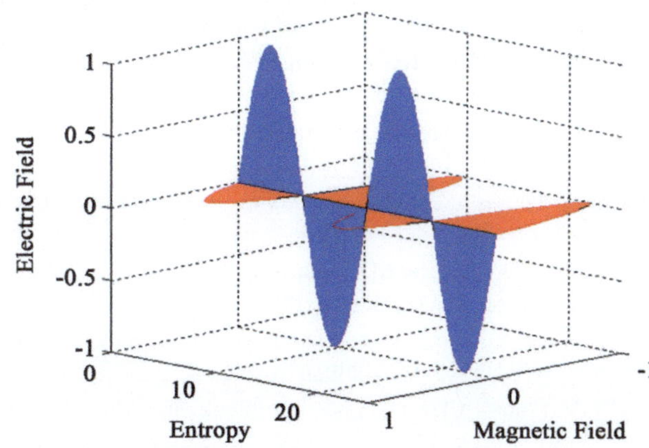

Figure 9. Flow of magnetic flux as E. M. waves of negative magnetic potential.

Such definition of the nature of magnetic flux as electromagnetic waves of magnetic potential found a plausible explanation of the MIT discovery of wireless power transmission by magnetic resonant coupling which was identified as "evanescent waves". Such waves can be plausibly defined as wireless transfer of normal magnetic flux, of wave-nature, through air by the high potential of the discovered magnetic resonant coupling [41]. Similarly, it helps to find a simple universal system of units that solves the complexities if the SI system of units in the field of electrodynamics [42].

8. Conclusion

Reviewing the natural flow of electric charges from high electric potential to low potential and the natural flow of magnetic flux from high magnetic potential to low potential which are similar to the natural flow of heat from high temperature to low temperature, it is possible to prove in this study that transfer of electric charge and magnetic flux represent sources of irreversibility and should be considered as a source of increase of entropy of the universe similar to the transfer of heat. Such results help through a followed entropy approach in defining the nature of the electric charge as electromagnetic wave of electric potential and the nature of magnetic flux as electromagnetic waves of magnetic potential similar to the heat flux as electromagnetic waves of thermal potential. Such postulated natures of electric charge and magnetic flux find plausible explanations of the Tesla dark waves, the MIT evanescent waves, solves the duality confusion and find a simple universal system of units that solves the complexities of the SI system of units in the field of electrodynamics. It also explains the speed of flow of the electric charges that approaches the velocity of light.

References

[1] Stevens, C.F. (1965) The Six Core Theories of Modern Physics. MIT Press, Cambridge.

[2] Yunus, A.C. and Michael, A.B. (2006) Thermodynamics: An Engineering Approach. McGraw-Hill Science Engineering, New York.

[3] Tschoegl, N. (2000) Fundamentals of Equilibrium and Steady-State Thermodynamics. Elsevier Science, Ltd., Amsterdam.

[4] Herbert, C.B. and Greene, R.F. (1952) On a Theorem of Irreversible Thermodynamics. *Physical Review*, **86**, 702-710.

[5] Rowe, D.M. (2006) Thermoelectrics Handbook: Macro to Nano. Taylor & Francis Group, 6000 Broken Sound Parkway.

[6] Riffat, S.B. and Ma, X. (2003) Thermoelectrics: A Review of Present and Potential Applications. *Applied Thermal Engineering*, **23**, 913-935. http://dx.doi.org/10.1016/S1359-4311(03)00012-7

[7] Heikes, R.R., Roland, W. and Ure, J. (1961) Thermoelectricity: Science and Engineering. Interscience Publishers Inc., New York.

[8] Van Herwaarden, A.W. and Sarro, P.M. (1986) Thermal Sensors Based on the Seebeck Effect. *Sensors and Actuators*, **10**, 321-346. http://dx.doi.org/10.1016/0250-6874(86)80053-1

[9] Jaydeep, V., Joshi, V. and Patel, N.M. (2012) Electricity Generation Using Thermopile System from the Flue Gases. *International Journal of Scientific Engineering and Technology*, **1**, 6-10.

[10] Ball, C., Jesser, W. and Maddux, J. (1995) The Distributed Peltier Effect and Its Influence on Cooling Devices. *Proceedings of the* 14*th International Conference on Thermoelectrics*, St. Petersburg, 27-30 June 1995, 27-30.

[11] Harman, T.C. and Honig, J.M. (1967) Thermoelectric and Thermomagnetic Effects and Applications. McGraw-Hill Book Company, New York.

[12] Buist, R.J. (1995) The Extrinsic Thomson Effect. *Proceedings of the* 14*th International Conference on Thermoelectrics*, St. Petersburg, 27-30 June 1995, 27-30.

[13] Weiling, L. and Shantung, T.U. (2004) Recent Developments of Thermoelectric Power Generation. *Chinese Science Bulletin*, **49**, 1212-1219. http://dx.doi.org/10.1360/04we0037

[14] Snyder, G.J., Fleurial, J.P., Caillat, T., Yang, R. and Chen, G. (2002) Supercooling of Peltier Cooler Using a Current Pulse. *Journal of Applied Physics*, **92**, 1564-1569. http://dx.doi.org/10.1063/1.1489713

[15] John IV, H.L. and John, H.L.V. (2008) A Heat Transfer Textbook. 3rd Edition, Phlogiston Press, Cambridge, MA.

[16] Serway, R.A. and Serway, J.W. (2010) Physics for Scientists and Engineers with Modern Physics. 8th Edition, Brooks/Cole, Boston.

[17] Callen, B. and Herbert, B. (1985) Thermodynamics and an Introduction to Themostatistics. John Wiley & Sons, New York.

[18] Zhang, X.X., Tejada, J., Xin, Y., Sun, G.F., Wong, K.W. and Bohigas, X. (2008) Materials Science—Poland, Magnetocaloric Effect. *Physics and Applications*, **26**, 807-814.

[19] Tishin, A.M. and Spichkin, Y.I. (2003) The Magnetocaloric Effect and Its Applications. Institute of Physics Publishing, Bristol. http://dx.doi.org/10.1887/0750309229

[20] Ryan, D.T. (2006) Toward a Cognitive-Historical Understanding of Michael Faraday's Research: Editor's Introduction. *Perspectives on Science*, **14**, 1-6.

[21] Haaiday, D., Resnick, R. and Walker, J. (2004) Fundamentals of Physics. 7th Edition, John Wiley & Sons, New York.

[22] Fujimtu, M. (2007) Physics of Classical Electromagnetism. Springer, Berlin.

[23] Hill, E.L. and Grünbaum, A. (1957) Irreversible Processes in Physical Theory. *Nature*, **179**, 1296-1297. http://dx.doi.org/10.1038/1791296b0

[24] Popper, K.R. (1965) Time's Arrow and Entropy. *Nature*, **207**, 233-234. http://dx.doi.org/10.1038/207233a0

[25] Abdelhady, S.A. (2010) Fundamental Equation of Thermodynamics that Embraces Electrical and Magnetic Potentials. *Journal of Electromagnetic Analysis & Applications*, **2**, 162-166. http://dx.doi.org/10.4236/jemaa.2010.23023

[26] Amiri, S.M., Pourazad, M.T., Nasiopoulos, P. and Leung, V.C.M. (2014) Power Quality Assessment by Entropy Harmonics and Quality of Power (ICHQP). *Proceedings of the IEEE* 16*th International Conference*, Bucharest, 25-28 May 2014, 380-384.

[27] Estrada, J.H., Valencia, R.S., Cortes, C.L. and Plata, E.A.C. (2013) Magnetic Flux Entropy as a Tool to Predict Transformer's Failures. *IEEE Transactions on Magnetics*, **49**, 4729-4732. http://dx.doi.org/10.1109/TMAG.2013.2260821

[28] Valencia, R.S., Estrada, J.H., Cortes, C.L. and Cano-Plata, E.A. (2013) Harmonics Detection in Transformers by En-

tropy of Electromagnetic Signals Radiated. *Proceedings of the* 2013 *Workshop on Power Electronics and Power Quality Applications* (*PEPQA*), Bogota, 6-7 July 2013, 1-5. http://dx.doi.org/10.1109/PEPQA.2013.6614967

[29] Fang, X. and Gao, S. (2015) Proof of Entropy Principle in Einstein Maxwell Theory. Physical Review D, **92**, 162-177.

[30] Lebowitz, J.L. (1993) Macroscopic Laws and Microscopic Dynamics, Time's Arrow and Boltzmann's Entropy. *Physica A*, **194**, 1-27. http://dx.doi.org/10.1016/0378-4371(93)90336-3

[31] Tolman, R.C. (1938) The Principles of Statistical Mechanics. Dover Publications, New York.

[32] Mayants, L. (1984) The Enigma of Probability and Physics. Springer, Dordrecht. http://dx.doi.org/10.1007/978-94-009-6294-1

[33] Haaiday, D., Resnick, R. and Walker, J. (1984) Fundamentals of Physics. John Wiley & Sons, New York.

[34] Parkinson, G. (2015) UBS: Tesla Powerwall Can Deliver 6-Year Payback in Australia. *Renew Economy*, Published Online.

[35] Eddington, A. (1981) The Nature of the Physical World. Cambridge University Press, Cambridge.

[36] Hunt, I. and Draper, W. (1981) Lightning in His Hand: The Life Story of Nikola Tesla. Omni Publications, Hawthorne.

[37] Franklin, A. (2013) Millikan's Measurement of Planck's Constant. *The European Physical Journal*, **38**, 573-594. http://dx.doi.org/10.1140/epjh/e2013-40021-3

[38] Abdelhady, S. (2012) A Thermodynamic Analysis of Energy Flow in Optical Fiber Communication Systems. *Applied Physics Research*, **4**, 22-29. http://dx.doi.org/10.5539/apr.v4n3p22

[39] Abdelhady, S. (2013) An Entropy Approach to Tesla's Discovery of Wireless Power Transmission. *Journal of Electromagnetic Analysis & Applications*, **5**, 157-161. http://dx.doi.org/10.4236/jemaa.2013.54025

[40] Abdelhady, S. (2011) An Entropy Approach to the Duality Property. *Journal of Electromagnetic Analysis & Applications*, **3**, 220-227. http://dx.doi.org/10.4236/jemaa.2011.36036

[41] Abdelhady, S. (2013) An Entropy Approach to Wireless Power Transmission by Magnetic Resonance. *Applied Physics Research*, **5**, 29-35.

[42] Abdelhady, S. (2010) An Approach to Universal System of Units. *Journal of Electromagnetic Analysis & Applications*, **2**, 549-556. http://dx.doi.org/10.4236/jemaa.2010.29071

26

Synoptic Analysis Clarifies Childhood Leukemia Risk from ELF Magnetic Field Exposure

Norbert Leitgeb

Institute of Health Care Engineering, Graz University of Technology, Graz, Austria
Email: norbert.leitgeb@tugraz.at

Abstract

In spite of 36 years epidemiologic research, there is still an ongoing controversy about a causal link between childhood leukemia (CL) and exposure to extremely low frequency (ELF) magnetic fields (MF). Public concern has been increased by the fact that ELF MF have been classified as possibly carcinogenic to humans (class 2B) while exposure limits still remain three orders of magnitudes above reported CL risk onset levels. In a new synoptic approach rather than few selected ORs, all reported epidemiological risk estimates (ORs) are analyzed, both pooled together as well as separated into sub-pools of different exposure metric as well as of high and low exposure levels. The results explain the worrying offset of ORs towards increased CL risk as well as the reported puzzling dose-response at low MF levels as an artifact caused by the small-number effect. The synoptic analysis clarifies that ORs critically depend on statistical power. With increasing statistical power ORs decrease and finally converge to and stay at zero risk. This is found consistently at the entire data pool as well as at all sub-pools related to investigated exposure parameters (wire code, distance to MF source, and magnetic field value). Former contradictory results can now be explained. The synoptic analysis provides convincing evidence that the risk of childhood leukemia is not increased by exposure to ELF magnetic fields. IARC's classification of ELF MF needs revision.

Keywords

Health Risk, Long-Term Effect, Carcinogenicity, Magnetic Field, Power Line

1. Introduction

Since 1979 when the first epidemiologic study has reported on an increased prevalence of childhood leukemia

(CL) in the presence of extremely low frequency (ELF) magnetic fields (MF) [1], in spite of many subsequent epidemiologic studies of the following 36 years the debate about a causal relationship is still ongoing. Based on merely epidemiologic evidence on a possible association with childhood leukemia (but not with other cancers neither in adults nor in children), the International Agency for Research on Cancer (IARC) of the World Health Organization (WHO) has classified ELF MF as possibly carcinogenic to humans, class 2B [2]. Based on reported risk estimates, the impact of an increased CL risk has been theoretically quantified to be about 1.5% - 2% of all incident childhood leukemia cases with annually 50 - 60 additional cases in the whole European Union (Croatia excluded) [3].

Health concern is based in particular on the puzzling fact that reported CL risk estimates (ORs) exhibit an offset towards increased risk and a seeming dose response in terms of an increase with increasing magnetic field exposure at low MF levels of 0.2 to 0.4 μT [4]-[8]. However, for the general public existing guidelines and regulations allow ELF MF exposures three orders of magnitudes above these values [9] [10]. The limit has even been doubled recently [9] without any idea whether and how the reported dose dependence should be extrapolated to exposures three orders of magnitudes above reported risk onset levels. This dilemma raises concern whether limits provide a sufficient safety level. Unfortunately, after several decades of epidemiological research, more of the same research cannot be expected to clarify the issue even in the future. Consequently, without a new approach risk believers and deniers are expected to further continue their controversy.

It is striking that so far only little use has been made of most of the already existing epidemiologic evidence. On the one hand (fortunately), it has become frequent practice that papers on epidemiological studies include tables with extensive numbers (up to about hundred) of ORs which have been calculated for a variety of different scenarios and exposure metrics. However, so far conclusions have been based only on few selected results.

In the new approach, the synoptic analysis makes use of the entire pool and selected sub-pools of available data rather than restricting to just a few selected results. This allows assessing consistency of data and dependencies of ORs on various parameters by joint evaluation. With this approach, it could already be shown that the reliability of reported ORs critically depends on the associated number of exposed cases [11] [12]. In the RF range, two quite different data pools with seeming controversial results could be identified. However, the synoptic analysis could clarify this discrepancy by showing that with increasing statistical power ORs of the contradictory data pools converge towards the same OR which, even more, turned out to be 0.8, hence, indicating decreased risk from mobile phone use [12]. In the ELF range, contradictory data pools could not be found. All reported data (except some few outliers) fitted well within one common pattern with ORs converging towards one single final value, in this case towards zero risk (OR = 1). This finding does not support the hypothesis of a causal relationship of childhood leukemia and ELF MF exposure [11].

However, this reassuring result can be challenged by the argument, that by pooling all reported ORs, an increased risk can have been masked by data from exposure scenarios with no risk, hence leaving the issue still open. Therefore, for further clarification, this paper concentrates on additional aspects, in particular on the worrying OR offset and the reported OR dose response. It will be shown that the synoptic approach allows generating convincing evidence and drawing consistent conclusions which can explain the reasons for the disturbing contradictions and controversial conclusions.

2. Method

For synoptic analysis, all available risk estimates published from 1979 until 2015 were pooled irrespective their statistical significance. Studies were identified by a literature search in the data bases PUBMED, FEMU, IEEE Xplore and MEDLINE with the keywords "childhood leukemia + magnetic fields + epidemiologic study". All identified studies were included without further selection such as due to their size or exposure metric. In case, ORs were given both unadjusted and adjusted for cofactors, the latter were used. However, for some analyses, ORs had to be accompanied by quantitative information on at least the related number of exposed cases or the quantitative exposure level. For quantitative analysis, the nominal scale of the "wire code" was converted into an ordinal scale by assigning numbers to author's verbally described classes, namely "very low" = 1, "ordinary low" = 2, "medium" = 3, "ordinary high" = 4 and "very high" = 5. In case OR values were given both unadjusted and adjusted for cofactors, the latter were included. The body of data was investigated with the Microsoft Excel software in particular:

1) at the overall data pool with regard to a dependence of ORs on the number of associated exposed cases N_{exp}

(as surrogate for statistical power) to check the acceptability of the null hypothesis (no MF-induced CL risk);

2) at sub-pools generated for different exposure parameters to check the consistency of results among different exposure metrics and the acceptability of the null hypothesis;

3) at sub-pools of different exposure parameters to investigate a potential dose response in terms of a the dependence of ORs on exposure levels;

4) at sub-pools of different exposure parameters related to high and low exposure to investigate the impact of the statistical power (N_{exp}) on a potential dose response.

3. Results

Overall, compared to the synoptic analysis of 2014 [11] some more studies could be included. The literature search resulted in 252 hits which were manually selected, ending up with 48 papers on childhood leukemia studies (excluding studies on devices and CL) [13]-[60]. More than 1500 reported OR values could be analyzed. It is important to note that inclusion of additional studies did not change the pattern of the overall data pool as previously presented. **Figure 1** shows the pattern of all ORs in dependence on the associated number of exposed cases (N_{exp}) including different exposure metrics and scenarios. It exhibits the already known funnel-shaped pattern with ORs scattered on either side of the zero-risk line (OR = 1). It can be seen that many ORs are based on few or even one case only. Consequently, it is no surprise that the variance of ORs is rather high at small N_{exp} and because of random influences extends to either side of the zero-risk line. It is obvious that ORs and conclusions based on only few exposed leukemia cases are not very reliable and hence exhibit a high variance. However, ORs exhibit an offset towards increased risk. Among other arguments this has been interpreted as indicator for increased childhood leukemia risk. However, note that this offset is mainly restricted to small N_{exp}. In addition, it is important to note that the offset decreases and finally disappears with increasing N_{exp}. As already shown in the previous analysis [11], the ORs decrease with increasing statistical power (N_{exp}). Moreover, ORs approach the zero-risk line from either side of the zero-risk line. Finally they converge towards zero-risk, and with some variance-remain there with further N_{exp} increase.

Interestingly, the same result can be found at sub-pools of different exposure parameters. In the data sub-pools of "wire code" (WC), "distance to MF source" (DMS) and "magnetic field immission" (MFI) with increasing N_{exp} ORs consistently show the same pattern, the similar decrease of OR offset, the similar decrease of

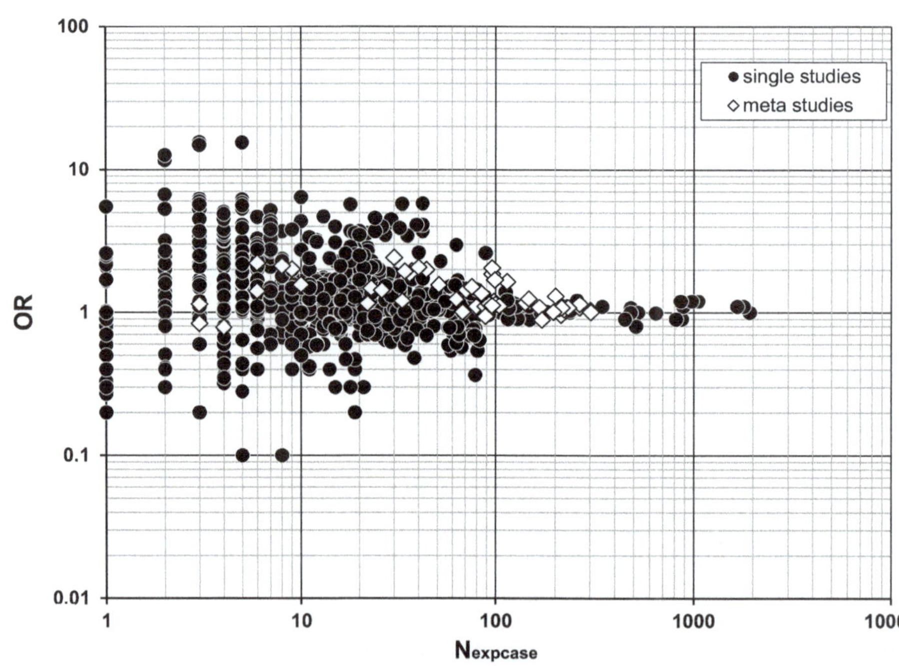

Figure 1. All reported odds ratios (OR) of 42 epidemiological studies on childhood leukemia and ELF magnetic field exposure in dependence on the associated number of exposed cases N_{exp}.

ORs and the similar convergence towards zero risk (**Figure 2**). This is not self-evident since this happens in spite of the quite different assessment strategies and different reliability of the exposure parameters.

The overall data pattern demonstrates that ORs critically depend on N_{exp} and decrease with increasing N_{exp}. However, this finding could be challenged by the argument, existing evidence for increased childhood leikemia risk at higher exposures could be masked by lacking risk at low exposure. To clarify this, a potential dependence of ORs on exposure level was investigated at different exposure parameters. The results at all exposure parameters are presented **Figure 3**. It shows that ORs increase with increasing exposure level even beyond 0.4 μT (note that MF immission increases with decreasing DMS). It is not self-evident that the increase is found at each of the exposure parameters. In the past, such a result had been interpreted as dose response and hence, as evidence for increased CL risk. This premature conclusion has become one of the main reasons for public concern.

However, such a premature conclusion ignores one important aspect, namely, that in **Figure 3** all ORs are given the same weight. However, this is not justified. **Figure 1** and **Figure 2** demonstrated already that the reliability of ORs significantly depends on the associated N_{exp}. Since highly exposed cases are rare compared to less exposed, it is no surprise that the data analysis showed that N_{exp} decreases with exposure (results not shown here, but can be verified by looking at **Figures 4-6**). This means that the statistical power and hence the reliability of ORs decreases with increasing exposure level. Therefore, the seeming dose-response shown in **Figure 3** needs to be interpreted with care. For further clarification, sub-pools were created for low and high exposure levels, respectively. The associated data distributions were analyzed in dependence of N_{exp} as a surrogate for statistical power.

Figure 4 shows the results for wire codes. The related data were separated into a sub-pool of low exposure (wiring configurations "very low" and "ordinary low") and a sub-pool of high exposure ("ordinary high" and "very high"), respectively. Wire coding stood at the beginning of epidemiologic investigations and later was substituted by quantitative immission assessment. Therefore, wire code data are less numerous and, hence, the sub-pool patterns less filled (note that there is still an overlap of similar data points, therefore not all available data points can be identified). It is important to note that in spite of fewer data, both sub-pools clearly exhibit the same generic properties as found in the overall data pools (see **Figure 1** and **Figure 2**), namely, ORs scattered around either side of the zero-risk line. In addition, both sub-pools show the tendency of ORs to converge towards zero risk, although less clear because of fewer data points. It can be seen that the range of N_{exp} associated with high exposure extends to much smaller numbers and ORs exhibit both larger variance and larger offset compared to the low-exposure sub-pool. That indicates a small-number effect probably due to selection bias. It needs to be noted, that there is no systematic difference between the patterns of high and low exposure which could indicate dose response in terms of a higher CL risk at high exposure. Consequently, with regard to WC data the seeming exposure-response shown in **Figure 3** is not supported by **Figure 4**.

Similar to WC data also from the data of the exposure surrogate "distance from MF source" two sub-pools were generated related to high ($d \leq 50$ m) and low exposure ($d > 100$ m), respectively. **Figure 5** shows that in spite of generic similarities both data sub-pools exhibit impressing differences. Again, it can be clearly seen that N_{exp} are much smaller at the high-exposure data with a majority of ORs based on less than 10 exposed cases while in the low-exposure sub-pool N_{exp} are about 10 fold higher. Consequently, the OR variance is much higher in the high-exposure sub-pool. In addition, there is again an obvious offset of ORs towards increased risk at small N_{exp}. However, with increasing N_{exp} it rapidly decreases, and ORs clearly converge towards zero risk with offset remaining. Interestingly, ORs of the low-exposure sub-pool do not have any offset from the zero-risk line: They are almost equally scattered around either side of the line OR = 1. Thus, DMS data are not indicating an increased leukaemia risk. The remaining OR variance at larger N_{exp} reflects methodological uncertainties in DMS assessment. It is much lower than at WC and MFI data. The patterns of the DMS sub-pools are in agreement with the overall pooled patterns (**Figure 1** and **Figure 2**) as well as with WC data (**Figure 4**). The differences between low- and high-exposure sub-pools stem from the different ranges of N_{exp}. Overall, there is no indication of a dose response and an increased childhood leukaemia risk from DMS data.

From the data related to quantitative MFI assessment (which includes measurements as well as numerical field calculation with quite different assessment strategies) sub-pools were generated for high (>200 μT) and low immission (≤100 μT), respectively. Results are presented in **Figure 6**. In the high-exposure sub-pool N_{exp} is much smaller with many ORs associated with one or two exposed cases only. The variance of ORs is large and decreases with increasing N_{exp}. The OR offset decreases with N_{exp} and disappears rapidly with increasing N_{exp}.

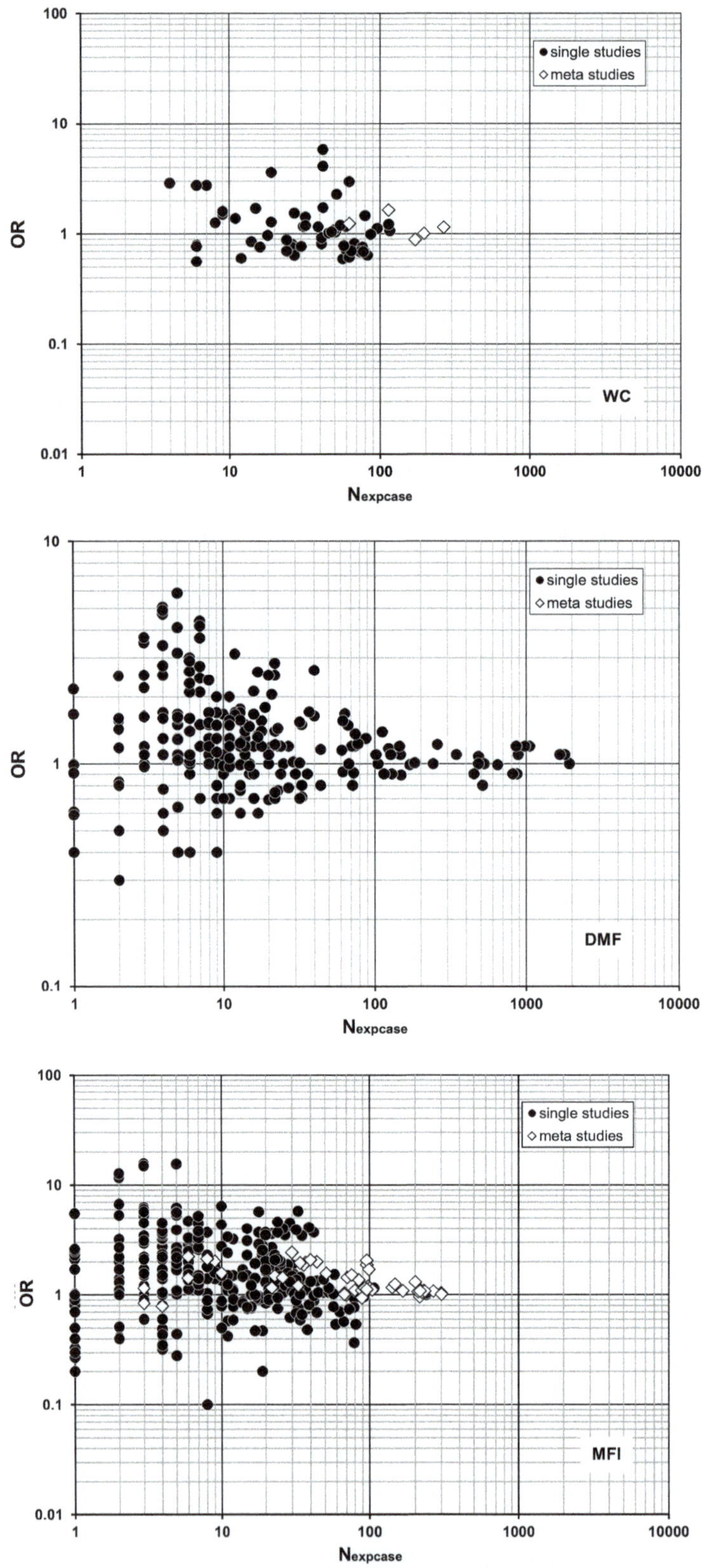

Figure 2. Reported odds ratios (OR) of epidemiological studies on childhood leukemia and ELF magnetic field exposure over the associated number of exposed cases N_{exp} for different exposure metrics; wiring code (WC, in ordinal scale, above), distance to MF source (DMF, middle), magnetic field immission (MFI, below).

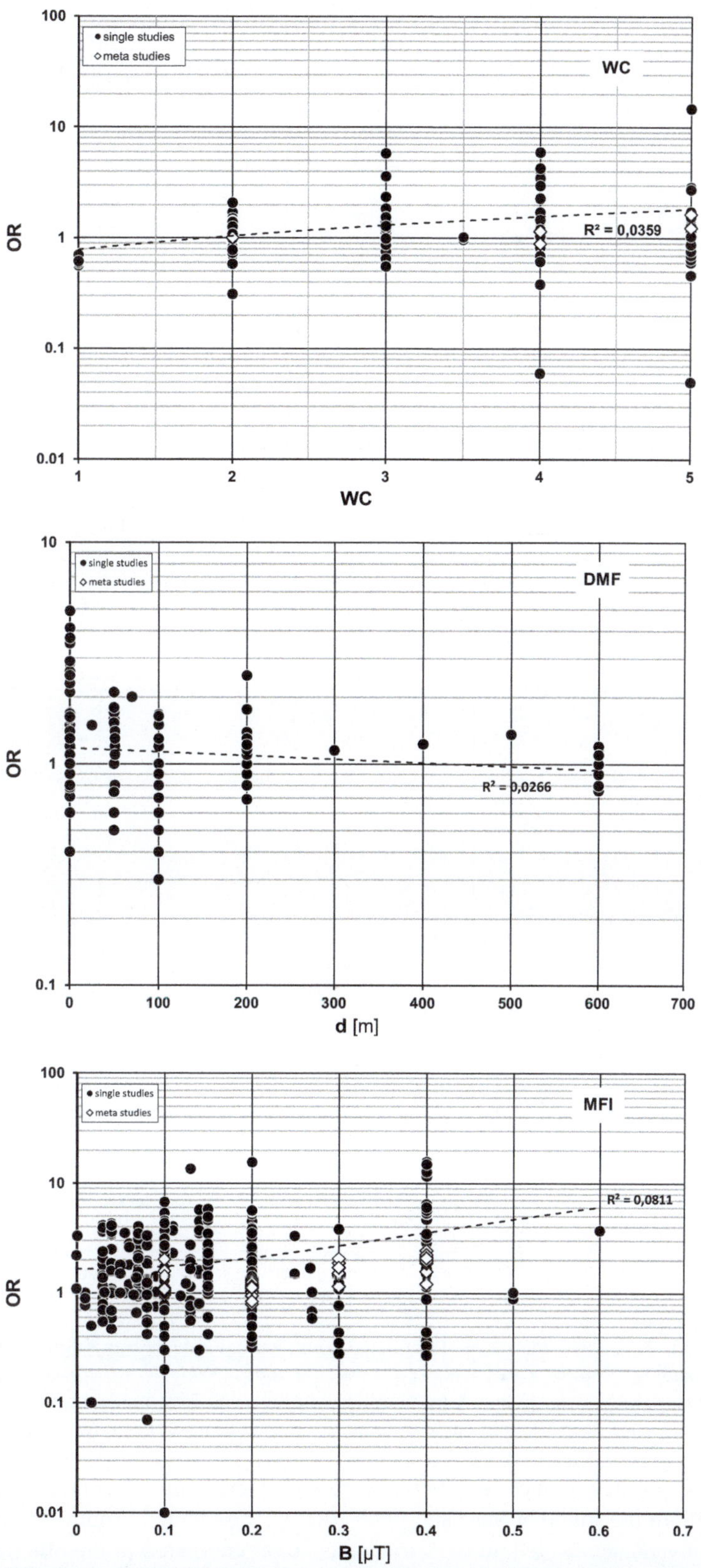

Figure 3. Childhood leukemia risk estimates (OR) over ELF magnetic field exposure level characterized by wiring code WC (in ordinal scale, above), distance to MF source d (DMF, middle), magnetic field immission B (MFI, below).

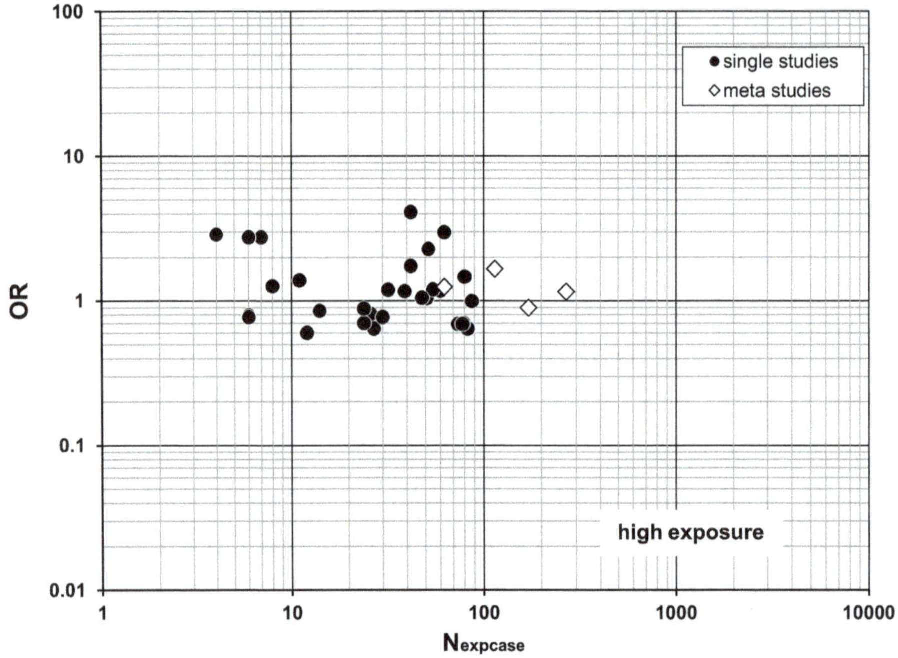

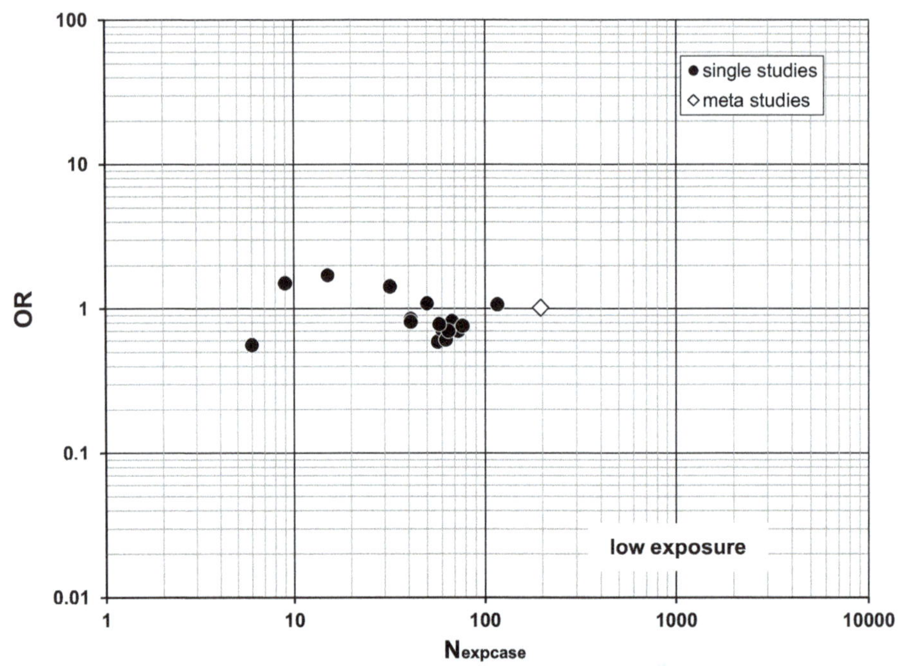

Figure 4. Reported odds ratios (OR) of epidemiological studies on childhood leukemia and ELF magnetic field exposure assessed by wiring codes (WC) over the associated number of exposed cases N_{exp} for data sub-pools related to high and low exposure, namely WC "ordinary high" and "high" (above) and "ordinary low" and "low"(below), respectively.

Similar to the other exposure parameters ORs decrease with increasing N_{exp} and converge towards the zero-risk line from either side. In the low-exposure sub-pool ORs are distributed around the zero risk line with neither offset nor further convergence. Note that the variance is larger compared to the other exposure parameters. This can be explained by the many different ways magnetic field values were derived. Overall, all generic features of MFI sub-pool patterns are in agreement with those of the other exposure parameters. There is no indication of a dose response and an increased CL risk with increasing MFI level.

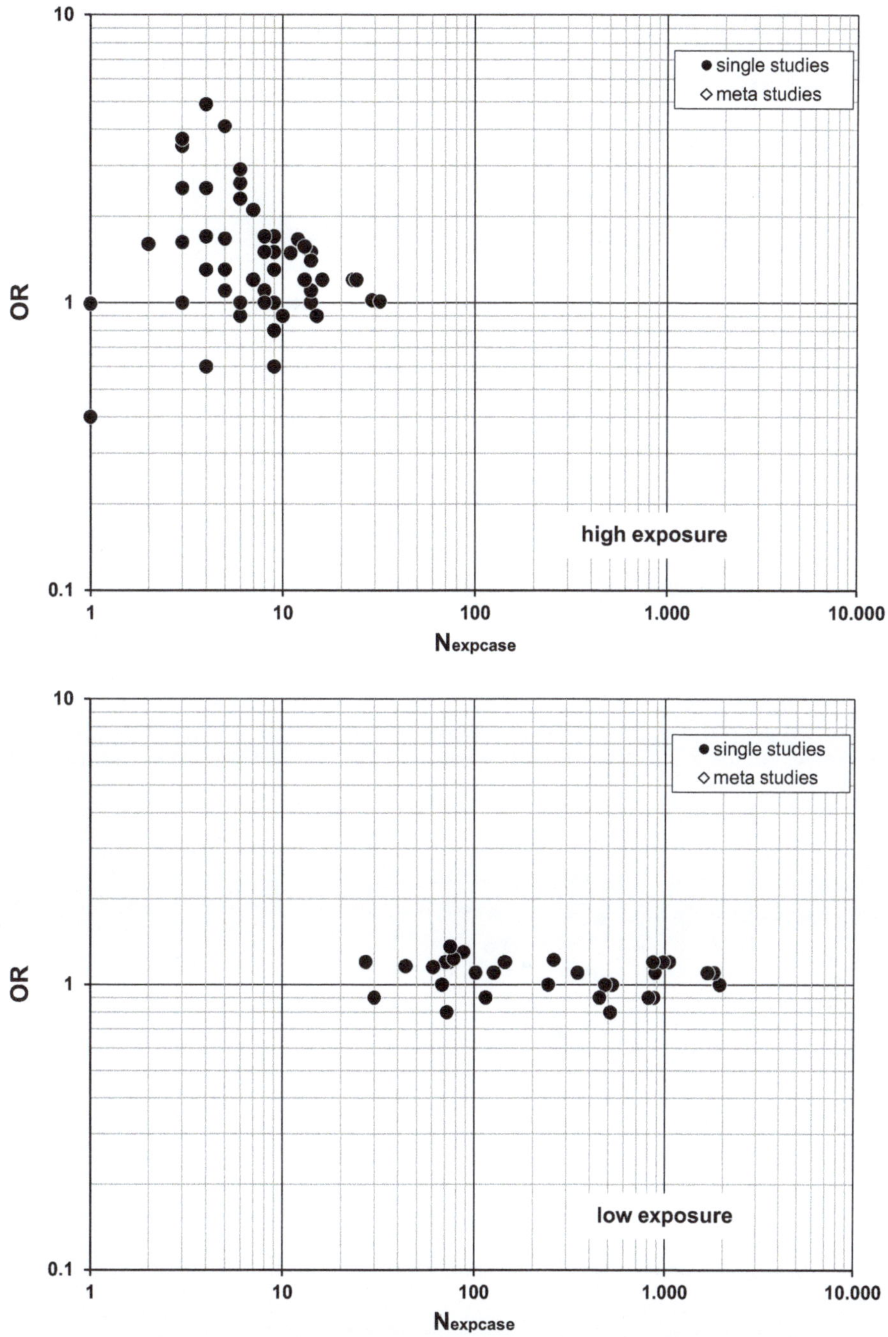

Figure 5. Reported odds ratios (OR) of epidemiological studies on childhood leukemia and ELF magnetic field exposure level assessed by "distance to magnetic field source" *d* over the associated number of exposed cases N_{exp} for data sub-pools related to high and low exposure, namely $d \leq 50$ m (above) and $d > 100$ m (below), respectively.

4. Discussion

The results show that the synoptic analysis provides additional information by including all available ORs rather than focusing at only one or few ORs selected by whatever consideration. It was decided to accept all available studies instead of rejecting studies based on exclusion criteria. Rejection would make it necessary to justify the dismissal (or acceptance) of studies and may be challenged by one or another stakeholder, by "believers" or "non-believers" of a MF induced health risks. In addition, selection may be prone to bias. This is demonstrated

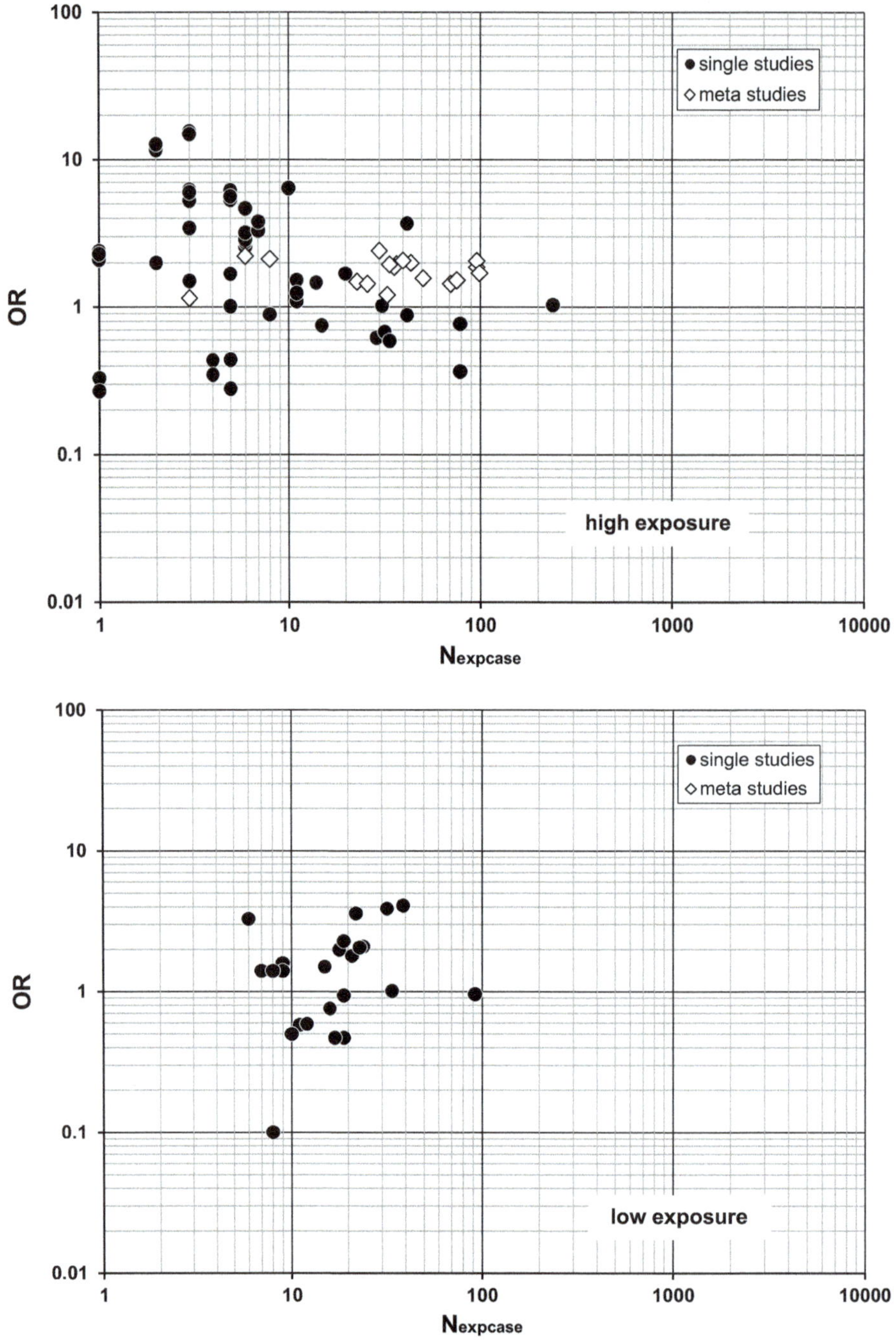

Figure 6. Reported odds ratios (OR) of epidemiological studies on childhood leukemia and power frequency magnetic field exposure assessed by the estimated magnetic field levels (MFI) over the associated number of exposed cases N_{exp} for data sub-pools related to high and low exposure, namely MFI > 200 nT (above) and MFI ≤ 100 nT (below), respectively.

by the fact that different groups such as ICNIRP and WHO versus Bioinitiative give quite different weight to the same studies. In contrast, not selecting does not require to justify selection criteria. However, this could also be challenged, this time by the argument, included weak studies could degrade the result. However, the advantage of the synoptic approach is to easily allow identifying whether studies fit into the overall body of evidence or suffer from methodical insufficiencies as those identified at mobile phone studies [12]. Likewise, this applies to outliers who are easily identified by their location aside the data pattern (see **Figure 1** and **Figure 2**). It is interesting to note that the pattern of data from meta-analyses differs strikingly from the general pattern. However, detailed discussion would go beyond the scope of this paper. It can be found elsewhere [61].

It needs to be noted that not all reported data could be included into the synoptic analysis namely those cases where needed OR-associated information such as on N_{exp} or exposure level were lacking. Therefore, even more data points could have been considered if published and if more authors would have refrained from restricting reporting on just few selected ORs only. However, the available body of data is already impressively large and stable. The inclusion of some more studies since the first synoptic analysis [11] did not change neither the data pattern nor the derived conclusions (note that due to overlapping ORs the added data points are hardly visible). Consequently, there is little reason to assume that further studies could lead to relevant changes. This is why synoptic analysis of the already available data allows drawing convincing conclusions in particular since some sparse data in sub-pools such as of the wire code are compensated by consistent support from other parameters.

The synoptic analysis allowed clarifying worrying issues and puzzling results. It is important to note that now the reported OR offset towards increased risk can be explained. The offset disappears if N_{exp} becomes large enough. This finding is supported by the fact that the same is seen at all exposure parameters irrespective their considerable methodical differences. This provides convincing evidence of a low-number artifact. It is obvious that a case-selection bias in terms of an unjustified selection of an exposed case has a larger impact if only one or few cases are used for OR calculation (note that a similar offset could also be seen at data of other childhood cancer [11]). The existence of a case-selection bias has already been discussed by others [62]-[65]. Therefore, there are convincing indication that the offset towards increased risk is an artifact caused by low-number effects which amplifies existing selection bias. It can no longer be taken as an indicator for an increased risk.

Another worrying issue could also be explained, namely the reported "dose-response". So far, it was puzzling to assume that ORs increased just within limited MF levels. On the one hand studies reported on a proportional OR increase with magnetic immission between 0.2 μT and 0.4 μT. However, on the other hand there was no idea how such a finding—if causal—could be extrapolated to higher exposures, in particular over three orders of magnitudes to the existing exposure limits. It is one merit of the synoptic analysis to be able to convincingly show that the conclusion on a "dose-response" is a consequence of the low-number aitifact. It can be explained by the coincidence of decreasing N_{exp} with increasing exposure which, as a consequence, causes a higher variance, and an increased offset of ORs due to the amplified selection bias at smaller N_{exp}. It is evident that such a bias does not endlessly increase. It is no surprise that it levels out. This explains why at higher MF levels a further increased leukemia risk could not be found. This is in agreement with the fact that the OR offset diminishes at higher N_{exp}. It is convincing that the same generic behavior was seen at quite different exposure parameters, namely, the decrease of OR offset with increasing number of cases and decreasing OR values with increasing N_{exp}. In addition, the reported "dose response" could not be confirmed at sub-pools of different exposure level and exposure metric. In contradiction to the dose-response hypothesis, in the exposure-related sub-pools ORs did not converge towards different endpoints. On contrary, irrespective the exposure metric ORs of high and low exposure sub- pools converged towards the same endpoint, which, in addition, consistently indicated zero risk (OR = 1). This finding does not support the hypothesis of a causal relationship between CL and ELF MF.

The synoptic analysis demonstrated that there is a remaining variance in ORs even if N_{exp} is large and the result stable. Moreover, the residual variance was different at different exposure parameters. This reflects methodical uncertainties and is in agreement with the fact that different exposure parameters are based on quite different approaches and metrics. Therefore, they suffer from different causes and amounts of uncertainty. It is interesting to note that the residual variance of ORs is highest at MFI which had been expected to perform better than the surrogates WC and DMS. The reason for this are the highly different assessment strategies associated with MFI levels such as measurement or numerical calculation at different spots within or outside homes, at different points in time (at diagnosis, before diagnosis, at study), instantaneous values or averaged values derived from different time intervals (daytime, nighttime, year) determined in different ways such as spot values, time averages, time arithmetic medians, time geometric means, time-weighted averages with different thresholds etc. The unavoidable residual variance is the reason why conclusions from single selected ORs even if based on high N_{exp} still suffer from uncertainty. This demonstrates the need of a synoptic approach.

It is one more argument against a causal relationship of childhood leukemia and magnetic field exposure that over decades of various attempts it was not possible to identify and agree upon a single reliable exposure metric.

The convincing evidence that the reported increase of ORs with MF levels does not indicate a dose-response nor a causal relationship now solves the dilemma of health risk assessment bodies and authorities, in particular of WHO who, on the one hand, concluded on a possible childhood leukemia risk from ELF MF exposure already at low environmental MF levels, and on the other hand stayed with exposure limits several orders of magnitude above reported risk onset levels.

The synoptic results allow terminating the long-term debate about a potential childhood leukemia risk from ELF magnetic field exposure with a reassuring message. The synoptic analysis provided convincing evidence that a causal relationship simply does not exist and risk of childhood leukemia is not increased by exposure to ELF magnetic fields. This is in agreement with the fact that in spite of intensive attempts the CL risk hypothesis could not be convincingly supported by an interaction mechanism, nor by in vitro or in vivo results or by explanations of the puzzling situation, that similar health risks could not be found in children nor at other cancer endpoints nor in even much higher exposed adults. The convincing evidence of lacking causality between childhood leukemia and exposure to ELF MF demonstrates that IARC's classification needs revision, the sooner the better.

However, the synoptic analysis of the entire body of available data from epidemiologic studies also leads to an important overarching conclusion, namely, that statistical power e.g. in terms of the number of exposed cases used for calculating ORs must not be neglected. Consequently, in epidemiology (not only on EMF) reported ORs need to be accompanied by associated N_{exp}. So far, this is already done by many authors but has not yet become general routine.

Another overarching consequence applies to reviews (and reviewers) which compare Ors from several studies. It is no longer justified to give all ORs the same weight. It is also inappropriate to list ORs in a manner which is not at all causally related with ORs such as in alphabetical order (e.g. of the first authors) instead of presenting ORs in dependence of a relevant parameter such as N_{exp}. Therefore, reviews that did not yet weigh ORs with the associated N_{exp} need to be interpreted with care.

5. Conclusion

The synoptic analysis is proofed to be an important and powerful tool. It avoids ignoring valuable information and is demonstrating the importance of weighing risk estimates (ORs) with their associated statistical power (N_{exp}). It allows clarifying the worrying offset of ORs towards increased risk and the seeming dose-response as an artifact caused by the small-number effect. It demonstrates that the indication of increased CL risk decreases with increasing statistical power until it diminished. This is consistently shown at the overall pooled data as well as at all investigated sub-pools of different exposure parameters. The results of the synoptic analysis allow terminating the long-term controversy about a potential childhood leukemia risk from ELF magnetic field exposure with a reassuring message. The synoptic analysis provides convincing evidence that a causal relationship does not exist and risk of childhood leukemia is not increased by exposure to magnetic fields. Consequently, IARC's classification of ELF MF as class 2B carcinogen needs revision.

Conflict of Interest

The author declares no conflict of interest.

References

[1] Wertheimer, N. and Leeper, N. (1979) Electric Wiring Configurations and Childhood Cancer. *American Journal of Epidemiology*, **109**, 273-284.

[2] IARC (2002) IARC Monographs on the Evaluation of Carcinogenic Risks to Humans. Vol. 80, Non-Ionizing Radiation, Part 1: Static and Extremely Low-Frequency (ELF) Electric and Magnetic Fields, IARC, Lyon.

[3] Grellier, J., Ravazzani, P. and Cardis, E. (2014) Potential Health Impacts of Residential Exposure to Extremely Low Frequency Magnetic Fields in Europe. *Environment International*, **62**, 55-63. http://dx.doi.org/10.1016/j.envint.2013.09.017

[4] WHO (2008) Environmental Health Criteria 238. Extremely Low Frequency Fields. WHO, Geneva.

[5] Ahlbom, A., Day, N., Feychting, M., Roman, E., Skinner, J., Dockerty, J., Linet, M., McBride, M., Michaelis, J., Olsen, J.H., Tynes, T. and Verkasalo, P.K. (2000) A Pooled Analysis of Magnetic Fields and Childhood Leukemia. *British Journal of Cancer*, **83**, 692-698. http://dx.doi.org/10.1054/bjoc.2000.1376

[6] Greenland, S., Sheppard, A.R., Kaune, W.T., Poole, C. and Kelsh, M.A. (2000) A Pooled Analysis of Magnetic Fields, Wire Codes, and Childhood Leukemia. *Epidemiology*, **11**, 624-634. http://dx.doi.org/10.1097/00001648-200011000-00003

[7] Kheifets, L., Ahlbom, A., Draper, G., Haghira, J., Lowenthal, R.M., Mezei, G., Oksuzyan, S., Schüz, J., Swanson, J., Tittarelli, A., Vincenti, M. and Wünsch-Filho, V. (2010) Pooled Analysis of Recent Studies on Magnetic Fields and Childhood Leukemia. *British Journal of Cancer*, **103**, 1128-1135. http://dx.doi.org/10.1038/sj.bjc.6605838

[8] Schüz, J. and Ahlbom, A. (2008) Exposure to Electromagnetic Fields and the Risk of Childhood Leukemia: A Review. *Radiation Protection Dosimetry*, **132**, 202-211. http://dx.doi.org/10.1093/rpd/ncn270

[9] ICNIRP (2010) Guidelines for Limiting Exposure to Time-Varying Electric and Magnetic Fields (1 Hz to 100 kHz). *Health Physics*, **99**, 818-836.

[10] EU (2013) Directive on the Minimum Health and Safety Requirements Regarding Exposure of Workers to the Risks Arising from Physical Agents (Electromagnetic Fields). *Official Journal of the European Union*, **L179**, 1-18.

[11] Leitgeb, N. (2014) Childhood Leukemia Not Linked with ELF Magnetic Fields. *Journal of Electric and Magnetic Analysis and Applications*, **6**, 174-183. http://dx.doi.org/10.4236/jemaa.2014.67017

[12] Leitgeb, N. (2014) Synoptic Analysis of Epidemiologic Evidence of Brain Cancer Risks from Mobile Communication. *Journal of Electric and Magnetic Analysis and Applications*, **6**, 413-424. http://dx.doi.org/10.4236/jemaa.2014.614043

[13] Salvan, A., Ranucci, A., Lagorio, S. and Magnani, C. (2015) Childhood Leukemia and 50Hz Magnetic Fields: Findings from the Italian SETIL Case-Control Study. *International Journal on Environmental Research and Public Health*, **12**, 2184-2204. http://dx.doi.org/10.3390/ijerph120202184

[14] Zhao, L.Y., Liu, X.D., Wang, C.P., Yan, K.K., Lin, X.J., Li, S., Bao, H.H. and Liu, X. (2014) Magnetic Fields Exposure and Childhood Leukemia Risk: A Meta-Analysis Based on 11,699 Cases and 13,194 Controls. *Leukemia Research*, **38**, 269-274. http://dx.doi.org/10.1016/j.leukres.2013.12.008

[15] Bunch, K.J., Keegan, T.J., Swanson, J., Vincent, T.J. and Murphy, M.F.G. (2014) Residential Distance at Birth from Overhead High-Voltage Power Lines: Childhood Cancer Risk in Britain 1962-2008. *British Journal of Cancer*, **110**, 1402-1408. http://dx.doi.org/10.1038/bjc.2014.15

[16] Sermage-Faure, C., Demoury, C., Rudant, J., Goujon-Bellec, S., Guvot-Goubin, A., Deschamps, F., Hemon, P. and Clavel, J. (2013) Childhood Leukemia Close to High-Voltage Power Lines—The Geocap Study, 2002-2007. *British Journal of Cancer*, **109**, 1899-1906. http://dx.doi.org/10.1038/bjc.2013.128

[17] Jirik, V., Pekarek, L., Janout, V. and Tomaskova, H. (2012) Association between Childhood Leukemia and Exposure to Power Frequency Magnetic Fields in Middle Europe. *Biomedical and Environmental Sciences*, **25**, 597-601.

[18] Wünsch-Filho, V., Pelissari, D.M., Barbieri, F.E., Sant'Anna, L., Oliviera, C.T., de Mata, J.F., Tone, L.G., Lee, M.L.D.M., de Andréa, M.L.M., Bruniera, P., Epelman, S., Filho, V.O. and Kheifets, L. (2011) Exposure to Magnetic Fields and Childhood Acute Lymphocytic Leukemia in Sao Paolo, Brazil. *Cancer Epidemiology*, **35**, 534-539. http://dx.doi.org/10.1016/j.canep.2011.05.008

[19] Malagoli, C., Fabbi, S., Teggi, S., Calzari, M., Poli, M., Ballotti, E., Notari, B., Bruni, M., Palazzi, G., Paolucci, G. and Vincenti, M. (2010) Risk of Hematological Malignancies Associated with Magnetic Fields Exposure from Power Lines: A Case-Control Study in Two Municipalities of Northern Italy. *Environmental Health*, **9**, 16. http://dx.doi.org/10.1186/1476-069X-9-16

[20] Kroll, M.E., Swanson, J., Vincent, T.J. and Draper, G.J. (2010) Childhood Cancer and Magnetic Fields from High-Voltage Power Lines in England and Wales: A Case-Control Study. *British Journal of Cancer*, **103**, 1122-1127. http://dx.doi.org/10.1038/sj.bjc.6605795

[21] Sohrabi, M.-R., Tarjoman, T., Abadi, A. and Yavari, P. (2010) Living Near Overhead High-Voltage Transmission Power Lines as a Risk Factor for Childhood Acute Lymphoblastic Leukemia: A Case-Control Study. *Asian Pacific Journal of Cancer Prevention*, **11**, 423-427.

[22] Maslanyj, M., Simpson, J., Roman, E. and Schüz, J. (2009) Power Frequency Magnetic Fields and Risk of Childhood Leukemia: Misclassification of Exposure from the Use of the "Distance from Power Line" Exposure Surrogate. *Bioelectromagnetics*, **30**, 183-188. http://dx.doi.org/10.1002/bem.20465

[23] Rahman, H.I.A., Shah, S.A., Alias, H. and Ibrahim, H.M. (2008) A Case-Control Study on the Association between Environmental Factors and the Occurrence of Acute Leukemia among Children in Klang Valley, Malaysia. *Asian Pacific Journal of Cancer Prevention*, **9**, 649-652.

[24] Schüz, J., Svendsen, A.L., LInet, M.S., McBride, M.L., Roman, E., Feychting, M., Kheifets, L., Lightfood, T., Mezel, G., Simpson, J. and Ahlbom, A. (2007) Nighttime Exposure to Electromagnetic Fields and Childhood Leukemia: An Extended Pooled Analysis. *American Journal of Epidemiology*, **166**, 263-269. http://dx.doi.org/10.1093/aje/kwm080

[25] Kabuto, M., Nitta, H., Yamamoto, S., Yamaguchi, N., Akiba, S., Honda, Y., Hagihara, J., Isaka, K., Saito, T., Ojima, T., Nalamura, Y., Mizoue, T., Iti, S., Eboshida, A., Yamazaki, S., Sokejima, S., Kurokawa, Y. and Kubo, O. (2006) Childhood Leukemia and Magnetic Fields in Japan: A Case-Control Study of Childhood Leukemia and Residential Power-Frequency Magnetic Fields in Japan. *International Journal of Cancer*, **119**, 643-650. http://dx.doi.org/10.1002/ijc.21374

[26] Draper, G., Vincent, T., Kroll, M.E. and Swanson, J. (2005) Childhood Cancer in Relation to Distance from High Voltage Power Lines in England and Wales: A Case-Control Study. *British Medical Journal*, **330**, 1290-1293. http://dx.doi.org/10.1136/bmj.330.7503.1290

[27] Mizoue, T., Onoe, Y., Moritake, H., Okamura, J., Sokejima, S. and Nitta, H. (2004) Residential Proximity to High-Voltage Power Lines and Risk of Childhood Hematological Malignancies. *Journal of Epidemiology*, **14**, 118-123. http://dx.doi.org/10.2188/jea.14.118

[28] Söderberg, K.C., Naumburg, E., Ander, G., Cnattingius, S., Ekborn, A. and Feychting, M. (2002) Childhood Leukemia and Magnetic Fields in Infant Incubators. *Epidemiology*, **13**, 45-49. http://dx.doi.org/10.1097/00001648-200201000-00008

[29] Schüz, J., Grigat, J.-P., Brinkmann, K. and Michaelis, J. (2001) Residential Magnetic Fields as a Risk Factor for Childhood Acute Leukemia: Results from a German Population-Based Case-Control Study. *International Journal of Cancer*, **91**, 728-735. http://dx.doi.org/10.1002/1097-0215(200002)9999:9999<::AID-IJC1097>3.0.CO;2-D

[30] Bianchi, N., Crosignani, P., Rovelli, A., Titterelli, A., Carnelli, C.A., Rossitto, F., Vanelli, U., Porro, E. and Berrino, F. (2000) Overhead Electricity Power Lines and Childhood Leukemia: A Registry-Based Case-Control Study. *Tumori*, **86**, 195-198.

[31] Skinner, J., Maslanyi, M.P., Mee, T.J., Allen, S.G., Simpson, J., Roman, E. and Day, N.E. (2000) Childhood Cancer and Residential Proximity to Power Lines. *British Journal of Cancer*, **83**, 1573-1580. http://dx.doi.org/10.1054/bjoc.2000.1550

[32] Auvinen, A., Line, M.S., Hatch, E.E., Kleinerman, R.A., Robinson, L.L., Kaune, W.T., Misakian, M., Niwa, S., Wacholder, S. and Tarone, R.E. (2000) Extremely-Low Frequency Magnetic Fields and Childhood Acute Lymphoblastic Leukemia: An Exploratory Analysis of Alternative Exposure Metrics. *American Journal of Epidemiology*, **152**, 20-31. http://dx.doi.org/10.1093/aje/152.1.20

[33] Kleinerman, R.A., Kaune, W.T., Hatch, E., Wacholder, S., Linet, M.S., Robinson, L., Niwa, S. and Tarone, R. (2000) Are Children Living near High Voltage Power Lines at Increased Risk of Acute Lymphoblastic Leukemia? *American Journal of Epidemiology*, **151**, 512-515. http://dx.doi.org/10.1093/oxfordjournals.aje.a010237

[34] McBride, M.L., Gallagher, R.P., Thériault, G., Armstrong, B.G., Tamaro, S., Spinellil, J.J., Deadman, J.E., Finchman, S., Robson, D. and Choi, W. (1999) Power-Frequency Electric and Magnetic Fields and Risk of Childhood Leukemia in Children. *American Journal of Epidemiology*, **149**, 831-842. http://dx.doi.org/10.1093/oxfordjournals.aje.a009899

[35] Green, L.M., Miller, A.B., Villeneuve, P.J., Agnew, D.A., Greenberg, M.L., Li, J. and Donelly, K.E. (1999) A Case-Control Study of Childhood Leukemia in Southern Ontario, Canada, and Exposure to Magnetic Fields in Residences. *International Journal of Cancer*, **82**, 161-170. http://dx.doi.org/10.1002/(SICI)1097-0215(19990719)82:2<161::AID-IJC2>3.0.CO;2-X

[36] Cheng, K.K., Day, N., Cartwright, R., Craft, A., Birch, J.M., Eden, O.B., McKinney, P.A., Petro, J., Beral, V., Roman, E., Elwood, P., Alexander, F.E., Chilvers, C.E.D., Doll, R., Taylor, C.M., Greaves, M., Goodhead, D., Fry, F.A. and Adams, G. (1999) Exposure to Power-Frequency Magnetic Fields and the Risk of Childhood Cancer. *The Lancet*, **354**, 1925-1931. http://dx.doi.org/10.1016/S0140-6736(99)10074-6

[37] Green, L.M., Miller, A.B., Agnew, D.A., Greenberg, M.L., Li, J., Villeneuve, P.J. and Tibshirian, R. (1999) Childhood Leukemia and Personal Monitoring of Residential Exposures to Electric and Magnetic Fields in Ontario. *Cancer Causes and Control*, **10**, 233-243. http://dx.doi.org/10.1023/A:1008919408855

[38] Dockerty, J.D., Elwood, J.M., Skegg, D.C.G. and Herbison, G.P. (1998) Electromagnetic Field Exposures and Childhood Cancers in New Zealand. *Cancer Causes and Control*, **9**, 299-309. http://dx.doi.org/10.1023/A:1008825220759

[39] Li, C.-Y., Lee, W.-C. and Lin, R.S. (1998) Risk of Leukemia in Children Living near High-Voltage Transmission Lines. *Journal of Occupational and Environmental Medicine*, **40**, 144-147. http://dx.doi.org/10.1097/00043764-199802000-00010

[40] Linet, M.S., Hatch, E., Kleinerman, R.A., Robinson, L.L., Kaune, W.T., Friedman, D.R., Secverson, R.K., Haines, C.M., Hartsock, C.T., Niwa, S., Wacholder, S. and Tarone, R. (1997) Residential Exposure to Magnetic Fields and Acute Lymphoblastic Leukemia in Children. *New England Journal of Medicine*, **337**, 1-8. http://dx.doi.org/10.1056/NEJM199707033370101

[41] Michaelis, J., Schüz, J., Meinert, R., Menger, M., Grigart, J.-P., Kaatsch, P., Kaletsch, U., Miesner, A., Stamm, A., Brinkmann, K. and Kärner, H. (1997) Childhood Leukemia and Electromagnetic Fields: Results of a Population-Based Case-Control Study in Germany. *Cancer Causes and Control*, **8**, 167-174. http://dx.doi.org/10.1023/A:1018464012055

[42] Petridou, E., Trichopoulos, D., Kravaritis, A., Pourtsidis, A., Dessypris, N., Skalkidis, Y., Kogevinas, M., Kalmanti, V., Koliouskas, H., Panagiotou, J.P., Piperopoulou, F., Tzortzatou, F. and Kalapothaki, V. (1997) Electric Power Lines and Childhood Leukemia: A Study from Greece. *International Journal of Cancer*, **73**, 345-348. http://dx.doi.org/10.1002/(SICI)1097-0215(19971104)73:3<345::AID-IJC7>3.0.CO;2-#

[43] Tynes, T. and Haldorsen, T. (1997) Electromagnetic Fields and Cancer in Children Residing near Norwegian High-

Voltage Power Lines. *American Journal of Epidemiology*, **145**, 219-226. http://dx.doi.org/10.1093/oxfordjournals.aje.a009094

[44] Coghill, R.W., Steward, J. and Philips, A. (1996) Extra Low Frequency Electric and Magnetic Fields in the Bedplace of Children Diagnosed with Leukemia: A Case-Control Study. *European Journal of Cancer Prevention*, **5**, 153-158. http://dx.doi.org/10.1097/00008469-199606000-00002

[45] Lin, R.S., Lee, W.-C. and Li, C.-Y. (1996) Risk of Childhood Leukemia in Households near High Power Lines. *Medical and Biological Engineering and Computing*, **34**, 131-132.

[46] Preston-Martin, S., Navidi, W., Thomas, D., Lee, P.-J., Bowman, J. and Pododa, J. (1996) Los Angeles Study of Residential Magnetic Fields and Childhood Brain Tumors. *American Journal of Epidemiology*, **143**, 105-119. http://dx.doi.org/10.1093/oxfordjournals.aje.a008717

[47] Wertheimer, N., Savitz, D.A. and Leeper, E. (1995) Childhood Cancer in Relation to Indicators of Magnetic Fields from Ground Current Sources. *Bioelectromagnetics*, **16**, 86-96. http://dx.doi.org/10.1002/bem.2250160204

[48] Lin, R.S. and Lee, W.C. (1994) Risk of Childhood Leukemia in Areas Passed by High Power Lines. *Review on Environmenal Health*, **10**, 97-103. http://dx.doi.org/10.1515/reveh.1994.10.2.97

[49] Fajardo-Guitérrez, A., Garduno-Espinosa, J., Yamamoto-Kimura, L., Hernández-Hernández, D.M., Gómez-Delgado, A. and Martínez-Garcia, M.D.C. (1993) Residencia cercana a fuentes electrical de alta tension y su asociación con Leukemia en ninos. *Boletín médico del Hospital Infantil de México*, **50**, 32-38.

[50] Feychting, M. and Ahlbom, A. (1993) Magnetic Fields and Cancer in Children Living near Swedish High-Voltage Power Lines. *American Journal of Epidemiology*, **138**, 467-481.

[51] Olsen, J.H., Nielsen, A. and Schulgen, G. (1993) Residence near High Voltage Facilities and Risk of Cancer in Children. *British Journal of Cancer*, **307**, 891-895. http://dx.doi.org/10.1136/bmj.307.6909.891

[52] Verkasalo, P.K., Pukkala, E., Hongisto, M.Y., Valjus, J.E., Järvinen, P.J., Heikkilä, K.V. and Koskenvou, M. (1993) Risk of Cancer in Finnish Children Living Close to Power Lines. *British Medical Journal*, **307**, 895-899. http://dx.doi.org/10.1136/bmj.307.6909.895

[53] Feychting, M. and Ahlbom, A. (1992) Magnetic Fields and Cancer in People Residing near Swedish High Voltage Power Lines. Institutet Miljömedicin Rapport 6/92, Karolinska Institutet, Stockholm.

[54] London, S.J., Thomas, D.C., Bowman, J.D., Soberl, E., Cheng, T.-C. and Peters, J.M. (1991) Exposure to Residential Electric and Magnetic Fields and Risk of Childhood Leukemia. *American Journal of Epidemiology*, **134**, 923-937.

[55] Myers, A., Clayden, A.D., Cartwright, R.A. and Cartwright, S.C. (1990) Childhood Cancer and Overhead Powerlines: A Case-Control Study. *British Journal of Cancer*, **62**, 1008-1014. http://dx.doi.org/10.1038/bjc.1990.428

[56] Savitz, D.A., John, E.M. and Kleckner, R.C. (1990) Magnetic Field Exposure from Electric Appliances and Childhood Cancer. *American Journal of Epidemiology*, **131**, 763-773.

[57] Coleman, M.P., Bell, C.M.J., Taylor, H.L. and Primic-Zakeli, M. (1989) Leukemia and Residence near Electric Transmission Equipment: A Case-Control Study. *British Journal of Cancer*, **60**, 793-798. http://dx.doi.org/10.1038/bjc.1989.362

[58] Savitz, D.A., Wachtel, H., Barnes, F.A., John, E.M. and Tvrdik, J. (1988) Case-Control Study of Childhood Cancer and Exposure to 60-Hz Magnetic Fields. *American Journal of Epidemiology*, **128**, 21-38.

[59] Tomenius, L. (1986) 50-Hz Electromagnetic Environment and the Incidence of Childhood Tumors in Stockholm County. *Bioelectromagnetics*, **7**, 191-207. http://dx.doi.org/10.1002/bem.2250070209

[60] Fulton, J.P., Cobb, S., Preble, L., Leone, L. and Forman, E. (1980) Electrical Wiring Configurations and Childhood Leukemia in Rhode Island. *American Journal of Epidemiology*, **111**, 292-296.

[61] Elwood, J.M. (2006) Childhood Leukaemia and Residential Magnetic Fields: Are Pooled Analyses More Valid than the Original Studies? *Bioelectromagnetics*, **27**, 112-118. http://dx.doi.org/10.1002/bem.20181

[62] Slusky, D.A., Does, M., Metayer, C., Mezei, G., Selvin, S. and Buffler, P.A. (2014) Potential Role of Selection Bias in the Association between Childhood Leukemia and Residential Magnetic Field Exposure. A Population-Based Assessment. *Cancer Epidemiology*, **38**, 307-313. http://dx.doi.org/10.1016/j.canep.2014.02.010

[63] Maslanyi, M., Slimpson, J. and Schüz, J. (2009) Power Frequency Magnetic Fields and Risk of Childhood Leukemia: Misclassification of Exposure from the Use of "Distance from Power Line" Exposure Surrogate. *Bioelectromagnetics*, **30**, 183-188. http://dx.doi.org/10.1002/bem.20465

[64] Mezei, G. and Kheifets, L. (2006) Selection Bias and Its Implications for Case-Control Studies: A Case Study of Magnetic Field Exposure and Childhood Leukemia. *International Journal of Epidemiology*, **15**, 397-406.

[65] Greenland, S., Schwartzbaum, J.A. and Finkle, W.D. (2000) Problems Due to Small Samples and Sparse Data in Conditional Logistic Regression Analysis. *American Journal of Epidemiology*, **151**, 531-539. http://dx.doi.org/10.1093/oxfordjournals.aje.a010240

27

A Comparative Study of Automated Segmentation Methods for Use in a Microwave Tomography System for Imaging Intracerebral Hemorrhage in Stroke Patients

Qaiser Mahmood[1,2], Shaochuan Li[1], Andreas Fhager[1,2], Stefan Candefjord[1,2], Artur Chodorowski[1,2], Andrew Mehnert[3], Mikael Persson[1,2]

[1]Department of Signals and Systems, Chalmers University of Technology, Gothenburg, Sweden
[2]MedTech West, Sahlgrenska University Hospital, Gothenburg, Sweden
[3]Centre for Microscopy, Characterisation and Analysis, The University of Western Australia, Perth, Australia
Email: qaiserm@chalmers.se, stephen.li1126@gmail.com, andreas.fhager@chalmers.se, stefan.candefjord@chalmers.se, atur.chodorowski@chalmers.se, andrew.mehnert@uwa.edu.au, mikael.persson@chalmers.se

Abstract

Microwave technology offers the possibility for pre-hospital stroke detection as we have previously demonstrated using non-imaging diagnostics. The focus in this paper is on image-based diagnostics wherein the technical and computational complexities of image reconstruction are a challenge for clinical realization. Herein we investigate whether information about a patient's brain anatomy obtained prior to a stroke event can be used to facilitate image-based stroke diagnostics. A priori information can be obtained by segmenting the patient's head tissues from magnetic resonance images. Expert manual segmentation is presently the gold standard, but it is laborious and subjective. A fully automatic method is thus desirable. This paper presents an evaluation of several such methods using both synthetic magnetic resonance imaging (MRI) data and real data from four healthy subjects. The segmentation was performed on the full 3D MRI data, whereas the electromagnetic evaluation was performed using a 2D slice. The methods were evaluated in terms of: i) tissue classification accuracy over all tissues with respect to ground truth, ii) the accuracy of the simulated electromagnetic wave propagation through the head, and iii) the accuracy of the image reconstruction of the hemorrhage. The segmentation accuracy was measured in terms of the degree of overlap (Dice score) with the ground truth. The electromagnetic simulation accuracy was measured in terms of signal deviation relative to the simulation based on the ground

truth. Finally, the image reconstruction accuracy was measured in terms of the Dice score, relative error of dielectric properties, and visual comparison between the true and reconstructed intracerebral hemorrhage. The results show that accurate segmentation of tissues (Dice score = 0.97) from the MRI data can lead to accurate image reconstruction (relative error = 0.24) for the intracerebral hemorrhage in the subject's brain. They also suggest that accurate automated segmentation can be used as a surrogate for manual segmentation and can facilitate the rapid diagnosis of intracerebral hemorrhage in stroke patients using a microwave imaging system.

Keywords

Magnetic Resonance Imaging, Automatic Segmentation, Microwave, Dielectric Head Model, Intracerebral Hemorrhage Reconstruction

1. Introduction

Worldwide, about 15 million people have a stroke each year and a third of these die [1]. Stroke is a disturbance to the intracerebral blood flow caused by either a clot blocking an artery (ischemic stroke) or bleeding as the result of a burst blood vessel (hemorrhagic stroke) [2]. As a consequence, the oxygen and glucose levels are reduced which in turn rapidly kills the affected brain tissue; e.g. an estimated loss of almost 2 million neurons per minute takes places for a large-vessel ischemic stroke [3]. Ischemic stroke is the most common type of stroke accounting for about 80% of cases (the proportion varies between countries) [4]. For ischemic stroke patients, thrombolytic treatment that dissolves the clot is very effective if given shortly after onset [2]. Unfortunately, only 1% - 8% of ischemic stroke patients today receive thrombolytic treatment within the necessary time frame (4.5 h) because of delays in seeking medical attention and making a diagnosis [2]. Before thrombolytic treatment can be given, intracerebral hemorrhage has to be ruled out because treatment may aggravate the bleeding causing further brain damage and even death.

Computerized tomography (CT) is currently the gold standard for stroke diagnosis, with MRI often being used when CT does not give a definitive answer [5]. The main drawback of these imaging modalities is that they are bulky and not portable. Therefore, stroke patients have to be transported to the hospital to be diagnosed. Although purpose-built large ambulances with CT have been trialed with promising results [6], this is an expensive solution and can only be considered in densely populated areas with well-developed traffic and telecommunications infrastructure. Ultrasound is a possible future portable alternative but at present is not effective for detecting intracerebral hemorrhage [7]. Thus there is a need for a fast, cost-effective and portable stroke detection imaging system that can be installed in ordinary ambulances.

To this end several microwave systems [8]-[11] have been proposed in recent years. The systems proposed in [8]-[10] were evaluated using numerical simulations whilst that proposed in [11] was used to obtain clinical measurements from stroke patients. Microwave imaging typically uses several microwave antennas positioned around the patient's head and measures the amplitude and phase for the propagated signal between all possible pairs of antenna combinations. Typically either a single frequency is used or multiple frequencies in an interval around 1 GHz are used. The dielectric properties of the brain tissues govern the microwave scattering [12]. The high contrast between blood and brain tissue means that a bleeding can be readily detected. Microwave imaging is based on reconstructing dielectric images that can be related to physiological changes in the brain.

At present, full 3D reconstruction algorithms typically need several hours to reconstruct an image. This is not feasible in ambulances where rapid reconstructions are needed. In principle fast reconstructions can be obtained using a patient-specific dielectric model of the head obtained from an MRI scan obtained prior to the patient suffering a stroke. An image reconstruction can then be calculated as an update to the prior head model, wherein the measured signals from the patient's head after they have suffered a stroke are compared with the corresponding simulations of the head model to localize changes in the tissue associated with the stroke.

The quality and fidelity of the dielectric head model necessarily impact on the accuracy of image reconstruction. The construction of a realistic dielectric head model involves, as a first step, the segmentation (delineation/labeling) of the patient's head tissues from MRI data obtained before the patient suffered a stroke. The segmentation is then used to construct a dielectric head model by assigning to each tissue type its corresponding dielectric property (conductivity/permittivity) value.

Accurate segmentation of brain tissues from MRI data is a challenging task because of several factors including: i) the complexity and variability of the underlying anatomy; ii) noise; iii) the bias field (an unwanted low-frequency signal caused by inhomogeneities in the magnetic fields of the MRI scanner); and iv) the low contrast between the skull, cerebrospinal fluid (CSF) and air in T1-weighted MRI data. Expert manual segmentation is today the gold standard but is time consuming, labor intensive and tedious.

In this paper we present a comparative study of several automated unsupervised (do not require training examples) segmentation algorithms for use in constructing a patient specific dielectric head model. The main purpose was to investigate how segmented head models could be used in a fast algorithm for reconstruction of hemorrhagic stroke. Another purpose was to investigate how the accuracy of the segmented model influenced the accuracy of the reconstructed image.

The algorithms were evaluated both directly in terms of segmentation accuracy and indirectly in terms of electromagnetic simulation accuracy and image reconstruction accuracy in a microwave tomography system for imaging intracerebral hemorrhage. The segmentation methods include the commonly used BET-FAST (brain extraction tool-FMRIB's automated segmentation tool) [13] [14] method, and five variations on a hierarchical segmentation approach (HSA) wherein the MRI data are initially partitioned into brain and non-brain, followed by a second step where individual tissues are segmented using either Bayesian adaptive mean-shift (BAMS), adaptive mean-shift (AMS), *k* means, fuzzy *c*-means, or FAST. The study was performed using both synthetic MRI data and real data from four healthy subjects. In the case of the synthetic data, the labeled tissues used to generate the MRI data were used as "ground truth" segmentation. In the case of the real data, expert manual segmentation served as "ground truth". The segmentations were done on the full 3D data. For computational reasons, the electromagnetic simulation and image reconstructions were done on a single 2D slice (located between the upper head and the eyes) from the 3D segmented MRI data.

The simulations were performed using a model of a microwave imaging system [15], consisting of 20 antennas, placed equidistantly around a 2D dielectric head model (constructed from the 2D segmented slice). Scattering measurements (amplitude and phase of the simulated scattering signals) were obtained from a dielectric head model constructed from each segmentation method independently, and for a dielectric head model constructed from the ground truth with a simulated intracerebral hemorrhage. The scattering measurements were used in the image reconstruction algorithm.

2. Materials and Methods

2.1. Proposed Stroke Diagnosis Procedure

This section describes the proposed stroke diagnosis procedure using the microwave imaging system (see **Figure 1**). It is assumed that brain MRI data exists for the patient prior to the onset of stroke; e.g. this might be routinely done for high-risk patients. The MRI data (**Figure 1**, top-left) is segmented (see Section 2.3) to construct a patient specific dielectric head model (**Figure 1**, bottom-left). This model is then used to simulate the scattering parameters (transmission and reflection coefficients) expected from the actual microwave measuring system (see Section 2.4).

The stroke patient is measured with the microwave imaging system (**Figure 1**, top-right). The simulated and measured scattering parameters are then compared (**Figure 1**, center-right) in the image reconstruction algorithm (see Section 2.4.1), wherein the reconstruction is generated based on the difference between these parameters. This reconstruction (**Figure 1**, bottom-right) shows the change in dielectric properties of the brain tissues associated with bleeding. Both the measurement and reconstruction are designed to be performed in an ambulance equipped with a microwave based stroke diagnostic system. Note that for the purposes of this study, we mimic this scenario by using simulations of an intracerebral bleeding in place of real measurements.

2.2. MRI Data

Synthetic multi-modal MRI data (*i.e.* data acquired using different MRI techniques) together with real multi-modal MRI data from four different healthy subjects were used in this study. These data are hereinafter referred to as the synthetic dataset, real dataset 1, real dataset 2, real dataset 3, and real dataset 4 respectively.

The synthetic dataset comprises T1-weighted, T2-weighted, and proton density-weighted (PD) MRI data, for 5% noise level and 20% bias field level, obtained from the BrainWeb simulated brain database [16].

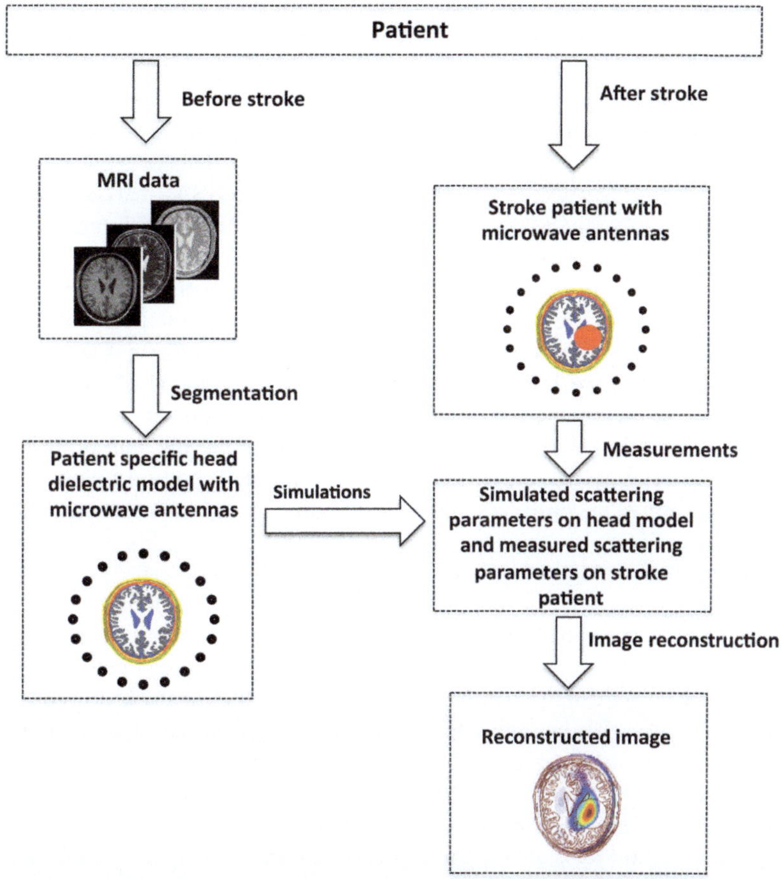

Figure 1. Schematic of the proposed stroke diagnosis procedure using the microwave imaging system.

Real dataset 1 comprises T1- and T2-weighted images of the head of a healthy subject. The data were acquired on a 3T (Tesla) Philips Achieva scanner.

Real datasets 2 - 4 comprise T1-, T2- and PD-weighted scans of the heads of three healthy subjects respectively from the IXI dataset (IXI040_Guys_0724, IXI121_Guys_0772, and IXI144_Guys_0788) [17]. The data were acquired using a Philips Medical Systems Intera 1.5T scanner. Details of the scan parameters and image and voxel dimensions for each MRI dataset are shown in Appendix (**Table A1**).

2.3. Segmentation Methods

2.3.1. Manual Segmentation (Ground Truth)

In the case of the synthetic MRI dataset the nine tissue classes—white matter, gray matter, CSF, skull, skin, fat, muscle, connective, and glial matter—defined in the BrainWeb simulated brain database [16] were taken to be ground truth segmentation. This was reduced to seven classes by merging tissues with similar dielectric properties. In particular the connective and fat tissue classes were merged, and the glial matter and gray matter classes were merged.

For the real MRI datasets manual segmentations, by an experienced radio-oncologist, into five tissue classes-white matter, gray matter, CSF, skull and skin—were taken as the ground truth. The decision to use five tissue classes represented a trade-off between the time needed to manually segment the images and the number of tissue classes. The manual segmentation took about 170 hours for 200 slices.

2.3.2. Automated Segmentation Methods

Six automatic (unsupervised) segmentation methods were compared in this study for use in constructing a dielectric head model of a subject, from MRI data acquired before the stroke event. Five are variations on a hierarchical segmentation approach and the last is a combination of the BET and FAST tools from the FMRIB Software Library (FSL).

1) Hierarchical Segmentation Approach

A schematic of the hierarchical segmentation approach (HSA) is presented in **Figure 2**. It takes as input multi-modal spatially-aligned MRI data of a subject (**Figure 2**, top-left). This data can be modeled as a single spatial head volume (V) with vector-valued voxels. In the first level of the HSA (marked level 1 in **Figure 2**), the head volume (V) is segmented into brain and non-brain tissue. For this the T1-weighted data is used to obtain both a brain mask and a whole head mask (**Figure 2**, top-center). The BET (Brain Extraction Tool) [13] is used to obtain the former and a simple whole head segmentation (WHSA) algorithm [18] [19] is used to obtain the latter. The set difference ("/") between these two masks then yields a mask of the non-brain head tissue (**Figure 2**, top-right). These masks effectively partition the head volume (V) into two disjoint sub-volumes: brain tissue (V_{BT}) and non-brain tissue (V_{NBT}) (**Figure 2**, center). In the second level of the HSA (marked level 2 in **Figure 2**), the brain and non-brain tissue are segmented into their individual tissue classes. This is achieved by applying a multi-tissue segmentation algorithm (MTSA) independently to the brain and non-brain tissue. The brain tissue is segmented into white matter, gray matter, and CSF tissue (**Figure 2**, bottom image on left side) while the non-brain tissue is segmented into skin, skull, fat, and muscle tissue or skin, skull tissue (**Figure 2**, bottom image on right side).

2) Multi-Tissue Segmentation Algorithms (MTSAs)

We used five different unsupervised multi-tissue segmentation algorithms (MTSAs): adaptive mean shift (AMS) [20], Bayesian adaptive mean shift (BAMS) [21], FMRIB's automated segmentation tool (FAST) [14], *k* means [22], and fuzzy *c*-means [23], independently in the hierarchical segmentation approach. AMS and BAMS (see [20] and [21] respectively for more detail) are variations on the mean shift algorithm. Mean shift [24] [25] is based on an adaptive gradient ascent approach to estimate the local maxima or modes of the probability density function underlying the feature space (features could be colors, texture, etc.). Ultimately each feature point is associated with a mode (local maxima of the probability density function corresponding to a dense region in the feature space) thereby defining clusters (groups of features points). The advantages of mean shift are that it doesn't require any initialization for clustering and performs well in the presence of noise and the bias field.

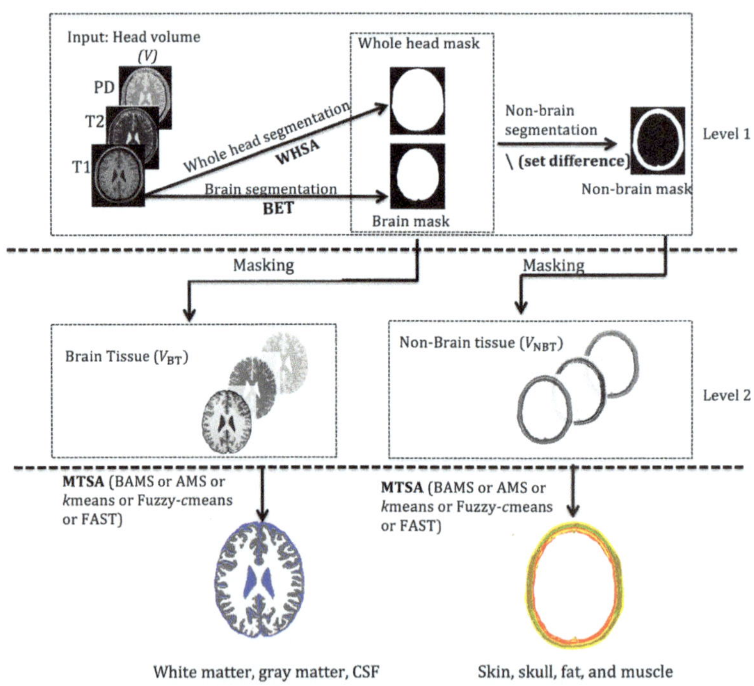

Figure 2. Schematic of the hierarchical segmentation approach (HSA) for automated whole head segmentation. In the figure, MTSA stands for multi-tissue segmentation algorithm.

The FAST tool [14] is commonly used for brain tissue segmentation. The underlying method is based on the hidden Markov random field model and associated Expectation-Maximization (HMRF-EM) algorithm [14].

The methods *k* means [22] and fuzzy *c*-means [23] are based on the distance function to partition the input data into clusters. These methods are frequently used in the pattern recognition field.

Hereinafter, the five variations of the HSA are denoted HSA-AMS, HSA-BAMS, HSA-FAST, HSA-*k* means and, HSA-fuzzy *c*-means.

3) FMRIB Software Library (FSL) tools: BET-FAST

The combination of the BET [13] and FAST [14] tools (BET-FAST) in FSL is commonly used for the creation of patient-specific dielectric models [26]-[28]. The BET-FAST segmentation approach takes the same inputs as the hierarchical segmentation approach (HSA), shown in **Figure 2**. The BET tool is applied to extract the brain-tissue mask, skull-tissue mask, and skin-tissue mask from the T1-weighted images (similar to HSA level 1 in **Figure 2**). The FAST tool is then used to segment the brain-only image (masked version of the T1-weighted images) into three tissue classes: white matter, gray matter, and CSF (similar to HSA level 2 in **Figure 2**).

2.3.3. Optimal Parameters Settings for the MTSAs and BET-FAST tool

In this study, the AMS and BAMS MTSAs were applied with the same parameter settings as described in [20] and [21] respectively. The *k* means, fuzzy *c*-means, and FAST MTSAs were initialized using the prior knowledge of tissue intensity ordering in the T1-weighted MRI scan. For example, in the T1-weighted MRI scan the highest intensity value was used to initialize the cluster for WM tissue, the lowest intensity value was used to initialize the cluster for CSF tissue, and the mean of the highest and lowest intensity values was used to initialize the cluster for GM tissue. Moreover, for the fuzzy *c*-means, the parameter m (controls the fuzziness of the resulting clusters) was set to 1.5 and for FAST, the parameter H (MRF beta) was set to 0.1. For the brain, skull and skin mask extraction, the BET tool was applied using the threshold parameter $f = 0.5$.

2.3.4. Computational Complexity and Time

Since the mean shift is an iterative intensive algorithm, the methods: HSA-AMS and HSA-BAMS have high computational complexity compared to the other competing methods. The approximate execution time for each method for the segmentation of the synthetic multi-modal 181 × 217 × 181 MRI head volume on a single desktop PC running 64-bit Ubuntu 10.04 with 8 GiB ram memory and Intel Core i7 CPU 870 at 2.93 GHz is presented in **Table A2** (see Appendix).

2.4. Procedure for Evaluating the Intracerebral Hemorrhage Reconstruction

In this study, the MRI datasets were independently segmented in 3D using each segmentation method. A single 2D axial slice (slice between the upper head and the eyes) from each segmented dataset was then independently used to construct a 2D dielectric head model (representing the situation where there is no hemorrhage) by assigning to each tissue class conductivity/permittivity properties obtained from [12]. A dielectric head model was similarly constructed from the ground truth segmentation for each MRI data set and manually modified to yield dielectric head models corresponding to intracerebral hemorrhages of different sizes and locations. Eight different hemorrhages were simulated by superimposing two different sized disks in four different positions (see **Figure 3**). The radii for the large and small intracerebral hemorrhages were 20 mm and 10 mm, respectively.

Each 2D dielectric head model without intracerebral hemorrhage was used in an electromagnetic simulation to compute scattering parameters (see **Figure 1**, bottom-left side). In principle these parameters can be

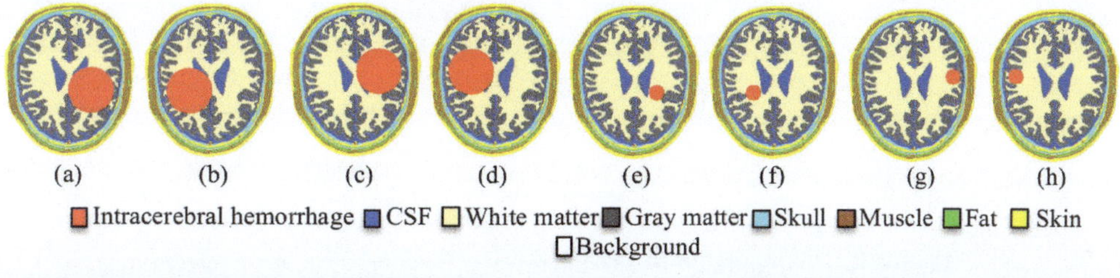

Figure 3. Dielectric head model with a large simulated intracerebral hemorrhage (a)-(d) and with a small simulated intracerebral hemorrhage (e)-(h) for four different positions.

pre-computed to facilitate real time reconstruction. Scattering parameters were similarly obtained for each dielectric head model corresponding to a hemorrhage (in the real world these would be obtained from the stroke patient using a helmet fitted with a microwave antenna array).

The antenna configuration for the microwave imaging system, used in this study for electromagnetic simulations, is shown in **Figure 4**. It comprises 20 antennas placed equidistantly around the dielectric head model. The antennas are used both as transmitters and receivers. For each tomographic measurement, one antenna at the time is used as transmitter while the others are used as receivers to measure the scattered signals. This process is continued until all antennas have been used as transmitters and all possible antenna combinations have been measured.

The electromagnetic simulations were performed with a 2D finite-difference time domain (FDTD) method [29] [30]. The FDTD method was used to simulate the electric fields inside the head as well as to simulate the scattering parameters from the head. The electric fields and scattering parameters for each pair comprising a head model without hemorrhage and the corresponding head model with a simulated hemorrhage were then used as inputs to the image reconstruction algorithm in order to reconstruct the intracerebral hemorrhage (see **Figure 1**, bottom-right).

Reconstruction Algorithm

The algorithm takes as input the simulated and measured signals, obtained from the head model without and the corresponding head model with a simulated hemorrhage respectively, and processes this data in order to determine the change in dielectric properties of the brain tissues associated with the intracerebral hemorrhage. In this study, we used a fast reconstruction algorithm, which is based on a Born approximation [31], wherein it is assumed that the scattering electric field due to the small perturbation (intracerebral hemorrhage) is negligible compared to the incident electric field.

Based on the Born approximation, the difference between the scattering parameters (obtained from the FDTD simulation as described above) simulated using the dielectric head model without hemorrhage and the scattering parameters simulated using the dielectric head model with hemorrhage (in practice, measured) [32] [33] is given by

$$\delta \boldsymbol{E}_s^{\text{total}}(\boldsymbol{r}) = \left(\frac{2\pi f}{c_0}\right)^2 \int \boldsymbol{G}(\boldsymbol{r}-\boldsymbol{r}')\boldsymbol{E}_0(\boldsymbol{r}')\delta\varepsilon \tag{1}$$

where $\delta\varepsilon$ represents the change in dielectric properties of the brain tissues due to the intracerebral hemorrhage, $\boldsymbol{E}_0$ is the incident electric field in the image domain and $\boldsymbol{G}$ *is* the Green's function that represents the electric field resulting from the pixels acting as single point scatterers in the image domain [34]. Both $\boldsymbol{E}_0$ and $\boldsymbol{G}$ are obtained using the dielectric head model without hemorrhage. For simplicity, Equation 1 can be expressed as

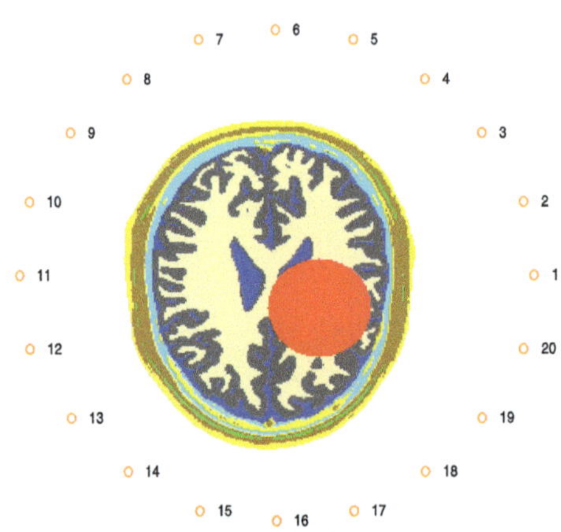

Figure 4. Configuration of the microwave imaging system for electromagnetic simulations (the figure shows an example 2D dielectric head model with simulated intracerebral hemorrhage).

$$b = Ax \tag{2}$$

where $b = \delta E_s^{\text{total}}(r)$, $A = \left(\frac{2\pi f}{c_0}\right)^2 G(r - r')E_0(r')$, and $x = \delta\varepsilon(r')$.

In order to construct the intracerebral hemorrhage, we solved Equation (2) for x. Equation (2) represents a system of ill-posed linear equations. To make it well posed, a Tikhonov regularization was used [35]. Equation (2) then becomes

$$\arg\min x = \|Ax - b\|^2 + \|\alpha Rx\|^2 \tag{3}$$

where α is the regularization parameter and R is the Tikhonov stabilization operator.

The solution of Equation (3) is then

$$\hat{x} = \left(A^+ A + \alpha R\right)^{-1} A^+ b \tag{4}$$

where $\hat{x}$ is the estimated change of dielectric properties of the brain tissues due to the intracerebral hemorrhage. In this study we chose R to be the identity matrix.

2.5. Quantitative Evaluation

The quantitative evaluation of each segmentation method for each data was done in terms of segmentation accuracy over all tissues, electromagnetic simulation accuracy, and reconstruction image accuracy for intracerebral hemorrhage.

Segmentation accuracy for each tissue type, each method and each dataset was evaluated in terms of the Dice score [36]. The Dice score measures the degree of overlap between the ground truth and the segmentation. The Dice score for the ith tissue is given by

$$DS^i = \frac{2V_{ae}^i}{\left(V_a^i + V_e^i\right)} \tag{5}$$

where V_{ae}^i is the number of voxels for i^{th} tissue the segmentation result and the ground truth have in common, and V_a^i and V_e^i denote the number of voxels for i^{th} tissue in the segmentation result and the ground truth respectively. The segmentation accuracy over all tissue types for each method and each dataset was then computed as the mean Dice score (*MDS*), given by

$$MDS = \frac{1}{N}\sum_{i=1}^{N} DS^i \tag{6}$$

where N is the total number of tissues.

The electromagnetic simulation accuracy was evaluated in terms of signal deviation (F). The signal deviation was defined as the sum of squared difference between the scattering signals obtained from the dielectric models with and without intracerebral hemorrhage, respectively. It is given by

$$F = \sum_{k-1}^{N} \sum_{t=1}^{n} \left\| S_{GTB}(k,t) - S_{SM}(k,t) \right\|^2 \tag{7}$$

where N is the total number of antennas, n is the total number of signal samples and $S_{GTB}(k,t)$ is the tth sample of the scattered signal received from the kth antenna, using the dielectric head model with simulated intracerebral hemorrhage and $S_{SM}(k,t)$ is the tth sample of the scattered signal received from the kth antenna, using the dielectric head model without hemorrhage.

The reconstruction image accuracy for both conductivity and permittivity of intracerebral hemorrhage for each dataset and each segmentation method was measured with respect to the ground truth in terms of relative error.

The relative error (*RE*) for the permittivity [29] is given by

$$RE = \frac{\iint_S \left|\varepsilon_{GT} - \varepsilon_{ASM}\right| \mathrm{d}S}{\iint_S \left|\varepsilon_{GT}\right| \mathrm{d}S} \tag{8}$$

where S is the region of intracerebral hemorrhage being reconstructed, ε_{GT} represents the permittivity profile of the intracerebral hemorrhage reconstruction using the ground truth, and ε_{ASM} is the permittivity profile of the intracerebral hemorrhage reconstruction using the automatic segmentation method. The relative error for the conductivity is similarly defined.

We also measured the accuracy of image reconstruction in terms of the Dice score by computing the degree of overlap between the actual (simulated bleeding in the ground truth brain) and reconstructed intracerebral hemorrhage.

Finally the mean signal deviation, mean relative error and mean Dice score were computed over the four different positions for both large and small intracerebral hemorrhages (see **Figure 3**).

Optimal Parameters Settings for the Reconstruction Algorithm

The ground truth from the synthetic data was used to empirically determine the optimal settings for the α and f parameters for the reconstruction algorithm. The optimal values of α and f for the reconstruction algorithm were obtained using a grid search over a predefined range of values. For each 2-tuple (α, f), the accuracy of image reconstruction for intracerebral hemorrhage was evaluated in terms of mean Dice score over the four different positions for two different sizes. The optimal 2-tuple was chosen to be that which yielded the highest mean Dice score.

We have previously shown that different sized objects are best reconstructed at different frequencies [30]. Therefore we have used different frequencies when reconstructing the large and the small intracerebral hemorrhage respectively. For the large hemorrhages $f = 350$ MHz was used. For the small hemorrhages separate reconstructions were made at $f = 350$, 550, 750 and 850 MHz. Averaging all the individual reconstructed dielectric images then generated the final reconstruction. The optimal value of α was empirically found to be $\alpha = 0.05$.

3. Results and Discussion

3.1. Qualitative Evaluation for Segmentation Accuracy

Figure 5 shows a segmentation example for each method including ground truth for a single axial slice from the synthetic dataset for 5% noise level with 20% bias field level. It can be seen that the methods HSA-BAMS and HSA-AMS yield fewer segmentation errors (every misclassified pixel is shown with a black dot) than the other methods. This suggests that HSA-BAMS and HSA-AMS are more robust to the noise and bias field. The methods: HSA-*k* means, HSA-fuzzy *c*-means, HSA-FAST and BET-FAST have a higher tendency to misclassify the gray matter as CSF. Moreover, HSA-*k* means and HSA-fuzzy *c*-means have a higher tendency to misclassify the fat as muscle tissue and the muscle as skin tissue while HSA-FAST has a higher tendency to misclassify the fat as muscle and skin tissue. However, in the BET-FAST method, BET is limited to segmenting the non-brain into two tissue types: skin and skull, wherein the fat and muscle are included in the "skin" class. Therefore,

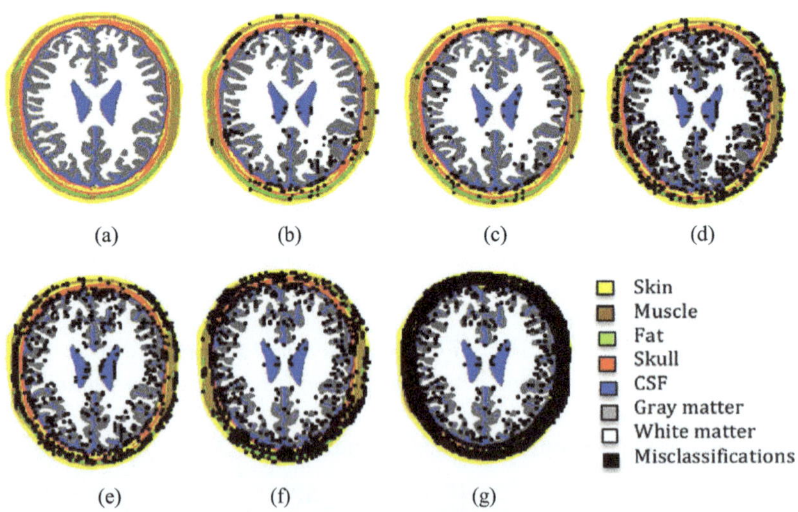

Figure 5. Segmentation results for the synthetic dataset for each segmentation method. (a) Ground truth; (b) HSA-BAMS; (c) HSA-AMS; (d) HSA-*k* means; (e) HSA-fuzzy *c*-means; (f) HSA-FAST; (g) BET-FAST.

BET-FAST has higher misclassification of skin tissue. Moreover, the BET-FAST method over segments the skull tissue.

Figure 6 shows a segmentation example for each method for a single axial slice from real dataset 2. It can be seen that the method HSA-BAMS yields fewer segmentation errors (shown in black dot) than the other methods. The methods: HSA-*k* means, HSA-fuzzy *c*-means, HSA-FAST and BET-FAST have a higher tendency to misclassify the white matter as gray matter and the gray matter as CSF. Moreover HSA-AMS as well as other competing methods (HSA-*k* means, HSA-fuzzy *c*-means. HSA-FAST and BET-FAST) over segment the skull tissue. For the other real datasets (dataset 1, dataset 3 and dataset 4), similar axial slices were selected for the evaluation and the qualitative results for these datasets were similar to those of dataset 2 (shown in **Figure 6**).

3.2. Segmentation Accuracy versus Electromagnetic Microwave Simulation Accuracy

The accuracy of image reconstruction depends on the accuracy of electromagnetic microwave simulation. In turn the accuracy of the electromagnetic simulation depends on the segmentation accuracy. Combined plots of segmentation accuracy (Dice score over all tissues) and electromagnetic simulation accuracy (signal deviation over all simulated hemorrhages) are shown in **Figure 7** for each segmentation method and each dataset.

The results in **Figure 7** show that the method HSA-BAMS has the lowest mean signal deviation and highest mean Dice score compared to the other methods. The results also show that the mean signal deviation measured using only the ground truth is very small compared to that measured using the automated segmentation methods. This suggests that tissue segmentation accuracy affects the simulated scattering signals more than the intracerebral hemorrhage.

3.3. Qualitative Evaluation for Intracerebral Hemorrhage Reconstruction

Example reconstructions of the permittivity and conductivity profiles for a large intracerebral hemorrhage with radius size of 20 mm from the synthetic dataset are shown in **Figure 8** and **Figure 9** respectively.

Example reconstructions of the permittivity and conductivity profiles for a small intracerebral hemorrhage with radius size of 10 mm from the real dataset 2 are shown in **Figure 10** and **Figure 11** respectively.

In each figure, the contours in the reconstructed images show the structures of the head tissues, wherein the actual intracerebral hemorrhage is shown as a circular contour. The color bar represents the range of intensity for the reconstructed images. In the reconstructed images, the white pixels correspond to background and the yellow and red pixels correspond to the intracerebral hemorrhage.

Figure 8 and **Figure 9** show that the reconstructions of the large intracerebral hemorrhage obtained using the

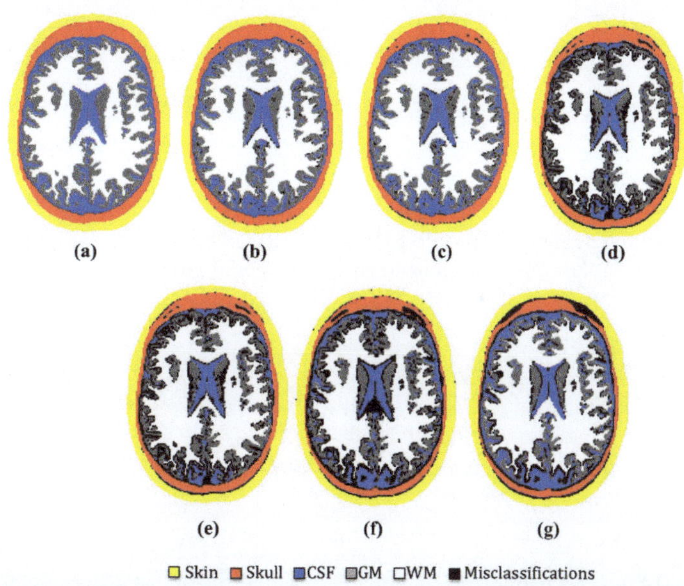

Figure 6. Segmentation results for the real dataset 2 for each segmentation method. (a) Ground truth; (b) HSA-BAMS; (c) HSA-AMS; (d) HSA-*k* means; (e) HSA-fuzzy *c*-means; (f) HSA-FAST; (g) BET-FAST.

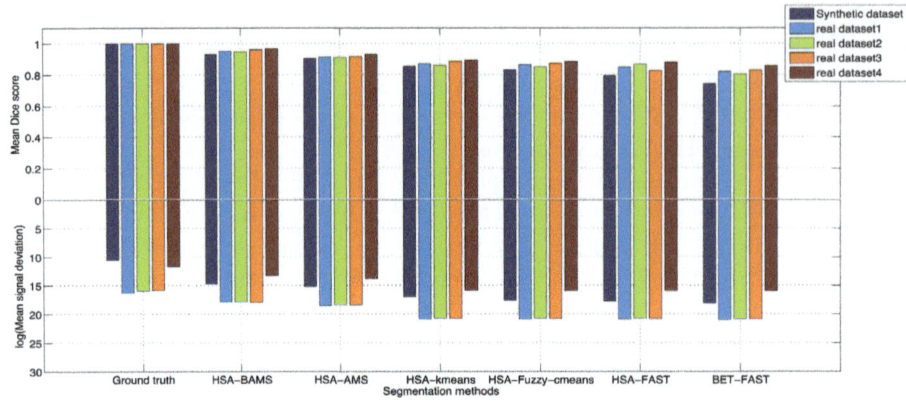

(a)

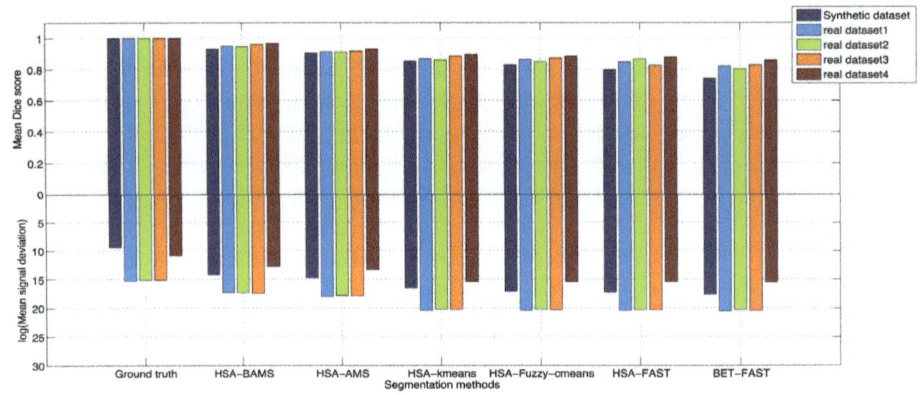

(b)

Figure 7. Mean Dice score and mean signal deviation (in logarithmic scale) for (a) large (b) small intracerebral hemorrhage. Results for different segmentation methods are shown along the x-axis, for every method a group of data is shown with different colors, corresponding to different datasets. Along the positive y-axis the histograms show the mean Dice score and on the negative y-axis the mean signal deviation in logarithmic scale.

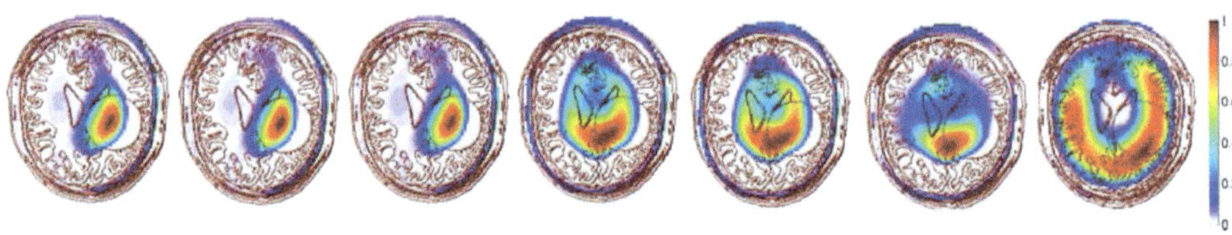

(a) (b) (c) (d) (e) (f) (g)

Figure 8. Reconstruction of the permittivity profile for a large intracerebral hemorrhage with radius size of 20 mm from the synthetic dataset obtained using the different segmentation methods. (a) Ground truth; (b) HSA-BAMS; (c) HSA-AMS; (d) HSA-*k* means; (e) HSA-fuzzy *c*-means; (f) HSA-FAST; (g) BET-FAST.

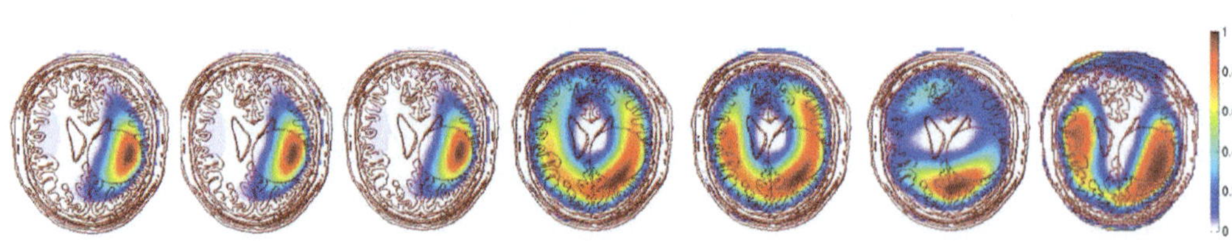

(a) (b) (c) (d) (e) (f) (g)

Figure 9. Reconstruction of the conductivity profile for a large intracerebral hemorrhage with radius size of 20 mm from the synthetic dataset obtained using the different segmentation methods. (a) Ground truth; (b) HSA-BAMS; (c) HSA-AMS; (d) HSA-*k* means; (e) HSA-fuzzy *c*-means; (f) HSA-FAST; (g) BET-FAST.

segmentation methods HSA-BAMS and HSA-AMS are better than that obtained using the HSA-*k* means, HSA-fuzzy *c*-means, HSA-FAST and the BET-FAST method, and are similar to that obtained using the ground truth.

Figure 10 and **Figure 11** show that the reconstructions of the small intracerebral hemorrhage obtained using the segmentation method HSA-BAMS are better than that obtained using the competing methods and are similar to that obtained using the ground truth. **Figure 10** also shows that there is no overlapping between the location of the actual intracerebral hemorrhage and the location of the reconstructed intracerebral hemorrhage obtained using HSA-*k* means, HSA-fuzzy *c*-means, HSA-FAST and the BET-FAST method.

3.4. Quantitative Evaluation for Intracerebral Hemorrhage Reconstruction

The reconstructed images for both normalized permittivity and conductivity are shown in **Figures 8-11**. In order to quantitatively assess the accuracy of the reconstructions, the first problem is to define what part of the reconstructed object defines the boundaries of the bleeding. It is not immediately straightforward how to do this and quantifying the levels of artifacts and reconstructions would require further investigations. We have therefore taken the following approach in the analysis, a threshold level "T" for both normalized permittivity and conductivity was defined. It was used to discriminate the intracerebral hemorrhage from the background. To exemplify, three different values T = 0.6, 0.7, 0.8 were selected and used for further analysis. The images were normalized to have maximum value 1, the permittivity/conductivity values below the threshold level, T were treated as background while the permittivity/conductivity values above T were considered to represent the intracerebral hemorrhage. The Dice score was then measured between the actual and reconstructed intracerebral hemorrhage for both permittivity and conductivity.

Figure 12 shows the combined plots for the mean Dice score for the threshold level 0.7 and the mean relative error for permittivity for both large and small intracerebral hemorrhages.

The plots for the synthetic data for both large and small intracerebral hemorrhages show that the methods HSA-BAMS and HSA-AMS have better reconstruction accuracy (higher mean Dice score and lower relative error) than HSA-*k* means, HSA-fuzzy *c*-means, HSA-FAST and BET-FAST. The plots for each real dataset show that, relative to the ground truth, HSA-BAMS has better reconstruction accuracy (higher mean Dice score and lower relative error) than all other segmentation methods. For the small intracerebral hemorrhage, it can also be observed that HSA-*k* means, HSA-fuzzy *c*-means, HSA-FAST and BET-FAST have zero mean Dice score; this indicates that there is no overlap between the reconstructed intracerebral hemorrhage and the actual intracerebral hemorrhage. This can be clearly seen in **Figures 10(d)-(g)**.

In the case of conductivity, similar results were obtained for the mean Dice score and mean relative error for both large and small intracerebral hemorrhages.

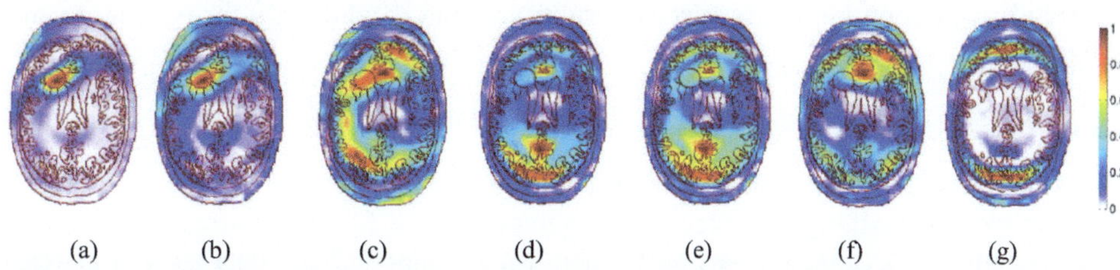

Figure 10. Reconstruction of the permittivity profile for a small intracerebral hemorrhage with radius size of 10 mm from the real dataset 2 obtained using the different segmentation methods. (a) Ground truth; (b) HSA-BAMS; (c) HSA-AMS; (d) HSA-*k* means; (e) HSA-fuzzy *c*-means; (f) HSA-FAST; (g) BET-FAST.

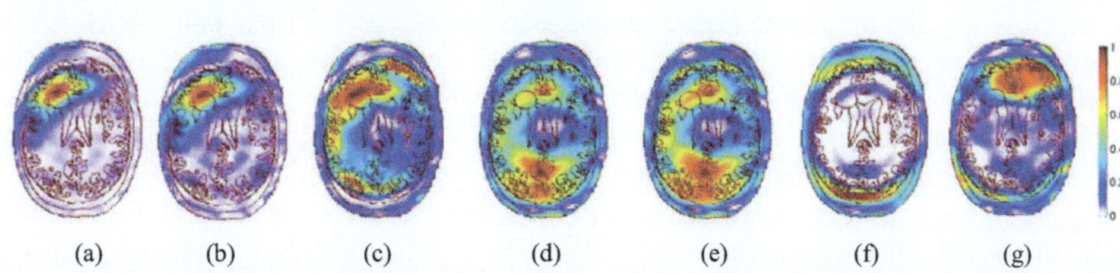

Figure 11. Reconstruction of the conductivity profile for a small intracerebral hemorrhage with radius size of 10 mm from the real dataset 2 obtained using the different segmentation methods. (a) Ground truth; (b) HSA-BAMS; (c) HSA-AMS; (d) HSA-*k* means; (e) HSA-fuzzy *c*-means; (f) HSA-FAST; (g) BET-FAST.

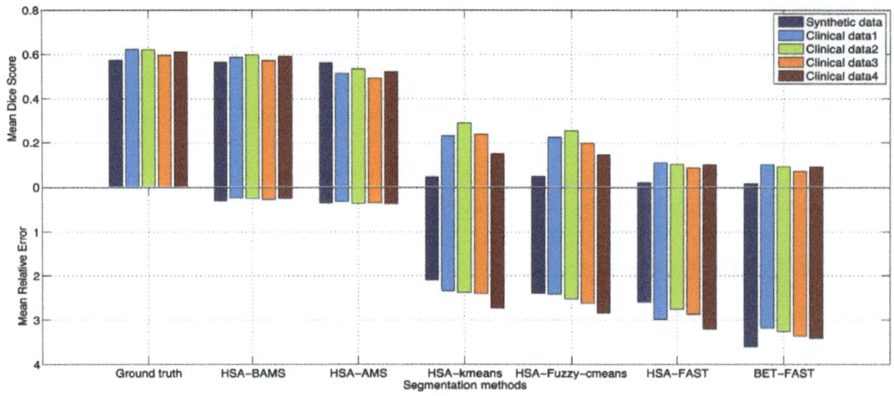

(a)

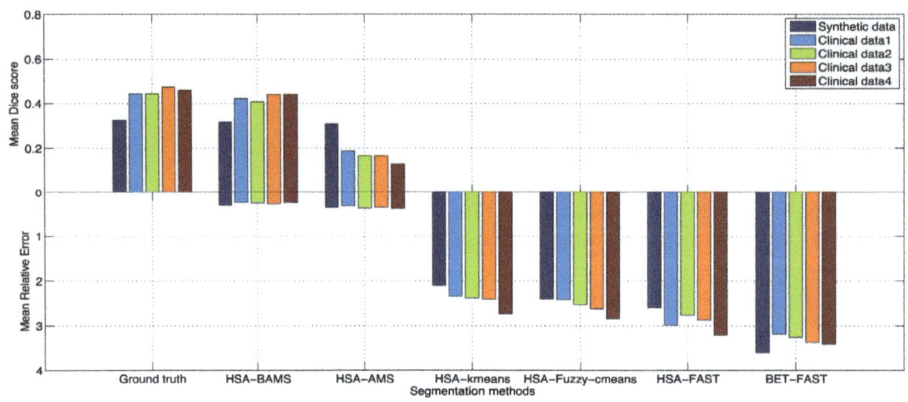

(b)

Figure 12. Mean Dice score and mean relative error for permittivity for (a) large and (b) small intracerebral hemorrhage for each method and each data.

The mean Dice score was also computed for the other threshold levels 0.6 and 0.8 respectively. For each method, similar trends were found in the results of permittivity and conductivity for both large and small intracere bral hemorrhages.

4. Conclusions

In this study, we presented an evaluation of six automatic segmentation methods for use in the construction of a dielectric head model for imaging an intracerebral hemorrhage in MR brain images using a simulated microwave imaging system. The evaluation was performed using synthetic MRI data as well as real MRI data from four different healthy subjects. The evaluation was done in terms of segmentation accuracy over all tissues, electromagnetic simulation accuracy, and image reconstruction accuracy for the intracerebral hemorrhage in the subject's brain for four different positions with two different sizes.

The results show that the automatic segmentation method HSA-BAMS has better performance (higher segmentation accuracy, lower signal deviation and lower relative error) compared with the other methods. The results also indicate that accurate segmentation of tissues leads to accurate reconstruction of intracerebral hemorrhage in the subject's brain. Overall the results suggest that automated segmentation (in particular HSA-BAMS) can be used as a surrogate for manual segmentation of the head tissues required for the image reconstruction algorithm.

This study represents a first step towards the visualization of intracerebral hemorrhage for stroke patients in near real-time using contemporary computing power in mobile devices suitable for field use. In the future, we aim to investigate the performance of the automatic segmentation method HSA-BAMS for the reconstruction of intracerebral hemorrhage in MR images using microwave-based measurements from stroke patients.

Acknowledgements

The author Mr. Mahmood acknowledges scholarship funding from the Higher Education Commission of Pakistan (HEC) and Chalmers University of Technology in support of this work.

References

[1] The Atlas of Heart Disease and Stroke. http://www.who.int/cardiovascular_diseases/resources/atlas/en/

[2] Fassbender, K., Balucani, C., Walter, S., Levine, S.R., Haass, A. and Grotta, J. (2013) Streamlining of Prehospital Stroke Management: The Golden Hour. *Lancet Neurology*, **12**, 585-596. http://dx.doi.org/10.1016/S1474-4422(13)70100-5

[3] Burns, J.D., Green, D.M., Metivier, K. and DeFusco, C. (2012) Intensive Care Management of Acute Ischemic Stroke. *Emergency Medicine Clinics of North America*, **30**, 713-744. http://dx.doi.org/10.1016/j.emc.2012.05.002

[4] Feigin, V.L., Lawes, C.M., Bennett, D.A., Barker-Collo, S.L. and Parag, V. (2009) Worldwide Stroke Incidence and Early Case Fatality Reported in 56 Population-Based Studies: A Systematic Review. *Lancet Neurology*, **8**, 355-369. http://dx.doi.org/10.1016/S1474-4422(09)70025-0

[5] Chalela, J., Kidwell, C.S., Nentwich, L.M., Luby, M., Butman, J.A., *et al.* (2007) Magnetic Resonance Imaging and Computer Tomography in Emergency Assessment of Patients with Suspected Acute Stroke: A Prospective Comparison. *Lancet*, **369**, 293-298. http://dx.doi.org/10.1016/S0140-6736(07)60151-2

[6] Walter, S., Kostopoulos, P., Haass, A., Keller, I., Lesmeister, M., *et al.* (2012) Diagnosis and Treatment of Patients with Stroke in a Mobile Stroke Unit versus in Hospital: A Randomised Controlled Trial. *Lancet Neurology*, **11**, 397-404. http://dx.doi.org/10.1016/S1474-4422(12)70057-1

[7] Hölscher, T., Dunford, J.V., Schlachetzki, F., Boy, S., Hemmen, T., *et al.* (2013) Prehospital Stroke Diagnosis and Treatment in Ambulances and Helicopters—A Concept Paper. *American Journal of Emergency Medicine*, **31**, 743-747. http://dx.doi.org/10.1016/j.ajem.2012.12.030

[8] Ireland, D. and Bialkowski, M. (2011) Microwave Head Imaging for Stroke Detection. *Progress in Electromagnetics Research M*, **21**, 163-175. http://dx.doi.org/10.2528/PIERM11082907

[9] Scapaticci, R., Di Donato, L., Catapano, I. and Crocco, L. (2012) A Feasibility Study on Microwave Imaging for Brain Stroke Monitoring. *Progress in Electromagnetics Research B*, **40**, 305-324. http://dx.doi.org/10.2528/PIERB12022006

[10] Irishina, N. and Torrente, A. (2013) Brain Stroke Detection by Microwaves Using Prior Information from Clinical Databases. *Abstract and Applied Analysis*, **2013**, Article ID: 412638. http://dx.doi.org/10.1155/2013/412638

[11] Persson, M., Fhager, A., Trefna, H.D., Yu, Y.N., McKelvey, T., Pegenius, G., *et al.* (2014) Microwave-Based Stroke Diagnosis Making Global Prehospital Thrombolytic Treatment Possible. *IEEE Transactions on Biomedical Engineering*, **61**, 2806-2817. http://dx.doi.org/10.1109/TBME.2014.2330554

[12] Dielectric Properties of Body Tissues. http://niremf.ifac.cnr.it/tissprop/

[13] Smith, S.M. (2002) Fast Robust Automated Brain Extraction. *Human Brain Mapping*, **17**, 143-155. http://dx.doi.org/10.1002/hbm.10062

[14] Zhang, Y., Brady, M. and Smith, S. (2001) Segmentation of Brain MR Images through a Hidden Markov Random Field Model and the Expectation Maximization Algorithm. *IEEE Transactions on Medical Imaging*, **20**, 45-57. http://dx.doi.org/10.1109/42.906424

[15] Fhager, A. and Persson, M. (2007) Using a Priori Data to Improve the Reconstruction of Small Objects in Microwave Tomography. *IEEE Transactions on Microwave Theory and Techniques*, **55**, 2454-2462. http://dx.doi.org/10.1109/TMTT.2007.908670

[16] Cocosco, C.A., Kollokian, V., Kwan, R.K.S. and Evans, A.C. (1997) BrainWeb: Online Interface to 3-D MRI Simulated Brain Database. *NeuroImage*, **5**, S425.

[17] IXI Dataset. http://www.brain-development.org/

[18] Otsu, N. (1979) A Threshold Selection Method from Gray-Level Histrogram. *IEEE Transactions on Systems, Man, and Cybernetics*, **9**, 62-66. http://dx.doi.org/10.1109/TSMC.1979.4310076

[19] Soille, P. (1999) Morphological Image Analysis: Principles and Applications. Springer-Verlag, Berlin, 173-174. http://dx.doi.org/10.1007/978-3-662-03939-7

[20] Mayer, A. and Greenspan, H. (2009) An Adaptive Mean-Shift Framework for MRI Brain Segmentation. *IEEE Transactions on Medical Imaging*, **28**, 1238-1249. http://dx.doi.org/10.1109/TMI.2009.2013850

[21] Mahmood, Q., Chodorowski, A., Mehnert, A. and Persson, M. (2012) A Novel Bayesian Approach to Adaptive Mean Shift Segmentation of Brain Images. *Proceedings of the IEEE International Symposium on Computer-Based Medical Systems* (*CBMS*), Rome, 20-22 June 2012, 1-6. http://dx.doi.org/10.1109/CBMS.2012.6266304

[22] Seber, G.A.F. (1984) Multivariate Observations. John Wiley & Sons, Inc., Hoboken. http://dx.doi.org/10.1002/9780470316641

[23] Bezdek, J.C. (1981) Pattern Recognition with Fuzzy Objective Function Algorithms. Kluwer Academic Publishers, Norwell. http://dx.doi.org/10.1007/978-1-4757-0450-1

[24] Comaniciu, D., Ramesh, V. and Meer, P. (2001) The Variable Bandwidth Mean-Shift and Data-Driven Scale Selection. *Proceedings of the IEEE International Conference on Computer Vision* (*ICCV*), Vancouver, 7-14 July 2001, 438-445. http://dx.doi.org/10.1109/ICCV.2001.937550

[25] Comaniciu, D. and Meer, P. (2002) Mean Shift: A Robust Approach toward Feature Space Analysis. *IEEE Transactions on Pattern Analysis and Machine Intelligence*, **24**, 603-619. http://dx.doi.org/10.1109/34.1000236

[26] Shirvany, Y., Edelvik, F., Jakobsson, S., Hedström, A., Mahmood, Q., Chodorowski, A. and Persson, M. (2012) Non-Invasive EEG Source Localization Using Particle Swarm Optimization: A Clinical Experiment. *Proceedings of the* 34*th Annual International Conference of the IEEE EMBS*, San Diego, 28 August-1 September 2012, 6232-6235.

[27] Rullmann, M., Anwander, A., Dannhauer, M., Warfield, S.K., Duffy, F. and Wolters, C. (2009) EEG Source Analysis of Epileptiform Activity Using a 1 mm Anisotropic Hexahedra Finite Element Head Model. *NeuroImage*, **44**, 399-410. http://dx.doi.org/10.1016/j.neuroimage.2008.09.009

[28] Lanfer, B., Scherg, M., Dannhauer, M., Knösche, T.R., Burger, M., Wolters, C.H. (2012) Influences of Skull Segmentation Inaccuracies on EEG Source Analysis. *NeuroImage*, **62**, 418-431. http://dx.doi.org/10.1016/j.neuroimage.2012.05.006

[29] Fhager, A. (2006) A Microwave Tomography. Doktorsavhandlingar vid Chalmers tekniska högskola, Gothenburg.

[30] Fhager, A., Hashemzadeh, P. and Persson, M. (2006) Reconstruction Quality and Spectral Content of an Electromagnetic Time-Domain Inversion Algorithm. *IEEE Transactions on Biomedical Engineering*, **53**, 1594-1604. http://dx.doi.org/10.1109/TBME.2006.878079

[31] Semenov, S.Y., Bulyshev, A.E., Abubakar, A., Posukh, V.G., Sizov, Y.E., Souvorov, A.E., *et al.* (2005) Microwave-Tomographic Imaging of the High Dielectric-Contrast Objects Using Different Image-Reconstruction Approaches. *IEEE Transactions on Microwave Theory and Techniques*, **53**, 2284-2294. http://dx.doi.org/10.1109/TMTT.2005.850459

[32] Slaney, M., Kak, A.C. and Larsen, L.E. (1984) Limitations of Imaging with First-Order Diffraction Tomography. *IEEE Transactions on Microwave Theory and Techniques*, **32**, 860-874. http://dx.doi.org/10.1109/TMTT.1984.1132783

[33] Pahomov, V., Semenchik, V. and Kurilo, S. (2005) Reconstructing Reflecting Object Images Using Born Approximation. *Proceedings of the IEEE* 35*th European Microwave Conference*, Paris, 4-6 October 2005.

[34] Larsen, L.E. and Jacobi, J.H. (1986) Medical Applications of Microwave Imaging. IEEE Press, Piscataway.

[35] Tikhonov, A.N. and Arsenin, V.Y. (1977) Solutions of Ill-Posed Problems. Winston and Sons, Washington DC.

[36] Dice, L.R. (1945) Measures of the Amount of Ecologic Association between Species. *Ecology*, **26**, 297-302. http://dx.doi.org/10.2307/1932409

Appendix

Table A1. Scan parameters and image and voxel dimensions for each MRI dataset.

MRI dataset	T1-weighted		T2-weighted		PD	
	Rows × columns × slices					
Synthetic	181 × 217 × 181		181 × 217 × 181		181 × 217 × 181	
	Voxel size (mm^3)					
	1 × 1 × 1		1 × 1 × 1		1 × 1 × 1	
	Rows × columns × slices					
Real 1	256 × 256 × 195		256 × 256 × 327		-	
	Voxel size (mm^3)					
	0.94 × 0.94 × 1		0.94 × 0.94 × 0.55		-	
	Scan parameters					
	TR (ms)	**TE (ms)**	**TR (ms)**	**TE (ms)**	**TR (ms)**	**TE (ms)**
	8.17	3.76	2500	233	-	-
	Rows × columns × slices					
Real 2-4	256 × 256 × 150		256 × 256 × 130		256 × 256 × 130	
	Voxel size (mm^3)					
	0.94 × 0.94 × 1.25		0.94 × 0.94 × 1.25		0.94 × 0.94 × 1.25	
	Scan parameters					
	TR (ms)	**TE (ms)**	**TR (ms)**	**TE (ms)**	**TR (ms)**	**TE (ms)**
	9.8	4.60	8178	100	8178	8.0

Table A2. Approximate execution time for each segmentation method.

Methods	Approximate execution time (minutes)
HSA-BAMS	85
HSA-AMS	70
HSA-*k* means	15
HSA-fuzzy *c*-means	20
HSA-FAST	45
BET-FAST	50

28

Antenna Performance Improvement in Elliptical Array Using RMI Method of Mutual Coupling Compensation

Aluru Venkata Lakshmi Narayana Rao[1], Nunna Bala Ankaiah[2], Dharma Raj Cheruku[3]

[1]Department of ECE, SSNCET, Ongole, India,
[2]ECIL, Hyderabad, India,
[3]Departement of ECE, GIT, GITAM University, Visakhapatnam, India
Email: narayanaavlrao@gmail.com, bala.nunna8@gmail.com, dharmarajc@yahoo.com

Abstract

The number of wireless electronic gadgets used in mobile communication, vehicle collision avoidance system, compact radars, etc. is extremely increasing at a rapid rate. Thus, the characteristics of the antennas involved in these gadgets are to be designed very stringently so as to avoid interferences & coupling and to improve compatibility, susceptibility, etc. Compact smart antenna with improved performance is highly essential to meet this challenging scenario. Mutual coupling between various elements of an array is one of the main factors which can be considered for improvement of performance of the antenna. Influence of mutual coupling on performance of the antenna is considered in this paper and various techniques to minimize this effect are presented. Effect of mutual coupling on radiation characteristics of the antenna can be compensated employing various methods like Conventional Mutual Impedance (CMI), Receiving Mutual Impedance (RMI). Analysis is presented as comparison between the two methods for different number of elements in the array. Analysis is also presented for different geometries of the array like circular and elliptical for improved performance. The results show performance improvement in the proposed array for parameters like SNR and Speed of convergence.

Keywords

Antenna Arrays, Elliptical (Oval) Arrays, Mutual Coupling, Compensation Methods, Beam Forming

1. Introduction

Wide spread application of wireless communication technology in daily life demands efficient and reliable signal transmission. Increased directivity, user capacity, battery life and source separation can be achieved by smart antenna to cope-up the demands. Smart antenna is popularly used in radar systems, wireless communications and in vehicle collision avoidance applications [1]. Antenna array is an integral part of smart antenna system. In smart antenna system, array of elements are designed to achieve required beam form for a given application.

A systematic arrangement of group of radiating elements is called an array. Size of the array and aperture proportionally increases with gain of the antenna [2]. Two methods to increase the aperture size of the array are to increase the number of elements and/or inter element spacing of the array [3]. Computational complexity increases with the number of elements. However, this leads to increase in grating lobes. Power efficiency reduces due to grating lobes [4]. A compromise between these two parameters can be achieved by reorienting the geometry and number of elements in the array.

Linear array is simple in structure and is limited to estimation of the elevation angle (one dimensional) of incoming signal only. In addition, the linear array can only estimate the range of azimuth angle limited by 180° (−90° < θ < 90°) [5]. Hence the accuracy of Direction of Arrival (DOA) estimation is greatly affected when the incoming signal is within the range of end-fire array (70° < |θ| < 90°).

The entire transmitted signal is equal to the sum and reradiated signal from the elements. Similarly, induced current on the antenna element reradiates electromagnetic field which would be received by the other neighboring elements in the array. Such mutual coupling effect is usually considered as a defect which degrades the performance of the array. Two types of coupling compensation methods: Conventional Mutual Impedance (CMI) and Receiving Mutual Impedance (RMI) methods are popular.

2. The Mutual Coupling Analysis

For an antenna array with N elements, receiving an incoming signal, being a plane wave source, the received terminal voltage at the k_{th} antenna element V_k can be expressed as

$$V_k = U_k + W_k \tag{1}$$

U_k = received voltage of k_{th} antenna element by the plane wave source alone

W_k = coupled voltage of the scattered fields from the other antenna elements in the array

The coupled voltage W_k in (1) can be written as [6]-[8]

$$W_K = Z_t^{k1} I_1 + Z_t^{k2} I_2 + \cdots + Z_t^{k(k-1)} I_{k-1} + Z_t^{k(k-1)} I_{k+1} \tag{2}$$

where Z_k is the receiving mutual impedance between the k_{th} and the i_{th} antenna elements and I_i is the terminal current at the i_{th} antenna element given by:

$$I_i = V_i / Z_L \quad \text{where } i = 1, 2, \cdots, N$$

with Z_L being the terminal load impedance of the antenna elements. Putting (2) and (3) into (1), we have:

$$V_k = U_k + Z_t^{k1} \frac{V_1}{Z_L} + Z_t^{k2} \frac{V_2}{Z_L} + \cdots + Z_t^{k(k-1)} \frac{V_{k-1}}{Z_L} + Z_t^{k(k-1)} \frac{V^{k+1}}{Z_L} + \cdots + Z_t^{kN} \frac{V^N}{Z_L} \tag{3}$$

The relationship between the uncoupled voltages U_k and the received voltages (*i.e.*, coupled voltages) V_k can be written in a matrix equation as:

$$\begin{bmatrix} U_1 \\ U_2 \\ \vdots \\ U_N \end{bmatrix} = \begin{bmatrix} 1 & \frac{Z_t^{12}}{Z_L} & \cdots & \frac{Z_t^{1N}}{ZL} \\ \frac{Z_t^{21}}{Z_L} & 1 & \cdots & \frac{Z_t^{2N}}{Z_L} \\ \vdots & \vdots & & \vdots \\ \frac{Z_t^{N1}}{Z_L} & \frac{Z_t^{N2}}{Z_L} & \cdots & 1 \end{bmatrix} \begin{bmatrix} V_1 \\ V_2 \\ \vdots \\ V_N \end{bmatrix} \tag{4}$$

2.1. Conventional Mutual Impedance Method

A circuit theory approach for reducing or compensating mutual coupling effect is discussed in this work. This method is called conventional mutual impedance method (CMI) [9]. Common mutual coupling can be easily measured directly or from S-parameters.

The relation between the terminal voltage and current can be given by

$$V_1 = I_1 Z_{K,1} + I_2 Z_{K,2} + \cdots + I_i Z_{k,i} + \cdots + I_K Z_{K,K} + \cdots + I_N Z_{K,N} + V_{OK} \tag{5}$$

Z_L being impedance, relation between voltage and current is given by

$$I_1^t = -\frac{v_i^t}{Z_L} \tag{6}$$

The relationship between the open-circuit voltages and terminal voltages can be written as

$$\begin{bmatrix} 1+\frac{Z_{11}}{Z_L} & \frac{Z_{12}}{Z_L} & \cdots & \frac{Z_{1N}}{Z_L} \\ \frac{Z_{21}}{Z_L} & 1+\frac{Z_{22}}{Z_L} & \cdots & \frac{Z_{2N}}{Z_L} \\ \vdots & \vdots & & \vdots \\ \frac{Z_{N1}}{Z_L} & \frac{Z_{N2}}{Z_L} & \cdots & 1+\frac{Z_{NN}}{Z_L} \end{bmatrix}_{N\times N} \begin{bmatrix} V_1 \\ V_2 \\ \vdots \\ V_N \end{bmatrix}_{1\times N} = \begin{bmatrix} V_{01} \\ V_{02} \\ \vdots \\ V_N \end{bmatrix}_{1\times N} \tag{7}$$

2.2. The Receiving Mutual Impedance

Another circuit theory approach for reducing or compensating mutual coupling effect is receiving mutual impedance method (RMI) [10]. The concept of receiving mutual impedance was suggested [11] as an alternative to the concept of conventional mutual impedance [12]. In fact, the need to handle mutual coupling problems differently in transmitting and receiving arrays has been noted before in mobile communications [13] and [14]. The basic difference between the receiving mutual impedance and the conventional mutual impedance lies in the fundamental causes of the coupling. In the conventional mutual impedance, the coupling is caused by the transmitting current distribution on an antenna while in the receiving mutual impedance; the coupling is caused by the receiving current distribution on an antenna.

The receiving mutual impedance between two antennas is the ratio of the coupled voltage across antenna1's terminal load Z_{L1} (due to the receiving current distribution on antenna 2) to the terminal current through antenna 2'sterminal load Z_{L2} when the array is excited by an external plane wave source.

$$Z_t^{12} = -(V_1 - U_1)/I_2 \tag{8}$$

i.e. Receiving mutual impedance with a receiving current on antenna 2 = coupled voltages across antenna 1's terminal load/receiving current through antenna 2's terminal load.

The definition of the receiving mutual impedance requires specifying a plane wave to excite the two antennas as shown in **Figure 1**. The coming direction of the plane wave is to be the same as the direction of the signal that the array is designed. In **Figure 1**, V_1 and V_2 are the received voltages across the terminal loads of antennas 1 and 2, respectively. The corresponding currents through the two antennas are I_1 and I_2. In equation (1), U_1 is the received voltage across the terminal load of antennas 1 when antenna 1 is excited by the external plane wave source alone (with antenna 2 removed from the array). U_1 is also called the isolation voltage on antenna 1.

3. Geometry and Array Factor

Two different geometries of the array are considered for comparison in this paper viz., Circular and Elliptical (Oval).

3.1. Circular Array

In circular array it is assumed that N equally spaced isotropic elements are placed on X-Y plane along a circular ring. Let the radius be "a". Circular array can able to scan 360° azimuthally. The geometry of N-element circular

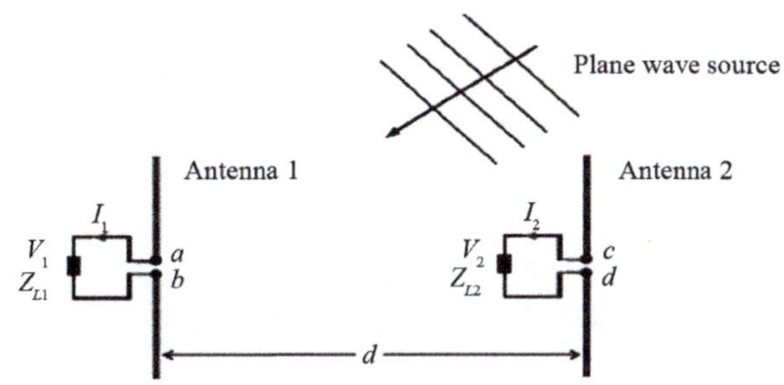

Figure 1. Mutual impedance between antennas.

array antenna is shown in **Figure 2** [15]. The array factor is given by the equation [16]

$$AF(\theta,\phi)=\sum_{N-1}^{N} I_n e^{j[ka\sin\theta\cos(\phi-\phi_n)+\alpha_n]} \tag{9}$$

where,

I_n = amplitude of excitation

α_n = phase of the n th element

θ = elevation angle from z axis.

Circumference = $2\pi r$

Area = πr^2

The radius of the array increases the directivity of uniform circular array and tends to a value N [17].

3.2. Elliptical Antenna Array

The geometry of the elliptical antenna array with origin as center is shown in **Figure 3** [18]

The array factor is given by equation (10) [19].

$$AF(\theta,\phi)=\sum I_n \exp\left(j\left[k\sin(\theta)\left(a\cos(\theta_n)\cos(\phi)+b\sin(\theta_n)\sin(\phi)\right)+\alpha_n\right]\right) \tag{10}$$

I_n=amplitude of excitation

α_n=phase of the n th element

θ=elevation angle from z axis.

$ø_n$=Azimuth angle measured from x axis for n-th element.

a, b=semi major and minor axises respectively.

e=eccentricity of elliptical array and is 0.5

$$\alpha_n=-k\sin(\theta_0)\left(a\cos(ø_n)\cos(\phi_0)+b\sin(ø_n)\sin(\phi_0)\right) \tag{11}$$

$\theta_0 = 90^\circ$, $ø_0 = 0^\circ$.

N=no of elements is 8 or 10.

$\text{Area}(A)=\Pi ab$

$\text{circumfcrcncc}(c)=\Pi\left(3(a+b)-\sqrt{(3a+b)(a+3b)}\right)$

Comparisons of area of circular and elliptical arrays are given below in **Table 1**.

4. Results

4.1. Conventional Mutual Impedance Method of Compensation

The objective is to analyze the response of elliptical antenna array with modified LMS (Least Mean Square) algorithm which optimizes the weight factor for DOA (Direction of Arrival) estimation. Mutual coupling is minimized by the CMI method in this case.

Example 1: Eight (Dipole) Element Array using CMI

In this example, an 8-element array using CMI is optimized with uniform spacing between the elements. It is assumed that the angle of signal and interference be 90° and 180° respectively.

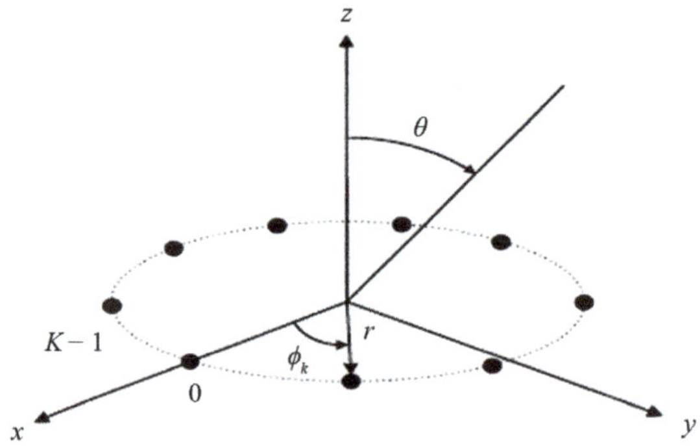

Figure 2. Geometry of circular antenna in XY-plane.

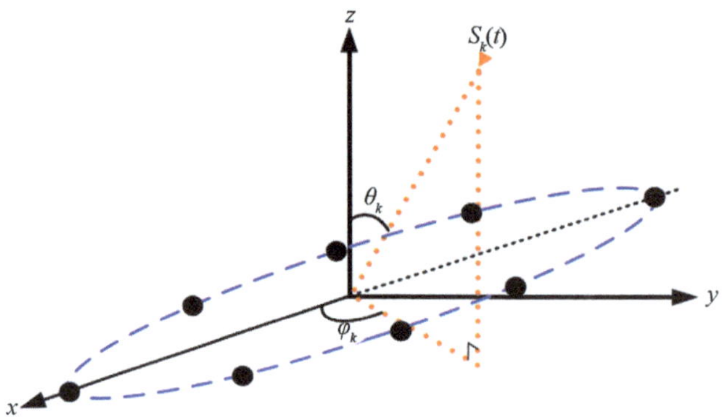

Figure 3. Geometry of ellipse in XY-plane.

Table 1. Comparision of area of circular and ellptical arrays.

Circumference (cm)	Circular and oval array area.				
	Circular array radius r (cm)	Circular array area A (cm)	Oval array major radius a (cm)	Oval array minor radius b (cm)	Area (cm^2)
1.00	0.16	0.08	0.20	0.10	0.07
1.50	0.24	0.18	0.30	0.15	0.15
2.00	0.32	0.32	0.40	0.21	0.27
3.00	0.48	0.72	0.62	0.31	0.60
4.00	0.64	1.27	0.84	0.42	1.07

It can be evident from **Figure 4** and **Figure 5**, that elliptical (oval) array with CMI method has reduced number of side lobes. Also, lower response of −20 dB observed in the interference direction, compared to circular antenna with −10 dB response.

Figure 6 and **Figure 7** compare the output noise in circular and elliptical antennas. Elliptical antenna has noise power which is less by 0.5 dB (negligible noise) for the iterations 10 to 100, whereas the circular antenna has considerable noise at low iteration rates below 20 for minimum of 0.5 dB. Hence it can be concluded that convergence is fast in elliptical array antenna than in circular for low noise.

Example 2: Ten (Dipole) Element Array using CMI

In this example, a 10-element array using CMI is optimized with uniform spacing between the elements. It is assumed that the angle of signal and interference be 90˚ and 180˚ respectively.

In **Figure 8**, iterations vs. SNR in circular array with 10-elements (uniform spacing between elements) using RMI Method is presented. **Figure 9** is similar plot for Elliptical array. The elliptical antenna leads with better SNR than circular antenna for iterations below 20. Circular antenna has −10 dB SNR and elliptical antenna has −2 dB which confirms the faster convergence of elliptical array.

Figure 10 shows circular array with CMI Method and without CMI Method which has deeper nulls if compensation is not applied. But elliptical array performance is good with null points covered by peaks as shown in **Figure 11**.

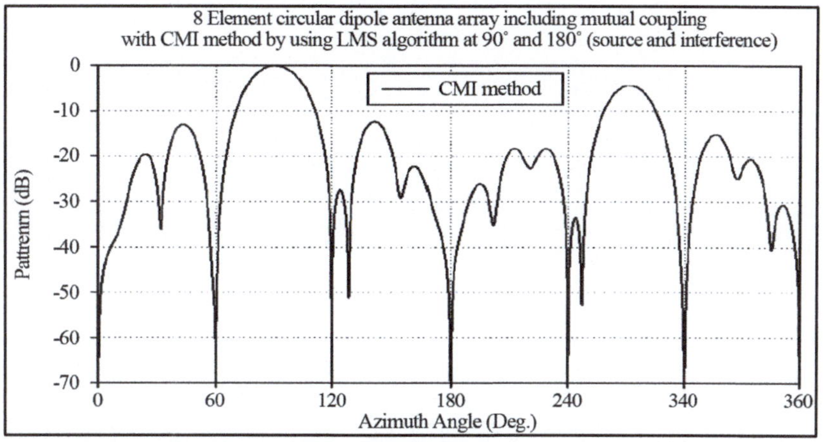

Figure 4. Circular array using CMI Method.

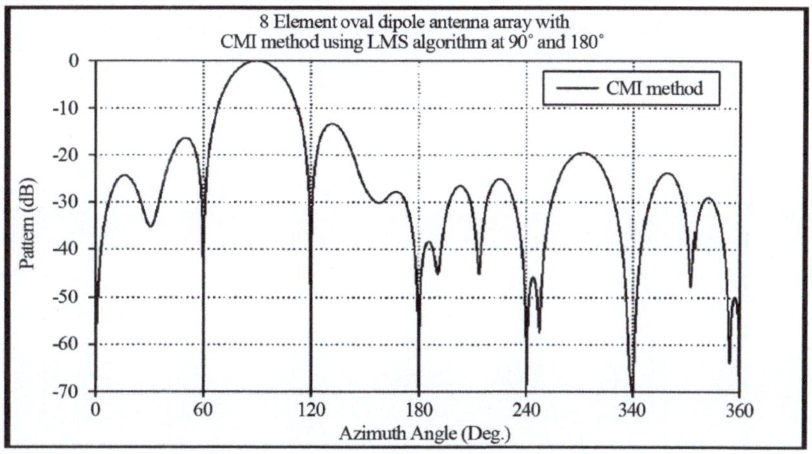

Figure 5. Elliptical (oval) array with CMI Method.

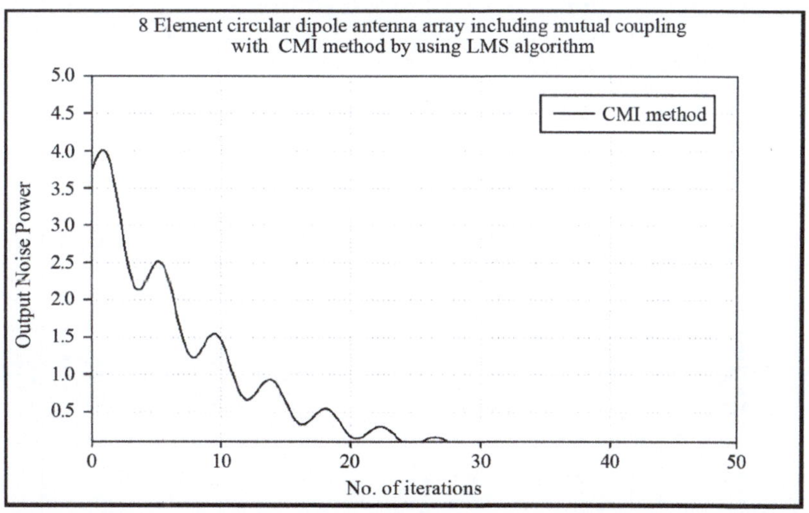

Figure 6. Iterations vs. output noise in circular array with CMIMethod.

In CMI compensation both 8 & 10 element arrays of circular and elliptical arrays are compared. Elliptical arrays show better in pattern generation, SNR and convergence speed. Advantage of elliptical antenna compared to circular antenna is its low area for same number of elements.

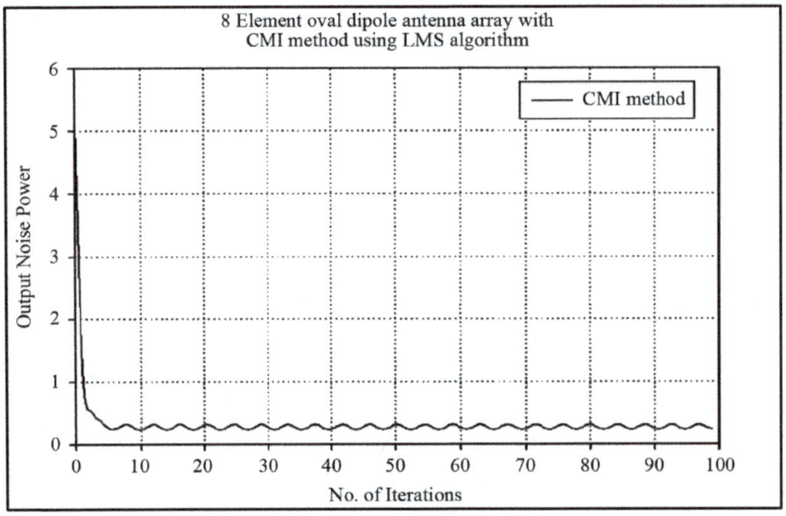

Figure 7. Iterations vs. output noise in Elliptical(oval) array using CMI Method.

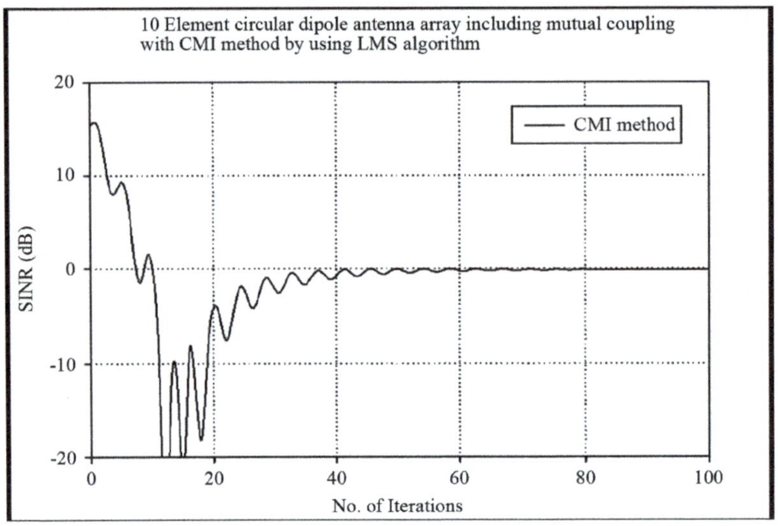

Figure 8. Iterations vs. SNR in circular array with CMI Method.

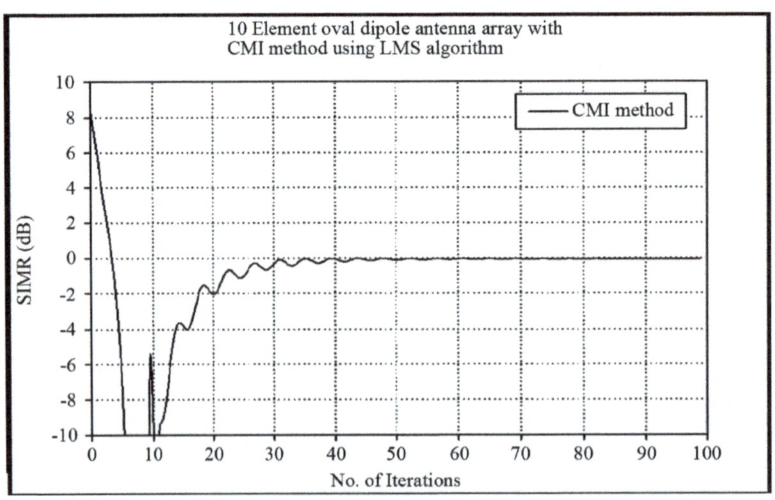

Figure 9. Iterations vs. output SNR in Elliptical array using CMI Method.

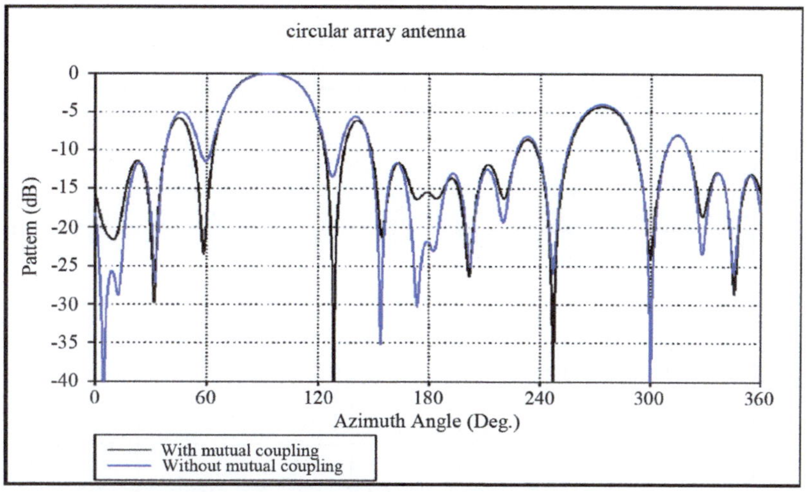

Figure 10. Circular array with and without CMI Method.

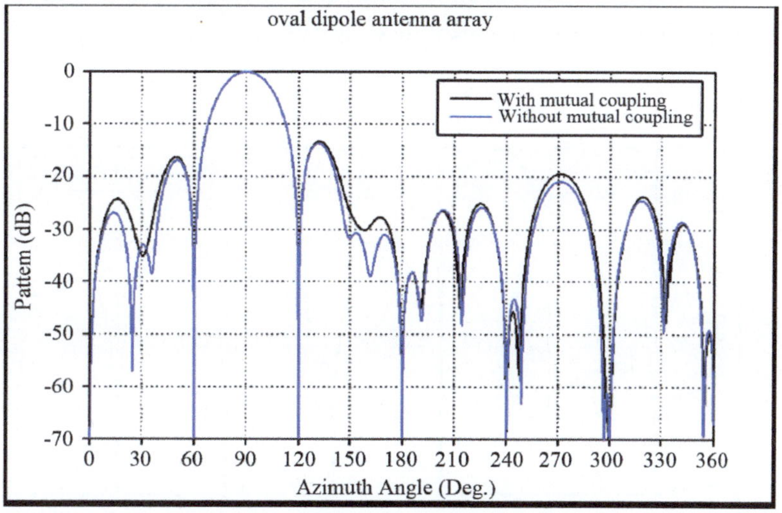

Figure 11. Elliptical array with and without CMI Method.

4.2. The Receiving Mutual Impedance Method of Compensation

In this RMI Method, the source and noise are assumed to be 90˚ and 180˚ of azimuth directions respectively. A modified LMS algorithm is used to optimize the weight factor for DOA estimation.

Example 3: Eight (Dipole) Element Array using RMI

In this example, an 8-element Circular array is optimized with uniform spacing between the elements.

The best amplitude value determined by the optimized technique is shown in **Figure 12**. From **Figure 13** it can be noted that the output noise power sustains minimum with the number of iterations above 10.

It can be evident that from **Figure 13** elliptical array with RMI Method has reduced side lobes and peaks compared with circular array with RMI Method in **Figure 12**. Elliptical antenna has lower response in the interference direction than circular antenna.

Figure 14 and **Figure 15** compare the output noise in circular and elliptical arrays. Elliptical antenna has negligible noise throughout the iterations from 10 to 100 below 0.5 dB, whereas the circular antenna has noise slightly more than that of elliptical antenna. Also below 10 iterations makes much difference in both antenna output noise. Convergence is faster in elliptical antenna.

Example 4: Ten (Dipole) Element Array using RMI

In this example, a 10-element array with uniform spacing between the elements is optimized using RMI. As per the **Figure 16** & **Figure 17** for Iterations vs. SNR in circular and elliptical arrays (no of elements 10) using

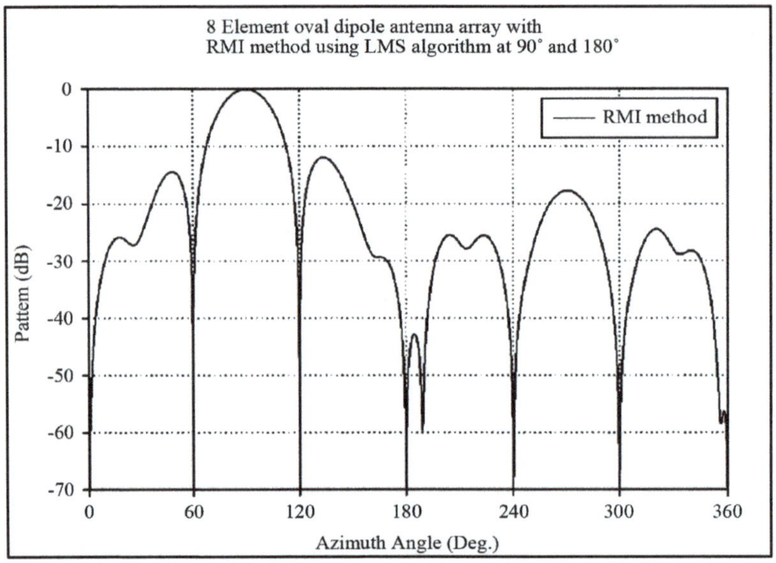

Figure 12. Circular array with RMI Method.

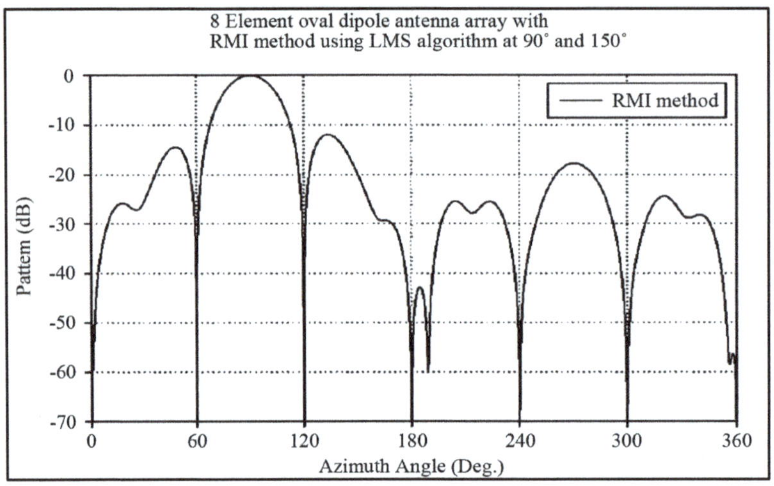

Figure 13. Elliptical (oval) array with RMI Method.

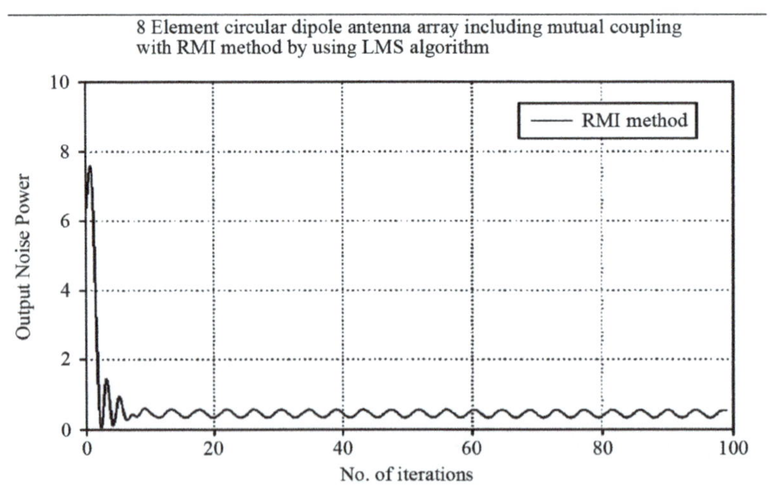

Figure 14. Iterations vs. output noise in circular array with RMI Method.

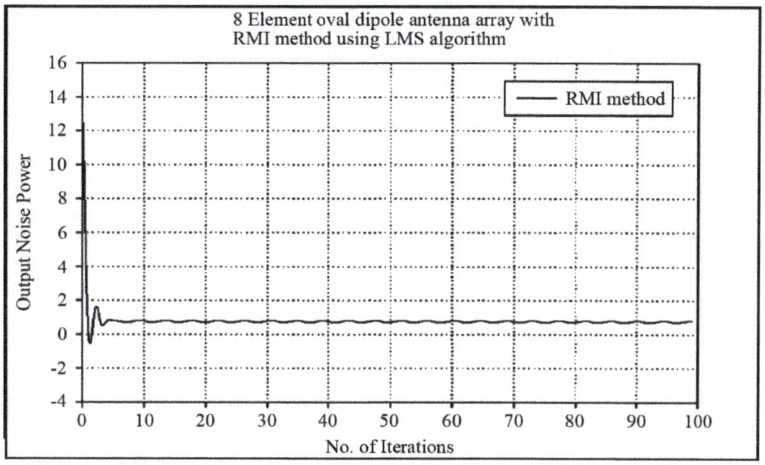

Figure 15. Iterations vs output noise in Elliptical (oval) array using RMI Method.

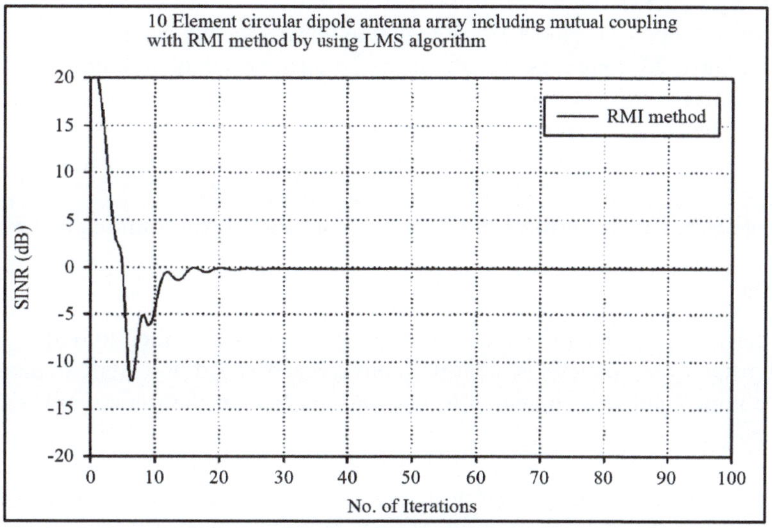

Figure 16. Iterations vs. SNR in circular array with RMI Method.

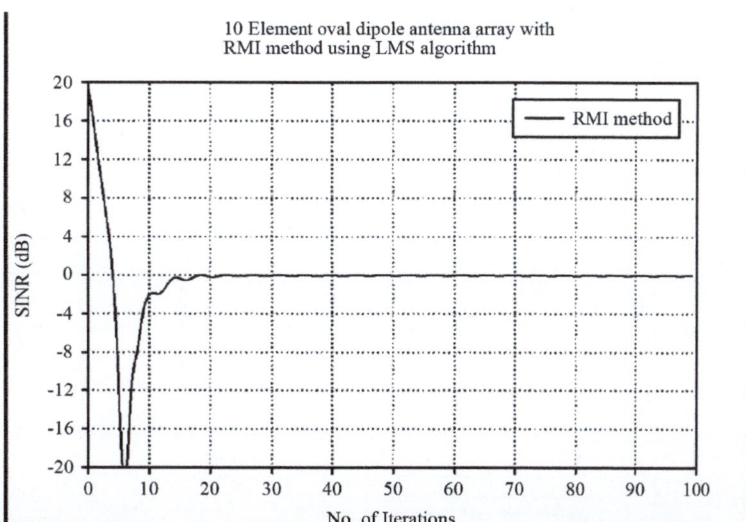

Figure 17. Iterations vs. output SNR in Elliptical array using RMI Method.

RMI Method, the elliptical antenna has better SNR below 20 iterations. Hence convergence is faster in elliptical antenna. For the iterations below 10, circular array has minimum of −12 db SNR.

Figure 18 and **Figure 19** presents the comparison between the patterns generated with RMI and without RMI method for circular array and elliptical antennas. Elliptical antenna beam minimized lobe in 120° to 240°, azimuth angle compared to circular antenna.

Example 5: Sixteen (Dipole) Element Array using RMI

In this example, an 16-element Circular & Elliptical arrays are optimized with uniform spacing between the elements. Simulations are shown in **Figure 20** & **Figure 21**. It observed that from **Figure 21** Elliptical (oval) array with RMI Method has reduced side lobes and peaks compared with circular array. Elliptical antenna has peak at −28 dB in the interference direction as circular antenna has −21 dB

In RMI compensation along with 8 & 10 element arrays of circular and elliptical arrays, 16-element arrays are also compared. Elliptical arrays shows better in pattern generation, SNR. Convergence speed is more improved. Geometrical advantage of compact size for elliptical antenna with same number of elements as in circular antenna is sustained.

More over RMI sanctions receiving current distribution on an antenna compared to CMI mode of compensation. It is additional advantage to recommend the RMI in elliptical arrays so that DOA estimation depends on Direction of receiving signal strength.

Example 6: Sixteen (Dipole) Element Array with non linear spacing using RMI

Figure 22 and **Figure 23** compares the output of 16-element circular & elliptical antenna with non linear spacing. Elliptical antenna has lower response at 210°, 270° and 330° Azimuth direction, whereas the circular antenna has response better than that of elliptical antenna. Also, side lobes are more in circular antenna.

Lobes between 180° - 240° & 300° - 330° are smoothening and shaped compared to circular array in elliptical antenna.

However, below 180° of angle uniform spaced array is seems to be performing better.

5. Conclusions

The simulations categorically show that CMI compensation method applied to elliptical array with 8 & 10 elements has better response in pattern generation, convergence speed and SNR compared with circular array. Compactness of elliptical antenna compared to circular antenna for the same number of elements is additional geometrical advantage.

Using RMI compensation method, a comparison is made for 8, 10 and 16 element array in circular as well as elliptical array. Elliptical arrays show better performance in pattern generation, SNR. Convergence speed is better compared to CMI method of compensation. Moreover, receiving current distribution on an antenna for RMI

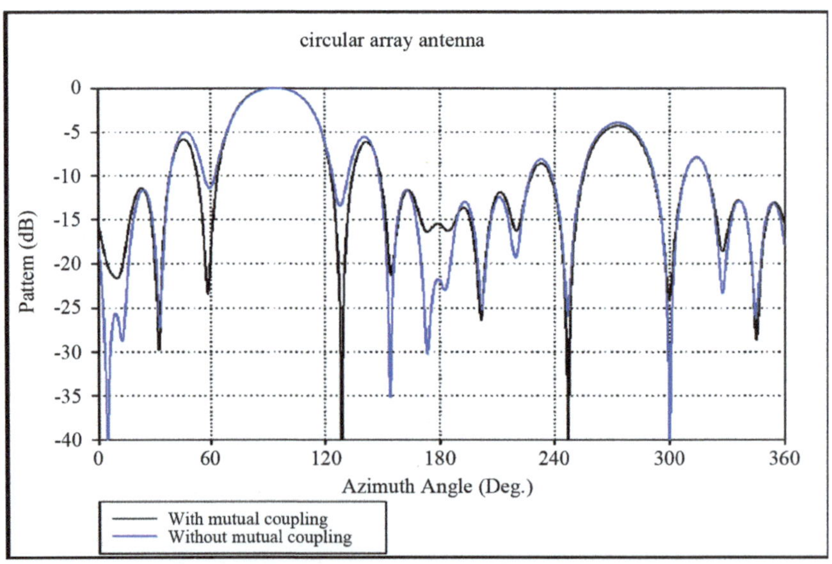

Figure 18. Circular array with and without RMI Method.

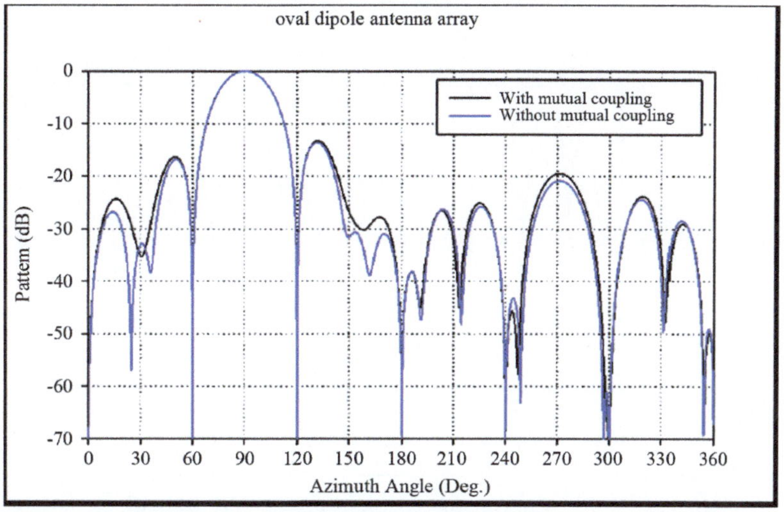

Figure 19. Elliptical array with and without RMI Method.

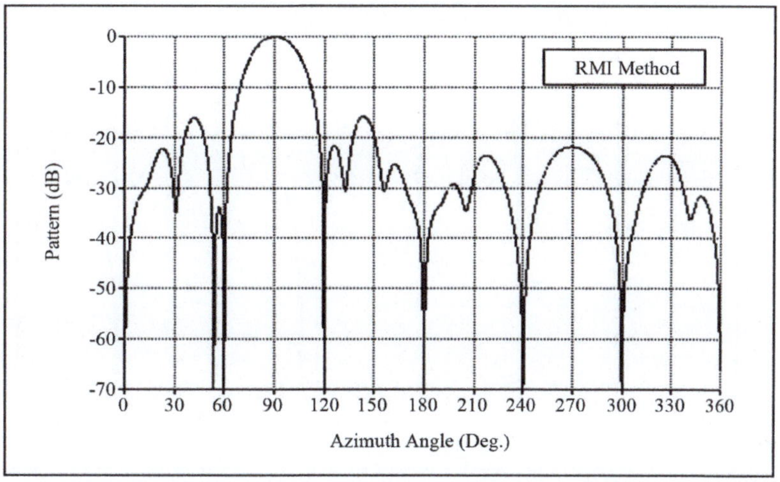

Figure 20. 16-Element Circular Dipole Antenna array with signal and interference at 90°and 180°.

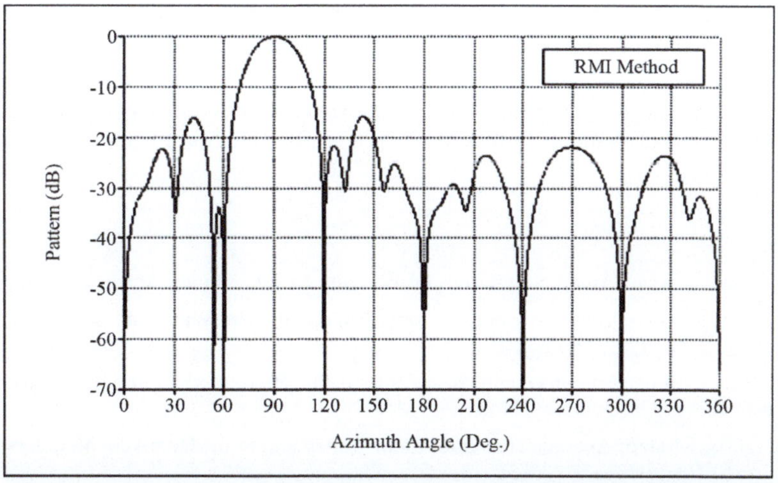

Figure 21. 16-Element elliptical (oval) Dipole Antenna array with signal and interference at 90° and 180°.

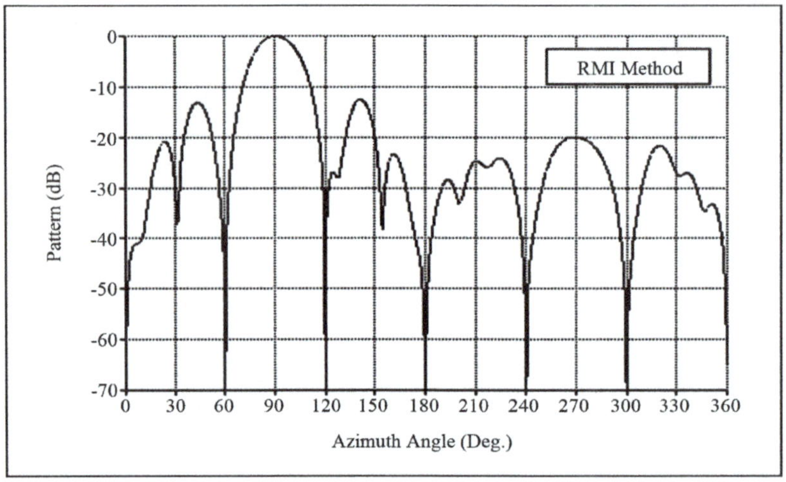

Figure 22. 16-Element Circular Dipole Antenna array with non-uniform signal and interference at 90˚ and 180˚.

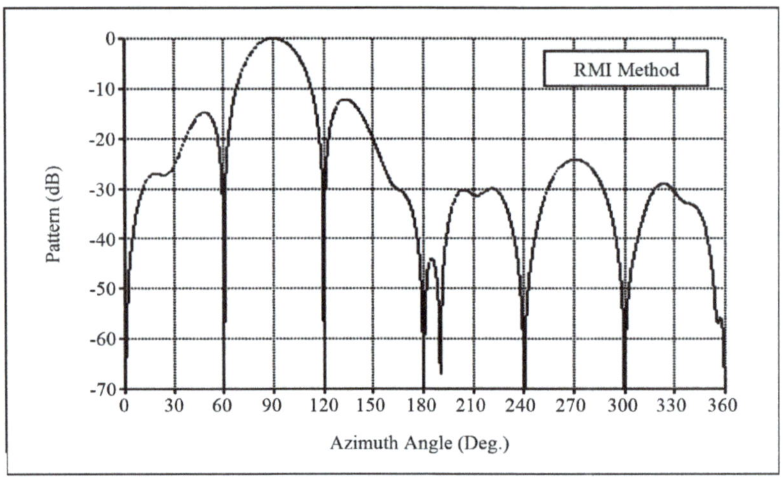

Figure 23. 16-Element Elliptical (oval) Dipole Antenna array with non-uniform signal and interference at 90˚ and 180˚.

method is better compared to CMI mode of compensation. It is additional advantage to recommend the RMI in elliptical arrays because DOA estimation depends on Direction of receiving signal strength.

Also, the simulation of non uniform spaced elements array in either circular or elliptical patterns confirms that below 180 degrees of angle, uniform spaced array performance is better. Hence, it can be concluded that elliptical array of uniform spaced elements with RMI compensation method of mutual coupling has the best performance in a smart antenna.

References

[1] Okamoto, G.T. (2002) Smart Antenna Systems and Wireless LANs. Kluwer Academic, Santa Clara.

[2] Haupt, R.L. (2010) Antenna Arrays: A Computational Approach. John Wiley and Sons, Hoboken, New Jersey. http://dx.doi.org/10.1002/9780470937464

[3] Fenn, A.J. (2008) Adaptive Antennas and Phased Arrays for Radar and Communications. Artech House Publishers, Massachusetts Institute of Technology, Norwood, MA.

[4] Balanis, C.A. (2008) Modern Antenna Handbook. John Wiley and Sons, Hoboken, New Jersey, 135. http://dx.doi.org/10.1002/9780470294154

[5] Ioannidis, P. and Balanis, C.A. (2005) Uniform Circular Arrays for Smart Antennas. *IEEE Antennas and Propagation Magazine*, **47**, 192-206. http://dx.doi.org/10.1109/MAP.2005.1589932

[6] Foutz, J., Spanias, A. and Banavar, M.K. (2008) Narrowband Direction of Arrival Estimation for Antenna Arrays. Morgan & Claypool, Arizona State University, Tempe, AZ.

[7] Lui, H.-S. Hui, H. T. and Leong, M. S. (2008) Investigation of Direction-of-Arrival Estimation Using Uniform Linear Arrays with Different Antenna Separations and Array Apertures. Presented at the Asia Pacific Microwave Conference.

[8] Balanis, C.A. (1997) Antenna Theory: Analysis and Design. Wiley, New York.

[9] Collin, R.E. and Zucker, F.J., Eds. (1968) Antenna Theory. McGraw-Hill, New York.

[10] Balanis, C.A. and Ioannides, P.I. (2007) Introduction to Smart Antenna. Morgan and Claypool, Arizona.

[11] Singh, H., Sneha, H.L. and Jha, R.M. (2013) Mutual Coupling in Phased Arrays: A Review. *International Journal of Antennas and Propagation*, **2013**, Article ID: 348123.

[12] Hopkins, G.D., Ratner, J., Traille, A. and Tripp V. (2007) Aperture Efficiency of Amplitude, Weighting Distributions for Array Antennas. 2007 *IEEE Aerospace Conference*, Big Sky, 3-10 March 2007, 1-9. http://dx.doi.org/10.1109/aero.2007.352856

[13] Cao, W.Q., Xiang, Y., Zhang, B.N., Liu, A.J., Yu, T.B. and Guo, D.S. (2011) A Low-Cost Compact Patch Antenna with Beam Steering Based on CSRR Loaded Ground. *IEEE Antennas and Wireless Propagation Letters*, **10**, 1520-1523. http://dx.doi.org/10.1109/LAWP.2011.2181316

[14] Chan, S.C. and Chen, H.H. (2006) Uniform Concentric Circular Arrays with Frequency Invariant Characteristics—Theory, Design, Adaptive Beam Forming and DOA Estimation. *IEEE Transactions on Signal Processing*, January 2007, 165-177.

[15] Tvrdik, J., Pavliska, V. and Habiballa, H (2008) Stochastic Self-Adaptive Algorithms for Global Optimization—MATLAB and C++ Library. http://www1.osu.cz/home/habibal/optimization/index.html

[16] Ammula, V., Rao, S.M. and Wentworth, S.M. (2010) Dual Excited Planar Circular Array Antennas for Direction Agile Applications. *42nd Southeastern Symposium on System Theory*, Tyler, 7-9 March 2010, 138-142. http://dx.doi.org/10.1109/ssst.2010.5442853

[17] Ahmidi, N., Neyestanak, A. and Dawes, R. (2008) Elliptical Array Antenna Design Based on Particle Swarm Method Using Fuzzy Decision Rules. *24th Biennial Symposium on Communications*, Kingston, 24-26 June 2008, 352-355. http://dx.doi.org/10.1109/BSC.2008.4563274

[18] Chou, J.Y.-L. (2002) An Investigation on the Impact of Antenna Array Geometry on Beam Forming User Capacity. Master of Science (Engineering), Queen's University, Kingston.

[19] Ioannidis, P. and Balanis, C.A. (2005) Uniform Circular Arrays for Smart Antennas. *IEEE Antennas and Propagation Magazine*, **47**, 192-206. http://dx.doi.org/10.1109/MAP.2005.1589932

29

Characterization of Breast Cancer Radiofrequency Ablation Assisted with Magnetic Nanoparticles: *In Silico and in Vitro Study*

Silverio Soto Alvarez[1], Lidia Flor E. Huerta[1], Alma Verónica Vargas[2], Jaime López[2], Jesus Gabriel Silva[1], César A. González[1,2*]

[1]Universidad del Ejército y FAM/EMGS, Laboratorio Multidisciplinario de Investigación, Mexico D.F., Mexico
[2]Instituto Politécnico Nacional, Escuela Superior de Medicina, SEPI, Mexico D.F., Mexico
Email: *cgonzalezd@ipn.mx

Abstract

The effectiveness of breast cancer ablation by radiofrequency (RF) has been associated to the capacity of concentrating the electromagnetic energy in the tumor region, our group has proposed that this condition could be modulated by appropriate RF exposure cycle times as well as modification of tissue electrical conductivity. The aim of this work was to evaluate analytical and experimental optimal exposure cycle times to induce tissue ablation by RF assisted with magnetic nanoparticles. The study was conducted both analytically by multiphysics simulation of the induced currents in cancer tissue given a magnetron source and experimentally by the observation of hyperthermic effects induced in agar phantoms by a magnetron device by the use of RF assisted with magnetic nanoparticles. The temperature showed significant increases in short periods of time, and was clearly higher under the condition with nanoparticles. Appropriate RF exposure cycle times as well as modification of tissue electrical conductivity by magnetic nanoparticles seems suitable factors to modulate the electromagnetic energy in tumoral regions.

Keywords

Breast Cancer, Radiofrequency Ablation, Multiphysics Simulation, Magnetic Nanoparticles

1. Introduction

Breast Cancer (BC) is the most frequent malignancy and the leading cause of death in women worldwide [1].

*Corresponding author.

The first therapy option for patients with localized breast cancer is radical mastectomy (surgery) and chemotherapy, radiotherapy and immunotherapy. In the last two decades minimally invasive alternative therapies have been developed, such as cryo-ablation, focused ultrasound high-intensity and radiofrequency ablation (RFA) [2] [3]. Radiofrequency ablation reached importance in recent years for treatment of cancerous tumors. Ablation therapy of tumors by hyperthermia is an application of magnetic nanoparticles, since they have the capacity to produce heat in response to application of external magnetic fields [4]. It is one of the procedures that have gained importance in recent years for the focal destruction of cancerous tumors in general, originally used in the treatment of liver tumors in early 1990s [5] [6]. His clinical potential has expanded significantly in the treatment of breast cancer; however, experimental results of several research groups indicate that the effectiveness of the RFA in the treatment of cancer appears to be associated with the potential ability to focus electromagnetic energy in a specific zone [5]-[7]. Tumors which are smaller than 3 cm in the greatest diameter are the best candidates for this type of ablation [5]. The aim of this work was to investigate theoretical and experimental optimal characteristics of magnetic fields to produce selective biodegradation effects in tumoral tissue with magnetic nanoparticles concentrations.

1.1. Biophysical Assumptions

The induction of currents in conductive materials by oscillating magnetic fields is explained by the induction Faraday's law in terms of general Maxwell differential equations:

$$\Delta \boldsymbol{B} \times E = -\partial B / \partial t \tag{1}$$

a magnetic field $\boldsymbol{B}$ variable in time induces an electric potential E in a conductive medium, this potential induces electric currents in the medium (Eddy currents). In addition, an induced current density J in a conductive medium is directly proportional to the electrical potential E and the electrical conductivity of the material as described following equation:

$$J = \sigma E \tag{2}$$

Equation above indicates that the absorbed energy by a conductive material is directly proportional to its electrical conductivity [5], and accordingly to the conservation charge law; the energy absorbed will be balanced as a heat dissipation, in this sense we have proposed the use of magnetic nanoparticles to increase the tissue electrical conductivity and concentrate the electromagnetic energy in the tumoral region, this procedure might be considered as focalized RF ablation technique.

1.2. Radiofrequency Ablation (RFA) and Magnetic Nanoparticles (MNP's)

Tumor ablation therapy by RF has the potential to be magnified by the use of magnetic nanoparticles (MNP's), since they have the ability to produce heat in response to application of external magnetic fields. The concept of RF tissue hyperthermia assisted by MNP's applies when MNP's are concentrated in the tumor site, and remotely apply heating by magnetic fields in order to reach temperatures up to 55°C. Ferromagnetic materials generate high heat output and allow the use of magnetic fields at lower frequency and intensity [8]. To the best of our knowledge, there is currently no selective technique that allows ablative process of cancer tissue. Taking advantage of the properties of electrical conductivity and relative permittivity of biological tissues, and based on physical theories of electromagnetism, a biophysical model is proposed to estimate the frequency and optimal exposure time cycles of a magnetic field to promote simulated ablation of cancerous tissue and experimentally tissue hyperthermia by magnetic induction assisted with MNPs.

1.3. COMSOL Multiphysics

The mathematical analysis was performed with COMSOL Multiphysics (COMSOL, Inc., Boston, MA, USA, 2005) which employs the finite element method to solve Equation (3) for the magnetic vector potential. Every structural component (subdomain) of the model was meshed in triangular elements. To define boundary conditions the inner electrical parameters of the surrounded sphere correspond to the air and represent electric and magnetic insulation. The process carried out for a modeling using COMSOL Multiphysics is defined by creating a geometry, mesh, physical specifications, the choice of solution and visualization results [9] [10].

1.4. Time-Harmonic, Quasi-Static and Electromagnetic Assumption

In the *time-harmonic* case we omit the displacement currents in the Ampere's equation [11]. To simplify initial conditions; we consider a non-moving geometry ($\boldsymbol{v}=0$) and no external electric potential gradient ($\Delta V=0$) on the border of the pixel model. In addition, we have assumed that the border of the sphere surrounding the system defined in above section is grounded. Thus, we can obtain:

$$\left(j\omega\sigma-\omega^{2}\varepsilon\right)\boldsymbol{A}+\nabla\left(\mu^{-1}\nabla\times\boldsymbol{A}\right)=\boldsymbol{J}^{e} \tag{3}$$

$$\nabla\times\left(\mu_{r}^{-1}\nabla\times\boldsymbol{E}\right)-\mathrm{K}_{0}^{2}\left(\varepsilon_{r}-j\sigma/\omega\varepsilon_{0}\right)\boldsymbol{E}=0,\varepsilon_{r}=n^{2} \tag{4}$$

where σ is the medium conductivity (siemens/meter), ω is the angular frequency (rad/s), $\boldsymbol{J}^{e}$ is a current density generated externally (ampere/meter^2), μ is the medium permeability (henry/meter), ε is the medium permittivity (farad/meter) and $\boldsymbol{E}$ is the electric field. In this study equation (3) was solved to obtain the magnetic vector potential ($\boldsymbol{A}$), a detailed deduction of (3) and its solution is shown in [12].

2. Materials and Methods

The study was designed to estimate the induced currents in cylindrical tissue phantoms by a RF magnetron device as well as the involved hyperthermic effect as follows: 1) Theoretical estimation by computer simulation (COMSOL Multiphysics 3.4) and 2) Experimental estimation of hyperthermic effects in agarose phantoms, both conditions by the use of a magnetron device as RF source. Estimations evaluate the presence of tumoral tissue in phantoms under the influence of MNP's. **Figure 1** shows a general view of the experimental design.

2.1. Multiphysics Simulation

Simulation was developed by the use of the RF module of COMSOL Multiphysics 3.4 (2.45 GHz frequency), and an external current density of 1000 A. Electrical conductivity as well as magnetic flux density were observed in two conditions: Phantom with tumoral tissue and Phantom with tumoral + MNP's. Subdomains description of the simulated model is as follows:

a) Subdomain 1: 29 × 21 × 30 cm cube. Simulates a RF exposure chamber and limits boundary conditions (air).
b) Subdomain 2: Cylinder 4 cm diameter and 7.4 cm length. Simulates phantom of healthy tissue.
c) Subdomain 3: Cylinder 1 cm diameter and3 cm length, simulates a RF antenna (magnetron).

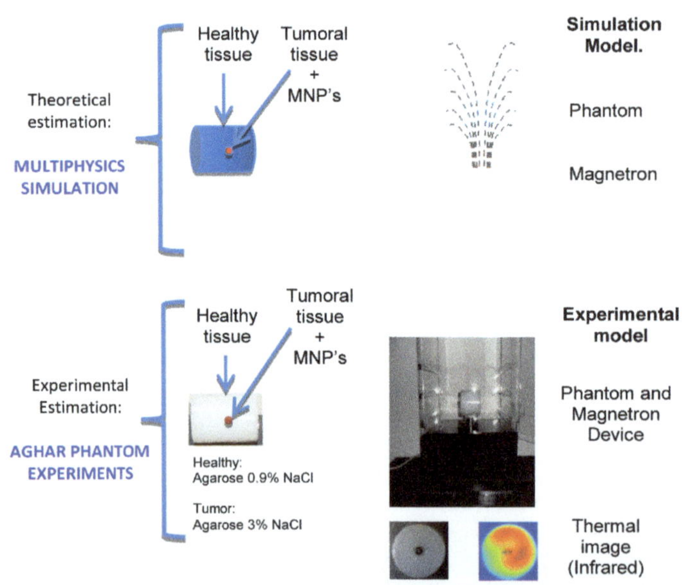

Figure 1. Experimental design for evaluating temperature increases in a cylindrical volume with electrical properties of biological tissues. The oreticaland experimental estimations.

d) Subdomain 4: Sphere 0.5 cm radius, simulates phantom of tumoral tissue.

e) Subdomain 5: Spheres of 0.04 cm radius, simulates concentrations of MNP's.

Respective electrical properties values such as conductivity (σ) and relative permittivity (ε) were assigned to every subdomain and Equations (3) and (4) were solved. **Figure 2** showed subdomains of the simulated model.

2.2. Agar Phantom Design and RF Experiment

Agar phantoms to emulate cylindrical healthy tissue volumes were designed and built with saline solution (C.S. PISA calve 3608), Agar (Invitrogen Cat No. 30391) and Sucrose (Gibco BRL Cat. No. 5503UA) accordingly to the technique previously described in [13]. In brief, a volume of 200 ml of NaCl 0.9%, Agar 5% and Sucrose 3% were homogenized at 90˚C and dropped in a cylindrical model. Additional cylindrical phantoms to emulate tumor volumes of 1 ml were developed with NaCl 3% (J.T. Baker, Cat. No. 3624-01) Agar 5% and Sucrose 3%, such cylinders were inserted in the posterior side of the cylindrical healthy phantom. Thus; phantoms in three conditions were prepared: 1) Healthy + Tumor + MNPs, 2) Healthy + Tumor and 3) Healthy (**Figure 3**).

An experimental RF device was designed on the basis of an industrial microwave generator type magnetron (Magnetron-Daewoo Model 2M218 JF, 2.45 1KW GHz) holding in a cubic acrylic chamber specifically designed to accommodate phantoms and minor species (**Figure 4**). The sequence of RF exposure was determined as 6 cycles of 25 seconds exposure/35 seconds relaxation. The phantoms RF exposure order was randomized and the experiment was developed by triplicate. Thermographic images of the phantoms were acquired with an infrared camera FLUKE® model Ti10, 76,800 pixels, which operates in the infrared spectral band of 7.5 to 14 μM with a thermal sensitivity ≤0.045˚C (Noise Equivalent Temperature Difference-NETD). The images were taken at the same time of day in an environment of temperature and relative humidity of 23˚C and 40% respectively, these parameters were monitored for the purpose of operating equipment of thermography under the same conditions only. Thermographic images were taken at a distance of 50cm with an isothermal surface background.

3. Results

3.1. Multiphysics Simulation

The results of multiphysics simulation are shown in **Table 1**. Increments of magnetic flux density and its induced currents in tumoral tissue as a function of the presence of MNP's are evident. **Figure 5** shows the graphic simulation results of magnetic flux density and electrical conductivity in the simulated tissue model.

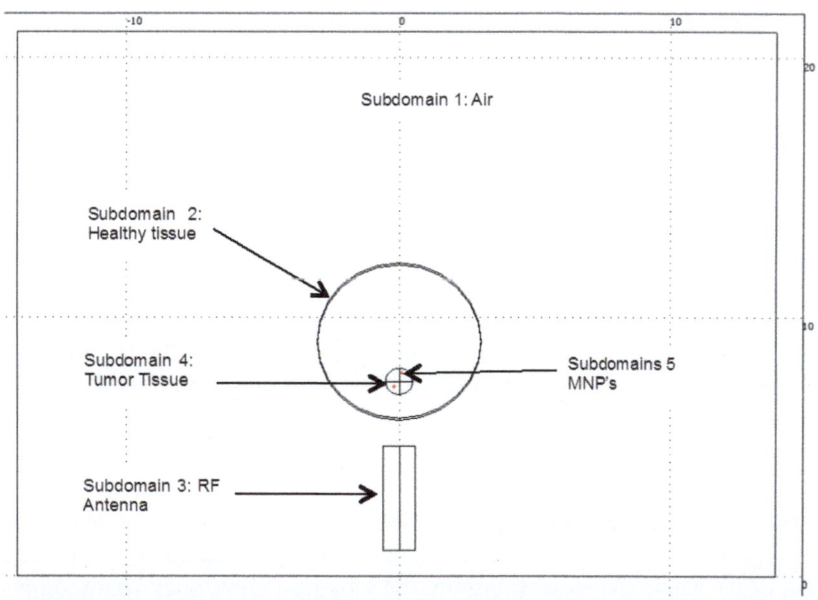

Figure 2. Simulation magnetron in COMSOL Multiphysics 3.4, with their respective subdomains, YZview.

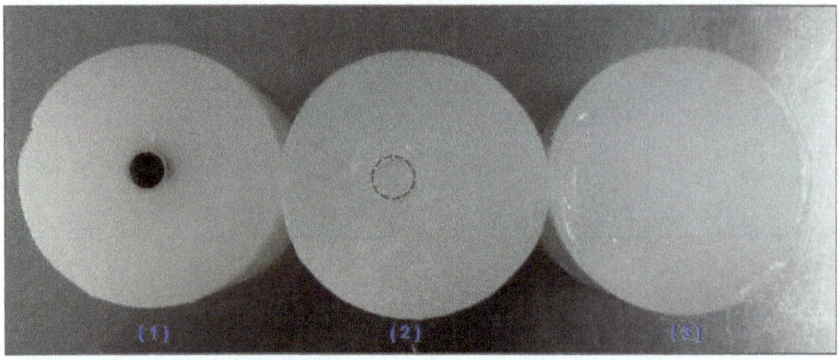

Figure 3. Phantoms prepared in three conditions: 1) Healthy + Tumor + MNPs, 2) Healthy + Tumor and 3) Healthy.

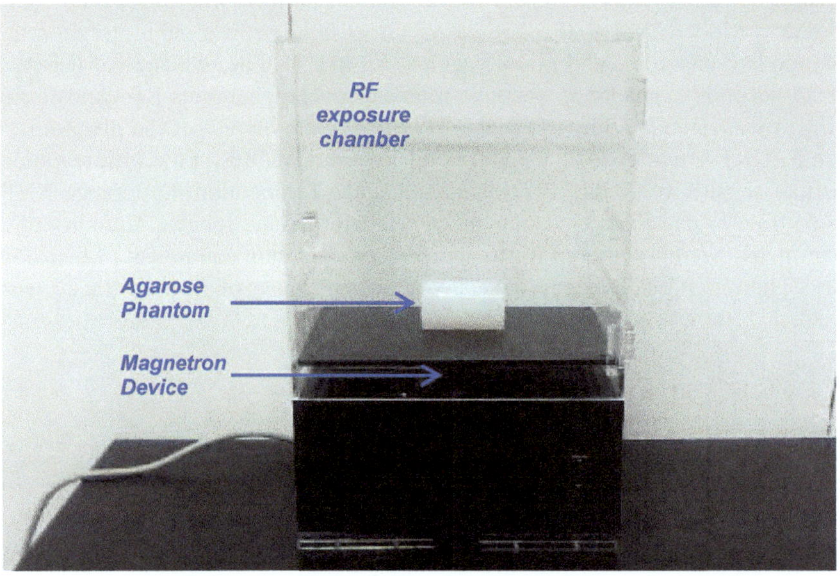

Figure 4. Magnetron device in acrylic holder with a representative agarose phantom.

Table 1. Magnetic fluxdensity estimation.

Without MNP's		With MNP's	
Healthy Tissue	Tumor Tissue	Healthy Tissue	Tumor Tissue
6.610202×10^{-7} T·m^3	1.622555×10^{-9} T·m^3	6.58401×10^{-7} T·m^3	1.744106×10^{-9} T·m^3

3.2. Experimental Simulation.

Figure 6 shows the experimental hyperthermia effect of the RF device in phantoms as a function of time (six cycles) and phantom conditions, relevant temperature increments for the Healthy + Tumor + MNP's condition are evident.

4. Discussion

The analytical estimations indicate a major concentration of the induced current density lines in the tumor region, the observations suggest that electrical conductivity increments as a function of the presence of MNP's promotes the concentration of the electromagnetic energy in the tumoral region, such condition is in agreement with Faraday's law as previously anticipated in the introduction section. The experimental findings confirm the primary hypothesis of this study, electrical conductivity increment given by the presence of MNP's in the tumoral region

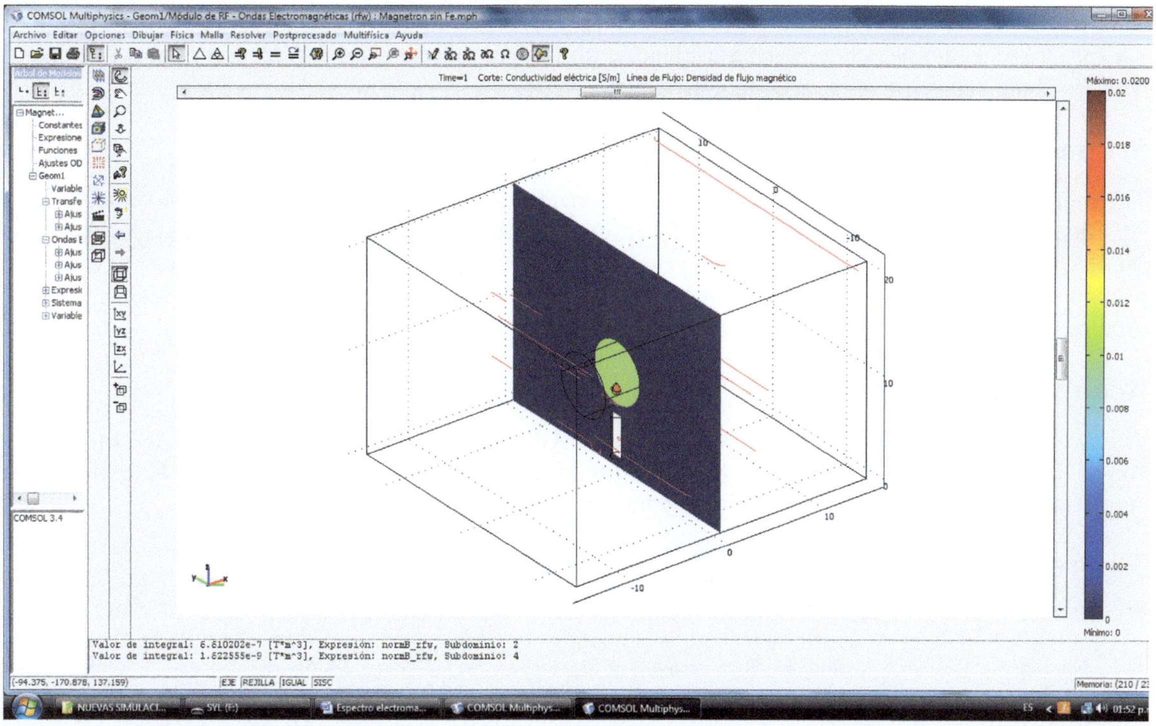

Figure 5. Induction of currentsin biological tissues (simulation), the electrical conductivity in colors, and themagnetic flux density through flow linesis observed.

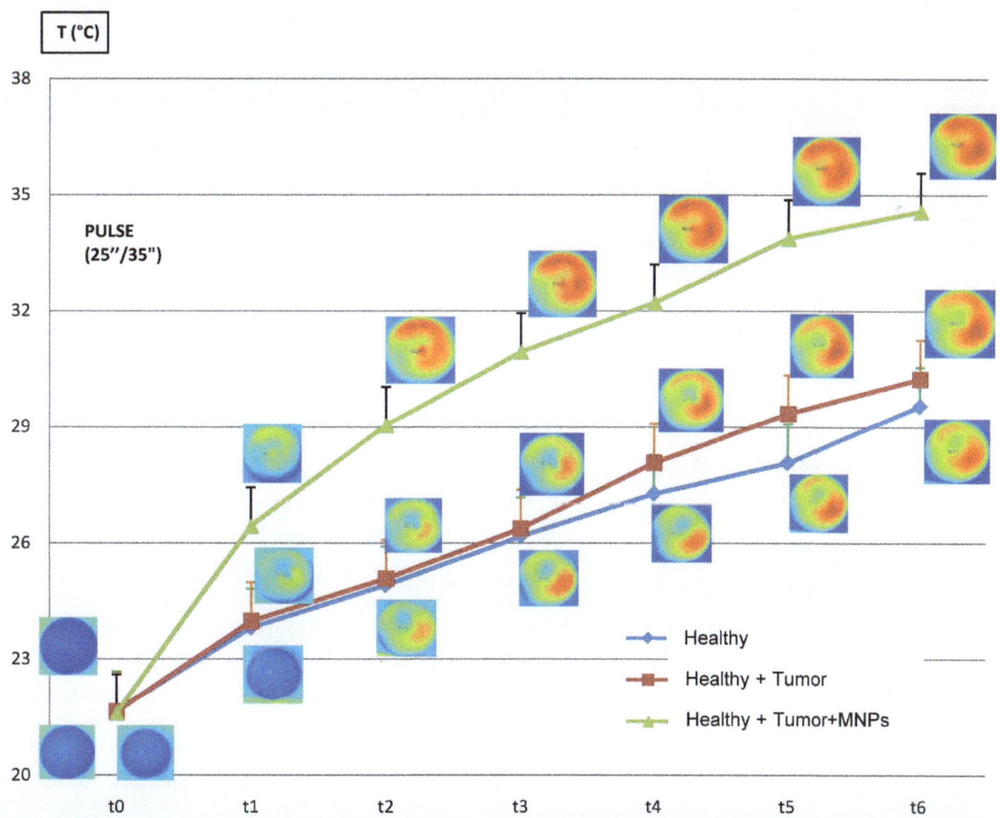

Figure 6. Temperature in crementsin experimental conditions 1) Healthy + Tumor + MNPs, 2) Healthy + Tumor and 3) Healthy. Significative increments in the condition with MNP's as a function of exposure cycles is observed.

is reflected as focal temperature increments, thus the explored RF exposure cycles (6 pulses of 25"/35" —exposure/relaxation) seem a feasible technique to produce selective biodegradation effects in tumoral tissue with magnetic nanoparticles concentrations.

5. Conclusion

The induced current densities and temperature estimated theoretically and experimentally respectively showed significant increments under the explored RF exposure cycles, and were clearly higher under the presence of MNP's. The modulation of RF exposure cycle times as well as modification of tissue electrical conductivity by MNP's seems a feasible technique to concentrate the electromagnetic energy in specific regions and eventually promotes selective ablation of tumoral tissue.

Acknowledgements

This work was performed in the "Escuela Militar de Graduados de Sanidad-Universidad del Ejército y Fuerza Aérea" (EMGS-UDEFA) Col. Lomas de San Isidro, Miguel Hidalgo, México D.F. C.P. 11620, México, and was supported by "CONACYT CB-2012 under grant No. 180536".

References

[1] González, C.A., Uscanga, M.C., Ibarra, C.D., Jiménez, M.E., Lozano, L.M., Silva, J.G., *et al.* (2012) Differentiation BIRADSI vs II by Magnetic Induction Spectroscopy: Innovative Potential Method for Early Detection of Breast Neoplasms. Research Article. *Mexican Journal of Biomedical Engineering*, **33**, 65-76.

[2] Zagoria, R.J., Hawkins, A.D., Clark, P.E., *et al.* (2004) Percutaneous CT-Guided Radiofrequency Ablation of Renal Neoplasms: Factors Influencing Success. *American Journal of Roentgenology* (*AJR*), **183**, 201-207. http://dx.doi.org/10.2214/ajr.183.1.1830201

[3] Hilger, I., Rapp, A., Greulich, K.-O. and Kaiser, W.A. (2005) Assessment of DNA Damage in Target Tumor Cells after Thermoablation in Mice. *Radiology*, **237**, 500-506. http://dx.doi.org/10.1148/radiol.2372041451

[4] Ramos, M. and Castillo, C. (2011) Biomedical Applications of Magnetic Nanoparticles. *Concyteg Ideas*, **6**, 629-646.

[5] González, D.C. (2012) Carcinogenic Ablation Breast Tissue through Radiofrequency Assisted with Bioconjugated Nanoparticles. Research Protocol. IPN School of Medicine Section of Graduate Studies and Research, Mexico D.F.

[6] Florez, C.D., *et al.* (2012) Radiofrequency Thermal Ablationand the Rapeutic Management of Patients with Lung Cancer. Case Report, Fundación Santa Fe de Bogotá, Bogotá.

[7] Goldberg, S.N., Gazelle, G.S. and Mueller, P.R. (2000) Thermal Ablation Therapy for Focal Malignancy: A Unified Approach to Underlying Principles, Techniques, and Diagnostic Imaging Guidance. *American Journal of Roentgenology* (*AJR*), **174**, 323-331. http://dx.doi.org/10.2214/ajr.174.2.1740323

[8] Jordan, A., Scholz, R., Maier-Hauff, K., *et al.* (2006) The Effect of Thermotherapy Using Magnetic Nanoparticles on Rat Malignant Glioma. *Journal of Neuro-Oncology*, **78**, 7-14. http://dx.doi.org/10.1007/s11060-005-9059-z

[9] www.taringa.net/posts/ciencia-educacion/6306379

[10] Mendoza, G.A. (2013) Introduction to the Software COMSOL Multiphysics. Thesis, Faculty of Mechanical and Electrical Engineering, Universidad Veracruzana, Mexico.

[11] Purcell, E.M. (1984) Electricity and Magnetism, Berkeley Physics Course. *Berkeley Physics*, **2**, 263-265.

[12] Gonzalez, C.A., Rojas, R. and Rubinsky, B. (2007) Circular and Magnetron Inductor/Sensor Coils to Detect Volumetric Brain Edema by Inductive Phase Shift Spectroscopy: A Sensitivity Simulation Study. *Proceedings of the* 13*th International Conference on Electrical Bioimpedance and* 8*th Conference on Electrical Impedance Tomography*, Graz, 29th August-2nd September 2007, 315-319. http://dx.doi.org/10.1007/978-3-540-73841-1_83

[13] Solazzo, S.A., Liu, Z., Lobo, M., *et al.* (2005) Radiofrequency Ablation: Importance of Background Tissue Electrical Conductivity—An Agar Phantom and Computer Modeling Study. *Radiology*, **236**, 495-502.

Permissions

All chapters in this book were first published in JEMAA, by Scientific Research Publishing; hereby published with permission under the Creative Commons Attribution License or equivalent. Every chapter published in this book has been scrutinized by our experts. Their significance has been extensively debated. The topics covered herein carry significant findings which will fuel the growth of the discipline. They may even be implemented as practical applications or may be referred to as a beginning point for another development.

The contributors of this book come from diverse backgrounds, making this book a truly international effort. This book will bring forth new frontiers with its revolutionizing research information and detailed analysis of the nascent developments around the world.

We would like to thank all the contributing authors for lending their expertise to make the book truly unique. They have played a crucial role in the development of this book. Without their invaluable contributions this book wouldn't have been possible. They have made vital efforts to compile up to date information on the varied aspects of this subject to make this book a valuable addition to the collection of many professionals and students.

This book was conceptualized with the vision of imparting up-to-date information and advanced data in this field. To ensure the same, a matchless editorial board was set up. Every individual on the board went through rigorous rounds of assessment to prove their worth. After which they invested a large part of their time researching and compiling the most relevant data for our readers.

The editorial board has been involved in producing this book since its inception. They have spent rigorous hours researching and exploring the diverse topics which have resulted in the successful publishing of this book. They have passed on their knowledge of decades through this book. To expedite this challenging task, the publisher supported the team at every step. A small team of assistant editors was also appointed to further simplify the editing procedure and attain best results for the readers.

Apart from the editorial board, the designing team has also invested a significant amount of their time in understanding the subject and creating the most relevant covers. They scrutinized every image to scout for the most suitable representation of the subject and create an appropriate cover for the book.

The publishing team has been an ardent support to the editorial, designing and production team. Their endless efforts to recruit the best for this project, has resulted in the accomplishment of this book. They are a veteran in the field of academics and their pool of knowledge is as vast as their experience in printing. Their expertise and guidance has proved useful at every step. Their uncompromising quality standards have made this book an exceptional effort. Their encouragement from time to time has been an inspiration for everyone.

The publisher and the editorial board hope that this book will prove to be a valuable piece of knowledge for researchers, students, practitioners and scholars across the globe.

List of Contributors

Mohammad Monirujjaman Khan, Ratil Hasnat Ashique, Badrun Naher Liya, Md. Mohsin Sajjad, Md. Anisur Rahman and M. T. Hasan Amin
Department of EEE, Primeasia University, Dhaka, Bangladesh

Bing Zhou
Petroleum Geosciences, The Petroleum Institute, Abu Dhabi, UAE

Graham Heinson and Aixa Rivera-Rios
Geology and Geophysics, The University of Adelaide, Adelaide, Australia

Ahmed Magdy and Hesham F. A. Hamed
Department of Electrical Engineering, Minia University, Minia, Egypt

Osama M. EL-Ghandour
Department of Electronics, Communications and Computer Engineering, Helwan University, Cairo, Egypt

Sabrina Macaire, Alexandre Catrain, Stéphane Tortel and Jean-Christophe Joly
CEA, DAM, Gramat, France

Sébastien Girard and Pierre Bonnet
Institut Pascal, Université Blaise Pascal, Clermont-Ferrand, France
CNRS, UMR 6602, Aubière, France

Alain Vian
ARCH-E. Université d'Angers, UMR 1345 IRHS, Angers, France

Douglas Letsholathebe and Kgakgamatso Mphale
Physics Department, University of Botswana, Gaborone, Botswana

Laurent Pecastaing, Marc Rivaletto and Antoine de Ferron
Laboratoire SIAME, Equipe Génie Electrique, Université de Pau et des Pays de l'Adour, Hélioparc Pau-Pyrénées, Pau, France

Romain Pecquois
Hi-Pulse, 300 rue des arts et métiers, Pont-de-Pany, France

René Vézinet, Jean-Christophe Diot and Stéphane Tortel
French Alternative Energies and Atomic Energy Commission (CEA), Gramat, France

Alejandro Barrientos García, Igor A. Sukhoivanov, José Amparo Andrade Lucio and Oscar Gerardo Ibarra Manzano
Department of Electronics Engineering, Universidad de Guanajuato, Salamanca, Mexico

Igor Guryev
Department of Multidisciplinary Studies, Universidad de Guanajuato, Yuriria, Mexico

Juan Carlos Hernández García
Department of Electronics Engineering, Universidad de Guanajuato, Salamanca, Mexico
CONACYT, Cátedras CONACYT, Ciudad de Mexico, Mexico

Gabriel Ramos Ortiz
Centro de Investigaciones en Óptica, A. C., GPOM, León, Mexico

Ahmed Fawzy and Hesham F. A. Hamed
Department of Electrical Engineering, Minia University, Minia, Egypt

Osama M. El-Ghandour
Department of Electronics, Communications and Computer Engineering, Helwan University, Cairo, Egypt

Altair S. de Assis
Universidade Federal Fluminense (IM-GMA), Niterói, Brasil

Hector Torres-Silva
Universidad de Tarapacá, Arica, Chile

Göran T. Marklund
Kungliga Tekniska Högskolan—KTH, Stockholm, Sweden

Seppo Mäkinen
School of Technology, Vaasa University of Applied Sciences, Vaasa, Finland

Bu S. Park, Leonardo M. Angelone and Sunder S. Rajan
Division of Biomedical Physics (DBP)/OSEL/CDRH, Food and Drug Administration (FDA), Silver Spring, MD, USA

Martin J. Lizak
Mouse Imaging Facility, National Institute of Neurological Disorders and Stroke (NINDS), National Institutes of Health (NIH), Bethesda, MD, USA

Gafur Gulyamov, Muhammadjon Gulomkodirovich Dadamirzaev, Nosir Yusupjanovich Sharibayev and Ne'matjon Zokirov
Namangan Engineering Pedagogical Institute, Namangan, Uzbekistan

Abdessalem Kaddouri, Mourad Aidi and Taoufik Aguili
Communication System Laboratory Sys'Com, National Engineering School of Tunis, University Tunis El Manar, Tunis, Tunisia

Ammar Armghan, Xinguang Hu, Shuai Yuan and Jinsong Xia
Wuhan National Laboratory for Optoelectronics and School of Optical and Electronic Information, Huazhong University of Science and Technology, Wuhan, China

Mohamad Y. Abou-Shahine, Youssef Nasser and Karim Y. Kabalan
Electrical and Computer Engineering Department, American University of Beirut, Beirut, Lebanon

Hilal M. El Misilmani, Karim Y. Kabalan and Mohamad Y. Abou-Shahine
Electrical and Computer Engineering Department, American University of Beirut, Beirut, Lebanon

Mohammed Al-Husseini
Beirut Research and Innovation Center, Lebanese Center for Studies and Research, Beirut, Lebanon

Volodymyr Grimalsky, Svetlana Koshevaya and Alvaro Zamudio-Lara
CIICAp, Autonomous University of State Morelos (UAEM), Cuernavaca, Mexico

Anatoliy Kotsarenko
Faculty of Engineering, Autonomous University of Carmen (UNACAR), Ciudad Del Carmen, Mexico

Michelangelo Rossetto
125 w 86 th st, New York, USA

Yuanhao Wang, Xiaoxi He, Jiangning Wang and Wolfgang Mathis
Institut für Theoretische Elektrotechnik, Gottfried Wilhelm Leibniz Universität Hannover, Hannover, Germany

Sergey Berezin
Applied Mathematics Department, Saint-Petersburg State Polytechnical University, St. Petersburg, Russia

Yahya S. H. Khraisat, Ahmad Bataineh and Malak Angor
Electrical and Electronics Department, Al-Balqa' Applied University/Al-Huson University College, Irbid, Jordan

S. Koshevaya, V. Grimalsky, Y. Kotsarenko and J. Escobedo-Alatorre
Autonomous University of Morelos State, CIICAp, Cuernavaca, Mexico

Norbert Leitgeb
Institute of Health Care Engineering, Graz University of Technology, Graz, Austria

Eduardo Javier Paez
Centre for Quantum Technologies, National University of Singapore, Singapore City, Singapore

Roberto Cessare Callarotti
Instituto Venezolano de Investigaciones Científicas (IVIC), Caracas, Venezuela and Universidad del Turabo, Gurabo, Puerto Rico

Yuande Sanchez
Fundaciôn Instituto de Ingeniería, Caracas, Venezuela

Rohani Bakar, Shumpei Fuse and Hiroyuki Arai
Department of Electrical and Computer Engineering, Yokohama National University, Yokohama, Japan

Yuki Inoue
Research Laboratories, NTT DOCOMO INC, Yokosuka, Japan

Keizo Cho
Department of Electrical, Electronics and Computer Engineering, Chiba Institute of Technology, Chiba, Japan

Salama Abdelhady
Faculty of Energy Engineering, Aswan University, Aswan, Egypt

Mohamed S. Abdelhady
Department of Mechanical Engineering, Beni-Suef University, Beni-Suef, Egypt

Norbert Leitgeb
Institute of Health Care Engineering, Graz University of Technology, Graz, Austria

Qaiser Mahmood, Andreas Fhager, Stefan Candefjord, Artur Chodorowski and Mikael Persson
Department of Signals and Systems, Chalmers University of Technology, Gothenburg, Sweden
MedTech West, Sahlgrenska University Hospital, Gothenburg, Sweden

Shaochuan Li
Department of Signals and Systems, Chalmers University of Technology, Gothenburg, Sweden

Andrew Mehnert
Centre for Microscopy, Characterisation and Analysis, The University of Western Australia, Perth, Australia

Aluru Venkata Lakshmi Narayana Rao
Department of ECE, SSNCET, Ongole, India

Nunna Bala Ankaiah
ECIL, Hyderabad, India

Dharma Raj Cheruku
Departement of ECE, GIT, GITAM University, Visakhapatnam, India

Silverio Soto Alvarez, Lidia Flor E. Huerta and Jesus Gabriel Silva
Universidad del Ejército y FAM/EMGS, Laboratorio Multidisciplinario de Investigación, Mexico D.F., Mexico

Alma Verónica Vargas and Jaime López
Instituto Politécnico Nacional, Escuela Superior de Medicina, SEPI, Mexico D.F., Mexico

César A. González
Universidad del Ejército y FAM/EMGS, Laboratorio Multidisciplinario de Investigación, Mexico D.F., Mexico
Instituto Politécnico Nacional, Escuela Superior de Medicina, SEPI, Mexico D.F., Mexico